EMERGENCY MEDICAL RESPONSE TO HAZARDOUS MATERIALS INCIDENTS

Online Services

This edition of Emergency Medical Response to Hazardous Materials Incidents includes an Online Companion World Wide Web Site featuring:

Correlating SOGs
Exposure Forms
Updated Material
Author Communication Link
Links to Related Sites
Question/Answer Forum
. . . And Much More!

To register just point your browser to: **http://www.firesci.com**

Online Services

Delmar Online
To access a wide variety of Delmar products and services on the World Wide Web, point your browser to:
http://www.delmar.com
or email: info@delmar.com

thomson.com
To access International Thomson Publishing's home site for information on more than 34 publishers and 20,000 products, point your browser to:
http://www.thomson.com
or email: findit@kiosk.thomson.com

A service of I(T)P®

EMERGENCY MEDICAL RESPONSE TO HAZARDOUS MATERIALS INCIDENTS

Richard H. Stilp

Armando S. Bevelacqua

Delmar Publishers

I(T)P an International Thomson Publishing Company

Albany • Bonn • Boston • Cincinnati • Detroit • London • Madrid
Melbourne • Mexico City • New York • Pacific Grove • Paris • San Francisco
Singapore • Tokyo • Toronto • Washington

NOTICE TO THE READER

Delmar Staff

Publisher: Robert D. Lynch
Acquisitions Editor: Mark Huth
Developmental Editor: Jeanne Mesick

Project Editor: Thomas Smith
Production Coordinator: Toni Bolognino
Art and Design Coordinator: Michael Prinzo

COPYRIGHT © 1997
By Delmar Publishers
an International Thomson Publishing Company

The ITP logo is a trademark under license

Printed in the United States of America

For more information, contact:

Cover photo courtesy of :
Baltimore County Fire Department AV Team

Delmar Publishers
3 Columbia Circle, Box 15015
Albany, New York 12212-5015

International Thomson Publishing Europe
Berkshire House 168-173
High Holborn
London, WC1V7AA
England

Thomas Nelson Australia
102 Dodds Street
South Melbourne, 3205
Victoria, Australia

Nelson Canada
1120 Birchmount Road
Scarborough, Ontario
Canada M1K 5G4

International Thomson Editores
Campos Eliseos 385, Piso 7
Col Polanco
11560 Mexico D F Mexico

International Thomson Publishing Gmbh
Königswinterer Strasse 418
53227 Bonn
Germany

International Thomson Publishing Asia
221 Henderson Road #05-10
Henderson Building
Singapore 0315

International Thomson Publishing - Japan
Hirakawacho Kyowa Building, 3F
2-2-1 Hirakawacho
Chiyoda-ku, 102 Tokyo
Japan

2 3 4 5 6 7 8 9 10 XXX 09 08 07 06 05 04

Library of Congress Cataloging-in-Publication Data

Stilp, Richard H.
    Emergency medical response to hazardous materials incidents / Richard H. Stilp, Armando S. Bevelacqua.
      p. cm.
    Includes bibliographical references and index.
    IBSN 0-8273-7829-7 (perfect)
    1. Hazardous substances. 2. Medical emergencies. I. Bevelacqua, Armando S., 1956–   . II. Title.
RC87.3.S75 1997
616.02'5—dc20
                                             96-34053
                                              CIP

To Doctor Robert Duplis who dedicated his time, love and, in the end, his life to those who provide emergency medical response in Central Florida. He was our friend, our confidant, and our mentor, providing insight and motivation to this project. His friendship and professionalism will be missed.

Robert Duplis MD
1947–1996

Contents

Foreword

The difference between an educated and uneducated man
is the same as between being alive and being dead.

Aristotle

The daily prevalence of emergency responses related to hazardous materials has grown tremendously since the inception of hazardous materials as a specialty response field. In fact, the sheer numbers and complexities of these hazardous agents is mind boggling. Just compare the first edition of the *Department of Transportation Emergency Guidebook* (Yellow Book) against the most recent edition now titled the *North American Emergency Response Guidebook* (NAERG 96) to appreciate the difference.

The emergency response community has been dealt a hand that can be won, and won safely, if the responders are well trained and well equipped. No longer are responses to emergency scenes "just another job." Short sightedness, tunnel vision, and a lack of education and equipment are perilous to responders. Emergency medical service providers have a doubly difficult task in protecting themselves from the harm of these agents and treating the victims of a release. The medical issues of hazardous materials emergency are numerous and complex. Organizations such as the National Fire Protection Association and the National Fire Academy have developed professional standards of operation and training programs to start the learning process about these potentially lethal events. But the learning process cannot stop there; it must continue in order to afford the members opportunities to safely discharge their duties while operating at a hazardous materials emergency.

Recently, events surrounding the discharge of chemical and biological weapons—weapons of mass destruction—have occurred in civilian populations around the world. Should some cowardly group opt to use hazardous materials as weapons, they could have a devastating effect in the United States. I don't say this to stir hysteria among our peers, but to point out the full ramifications of what hazardous releases, both accidental and deliberate, can yield.

Many emergency medical service providers as well as hospital emergency department staff members have had little or no exposure to hazmat response. The importance of the information contained in this book cannot be overstated. In fact, when utilized as a resource this document will enhance your understanding of hazmat response and assist you in conducting your duties to the fullest expectations and in the safest manner possible. To paraphrase Aristotle, I'd rather be educated.

Paul M. Maniscalco BS EMT/P
Deputy Chief, NYC*EMS
Commanding Officer
Special Operations Division

Preface

Emergency response work involves a defined role of safe, immediate, and effective action. Precise thinking is needed to safely mitigate the situation at hand. For years emergency medical service and fire response personnel have been using an action-oriented approach to handling emergencies. However, when dealing with a hazardous material the approach must be much slower, more defined, and initially precise, without any mistakes or gut feelings that may affect the plan. All action must be previously established in order to avoid sudden exposure, serious injury, or even the death of the emergency worker. Such a predetermined response—one that has been worked out prior to the event (preplanning), allows for all the players to know, to understand, and to be responsible for their roles within the system, that is, following organized standard operating procedures.

Only recently has the magnitude in which hazardous materials affects our lives been recognized. Fire departments, medical facilities, police departments, ambulance services, and first response agencies, must address the problem. They must understand their purpose within the system and adopt the appropriate action plan to prepare their agencies for the emergency. In August 1995, two environmental groups, the National Environmental Law Center and the U.S. Public Interest Research Group, found that forty-four million Americans are at risk of injury or death from toxic chemical release, explosions, or plant fires. This number is astounding but defines the need for better education and preparation for chemical disasters.

We all have read about or witnessed industrial plant fires, large hazardous materials spills, or the single-patient suicide using carbon monoxide as the chemical of choice. We have probably seen the effects of limited malathion insecticide exposure without even taking note of signs and symptoms (SLUD—salivation, lacrimation, urination, and defecation). The incidence of hazardous materials responses that results from exposure is higher than one could reasonably imagine. According to a recent literature; the U.S. Environmental Protection Agency has *estimated* that approximately 260 million tons of hazardous waste are generated each year (this figure is extremely conservative according to other sources). This number represents waste, not the generation of new products. 63,000 chemicals were used within our society. On an average, we are seeing approximately 150,000 cargo shipments of compressed gas on a weekly basis, 200,000 shipments of hazardous commodities every year, 20 million metric tons of hazardous commodities bought and sold every week, and 3 or 4 new chemical compounds placed on the market every day.

The statistics are worrisome. Federal, state, and local governments see the problem that has stimulated a growth of hazardous material response teams, hazmat brigades, regional response teams, and the medical community's involvement. The chances of you, the emergency responder, becoming involved with a hazardous materials incident during your employment is very high.

In the past it was assumed that the medical response team would have the expertise to handle such an incident. How this false assumption came about is unknown. It is frankly quite remarkable that an industry that prides itself on being ready for all emergencies, has for the most part, overlooked such an important role within the emergency services. And, unfortunately, through the transition of hazardous materials response, the topics of medical control and engineering controls for the emergency services have not gained much attention. Most of you have listened to lectures on hazardous materials that identify a need for the medical response, yet no one has outlined and identified the roles and responsibilities for the medical personnel. This book has been assembled to organize the information in a manner so that all emergency services can benefit. It is our aim to present a general discussion of toxic exposures and their effects. Current policy, procedures, standards, governmental rulings, guidelines, recent trends, and theories, all interrelated to the hazardous materials medical discipline, are covered to establish a standard of care during the medical response to hazardous materials incidents.

While conducting the research for this book, we discovered a serious lack of information available for emergency medical responders. Most of the material found on the subject came from physician-based subject matter that was difficult to understand and was written on a scientific level. Our goal for writing this book was to present the reader with information pertinent to performing the job of hazardous materials medical response. The information ranges from basic concepts of hazardous materials response to general chemistry and pathophysiology. Also included in the text are rules, regulations, and standard operating guidelines to give emergency medical personnel information to assist them in performing their jobs within this technical field.

Patients suffering from exposure and resultant injury are carried through a loop that includes the initial hazardous materials response, medical care, transportation, definitive hospital care, and release from the hospital. The loop includes members from both the public and private services, including fire departments, medical transport agencies, hospital staff, and even police departments, who must be educated to the activities involved in an incident so that they can accomplish goals consistent with their agency's mission as it relates to patient care. It is our intent to present information that can be used to develop, implement, and manage the medical concerns present at hazardous materials scenes.

As each section is presented, practical information, safety issues, areas of caution, and considerations are discussed, both in theory and application. These issues are presented in a stepping-stone approach, building and broadening the knowledge base as the reader progresses through the text. It is *not* our intention to make the reader a toxicologist or physiologist, although it is our intention to set forth ideas and procedures that will give the reader tools to be an effective practitioner when dealing with patients exposed to hazardous materials. The citizens of your city or county, your hazardous materials team members, and your crew depend on you and your knowledge.

The book is divided into five parts. Section 1 discusses the fundamental principles of hazardous materials response including basic referencing, identification, and preplanning.

Hazardous materials scene organization is also discussed and the steps needed to mitigate a typical incident are addressed.

Section 2 gives the reader a detailed overview of chemistry and toxicology. These principles are interrelated and complex. This section is intended to aid the emergency worker in identifying the hazardous material. The information is directed toward referencing chemical materials as it relates to medical applications and overall scene management.

Section 3 presents the physiology of the involved body systems, and principles of treatment as they relate to the hazardous materials environment. The section then details some of the more common poisonings and antidotes or specific treatments for them.

Section 4 identifies medical controls and presents sample guidelines that may be adopted for use or modified to meet specific needs. These guidelines were written for the emergency provider's use.

Section 5 includes areas of special interest, such as clandestine drug laboratories and air monitoring, but that may not be needed within every system. The selected topics tie in with the previous chapters to form a well-rounded knowledge base for the emergency medical responder.

Each chapter is supported in the student manual through the use of questions, scenarios, and practical tabletop exercises. The instructor's manual highlights points to be stressed and teaching tools to be used in presenting the material.

In utilizing these publications, we suggest that the instructional staff organize a meeting with local medical control so that all treatments can be reviewed and approved for the agencies providing the care. Most of the text can be delivered to both emergency medical technicians (EMT) and paramedic level response personnel. Although chemistry, toxicology, and physiology are complex subjects, none are presented at a level that cannot be grasped by both EMT and paramedic personnel. Invasive treatments identified in the book are the only subject matter that cannot be pursued at the EMT level. Emphasis is placed on acute injuries requiring immediate treatment, although in some cases chronic exposures and long-term effects are discussed. The treatment modalities are directed toward those jurisdictions that provide advanced life support as a part of their normal response criteria.

The pathophysiology of the injuries and treatments are based on researched and documented case studies, although, in some cases, disagreements may exist within the medical community. These disagreements are usually due to a lack of complete understanding when dealing with the toxic effects on humans. Furthermore, there may be controversy between medical control and the roles and responsibilities of responders to an incident. Most communities have either not recognized the problem or the infrequency of such an incident leads to complacency. All of the treatments presented in this text have been thoroughly researched and are recognized as acceptable therapy.

As in any type of science or medicine, new and greater understanding may be just around the corner. We urge you to discuss any of these modalities with your respective medical authority before instituting them within your jurisdiction. Because of the complexity of the information presented here, we urge you to arrange training that combines the efforts of the hazardous materials team, emergency medical units, medical director, and hospital staff. This approach will assist those who are not knowledgeable of hazardous materials concepts to become indoctrinated into the full loop of care associated with chemically injured patients.

This discipline is relatively new, with few cases actually documented, therefore the authors would be interested in any unusual cases, new treatment, or management techniques found by the readers of this text. By working together we can form the future of this highly technical, exciting, yet necessary discipline.

Acknowledgments

We would like to extend our appreciation to a number of individuals who made this project possible. It all started sometime in 1989 with some ideas scratched out on a table napkin and later presented to our local medical director. After months of personal research, and a formal presentation, Fire Chief Robert Bowman, and the late Medical Director Dr. Robert Duplis endorsed the project as one that should be taught. With the cooperation of local training facilities and Valencia Community College, the first classes were presented to the Orlando Fire Department and other surrounding fire departments and transport agencies. All the work contained herein is a reflection of many long off-duty hours of research and class presentations. As so often happens, the instructors learned more about the subject from each class, which eventually led to much of the material found in this book.

We are proud to say that this book was developed by street practitioners who were motivated through personal experience, a quest for knowledge, and a dedication to the job. The following individuals greatly supported our writing and research endeavors. Through their encouragement and support this book has become a reality. Jeanne Mesick, Paul Shepardson, and their staff for the continued support from Delmar. Chief R. Bowman and District Chief R. Huggins of the Orlando Fire Department, for their enthusiasm in establishing one of the first Advanced Life Support Hazardous Materials Response Teams. Chief Lee Newsome of Ocala Fire Department for his input and review. Chief Paul Maniscalco of New York City EMS for his review and words of encouragement. To the staff of the Orange County Department of Emergency Medical Service and the Orlando Fire Department Hazardous Materials Team, Orlando, Florida, for their implementation of procedures and protocols. Special thanks to Lt. Don (Doc) Adams for assistance in writing Chapter 9, Biohazard Awareness, Prevention, and Protection.

The acknowledgments would not be complete without the recognition of those who are close friends, peers, and family, who have been a constant source of encouragement throughout the writing and rewriting of this manuscript. Much gratitude to the following who reviewed the project in its entirety under very constrained deadlines:

Clinton Smoke
Northern Virginia Community College

Kevin Coleman
FDIC Instructor of the Year 1995
ABI Consulting

Christopher Hawley
Baltimore County Fire Department Hazardous Materials Team

James Angle
Palm Harbor Fire Department
University of Concinnati

Lee Cooper
Wisconsin Indianhead Technical College

Chief R. C. Dawson
Mechanicsville Fire Department

James Madden
Lake Superior State University

James Barnes, Jr.
Baltimore County Fire Department

Michael Gilroy
Nassau County Fire Service Academy

Michael McKenna
Mission College

Jay Franey
Aims Community College

Robert Laeng
Training for Life Safety

Johnny Mack
South Puget Sound Community College

About the Authors

Richard H. Stilp has been a member of the Orlando Fire Department since 1976 and a member of the hazardous materials team since its inception in 1982. In 1986, Mr. Stilp graduated from Valencia Community College as a registered nurse. He currently holds a bachelor of arts degree and is attending classes toward an MBA.

He serves as a lieutenant paramedic with the Orlando Fire Department, Orlando, Florida, where he is responsible for hazardous materials response and is a member of the fire department's dive team and high-angle tactical response team. Mr. Stilp is a hazardous materials technician, and certified and nationally recognized hazardous materials instructor, and an adjunct instructor for the National Fire Academy and the International Association of Firefighters. He teaches in local colleges in both the paramedic and hazardous materials disciplines.

Mr. Stilp participates regionally with hazardous materials issues and is involved with the local emergency planning committee and local hospitals. In the hospitals where he has worked, Mr. Stilp was responsible for the establishment of hospital preparedness for hazardous materials.

Armando S. Bevelacqua has been involved with prehospital care since 1975. During this time, he was a flight medic, supervisor of an ambulance service, and technician in an emergency room and cardiac intensive care unit.

He currently serves as a lieutenant paramedic with the City of Orlando Fire Department, Orlando, Florida, where he is responsible for hazardous materials response and high-angle tactical rescue. He has developed many educational programs for the Orlando and surrounding fire departments. He has been actively involved with quality assurance, documentation review, and is an EMS and hazardous materials training officer.

Mr. Bevelacqua teaches at local colleges, instructing EMS and fire classes. As a freelance writer, he recently published a book on report writing for EMS providers. He has lectured nationally on several issues and is an adjunct instructor for the National Fire Academy and International Association of Firefighters.

He is involved with many emergency planning entities including the local emergency planning committee. Mr. Bevelacqua holds a bachelor's degree in fire safety engineering from the University of Cincinnati and is preparing to enter an MBA program.

INTRODUCTION TO HAZARDOUS MATERIALS

Section 1 is an overview of the typical hazardous materials (hazmat) response. Before examining emergency medical response, we must understand what else is taking place at the scene. It is also important to identify why emergency responders are there in the first place. Once an understanding of the scene is established, the medical responder can better fit into the chain of events that transpire at these scenes.

Chapter 1 addresses the issues concerning preparation and response to hazardous materials. Many of the issues are not directed toward only medical responders but to the response agencies in general. To understand the medical aspects of hazardous materials emergencies the responders must have a general understanding of the incident as a whole.

Both before and during the incident, information is gathered that guides emergency responders in the completion of their task. Knowing where to go to retrieve information is an important part of the job. As Samuel Johnson once said, "The next best thing to knowing something is knowing where to find it." Readers are guided as to where the initial referencing should start and instructed where to go for more information.

Chapter 2 discusses the priorities and guidelines involved in mitigation of a hazmat incident. Each individual has a job to do on the scene and if all participate by completing their part, the scene will be handled in a logical and safe manner. In some cases the subjects may be taken to the extreme. This is done to build a solid foundation from which to work. Emergency response, by its nature, is already hazardous. It is prudent to be overcautious and err on the side of safety.

Chapter

1

Hazardous Materials Concepts

Objectives

Given a hazardous materials incident, you should be able to describe the general roles, responsibilities, and hazards presented to the responder and victims, appropriately describing the importance of a planned response.

As a hazardous materials medical technician, you should be able to:

- Identify the responsibilities and roles of the Local Emergency Planning Committee (LEPC) as they relate to the hazardous materials event.
- Describe the functions of the emergency dispatch center.
- Define size-up as it relates to the hazardous materials incident.
- Identify basic referencing techniques, discussing the pros and cons of these referencing techniques.
- Describe the DOT placard, identifying its strengths and weaknesses.
- Identify and define NFPA 704.
- Describe the four identifiers of the placard and the DOT hazard classes.
- Identify the zones and describe the functions within each zone.

- Describe the three components necessary to create a plan.
- Identify and define the three levels of a hazardous materials incident.
- List the internal and external elements of resource management.
- Define the standard of care as it relates to a hazardous materials incident.
- List the four elements of negligence and define each element.

HISTORY OF HAZMAT RESPONSE

Since the late 1960s there has been much discussion about the effects suffered when chemicals are released or spilled into the environment or when one is exposed to many of the chemicals used in industry. With this newfound knowledge, government action soon followed. Throughout this book there are references to the Occupational Safety and Health Association (OSHA) and the Environmental Protection Agency (EPA). Although OSHA was established in 1974, it was not until 1986 that both of these organizations received their power through a legislative action. The Superfund Amendments and Reauthorization Act (SARA) was passed mandating regulation in the storage, transportation, use, and disposal of chemicals into the environment. Within this document several provisions were established, of which Title I and Title III are the most important to emergency responders.

Title I mandates that OSHA and EPA establish regulations on training, emergency response, safety, and associated hazardous materials activities. Within this title OSHA 29 CFR (Code of Federal Regulations) 1910.120 and EPA 40 CFR 311 were established. These federal regulations outlined training standards and mandated written standard operating procedures (SOPs) for hazardous materials incidents.

Title III sets requirements for industry to report the chemicals used or stored in the workplace. Tier 2 reports are required by this title. These reports supply emergency responders with an inventory of what chemicals may be found in an industrial setting. Part of the reporting requirements require businesses to supply a material safety data sheet (MSDS) for each chemical that meets the reporting requirements. The MSDS supplies the responder with valuable data about a particular chemical. Title III further requires planning for hazardous materials emergencies on the state and local level. To ensure that the planning was completed, the federal government, through the 1986 SARA legislation, directed the states to form state emergency response commissions (SERC), which implements the SARA requirements, and the local emergency planning committees (LEPC), which identifies needs, resources, and planning for the community. Through SERC and LEPC, emergency response could be coordinated between multiple agencies. LEPC's responsibilities include:

- Identifying facilities that use, store, or transport hazardous materials
- Designating emergency response coordinators from the community and facilities

- Preparing response procedures facilities, emergency responders, and medical personnel
- Maintaining an inventory of essential response equipment
- Providing notification of accidental release
- Assisting in the development of evacuation plans, training programs, and exercises

Recent years have seen the passage of laws regulating the storage, transportation, and use of hazardous materials. These laws have also addressed planning for the incident, training standards, and skill competencies, but the medical aspects of a hazardous materials emergency have been mostly overlooked. OSHA identifies several levels of training in the CFR 1910.120 document. Among these levels are Awareness, Operations, Technician, Specialist, and Incident Commander. In all of these levels the medical issues are touched on but not truly explored. The National Fire Protection Association's (NFPA) 473 document, Professional Competence for EMS Personnel Responding to Hazardous Materials Incidents, was the first to identify training specifically for emergency medical personnel. This document has great merit but still falls short of what should be required for the medical responder.

ASSESSING THE PROBLEM

Hazardous materials pose many problems: One of the obvious is determining how the emergency medical services (EMS) function fits into a hazardous materials response. Many agencies are involved with the handling, use, and the problems associated with hazardous materials. Each of these agencies has identified hazardous materials as it relates to their realm of service. Several examples of these identifiers are listed below:

hazardous material
according to the EPA, a material that may be potentially harmful to the public's health or welfare if it is discharged into the environment and according to the DOT, any substance or material in any form or quantity that poses an unreasonable risk to safety, health, and property when transported in commerce

- The EPA defines a **hazardous material** as a substance that may be potentially harmful to the public's health or welfare if it is discharged into the environment.
- The Department of Transportation (DOT), in an act dating back to the mid 1970s, defines a hazardous material as any substance or material in any form or quantity that poses an unreasonable risk to safety and health and to property when transported in commerce. DOT further identifies the hazardous materials in classifications discussed later on in the chapter.
- The American Conference of Governmental Industrial Hygienists (ACGIH) does not define hazardous materials but has established levels of chemical substances that a person can be exposed to without sustaining permanent injury. These values guide industry in setting exposure limits on particular chemicals commonly found in their environment.
- OSHA and the National Institute for Occupational Safety and Health Administration (NIOSH) view a hazardous material from the standpoint of

potential hazard. They rate conditions that may cause injury or death as they are found in the working environment, whether they are obvious or not.

In each of these statements health issues are addressed and appear to be the main concern. As health-care workers, we have been taught that the primary objective is to save life and property, which many times requires immediate action. We may not be able to act immediately at a hazmat incident. A risk versus benefit analysis must be done in order to manage the scene appropriately. As difficult as it may be, no offensive action may be the most appropriate action to take. The more typical approach to this type of situation may be an indirect one. Evacuation of surrounding areas or other defensive measures may be the appropriate life-saving technique.

The problems encountered at a hazmat incident are many. The primary threats involve injury to the emergency worker and harm to the community. Without emergency personnel, the situation cannot be handled in a safe or timely manner. Unfortunately, sometimes emergency response personnel do not view the safety of themselves or even their peers as the primary consideration. When dealing with this type of incident, personal safety should *always* be the primary concern.

Only through training and education will this idea become firmly seated in the minds of responders. Responding agencies must have the forethought to develop standard operating guidelines that further define their roles and responsibilities. Such guidelines will not only help protect those responding to these emergencies, but also peers, family, and the communities for whom they serve.

PREPLANNING

Unfortunately, many emergency response personnel believe that when an emergency occurs, their training alone will allow them to complete the needed tasks safely. Although there is no replacement for good training and education, preplanning has been proved to reduce the accident rate and to better prepare the responders for their job. Preplanning is the only tool emergency responders have to change a typical reactive response into a proactive process for mitigating the emergency.

Fire departments and emergency medical providers have used preplanning for years. Fire departments preplan for fires in high hazard occupancies. Areas addressed by preplans include water supplies, predicting the spread of fire, and structural hazards, along with many other safety factors. Emergency medical providers use preplans to prepare for multiple-patient incidents that tax their resources. Preplanning must also be used by the hazardous materials response community to determine the danger to life, property, and environment. These preplans should involve identifying the areas where hazardous chemicals are stored, used, and transported so that the type of injuries or damage that would occur if an accident happened can be predicted.

! SAFETY
Personal safety should always be your primary concern.

■ NOTE
Responding agencies must develop standard operating guidelines to define their roles and responsibilities.

■ NOTE
Preplanning is the only tool emergency responders have to change a typical reactive response into a proactive process for mitigating the emergency.

■ NOTE
Preplans should involve identifying the areas where hazardous chemicals are stored, used, and transported so that the types of injuries or damage that would occur if an accident happened can be predicted.

Planning may be the first step toward exposure prevention. By analyzing a potential hazardous materials emergency, the impact of such an emergency may be forecast. The preplanning process identifies hazards, predicts accidents, then addresses appropriate solutions. Only when this process is complete is a responder confident that the emergency can be handled in the most efficient and safe manner.

The goal of preplanning is to safeguard against the chemical emergency or natural disaster before the incident occurs. Preplanning can be accomplished on an informal basis or completed formally through the institution of department policy or community plans. Under the LEPC, advisory councils, community councils, and prevention bureaus are established to monitor, analyze, and plan for an accidental release. They also must assess the location and quantities of hazardous materials found within or transported through their community. The LEPC must oversee community-wide objectives written to identify the relationships between emergency management, emergency services, and the hospitals. Each plays a role in the system as it pertains to medical management during a major chemical release.

A hazardous materials event may impact the emergency system as a whole and even the infrastructure of a community. Even on a small scale, the impact is greater than one might think. In many cases the medical aspect of hazardous materials has not been adequately addressed by medical responders, emergency transport agencies, or the hospitals.

INCIDENT OVERVIEW

Receiving the Call and Identifying a Hazardous Materials Emergency

Appropriate scene control and incident management begins within the emergency dispatch center (EDC, also known in different areas under other terminology). Dispatch centers across the country are answering calls for help at an alarming rate. Each time the phone rings, the individuals within these centers must be alert to the possibility of a hazardous materials incident. Before an incident occurs, each dispatch center employee or volunteer must know and understand the local standard operating guidelines (SOG). Education and training of these important participants can be ensured through practice drills and annual continuing education.

Written policies must be in place. A standard hazardous materials incident reporting form should be adopted so the appropriate questions are asked without delay. This is not only true for fire department dispatch centers, but also for emergency medical dispatch centers. Many times the medical centers are called first because of an injury related to the release of a hazardous material. The form should stimulate the call taker to ask questions concerning the type facility where the accident took place, how much material is spilled or leaking, the type of container involved, and if anyone is injured or exposed.

These questions assume that the caller identifies the incident as involving hazardous materials. When the scene is not immediately identified as involving hazardous materials, the dispatcher must acquire pertinent information by asking specific questions. The dispatch center should be aware of other clues. Receiving numerous requests for assistance from the same address, all with similar complaints would be highly coincidental. Multiple patients complaining of difficulty breathing, chest pains, excessive salivation, diaphoresis, or decreased level of consciousness all warrant aggressive questioning from the call taker.

Once a hazardous materials incident has been identified and confirmed, additional considerations can be addressed by the dispatcher. A variety of overall information is needed for the responding units, along with specific information for the incident commander. Wind direction and speed may indicate the necessity for evacuation. Humidity, dew point, and chemical information indicate the need for further protective equipment. Additional information as to the type of chemical, its present state of matter, and its normal state of matter are all helpful. Many times the caller has specific knowledge of the chemical involved and is able to assist with information even before the responding units arrive on the scene. If preplans are available to the dispatcher, they can be referenced to determine known hazards and what special precautions should be taken at the incident.

Initial Size-Up

Size-up involves analyzing the information about the scene. This process may start during preplanning and should continue as information about the incident is received. The gathered information should include the location of the incident so population densities can be predicted, and the time of day to further assess life hazard. If the incident involves transportation of a material, it is important to know if it exists on an interstate, city/county road, rural setting, or railway system. Each of these settings presents a different set of considerations for the command staff and emergency response teams.

Building types and occupancies are important information for the responding units. Is the incident in an industrial park, or is it light industry within a business area? School and university chemistry laboratories contain large varieties of chemicals with varying hazards. Public and private sewer treatment plants, recreational areas, and waste sites also store or house large volumes of chemicals. Hospitals, clinics, and doctors offices have a variety of hazards (biological, radiological, laser emissions, and chemical agents) that under the right condition may present negative consequences.

If the incident is confirmed and injuries resulted, certain questions must be answered:

To what extent are the victims injured?

Do the victims need decontamination?

Are the local hospitals capable and ready to receive patients?

■ **NOTE**
If an incident is confirmed and injuries resulted, certain questions must be answered concerning the extent of the injuries, the need for decontamination, and the preparedness of local hospitals.

These and many other questions must be answered as the incident progresses.

Response

As dispatch is gathering and assembling information, the responding units also have a variety of information available to consider. Responders who have an extensive knowledge of the accident area may help assess the particular needs of that area. From the information given at the time of dispatch, several items can be evaluated for the initial scene setup. If the product is airborne, wind direction and speed will provide clues on the direction of travel. When the location is identified, preplans should be reviewed and examined for the resources that may be required. From the initial information, identification of the chemical and recognition of the hazards become an early consideration. After this is accomplished, isolation and evacuation may take place. It would be ludicrous to say that you know or are able to read all of the information on hand about the chemical in question. However, if you give yourself a brief overview of the reference material, size-up will become a much easier task.

Preplanning of certain industrial occupancies, schools, shops, or other high-hazard areas can give exact evacuation and hazard zone parameters. By preplanning, a determination to defend in place may be considered in lieu of a full evacuation. It is commonly thought that the two populations presenting th greatest challenge during evacuation are children and the elderly. If these two groups are found in high number within the hazardous area, defending in place may be a real alternative to evacuation. If there is a likelihood of exposed patients, hospital notification via cellular phone or medical control radio must be made. The hospital can then start to prepare its operation in order to meet the demands of the incident.

Arrival on the Scene

Once the units arrive on the scene, the true work begins. Both physical and mental demands will continue until the termination of the incident. The first unit on the scene must continue a complete and accurate size-up using all of the available information. Once you arrive on the scene, the information gathered there should be added to the information gained through preplanning and the dispatch process. The accuracy of the information gathered by the first arriving units is critical in conducting a safe and efficient incident.

Upon arrival stop, look, and listen.

Stop and stage far enough away to be out of possible danger, yet close enough to assess the situation. This distance may not be the true safe distance; rather it is a distance that will give you time to react, if the situation deteriorates. (Remember the rule of thumb! Staging for the incident should be initially set up at a distance

! SAFETY
Upon arrival stop, look, and listen.

that allows you to cover the view of the incident by holding your arm straight out with your thumb up. This is usually a safe initial distance from the scene.)

Next, *look* for smoke, clouds of vapors or gases, and most important, indications of fire or fire potential.

Finally, *listen* for high-pitched noises, an indication that a fluid is creating a gas or a gas is under pressure and is being released. Remember to *stop*, *look*, and *listen* during initial size-up.

Information gathering is a key component in all four steps of the process used during hazardous materials emergencies. Those four steps are analyzing, planning, implementing and evaluating (reevaluating once you have gone through the processes). This process is ongoing until termination of the incident.

Identification

Once on the scene, true identification should be made. DOT numbers, CAS (Chemical Abstract Service) numbers, placards, and labels will give clues to the product involved. Referencing the product in question should be one of the initial goals. All procedures and considerations hinge on this one function. Because of the great importance of referencing the material involved, using three different resources to verify information is a good rule to follow. Once the product is identified further referencing can take place. From the medical standpoint, information about the involved material must be researched from a chemical, physical, and medical view.

The North American Emergency Response Guidebook (NAERG)96 helps identify the product classification of the chemical in question. This reference book gives you basic information. The NIOSH book provides additional and more technical information including the threshold limit values, target organs affected, personnel protective equipment, and chemical and physical properties. Preincident planning, shipping papers, material safety data sheets, placards, labels, and container shapes can all give clues to the product, dictating the appropriate procedures to follow. Many other excellent reference books are available and should be used during this process (see Appendix A).

First Response Reference   The single most important function at the scene of a hazardous materials incident is identification. The ability to identify the potential of an incident, its products, and victim possibility takes education and practice. This degree of training and experience is not generally acquired at the basic responder level and is beyond the scope of what is generally taught to the emergency medical services during the first response segment. However, because of growing awareness about hazardous materials and the increased frequency of transportation of these materials, it is imperative that first responders become educated in this skill (Fig. 1-1).

Most hazardous materials incidents require a slow premeditated approach. Unlike most incidents handled by the emergency services, hazardous materials

Figure 1-1 *The ability to identify the chemical takes expertise and practice. The reference officer must be familiar with the materials available to him or her in the research process.*

should not be handled through a reactionary approach. Rather, it should be a well-informed, logical progression toward involvement. The questions that must be asked are: What are the ramifications of your initial actions and, in what order should these actions be taken? The information gained while referencing the materials involved, can answer these questions in an informed and intelligent manner, insuring a safer mitigation to the incident. Occasionally a rapid action is needed to rescue victims. This rapid action should always be instituted after the scene has been assessed and the level of danger has been determined to be acceptable. This type of action is not the typical response but is instead an exception to the rule.

■ NOTE
Emergency responders must have a basic understanding of the rules and regulations involved in placarding, labeling, and identifying of fixed-storage facilities.

Placards and Labels   Emergency responders must have a basic understanding of the rules and regulations that are involved in placarding, labeling, and identifying of fixed storage facilities. Federal law mandates placarding of vehicles that transport hazardous substances (DOT 49 CFR) (Fig. 1-2). Reviewing federal laws is beyond the scope of this text, but it is worth mentioning that not all who transport or store hazardous materials clearly state the substances they have. Only an estimated 50% to 60% of all placarding is correct. In some cases one may also notice placarding of warehouses and businesses, referred to as *fixed-site facilities*. In most jurisdictions, providing fixed-site placards is a voluntary standard. Some localities have adopted the standards as part of the local code, but they are not mandated by federal law.

It would be nice if all transporters and occupancies had a warning sign, but they do not. Rules and regulations are only as good as user compliance which, as noted, is quite low. The laws themselves also are inconsistent. Some hazardous

Figure 1-2 *Placards are mandated by law on vehicles transporting hazardous substances. There are exceptions to the rule and violations of the law, but a placard can clue the responders as to what hazards may be present.*

materials are identified under a different class than one might expect. For example, anhydrous ammonia has a flammable range of 16% to 25% but is placarded as a nonflammable gas. DOT says that for a gas to be qualified as flammable, it must have a lower flammable limit of no higher than 13% with a flammable range of greater than 10%. Furthermore, if the quantity is low enough, some hazardous materials may not be identified at all.

Probably the most commonly seen hazardous materials markings are the placards used on vehicles. Two concepts of regulations come into play when transporting a hazardous material, and all placard regulations are based on these two concepts.

1. A hazardous material is defined as a substance, material, or compound that has been determined by the DOT to have the capacity to do harm to the environment and human and animal populations. It is or has an unreasonable risk to the safety, health, and property upon contact.

2. Quantity dictates if most materials will be placarded at all.

Those materials determined to be the most hazardous according to the CFR are called **table I materials** and must be placarded any time they are transported regardless of the weight. In this category are:

Explosives in class numbers 1.1, 1.2, 1.3

Poison gas class 2.3 and Poison 6.1 inhalation hazard

Dangerous when wet class 4.3

Radioactive III materials

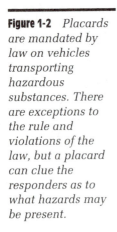

table I materials
materials determined to be the most hazardous according to the Code of Federal Regulations

All other hazardous materials are found in table II. Placards for table II materials are only required when the amount of the chemical exceeds 1,001 pounds. In other words, 990 pounds of a hazardous material may not require a placard at all, but at 1,001 pounds it will. From a health hazard point of view, does 11 pounds create a safer situation?

Placards are signs made of a durable material so that they will not deteriorate under normal environmental conditions. This does not mean that when the vehicle is on fire or if the chemical is leaking the placard will not be destroyed. All vehicles that require placarding must display the placard in four locations—the front, the rear, and each side—as outlined under the DOT guidelines. Placards are square, 10 3/4 inches by 10 3/4 inches, and placed on a point giving them the look of a diamond. Each placard has four general characteristics to assist in identifying a substance:

Color. Different colored backgrounds identify the classifications of hazardous material.

Symbol. Specific symbols identify the type of hazard.

United Nations (U.N.) identification number. A four-digit identification number found in the middle of the placard or in an orange panel located above or below the placard must be displayed. The number identifies the chemical or group of chemicals contained within the vessel. The *North American Emergency Response Guidebook* contains a list of chemicals that corresponds to these numbers and can be referenced quickly to give the responder general information. (Some placards include the hazard name.)

Classification number. A subclassification number to identify the sub-category of the substance and compatibility codes.

Labels look very similar to placards and contain much of the same information. Labels are placed on two sides of the container and are normally 4 inches by 4 inches.

Under the DOT transport placard and labeling system, chemicals are categorized under nine hazard classes and two word classifications:

Class 1: Explosives. This classification placard shows a symbol of an exploding object (bursting ball) in the top corner, and has an orange background. There are six subdivisions within this hazard class:

1.1 Materials that have a mass explosion hazard

1.2 Materials that have distinction through projection hazard

1.3 Fire hazard materials

1.4 Materials with a minor explosive hazard

1.5 Materials that are considered insensitive

1.6 Materials that are dangerous but are considered extremely insensitive

Class 2: Gases. This classification includes a variety of hazards with the hazard denoted by a symbol representative of the danger. A symbol of a fire against a red background is used for the flammable gases (2.1). A cylinder against a green background represents the nonflammable gases (2.2). For a poisonous gas, the skull and crossbones symbol is used on a white background.

2.1 Flammable Gas

2.2 Nonflammable Gas (lower flammable limit of no higher than 13% and a flammable range of greater than 10%)

2.3 Poisonous Gas

2.4 Corrosive Gases

Class 3: Flammable Liquids. In this classification, a fire symbol against a red background is used. The liquid can either be combustible (those liquids with flash points above 140° F) or flammable, which is rated against the material's flash point.

3.1 Materials that are flammable with a flash point less than 0° F

3.2 Materials that are flammable with a flash point between 0° and 72° (equal to 0° but less than 73°) F

3.3 Materials that are flammable with a flash point between 73° and 141° F

Class 4: Flammable Solids. This class includes a variety of placards, each designed to identify the hazard classification. The red vertical-striped background with a flame symbol at the top denotes the flammable solid (4.1). The spontaneously combustible solids (4.2) are denoted by the white top and red bottom placard with the symbol of a flame at the top. The Dangerous when wet (4.3) is a blue placard with a flame symbol at the top.

4.1 Flammable Solids

4.2 Spontaneously Combustible

4.3 Dangerous When Wet

Class 5: Oxidizing Substances. Those substances that support combustion are symbolized by a ball on fire with a yellow background. The words oxidizer (5.1) or organic peroxide (5.2) are across the middle. The subdivision numbers 5.1 and 5.2 must also be placed on the placard.

5.1 Oxidizers

5.2 Organic Peroxides

Class 6: Poisonous* and Infectious Substances. These placards have the symbol of the skull and crossbones with a white background for the poison and a diagram of the biohazard symbol for the etiologic materials.

6.1 Poisons*

6.2 Infectious Material

Class 7: Radioactive Materials. This class of materials emits ionizing radiation. There are three subdivisions to this classification, each denoting the level of ionization potential. Each placard shows the radioactive propeller on a yellow background and the word radioactive on the bottom half on a white background. Class 1 is all white and is the lowest in hazard (5 microsievert per hour). The middle hazard is class two with a yellow top half and a white bottom (5 microsievert to .5 millisievert per hour). Class three is the most hazardous and has emissions above the .5 millisievert per hour range.

Class 8: Corrosive Material. This class contains those materials that are extremely basic or acidic. The placard is white on top with a black bottom. The top diagram area shows a hand and an object being destroyed by a liquid that is being poured out of a test tube, denoting the corrosivity.

Class 9: Miscellaneous Materials. This classification denotes hazards that do not fit into any of the above definitions but that present a significant danger. This placard has vertical black lines on the top half and white on the bottom half.

9.1 Miscellaneous Dangerous Materials**

9.2 Environmentally Hazardous**

9.3 Dangerous Waste**

ORM-D: Consumer Commodity. These are mostly labels with ORM-D (other regulated material, classification D) in the middle. They are for household commodities such as bug spray, drain cleaner, and the like. The letter D is from an old classification. Class nine was the ORM classification. There were five subdivisions: A, B, C, D, and E. The present-day miscellaneous materials are the old ORM A, B, C, and E. The new class of ORM-D are those commodities from the old class nine ORM-D.

DANGEROUS: Mixed Loads. When the total weight of two or more Table II materials is 1,001 pounds or more, they are identified by this sign.

Warehouses and businesses, referred to as fixed-site or fixed-storage facilities also may be placarded. The NFPA 704 fixed storage placarding system (Fig. 1-3) is

---

* Poison, poisonous, and toxic are all synonymous.

**Canada's classification.

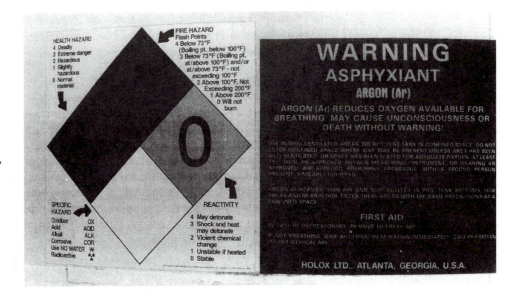

Figure 1-3 *The NFPA 704 placard is colorful and easy to see. Each hazard class is assigned a color and a number indicating the level of hazard.*

designed specifically for storage facilities. Many communities have adopted the standard as a local ordinance to alert emergency responders to the dangers inside, but it is voluntary, not mandated by federal law. The Building Officials and Code Administrators (better known as the BOCA code) adopted by many localities requires the use of the 704 system. The 704 system consists of a large diamond-type placard that is colored in four quadrants: blue, red, yellow, and white. Each color represents a different type hazard: health, flammability, reactivity, and special information respectively. Each of these quadrants contains numbers on a scale from 0 to 4. Zero indicates no hazard for that color and 4 indicates the worst possible hazard. In the special information (white) quadrant, two symbols, W̶ for do not use water and OXY for oxidizers, may indicate additional hazards.

The following symbols or abbreviations have also been used and may be seen in your area:

    FOR RADIOACTIVITY

    FOR ETIOLOGIC AGENTS

COR   FOR CORROSIVES

    FOR LASER EMISSIONS

EXP   FOR EXPLOSIVES

Although this system provides much-needed information, it is still limited as it does not identify the chemical found at the structure, only the potential hazard. The occupancies that actively use the marking system provide important information to those responders who may be faced with entering such a facility during an emergency.

Another type of identification label is the Hazardous Materials Information System (HMIS) label. This label utilizes the same coloring scheme as the NFPA 704 system, but the numbering may not be the same. The HMIS system was not designed for the emergency responder so it is rarely seen during transport. It is primarily used in storage and manufacturing facilities so that employees can quickly reference the hazards associated with the chemicals they are handling. This label helped industry meet portions of the hazard communication standard. Most HMIS labels also contain information on personal protective equipment needs associated with the use of that chemical.

Shipping Papers   Another source of information available on the scene are the shipping papers. These papers can provide an excellent starting point for your chemical research. The shipping papers have the chemical name and the quantity being shipped, and they identify the manufacturer. Most companies that produce chemicals have an emergency contact number and some even have response teams or technicians to assist with the technical questions about the chemical that you are dealing with.

During some hazardous materials incidents the papers may be left in the hot zone or contaminated prior to removal from the vehicle. In these cases the papers should be placed in a large plastic food storage bag. If needed the storage bag can be decontaminated for safe handling in the cold zone. The bag will prevent the possibility of secondary contamination and allow the papers to be safely read in the cold zone.

Material Safety Data Sheets   MSDS must be maintained on the premises of all hazardous materials sites—a fixed-storage facility, an industrial setting, or anywhere hazardous materials are kept and used (they are usually not available in transit). They contain a wealth of information, including the chemical's name, synonyms, chemical properties, physical properties, health and safety issues, fire potential, flammability, and reactivity, are but a few items addressed on these documents.

Container Shape   The shape of the container may provide responders with clues to the general type of hazard. Cylinders that have a pressure cap, like those found on an oxygen cylinder, may indicate a compressed gas or poison. Larger containers that have an arrangement of valves and input gates should alert the responder to the possibilities of a compressed gas and/or a cryogenic product (a product usually existing in a gas form but placed in liquefied form through pressure and refrigeration). Large cylindrical objects stacked lengthwise on flatcars or semi-tractor trailers may indicate high-pressure compressed gas.

Emergency responders have a limited time to gain information on an emergency incident. The primary goal is to quickly identify the product and to gather the appropriate information for mitigation. The *North American Emergency Response Guidebook* is a good start (the *North American Emergency Response Guidebook* is covered in detail in Appendix A). Several other resources can also be relied upon. CHEMTREC, a 24-hour service offered by the Chemical Manufacturers Association, can give general recommendations and contacts. The National Response Center (NRC) is manned by the United States Coast Guard and can provide you with a variety of informational contacts. Each of these emergency sources give general information and contacts, but as a rule they do not tell you what to do.

To gain more information involving toxicology and specific field treatment your local poison control center can provide valuable information. This reference in particular is addressed in Chapter 8. Listings of local poison control centers can be found in a *Physician Desk Reference* or by contacting a local hospital.

Identifying Zones

The next goal is to determine the hazard and evacuation zones. Data gained from reference material will aid in identifying the hazard zones and determining if evacuation or defending in place is needed. The universally accepted zone terminology is the hot, warm, and cold zones. Each has a specific meaning and use.

The **hot zone** (also called the *exclusion zone*) is the area where the hazardous material was released. The boundaries of the zone are established using reference data that predicts the distance that contamination can be expected to be found. The hot zone can also be established by using instruments to establish the areas of contamination then marking those areas as the hot zone. Only individuals in proper personal protective equipment are allowed to enter the hot zone; all others are excluded.

The **warm zone** is a buffer area surrounding the hot zone. Because the decontamination process takes place within the warm zone, it is often referred to as the *decontamination zone* or the *contamination reduction zone*.

The **cold zone** (*safe* or *support zone*) is where equipment and personnel directly involved in the incident are located. The cold zone is a safe zone, free of contamination (or as free of contamination as possible) where the hazardous materials team members function, medical assessment takes place, and rehabilitation is located. Bystanders and press are kept out of this zone unless accompanied by an emergency worker and then are allowed in only for specific reasons.

Most books picture the zones as perfect circles around the release site (Fig. 1-4), but this is not the case. Many times the zones are located between buildings, using automobiles, structures, or other landmarks to border the zones. Where open spaces are involved, the topography and wind both play major roles in determining the zones' shapes and sizes.

■ NOTE

A hazard zone is characterized by 3 zones: the hot, or exclusion zone; the warm, or contamination reduction zone; and the cold, the safe or support zone.

hot zone
the area where a hazardous material was released

warm zone
a buffer area surrounding the hot zone

cold zone
also called the safe zone or support zone; the area where equipment and personnel directly involved in an incident are located

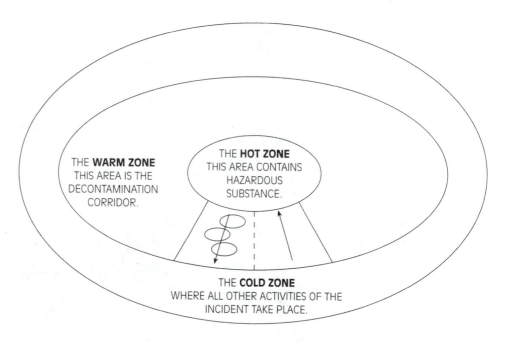

Figure 1-4 *The hot, warm, and cold zones may not be perfectly round circles as shown here. Wind, terrain, and structures can all effect their shapes and sizes. Landmarks can be used to help define them.*

Cones, banner tape, or other easily recognized materials should be used to identify the entrance and exit of the decontamination corridor located in the warm zone (Fig. 1-5). Banner tape may also be used to isolate the entire hazard area from the public.

If evacuation is indicated, a loudspeaker or the police department may be utilized. Care must be taken so unprotected workers are not placed in a dangerous position downwind. Under large-scale conditions where there are enormous geographical distances to cover, the media can be used for evacuation notification. Although evacuation is mentioned several times in this text it is an extremely difficult task to complete successfully. If large areas must be evacuated the resources to accomplish this task are enormous.

Treatment and Transportation Considerations

Emergency medical providers may be faced with several scenarios on a hazardous materials incident, including civilians or emergency response personnel being exposed or a hazardous materials entry team member being exposed because of a failure of personal protective equipment.

Persons who are found in the hot zone upon the arrival of emergency response units present several challenges for the responders. If the person is ambulatory he/she can be directed to an area where emergency decon can take place if needed. In this area responders in appropriate protective equipment can

Figure 1-5 *The zones should be clearly marked with banner tape or cones, which allows those participating in the incident and bystanders to understand the boundaries.*

assess the medical status of the patient and determine his or her medical needs. Many times the action-oriented nature of emergency responders causes them to enter unsafe areas resulting in injury or contamination, further complicating the incident. These contaminated responders must be treated the same as a contaminated civilian and be assessed and decontaminated if warranted.

Unconscious patients present a more critical situation. Emergency entry (entry that is done quickly and with less than normal preparation to remove a viable victim from the hot zone) is a dangerous practice with the potential to injure emergency responders. Guidelines for emergency entry should be followed, but even with guidelines the results of an entry of this type could be devastating. The decision to make a rapid entry into the hot zone for the purpose of rescue should be made only after the dangers are known and it is determined that the benefit outweighs the risks.

Another scenario is that of an entry team member in full protective equipment becoming incapacitated in the hot zone. When an entry team member becomes unconscious or unable to remove himself from the hot zone, several backup plans should go into effect. A plan to remove this team member from the hot zone is of vital importance. The medical responder must then be advised of when appropriate treatment can start. It is difficult if not impossible to determine the cause of the team member's difficulty until the protective equipment is removed and assessment performed. The most common injuries to entry team members usually involve a failure of the protective equipment or heat-related injuries.

! SAFETY
Emergency entry (entry done quickly and with less than normal preparation to remove a viable victim from a hot zone) is a dangerous practice.

In the above-mentioned cases, decontamination must be set up to clean the contaminated patients. When patients have been grossly contaminated, high-level personnel protection should be utilized in decon and considered even during transportation. Emergency medical providers and transportation personnel must realize that it is usually impossible to completely decontaminate a patient. In some instances, even after a washing process, the chemicals may still be issuing from the patient's skin or body fluids in high enough concentration to injure emergency personnel. For this reason the term *contamination reduction* is becoming an alternate name for *decontamination*. If it is determined that the patient may still pose a danger, additional protective measures to prevent secondary exposures should be practiced. In addition, preparing the ambulance by covering the interior walls with plastic sheeting and removing nonessential equipment may also be considered.

If plastic sheeting is used to cover the interior of a transport vehicle, then the attendant must be protected from contamination. This process makes decontamination of the interior portion of the unit easier. However, the space is now confined, and the risk of secondary contamination has increased. This may expose the treatment crew to the same level of contaminates that you are trying to keep off the vehicle.

If this level of protection is determined to be necessary, then the same level will also be needed in the hospital setting. A determination of the hospital's capabilities is necessary even before the patient leaves the scene. If the hospital cannot provide the same level of protection, and indeed the patient presents a high hazard, then the hospital staff and other patients may be placed in danger.

● CAUTION

Decontamination must occur prior to aggressive patient treatment, and full treatment should only be performed after decontamination is completed.

If decontamination is necessary, it *must* occur prior to aggressive patient treatment. If necessary, lifesaving treatment may start during the last phases of decontamination. Full treatment should only be performed after decontamination is completed. At no time should a grossly contaminated patient (one that has not received any decontamination) be transported to the hospital. This practice will contaminate the transport crew, nurses, doctors, technicians, and other patients at the hospital. It is vitally important to leave the hazardous material at the scene!

! SAFETY

● A grossly contaminated patient (one that has not received any decontamination) should never be transported to the hospital.

Most chemicals can be reasonably decontaminated at the scene and treatment of the patient can ensue while at the incident and en route. Some chemical exposures, such as organophosphate insecticides, mandate the highest level of transport protection. Organophosphates and other cholinesterase inhibitors, trigger parasympathetic stimulation. This stimulation causes the patient to loose many types of body fluids. These fluids may be highly contaminated with the chemical, causing a continued threat to the medical care givers. In this case, and others like it, the emergency workers must be protected from potential secondary exposures.

Research of the product involved is the only way one can become informed of the medical effects that results from an exposure. If medication is needed, including antidotes, it should be administered as early in the event as possible. Care for the patient must be provided aggressively with knowledge and understanding. The conditions presented by a patient injured at a hazardous materials emergency are not so different than other medical or traumatic emergencies. Care for cervical spine, airway, breathing, and circulation are of primary concern. The whole process of patient care involving a chemically exposed patient involves a delicate balance between hazardous materials concepts and medical treatment.

Transportation of Victims

Transportation of hazardous materials victims should be done to a predetermined, prenotified medical facility capable of handling this particular type of patient. Because of the unique conditions presented by a hazmat patient, transportation to an uninformed hospital should *never* occur. Alerting the hospital should be a part of the initial stages of the incident. Once dispatch has been notified of a hazardous materials incident, the hospital should be alerted to the possibility of chemically exposed patients.

● **CAUTION**

Because of the unique conditions presented by a hazmat patient, transportation to an uninformed hospital should <u>never</u> occur.

Transportation of these victims should be accomplished by ground ambulances only. Ground units can be properly protected against contamination, but if contamination occurs decontamination of an ambulance can be accomplished without specialized equipment with reasonable cleanliness assured. As a general rule, helicopters should never be used to transport a chemically contaminated patient. Helicopters cannot be adequately protected against these patients and, if the chemical affects the pilot, the results could be tragic.

Fortunately, the type of incidents requiring this level of protection are rare but they do exist. Incidents of this magnitude have resulted in emergency medical care givers becoming patients themselves. Secondary contamination has forced the closing of emergency departments. Because of the increased use of dangerous chemicals, the potential for these situations are increasing daily. Preparation through training and education for an incident of this magnitude is the only way to lessen the effects it will have on the community as a whole. Emergency service, the fire department, ambulance service, police department, and hospital staff must be prepared. Preplanning on the part of each agency and the local emergency planning committee is a must. All affected agencies should be involved in planning, training, implementing, and evaluating the system.

REGIONAL PLANNING

Planning is not only the job of each emergency response agency but is also the responsibility of the local, state, and federal governments. State and local sponsored committees assist in preparing and planning for natural and human-caused emergencies. Together, these groups comprised of emergency agency managers and public representatives, make up local emergency management. The job of emergency management involves the education of the public as well as planning for emergencies. From the hazardous materials standpoint, the community's emergency management service helps identify prospective chemical problem areas and formulate solutions. It also aids in public awareness by highlighting the public safety regarding health issues and identifying emergency evacuation procedures. By informing the media, school officials, local businesses, and the public at large, unfounded public perception can become public understanding.

Planning must be an interdisciplinary effort with its team members derived from different backgrounds. Within this team, community leaders must employ competent specialists from appropriate disciplines. Each member must bring a full understanding of his or her respective role and responsibility in the planning process.

The local emergency planning committee under the direction of the state emergency response commission has many planning responsibilities, including:

- Identifying local users and transporters of hazardous materials
- Designating facility and community response coordinators
- Preparing emergency response procedures for facilities, emergency responders, and medical personnel
- Providing notification procedures
- Maintaining inventory for emergency response equipment
- Assisting in developing evacuation plans, training, and exercises

Components of Creating a Plan

The actual planning activity should consist of small groups, each working on isolated issues. Once all issues have been addressed, then the plan will start to take form. In the general sense there are three basic components to preplanning, each of which have areas of overlap that can be thought of as a loop of planning activity. These three concepts are:

1. Controlling the risk
2. Managing resources
3. Collecting adequate information before the incident

Controlling the Risk

The hazardous materials event affects all agencies responsible for emergency care. Fire suppression, law enforcement, emergency medical services, and hospitals all have a high probability of involvement in a hazardous materials incident. Because society is dynamic, what could occur tomorrow has the potential of occurring today, therefore, planning is critical. By identifying the problems and evaluating the solutions, planning can define the roles and responsibilities of the responding and receiving agencies. With knowledge comes the awareness of potential needs. Five steps are involved toward achieving this goal of risk control and needs identification.

tier II reports
forms that describe certain chemical quantities and locations within an occupancy

1. *Identify the risks.* As a preplanning component the community must be realistically evaluated. This evaluation examines major roadways and railroads, **Tier II reports** (forms that describe certain chemical quantities and locations within an occupancy), and general industrial processes within the community. Occupancy, location, and manufacturing process, are but a few areas of concern. All of these should be identified and assigned a risk level.

Risk levels are hazard categories. For example Hazard Type A are transportation accidents that could involve large quantities of hazardous materials. Railroads, roadways, waterways, and airways all can carry bulk loads. Type B are occupancies that contain a quantity of materials that could represent a potential problem. Manufacturing plants and industries are good examples. Type C are those areas that may not contain a large quantity of material, however the high potential for risk is apparent. School laboratories, university laboratories, industrial chemical laboratories and other occupancies that may carry low nonreportable quantities are good examples. Type D involve common household, over-the-counter chemicals. These include incompatibles, small leaks, and simple chemicals. Tier II reports are a good place to start when identifying hazardous business occupancies. Maps, demographics, and geographic information all aid the planning effort.

2. *Evaluate resource needs.* Each risk level requires different levels of resources, and the available resources must be evaluated. These resources are typically based on the vulnerability and risk assessment within a community.

Vulnerability is an evaluation of the possible damage that could occur within a community. The simplest evaluations—life, environment, and property—are the priorities. Population densities, geographical layout, weather, and topography all play into the vulnerability assessment.

Risk assessment is an outlined approach toward what could happen "if." It is the process of evaluating plausible scenarios based on known facts and information. These anticipated events can be placed into several levels of resource requirement. The levels to which an incident can escalate are defined as:

level I incident
an occurrence that is not life threatening

- **Level I incident** is one that is not life threatening. The first responding agency or local system can handle the event, which may include such activities as decontamination, and treatment. The responding agency's operations would not be hindered. There is minimal effect to human life, the surrounding community, and the environment. The amount of resources required are equal to or less than any other alarm that may be handled. Examples of daily resources that are used are fire department, emergency medical services, law enforcement, poison control centers, CHEMTREC, and area hospitals. This event usually involves only one or two minor patients not capable of overtaxing the responding agency.

level II incident
an occurrence that poses a potential hazard to a large population segment

- **Level II incident** is the start of a potentially hazardous situation involving a large population segment. The condition is either a public health hazard or has the potential to become a danger to life and the environment. Here the responding agency has reached its resource capability or the incident has the high potential to exceed the agency's capabilities. Both internal and external resources will be necessary at some level for mitigation to take place. These resources are outlined under Resource Management.

level III incident

an occurrence that has a high probability of becoming a serious health hazard to human life, or that will adversely affect the environment

- **Level III incident** has the high probability of becoming a serious health hazard to human life or will adversely affect the environment. The resource management outlined in the next section will be used to mitigate the incident. These incidents are usually large-scale, multijurisdictional, and affect communities infrastructure.

First, the level versus the type of incident must be identified, then the resources that are required to handle the incident must be determined. This may take the brainstorming of task teams that are knowledgeable in certain aspects of the system. In the level II and level III incident, many agencies must become involved. Each has a resource that another agency can use. This sharing of allocated resources requires planning before the event occurs.

At each level, manpower must be realistically evaluated. For example, within a level I hazardous materials response where the responding agency has the available resources, the local ambulance service can be used in a supportive role within the medical sector. This one activity can make available from two to four hazardous materials team personnel for direct mitigation tasks. Each agency has a cadre of personnel that can be utilized within an incident if properly trained and preplanned. By training and preplanning the event, everyone knows to what level their respective jobs will take them. Thus, the incident will move predictably toward termination in a more efficient manner.

Equipment resources must also be evaluated. As the incident becomes more involved, the amount of equipment also escalates. For example, at a level I incident the amount of time on the scene may become extended. If the time frame toward termination approaches 4 hours, food, toilets, and the rotation of crews, must be provided for the responders and the supportive staff. Deciding who will provide these facilities must be preplanned.

3. *Identify the entities needed.* As the incident progresses the need to utilize other resources increases. We term these *inside* and *outside* resources. The inside resources are those that are already maintained by the county or city government, such as police departments, building department, or water department. Outside resources include private companies and require financial compensation for services. These internal and external entities must be identified and contacted for their implementation within the system. Contingency planning with retaining fees for the private contractor may be a consideration within the planning process.

4. *Develop a plan.* Develop a workable plan encompassing short-term and long-term goals. The plan must be realistic in its endeavor to deliver service.

System delivery is based on the local agencies' ability to serve and protect while maintaining a budget. Politics can play an influential role in the delivery of service. The county or city government allocates the money and expects the fire service to maintain that budget. As we have already seen across the country, budget cuts and personnel accountability have become the norm of local government. Basically, local government's ability to maintain a level of service that the community is expecting is based on acceptable risks that the locale is willing to take.

Short-term goals are those that are required within the first few minutes to an hour. These goals are met through standard operating procedures. All goals and objectives must be identified within these standard operating procedures. Each agency should have a written policy identifying the use of equipment and the personnel on the scene. Furthermore, the policy should outline tactical and operational procedures.

Long-term goals are those that are required after the 1-hour time frame and extend to recovery. These are addressed in policy-oriented procedures or contingency plans. If long-term considerations are identified in the plan, when the equipment or manpower is needed, it can be summoned quickly.

By establishing short- and long-term goals and combining the resources of several agencies, even with budgetary concerns, a plan for the good for all, can be accomplished. For example, the fire department may need supplied air respirators (SARs), breathing apparatuses that are plumbed into a large reservoir of air, however, it cannot justify the purchase of the equipment. The need was identified during a hazardous materials incident or confined space operations, however the number of these incidents occurring within the region is limited. The level of accepted (and assumed) liability is high.

On the other hand, the sewer department has SARs for its daily operation however, it needs the annual training that is required of all personnel who uses a respirator in their job. A happy marriage between the departments can occur. The fire department can provide the intensive annual training, while the sewer department can provide the SARs for fire departments needs. The budget in both agencies has been maintained without compromising the level of service within either department.

5. *Implement the plan with annual revision.* In order to be successful the planning process needs a committed task force from the involved departments during the planning and revision process. By utilizing interested individuals, solutions will be identified on the task level. Allow task operators to take the lead in solution development, implementation, and review. Task operators understand the details of the operation. These operators will identify solutions that command staff may not think of. On the other hand, command functions implemented on the command level should be presented to the task operators to see if they are realistic. If the commander does not understand the problems encountered within a Level A encapsulating air-tight suit, unreasonable tasks may be planned and ordered.

Resource Management

The short- and long-term plans identify the needs of both a single jurisdictional event as well as a multijurisdictional event. Within any plan, levels of involvement from several agencies must be identified. Each agency possesses interacting elements, without each other none can exist at the hazardous materials event. Within this structure of management internal, or government supported, and external, or privately owned, elements exist.

Internal Elements

- Building department. This department can assist in planning for the safe storage of hazardous materials in new facilities. In addition structural plans, HVAC systems, drainage, and water supplies in the building can be readily identified. Building departments may also be responsible for assisting in design modifications to make existing facilities safer.

- Mutual aid agreements. For the level II and III incidents emergency response groups from surrounding communities must be placed within the plan. Equipment and manpower can be shared by several communities, while providing service to their individual municipalities or townships. Several areas around the country have taken this concept a step further. In these adjacent communities, each agency trains and maintains a different aspect of the hazardous materials response. For example, one department may support and maintain the decontamination and medical requirement, while the next community has entry team equipment and personnel, and yet another department manages the support staff and reference sector. Each benefits from each other, however, as individual departments, costs are reduced. All train and work as one team, further reducing the cost.

- Water and sewer departments and public works. These departments can assist in the placement of water supply, drainage areas, and heavy equipment. They maintain a variety of equipment that can be used for hazardous materials mitigation. Through planning, additional equipment can be obtained without incurring costs to individual departments.

- Police department. The law enforcement agencies will be needed to block off areas of involvement, aid in evacuation routes, and maintain the orderly conduct of the community as a whole. To efficiently function in this role, the police department should have training equivalent to the operational level. They are on the road more often than any other emergency response group, however they often receive the least amount of hazardous materials training. Their role within the system must be addressed and identified as an integral part of all responses. The paramedical support teams of police departments must also be educated in the medical aspects of hazardous materials. The police, fire, and ambulance departments must have the necessary education in medical support required for such an event.

- Schools. Schools can be used to educate the public or house evacuees. Schools and their available facilities can provide temporary shelters when an evacuation order is given. In some cases they are themselves a hazardous materials occupancy. Contingency plans should identify which schools to evacuate and which schools to use as a safe haven during a hazardous materials release.

External Elements

- Contractors. Many private firms within a community may be of assistance to the fire, police, and ambulance service in times of emergencies. However, financial considerations must be addressed before the event. Heavy equipment, portable toilets, food services, housing, transportation, and communications are but a few services that private business can provide.

- Hospitals. Although these occupancies contain their own hazardous materials potential, when the event is outside of the hospital, these institutions may receive the injured from the hazmat incident. As a part of the emergency response loop, the hospital must be educated in the aspects of hazardous materials. Hospitals have the potential for receiving walk-in contaminated patients. These patients may expose the staff, patients, and visitors. Dealing with this class of patient can only be safely accomplished through preplanning.

- EMS agencies. Unfortunately many of these groups have not been trained to the level that is required for functioning at hazardous materials incidents. These agencies can serve a vital role within the hazardous materials response, even without any patients. Once an incident produces patients, then the EMS agencies become part of the hazardous materials scene. If the agency is only involved with transport, they should be educated in the aspects of medical management of a chemically exposed patient. At a minimum, all EMS personnel should be educated to the operational level and in the medical management of hazardous materials exposures.

- Chemical engineers, industrial hygienists, and toxicologists. These scientists can provide the reference sector and the planning committees with insightful information. Opinions and possible solutions based on the chemical and toxicological properties can be interpreted and disseminated to the appropriate staff member.

- Fire brigades. These entities will probably be the first response to an in-house chemical problem. Evacuation, isolation, and gathering information can all be provided prior to the responding agency's arrival. Industrial businesses and hospitals can both benefit from these first responders. The outside agency hazardous materials response team must play a part in their training so that goals and objectives of the community can be shared.

Collecting Adequate Information before the Incident

Preplanning allows responders to be ready today for the incident of tomorrow. Another means of gaining insight into incidents is to examine what has already happened. Learn from the mistakes of others. No one will handle an incident the same way. Each of us can have the opportunity to enhance our systems by placing another community's unfortunate mishap within the context of our response area. What could we do if this occurred here?

Most communities store or use chemicals that are commonly misused or involved in accidents. Using the information gathered from the incident and, based on your own system structure, work through the problem. By working through the problem (tabletop scenario) you will gain experience in that incident, the resources that may be needed, and the realistic availability of those resources.

In order to understand what problems may be present, the chemicals stored, handled, and transported through the district must be identified. From Tier II reports one can identify quantity, location, and chemical type. Many occupancies use the same chemical for different or similar processes. Once the chemical has been identified, the same prehazard analysis can be utilized for a variety of incidents and occupancies.

LEGAL ISSUES

Law

The business of hazardous materials has become entwined in law, legal premise, and standards of care. The word *law* describes a set of rules that are prescribed by a governing body. These rules are mandated to all as a role of action or conduct. Within this general definition, several areas are not truly law, but follow the basic premise of law.

You may be familiar with the impact that the 1986 SARA had and is still having on the emergency response community. SARA established requirements for the management of hazardous materials incidents. It provides a framework for training, planning, and response that emergency providers must follow. SARA also provides a structured division of federal, state, and local involvement, by utilizing emergency planning committees.

OSHA was brought into the picture as an arm of government that could establish training requirements, scene responsibilities, and enforcement of these objectives (29 CFR 1910.120). Some states adopted the OSHA plans. Other states chose not to and were subsequently covered under a mirror regulation to 29 CFR 1910.120, called 40 CFR 311. These two regulations ensured that all responders would have to abide by the law.

What is the difference between law and federal regulation? SARA is law; it was enacted by legislation. From this legislation, regulations are promulgated. That is, the law states what must be accomplished; the CFRs state how it should be accomplished. Because the CFRs are established to meet the law, they hold the weight of law. The CFRs are not legislative law, however the OSHA or EPA standard applies with the same intent.

Standard of Care (SOC)

Much information describing what should be done prior to, during, and after a hazardous materials release can be found within standards, laws, and regulations. These standards, laws, and regulations combined with experience make up what is referred to as the *standard of care*.

The SOC is the level of service that one provides as a minimum competence level during the duty or service of the provider. So we can think of these regulations as the standard of care for hazardous materials. Although the standards within the federal regulations are considered federally mandated procedures, other documents also provide a base to the SOC definition. Under standard of care there are internal components (those found inside the agency), and external components (those found outside the agency).

The internal components directly affect the operations of the agency in question. These are items such as roles and responsibility, rules and regulations, agency protocol, local custom, directives, memorandums, and standard operating guidelines. All provide a framework for the operations of the responding agency. Therefore, the SOC can not only involve federal regulations but also may involve regional or local specific standards as part of the definition.

The CFRs, legislative laws, and other documents apply to the understanding of external SOC. Federal guidance documents (National Response Team Guidelines, Civil Preparedness Guide, etc.), Consensus Standards (NFPA), training manuals, and trade magazine articles all may apply to the definition of external SOC.

The standard of care is not written in any document but instead is defined through regulations, laws, local protocol, local customs, common sense, and even in weighing what prudent emergency responders would do if faced with a similar set of circumstances. This is providing that a request for service was enacted and the agency responding had a reasonable level of skill to actually act upon the emergency. This central idea revolves around the standard of care and is the focus of negligence.

■ NOTE

As a prudent, responsible emergency responder, you are expected to act appropriately in regard to your training and education under the standard of care.

Negligence

Negligence has four separate yet interrelated elements. Each must be proved before a suit of negligence is applied. It assumes that as a prudent, responsible emergency responder, you will act appropriately under the standard of care. The four elements of negligence are:

1. Duty to act—given a set of circumstances a prudent individual would perform
2. Breach of the duty—failure of actions that are clearly a part of the SOC
3. Damage
4. The failure to act resulted in damage—causative

Duty to Act   There is an assumption that while employed by an emergency agency, providing that assistance was requested, emergency providers have a duty to perform. Once on scene the duty continues until an equally or higher qualified agency arrives. At this point transfer of the incident can occur or the scene can be effectively terminated without harm to the community, person, or environment.

Breach of Duty   Failure of actions or breach of duty occurs when the acceptable standard of care is violated. This assumes that given the same set of circumstances any prudent, responsible individual would handle the incident in the same manner. The test for this usually lies with the federal or state government (external standards). In basic terms, the standard of care was not observed.

Damage   Damage indicates that punitive, compensatable harm has occurred. An injury was suffered such that reconciliation must take place. Further damages against the defendant can be placed as a deterrent to others and/or the defendant, in a hope that the incident may not happen again.

Failure to Act   The failure to comply to the standard, or the act of failure that resulted in injury, foreseeable by the defendant constitutes failure to act. As a protection from this, adequate training, continuing education, and proper documentation of the incident is the best defense.

Negligence claims can be classified in two ways. *Gross negligence* occurs when there is a willful failure to meet the standard of care. For example, failure to provide training, not enacting an incident command system, or declining to provide emergency medical service and transportation, all would be considered as gross negligence. Not having the appropriate level of funding to meet the standard of care is not a good plea; this too may be considered gross negligence. *Malfeasance* is the intentional action to harm others while violating the standard. For example, ordering a transporting agency to take a patient to the hospital without proper decontamination and without the appropriate level of personnel protection could be malfeasance on the part of the officer ordering the patient transport. *Nonfeasance* is the failure to act when it was required. In either the case of malfeasance or nonfeasance, the standard of care has been breached or violated.

It is important to discuss negligence in the context of what should be done. The importance of these points are to develop a deliberate approach toward hazardous materials mitigation. Levels of involvement will vary among emergency responders. However, if a conscientious effort is provided toward response plan development, training activity, safety officer designation, standard operating procedures, incident command system (ICS) establishment, and EMS involvement/ training, the incidence of a negligence claim may be reduced. But the limitless level of subjective reasoning, along with broad legal applications, will only give you an appreciation for what *may* occur.

Consider this: Some legal authorities suggest that as a society recognizes certain acts or conditions, these acts become as much a part of life as the considered "normal" activities. For example, an agency maintains a local hazmat team, and this team has technician level training and has had for the past 10 to 15 years. If the agency terminates the team's activities, can the agency be held liable? Of course, this is up to severe subjective scrutiny. However, if the community has learned to accept it as the norm, then the act of abandonment could be interpreted as a conscious effort to reduce the standard of care to that community and society, in effect having a knowledgeable cause of damage before the event. The concern is that a standard of care has been established. If the social norm is to have an agency that can isolate, or mitigate, an incident and the majority of surrounding agencies have such a capability, then a standard of care has emerged.

Responding agencies that become involved with an incident if not trained to the appropriate level could be held liable. In other words, unless the responding agency has legal authority or has been given legal authority, this agency could be held liable for any negligent acts when dealing with the incident. For example, a fire department has the authority within a said community to act upon any and all hazardous situations, and for the most part it is assumed how the department will act. However, if a policy and procedure is written by the fire department authorizing a neighboring ambulance to respond to all hazardous materials incidents and the crews are not properly trained, the community and fire department may be held liable.

In the Code of Federal Regulations and within SARA, there is a strong recommendation for emergency medical services to respond to hazardous materials incidents. In the consensus standards (NFPA 473), again we see the requirements of knowledge for all EMS providers to hazardous materials emergencies. However, these documents do not provide all the knowledge that an EMS provider must possess. It represents only the tip of the iceberg.

Cost Recovery

A common incident may have the potential to create an unreasonable cost for the responding agency. For this reason known hazards can be preplanned and the cost involved in handling the incident addressed. Cost recovery is one mechanism that communities can use to regain the expenditures of an incident from the responsible party. However, cost recovery must be a part of local or state statute. Ordinances that provide for cost recovery by the response agency must be in place prior to the event.

The Federal Regulation 40 CFR 310 defines the regulations for which the intent and the process of cost containment or reimbursement to local government occurs. The regulation provides for reimbursement of the temporary emergency costs. State statutes may have to be researched for direction or compliance. Cost recovery has been a difficult issue for many departments. Most emergency services are supported by local government and have never charged for services of any kind. Therefore, it may be difficult for local government officials to recognize that charging the responsible party is one way to defer the high cost of hazmat response. Laws are difficult to wade through and can be interpreted in many different ways, making enforcement difficult. Local ordinances must support state and federal laws and regulations for the system to work.

Some areas around the country have experienced litigation against the emergency responders and in defense of the responsible party, attempting to recover expenses. Attorneys for the responsible party have charged hazardous materials teams with improperly controlling leaks or expanding spills, incurring higher costs to the responsible party. If proved then the local emergency response agency must cover the cost difference between the level of containment or confinement that occurred and the level of confinement or containment which according to "standard" tactical operations should have occurred.

● **CAUTION**

Local emergency response agencies may be required to cover the cost difference between the actual and expected levels of confinement or containment, as defined by "standard" tactical operations.

Summary

The business of emergency response is usually thought of as taking quick and decisive action. However, when discussing hazardous materials, a more deliberate and thought-out approach must take place. The problem with hazardous materials response is that one must not only become knowledgeable in chemical behavior, but also in the responding agencies' capabilities, available resources, and preplanning. This chapter was intended to refresh the memories of those familiar with the hazmat response, and to educate those whose knowledge of the hazmat response is limited.

Hazardous materials mitigation on the most basic level revolves around research, common sense, knowledge, and legality. A large component of this is the process of recognition and identification. Understanding how resources within the community can assist you during a hazardous materials event will enhance your capabilities at the time of need. The most important aspect of preparing for a hazmat event is the act of preplanning.

SCENARIO

It is 5:00 P.M. on Saturday and you have just received a call about a suspicious odor in the warehouse area of your response district. The alarm was phoned in by a security guard working in a food distribution warehouse. She stated that the odor was unpleasant and caused an irritation in her throat. She first noticed it when she went outside to smoke a cigarette. It appeared to be coming from the buildings to the east of the food warehouse. The dispatcher further explained that it became more difficult to understand the caller because she was coughing so violently. Because your crew had previously preplanned many of the warehouses in that area, you decided to reference the copies of the preplans while en route to the location. There are two warehouses to the east, one housing lumber and building supplies, the other a pool supply and recreation warehouse.

The dispatcher updates you on weather conditions while you are en route. It is 84° F, humidity is 66%, and the wind is out of the southeast at 12 mph. You arrive on the scene placing your unit upwind and uphill of the situation. A scan of the scene indicates what appears to be a white to light yellow vapor blowing across the parking lot to the south side. A closer look, using binoculars, indicates the vapor is seeping through the doors of the pool supply warehouse.

Scenario Questions

1. What should be your next function once you have pinpointed which warehouse the vapors are issuing from?

2. Knowing that a patient exists in the warehouse to the west, when should rescue be considered and how should it be accomplished?

3. To help identify the chemicals involved, what clues can you look for without placing yourself in danger?

4. To establish an initial hazard zone, what factors must be used to determine the size of the incident?

5. Several box truck trailers are in the fenced parking area and, with binoculars, you can see a yellow placard and a green placard. A red dangerous placard is also found on one truck. What might these placards indicate?

6. In addition to the DOT placards found on the trailers, the building has a multicolored placard with numbers in each color. The blue quadrant indicates a 4, the red a 2, and the yellow a 3. The white section does not have any marks. What is this placard, and what degree of hazards are present?

7. You were just advised by your dispatch that an ambulance crew was dispatched to the location of the security guard. They approached the scene from downwind and are now inside the warehouse to the west and have advised their dispatcher that they to are experiencing difficulty breathing, coughing, and tearing eyes. They are unable to drive their vehicle and are requesting that the fire department hazardous materials team respond. What do you do to help coordinate this situation?

8. Since you now know that at least three patients exist who should be informed?

9. Are there any special considerations for the care of these patients?

10. You determine that immediate treatment and transportation is a must since the closest hospital is 30 minutes away. You have been advised that a medical transport helicopter is en route and will be on the scene shortly. Is the use of a helicopter a viable solution for the long transport time?

Chapter 2

Scene Organization and Standard Operating Guidelines

Objectives

Given a hazardous materials incident, you should understand the organization of the hazardous material scene and identify the stages of the event; the competencies necessary to safely work within the command structure of an incident in which patients are exposed to a chemical; and your obligations to the victim, to the team, and to command.

As a hazardous materials medical technician, you should be able to:

■ Describe the overall priorities of a hazardous materials incident and the special considerations of each.

■ List the three functional groups involved in most hazardous materials incidents.

■ Describe the concerns of transporting a victim of a hazardous material exposure and identify the options for dealing with them.

■ List the different types of decontamination and define each.

■ List the steps that should be taken during site termination.

■ Identify the differences between A, B, C, and D levels of protection.

■ Define the three types of suit failure.

■ List and define the seven stages of a hazardous materials incident as described in the chapter.

HAZARDOUS MATERIALS MEDICAL RESPONSE

Hazardous materials scenes have elements of medical concerns. In fact the definitions given to hazardous materials all identify the medical effects of a hazardous material. Providing safety to the public and responders is vitally important. Some hazardous materials response teams still do not recognize that participation by medical personnel is not only essential but implied by laws and standards. Medical responders from private and public service, whether third service or fire department based, must become familiar with scene organization and incident command. Through coordinated efforts, complex and dangerous scenes can be handled safely and without significant damage to personnel, property, or environment.

Incident Management System

Coordination of an incident of any magnitude requires a structure that divides the scene into smaller more manageable segments. Although several systems exist throughout the United States today, all were conceived in the early 1970s during the wildland fires in Southern California. Since then many systems have been written about and adopted not only by fire services but emergency agencies of all kinds. These systems go by many names including incident command system (ICS), fire command, and even incident management system (IMS). The term *management* denotes to control or guide by making decisions based on information gained about the situation. *Command*, on the other hand, is a term adopted from the military and means to order and direct authoritatively. We chose IMS because coordinating a complex incident requires good management, not necessarily overall authority.

Throughout this book you will also note the use of the term *standard operating guidelines* instead of *standard operating procedures*. This terminology was chosen because legally, procedures are meant to be followed exactly. In hazardous materials response, each emergency is handled differently and conditions may dictate the need to violate written procedure. Therefore we are using the term *guidelines*. Guidelines are not intended to be followed to the letter but instead used to guide responders in the decision-making process.

All emergency incidents have a degree of complexity. Consider the single-unit response where an intoxicated patient falls down a flight of stairs. Although this situation is simple as compared to other emergency events, it still may contain elements that, if personnel are not properly trained, are highly complex. For the emergency responder this type of call may be classified as routine. However, take this same incident and increase the number of patients to six or seven. The scene has now become too complex for the single unit to handle. Triage must be performed, assignments completed, and treatment provided as other units arrive. The incident becomes complex in terms of resource management. This management of resources and equipment is the basis behind the incident management system. The management of emergency incidents is a necessity if resources, equipment, and manpower are to be utilized in an efficient manner.

The incident management system enables the responder to organize and control any size incident that he/she may be presented with. For the system to work, standard operating guidelines must have been developed and implemented, following the local command structure. It is a modular system that allows the manager to add needed aspects and delete unnecessary ones. Only those areas that are required for the functioning of the incident need be enacted. The system also creates a communication link between all of the necessary participants and the resources that they control. Information is disseminated in a timely manner to only those sectors that need it while others are not burdened with it. Decisions that require immediate information (intelligence) are given priority as the information is gained. The system controls, organizes, and directs the resources that are needed within the appropriate communication network (Fig. 2-1).

The system also divides tasks into functional groups, enabling these groups to concentrate on the problem at hand. For the system to properly function, all personnel within the response system must be trained in its use and activation, understanding their roles and responsibilities within the system.

Many hazardous materials response laws and standards identify the use of an incident management system. OSHA 29 CFR 1910.120 and 40 EPA 311 mandate the use of such a system. NFPA 1500 Standard on Fire Department Occupational Safety and Health Program, NFPA 1561 Standard for Fire Department Incident Management System, NFPA 1521 Fire Department Safety Offices, NFPA 471 Recommended Practice for Responding to Hazardous Incidents, NFPA 472 Standard for Professional Competence of Responders to Hazardous Materials Incidents, and NFPA 473 Professional Competencies for EMS Personnel Responding to Hazardous Materials Incidents all recommend or suggest a similar management system.

The benefit of the IMS is that it can be used for any emergency including multiple casualty incidents, fires, high-rise emergencies, hazardous materials releases, or basic medical emergencies. The basis of utilizing a command structure is to limit the span of control for any supervisor to between five and seven functions or individuals. By limiting the span of control, the safety of an incident is greatly increased.

The command structure starts with the first arriving unit and builds as the incident escalates. The first arriving company officer takes command and assigns duties as needed. If resources are required, the company commander requests these resources. As the incident grows, so does the command structure.

■ NOTE
The same principle used to manage multiple-casualty incidents or to control a large fire can be used on a hazardous materials scene.

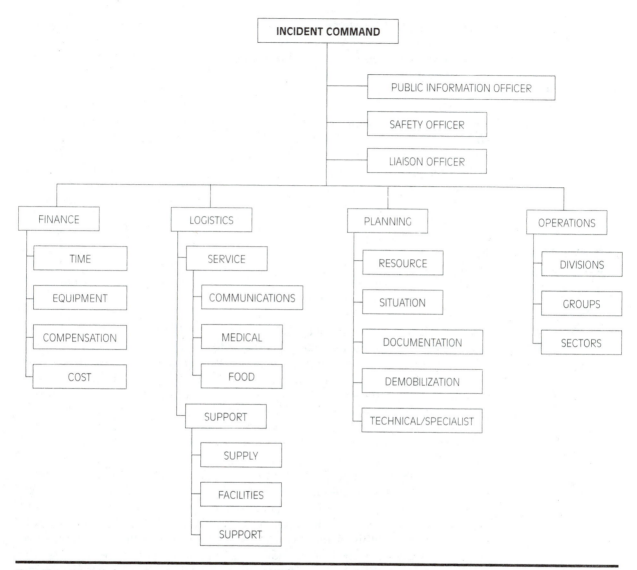

Figure 2-1 *The command structure may be complex, involving many sections, divisions, and functional groups.*

Hazardous Materials Incident Priorities

As with any emergency incident, a hazardous materials incident has priorities. These priorities are life safety, incident stabilization, property conservation, and protection of the environment (an acronym commonly used is **LIPE**). The incident commander must address these priorities early and in the order of importance.

Life Safety—Dealing with Victims   Because the threat to life is of utmost importance, one of the first items to consider is that of victims or potential victims. Establishing from a reliable source early in the incident whether people are still in the area of the chemical release is vital. If so, are there injuries to these persons? Are they unaware of the danger or unable to evacuate? Establishing further information about the condition of the potential patients is essential.

On a hazardous materials incident two basic types of victims may result from the release of a chemical. Those affected by the chemical itself (exposed and contaminated) and those suffering from injury as a result of the incident. Trauma such as concussion injuries, thermal burns, or other traumatic injuries associated with the chemical release may not be caused from the chemical itself.

Because of the magnitude of some hazardous materials incidents, victims in the hot zone may be secondary to those potential victims downwind or in close proximity to an expanding incident. Remember, in a disaster situation, responders are there to do the "greatest good for the greatest number." All rescues should be weighed on a risk versus benefit scale. A high-risk situation with questionable benefit (removal of victims who have little chance of survival) does not warrant the entry into a dangerous situation.

❗ SAFETY
All rescues should be weighed on a risk versus benefit scale: A high-risk situation with questionable benefit does not warrant the entry into a dangerous situation.

Incident Stabilization   It is the incident commander's responsibility to establish the tactical goals and strategies when dealing with an incident. The incident commander may be the first arriving ambulance attendant or the first fire commander to arrive. In any case as the incident grows the command is passed to the higher ranking official or the most qualified official on the scene. The functions performed to stabilize the scene are normally completed at the other end of the command structure. Depending on the size of the incident and the resources needed the command structure can become quite large. The final hands-on work is still done by the lower levels of the structure.

Both EMS personnel and firefighters have been ingrained with the lessons of speed and aggressiveness. EMS responders are taught from day one the importance of hurrying to an emergency scene. They are all preached to about the significance of the "Golden Hour" and most are limited to 10 minutes on the scene of traumatic injuries and 25 minutes on the scene of a medical emergency. Fire departments are no different. Standard operating guidelines all over the country indicate minimum time frames of less than 5 minutes to search an occupancy for victims and report an "all clear." Firefighters learn from the first day of training the importance of speed. They learn to don the air packs in less than a minute, catch hydrants in only seconds, pull attack lines and get water on the fire within moments. Using the time limits and aggressiveness on hazardous materials incidents can have dangerous and deleterious results. It cannot be stressed enough that a more deliberate and almost always slower pace must be taken during these emergencies.

Occasionally, a rapid response is necessary. We have personal experience with incidents in which multiple injured and contaminated persons needing help rushed the emergency vehicles as soon as they pulled up to the scene. In these cases the initial scene organization is lost and chaos is the result. It becomes somewhat of a chore to regain control, but it is the most important priority.

Establishing written guidelines will assist in the on-scene organization. The guidelines should be just that, guidelines. Each and every hazardous materials emergency is different, necessitating the need to modify the guidelines to meet the demands of that particular emergency.

Property Conservation   Because it is not regarded as a high priority, many times property conservation is neglected. However, in a hazardous materials event the property that is neglected may be the highway, railway, or community area. All of these areas can affect the infrastructure of a community to one degree or another. In any case, the effect may have far-reaching social, economic, and health-related effects. The appropriate action by the incident commanded is to impose actions in a timely manner to reduce the long-term effects of the incident.

Environmental Considerations   Recently concerns for the environment have become a top priority. If the order of priorities follow the acronym of LIPE—life, incident, property, and environment—the last item to be addressed is the environment. However, at all hazardous materials incidents the environment should be an early consideration. If the life hazard has been controlled and the incident stabilized, then both the loss of property and damage to the environment will be contained.

Many new products, equipment, and techniques are geared toward environmental conservation. What was acceptable practice on a hazardous materials incident 10 years ago is today not responsible practice. Reporting spills of even small quantities to environmental protection departments has become an important part of hazardous materials response.

ORGANIZING THE SCENE

An involved hazardous materials incident can be complex, where responsibilities must fit together for the incident to be safely completed. The functions are divided into sectors, sections, divisions, or groups, each containing one main task to complete. The task may be simple or complex and may be completed early in the incident or continue throughout the incident to mitigation.

Any hazardous materials scene must follow an organized and intentional plan involving personnel assigned to specific duties that must be carried out in a particular manner to mitigate the incident. These duties may be only a small part of a large complex scene or may make up the scene as a whole. Three overall functional groups must exist at any hazardous materials incident :

1. Hazardous Materials Operations
2. Hazardous Materials Safety
3. Hazardous Materials Medical

Hazardous Materials Operations

The complexity of a hazardous materials incident can hinder the effectiveness of any staff officer. The creation of smaller manageable sectors, groups, or divisions will assist an officer in performing specific functions. In general the tactical priorities are the primary concern of the operations officer (life safety, incident stabilization, property conservation, and environmental control).

In order to set reasonable priorities, the reference section must attain and provide the appropriate level of intelligence to the operations officer. The reference section is situated in the middle of the command structure so that all sectors needing intelligence from reference can readily obtain it.

Public Information Officer (PIO)   Utilizing a PIO to distribute information keeps the news media informed about dangers or precautions concerning the incident. The news media can in turn provide concrete information to the community to lessen fears generated by these emergencies. If plans for evacuation have been made, the PIO can alert the community through the media. As the operation progresses, the media can update the public with accurate news reports.

Reference   The success of a hazardous materials operation depends on the completion of many tasks, some conducted simultaneously (Fig. 2-2). Each task is dependent on the reference (also called science, research, or intelligence) section gaining accurate reliable information, sometimes rapidly. Within the reference sector vital information about the incident is researched based on known vari-

Figure 2-2 *Gathering information can start enroute but must continue throughout the incident.*

ables. If the incident involves a fixed-site facility, Tier II reports should be available. These printed materials can be compared to other reference texts. If the incident involves a means of transportation, container shape, size, placards, and U.N. numbers can be analyzed to identify the hazards presented by the incident. Most of the tactical and strategic goals are based on the information that is found within this section.

The roles and responsibilities of reference include information gathered in conjunction with analytical research (see Chapters 3 and 4). This section assists in the appropriate identification of the unknown materials based on visual assessment or other intelligence sources. These sources may involve video, still pictures, or first-hand accounts of the incident. Once the materials involved are identified, the hazards of those materials can be researched. The hazards may include but are not limited to fire potential, health hazards, reactivity, and incompatibilities. A rule of thumb for researching a chemical is to access information from at lease three sources.

Equipment Officer   The equipment officer identifies the material needs for the operation by consulting with the reference and operations officer. Suits, tools, monitoring, and decontamination equipment are the areas of responsibility. A prime concern of the equipment officer is the compatibilities of the materials entering the hot zone with the chemical in question. The equipment team also prepares the entry team, their equipment, and establishes the decontamination corridor. Chemical compatibility and incompatibility is a major issue when dealing with these sections. Appropriate suits and equipment must be selected that will not hinder the mitigation process and decontamination.

The Entry and Backup Team   Any time a team enters a hazardous environment, a backup team must be ready for entry, in the event that the entry team needs immediate help. This is termed the "guardian angel" concept.

Each entry team member is assigned a support person. This individual assists in dressing, ensuring that all of the needed equipment is utilized and in working order. The use of a checkoff sheet places all of the functions in order and ensures that no portion of the dress out process is missed. Dressing of the entry crews is conducted simultaneously with medical surveillance and tactical briefing. Once the process is completed the equipment support person reports the readiness of the entry personnel to the safety officer. The safety officer then ensures that the equipment used is appropriate for the incident. This system of checks and balances allows the entry personnel to enter the hot zone with a reasonable degree of safety.

The level of protection and the type of suit utilized by the entry, backup, or decontamination teams is determined by information generated from the reference section. It is the responsibility of the reference sector to watch for compatibilities and incompatibilities in suit selection.

Suit Selection   The levels of protection as defined by EPA are outlined below. Generally four levels, from Level A, the highest protection, to Level D the lowest are recognized, although other specialized equipment may be used.

- Level A consists of a fully encapsulated chemical protective suit with self-contained breathing apparatus. This suit enables the wearer to enter atmospheres that can cause injury by absorption, inhalation, and ingestion. It offers limited fire protection and therefore should not be used in a flammable atmosphere.
- Level B consists of hooded chemical-resistant clothing including gloves and boots with a self-contained breathing apparatus. Level B protects the responder against inhalation and ingestion and, to a lesser degree, the absorption of hazardous materials. Level B is not fully encapsulating and therefore some areas of skin are exposed.
- Level C protection includes the same level of skin protection as level B but a lower level of respiratory protection. The skin protection protects against absorption from a liquid or a solid. The respiratory protection consists of an air purifying respirator that provides a filtration of air through an air purifying canister face mask specific for the chemical involved. The user breathes the air from the surrounding atmosphere and therefore the air must be at least 19.5% oxygen.
- Level D is the work uniform, chemical-resistant boots, and safety glasses. There is no provision for respiratory protection. This level is used where there is a very low level of contamination and no known hazard. The responder wearing Level D should not be subject to splash, immersion, or inhalation of a hazardous material.

Chemical protective clothing is designed to protect the user against chemical contamination. However, no garment can protect the user against every chemical. Most protective garments are specific for a certain type of chemical. Sometimes multiple layers of materials are used to afford the widest range of protection. Because of this layered protection, the garments become bulky and reduce the user's dexterity and vision. These garments not only keep hazardous chemicals out but also keep heat and moisture in, additionally limiting the user.

Three major concerns are evaluated when selecting the most appropriate suit: Penetration, degradation, and permeation. The equipment officer and reference must be cognizant of these principles.

Penetration is the physical movement of a chemical through the natural openings of the suit. Zippers, exhalation ports, seal around the face shield, and connections can provide a route for chemicals to enter the suit. Abrasions, punctures, or tears can contribute to penetration. Temperature gradients (the temperature of the suit and the atmosphere) can greatly affect the penetration values of the suit. The most common form of matter that enters the suit in this fashion is liquid material, however, finely divided particles (solids) and gases (under pressure) can also penetrate.

Degradation can be caused from the physical destruction of the suit by temperature, concurrent chemical exposure, inappropriate storage environment, and incompatibility with a chemical. It is marked by discoloration, bubbling, chaffing, shrinkage, or any visible signs of destruction of the protective material. Unfortunately, most of the protective materials are tested at approximately 70° F, without considerations of repetitive chemical exposures, temperature fluctuations, or more than one exposure over time.

Permeation is the movement of the chemical through the protective material at the molecular level. As is discussed in Chapter 3, atoms have "space" between the nucleus and the electron and space between bonded atoms. Chemicals can permeate the material in these areas. Chemicals can also react on the molecular level causing permeation. It is thought that because atoms are constantly trying to attain a more stable state, some chemical exposures to the suit fabric will alter or absorb a percentage of the assaulting chemical. As time continues the absorption produces a "layer" of the toxic material just under the fabric's surface. When the concentrations build, either through prolonged exposure, repetitive exposure, or incompatibilities with the chemical, a diffusion of the hazardous material takes place, moving through the outside of the fabric. Eventually, the hazardous material breaks through the fabric (called *desorption*), leading to an exposure of the individual within the suit. When referencing suit compatibility, the breakthrough time should be referenced and reported to the equipment officer. Breakthrough time is based on a maximum of 8 hours or less (480 minutes) with the movement measure in microgram per centimeter squared per minute.

Hazardous Materials Safety

The safety officer is directly in control of those areas of the operation entering, within, or exiting the hot or warm zone (Fig. 2-3). This individual oversees the critical areas of potential secondary contamination or other safety concerns. If at any point during the operation, the safety officer deems any operation to be unsafe or conducted in an unsafe manner, this individual must have the full authority to stop the operation. Responder safety must be in the forefront of the safety officer's mind.

Hazardous Materials Medical

The hazardous materials medical section has many responsibilities including overseeing medical readiness for entry, rehabilitation, monitoring the effects of heat on the crews, treating patients affected by a hazardous material, and overseeing medical staging and transportation.

Rehabilitation   Rehabilitation of personnel is often overlooked or not addressed to an appropriate level. Too often it is assumed that emergency responders who become fatigued will request to rest when unable to work. But they cannot be relied upon to decide when a rehabilitation break is needed. Most emergency

Figure 2-3 *The hazardous materials safety officer is in direct control of all duties in the hot zone and warm zone.*

responders will work to total exhaustion and then be unable to work any longer. The rehab time needed for the totally exhausted individual may span hours or even days. By working the responders to full exhaustion, the incident commander (IC) will eventually lose his/her entire workforce.

Medical surveillance (Fig. 2-4) should be applied to any task worker at a hazardous materials scene to establish baseline health criteria. The cursory entry medical and exclusion criteria for hazardous materials team workers is outlined in Chapter 7. The medical officer should ensure that procedures are followed that help to enhance the productivity of the team, while maintaining health status.

Treatment   Treating chemically contaminated patients is a challenge. The medical treatment personnel should have a working knowledge of hazardous materials incidents, patient removal, and decontamination and chemical properties so that they can better protect themselves and their equipment. Legal considerations must also be examined when the treatment personnel are not trained in hazardous materials response. It is generally interpreted that persons trained to a minimum of operations level are allowed to function in a contaminated environment but only when they have received additional training in specific functions like decontamination and air monitoring. Patients who have been through the decontamination process can be treated by persons who have not been trained in hazmat response. Utilization of these personnel may be risky because, as men-

■ **NOTE**
Since most emergency responders will work until they are unable to work any longer, and the rehab time needed for a totally exhausted individual may be hours or even a day, by working responders to full exhaustion the incident commander will eventually lose his/her entire workforce.

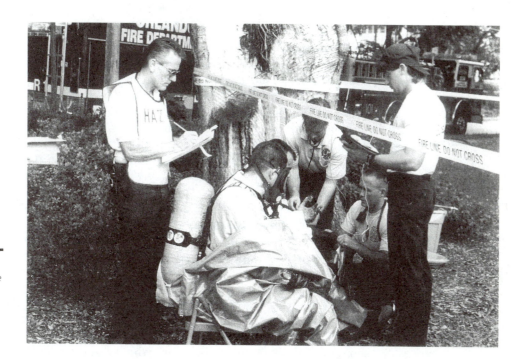

Figure 2-4 *Medical surveillance is done on all entry personnel before entry and after exit from the hot zone.*

● CAUTION

The medical treatment personnel should understand that exposures can have a long-lasting effect, causing damage to the body systems days, weeks, or even years later

secondary contamination

contamination from a previously contaminated person or object that occurs away from the initial scene

tioned throughout this book, contaminants can never be completely removed and may still pose a danger. The medical treatment personnel should understand that exposures can have a long-lasting effect, causing damage to the body systems days, weeks, or even years later.

Transportation   Transportation of victims who have been contaminated involves many considerations. The complete removal of contaminants may not be possible. Prior to transport, a determination must be made as to the possibility of **secondary contamination** (contamination from a previously contaminated person or object that occurs away from the initial scene). If secondary contamination is a hazard either from the chemical emitting from the skin, respiratory system, or body fluids, protection of equipment and the ambulance itself may be necessary. Although complete protection of an ambulance is not common, occasionally it is the safest and least expensive way to protect costly equipment.

Providing protection to the interior of an ambulance can be a time-consuming job, but can be easy to accomplish. If possible, a reserve ambulance should be used so that in the event of serious contamination it can remain out of service for decontamination and not hinder normal department functioning. Fortunately most chemicals can be sufficiently reduced through proper decontamination. Draping the interior of a transport unit is an option. Another means of transporting patients who still pose a danger after decontamination is by encapsulating the victim. These two methods are called the Reverse Isolation Method and the Radiation Model. Each has benefit in differing situations but both have drawbacks.

Reverse Isolation Method. This method provides a barrier between the rescuer and the patient. The process is accomplished by first placing a piece of plastic under the stretcher and rolling the edges in to form a basin under the stretcher. This provides a means for catching any runoff or body fluids under the stretcher. Next, place another piece of plastic sheeting under the patient and then wrap the same piece over the top of the patient. The body is separated from the outside environment. It was once acceptable to place patients in body bags to accomplish the same isolation. Although this practice provides an excellent barrier between the rescuer and the patient, it has proved to be detrimental to the patient in several ways. First, the patient is continuously exposed to the chemical trying to escape the victim's body. Next, some chemicals, such as solvents, can cause central nervous system stimulation. This stimulation increases the body temperature and when confined in this type of surrounding may lead to heat stroke or other related injuries.

Radiation Model. This isolation model is more widely used as it allows the caregiver to fully assess the patient while en route. With this model the entire patient compartment is draped in plastic in order to provide an "isolated" environment. Once the compartment is sealed, the rescuer and the patient may need some form of respiratory protection, which can come from air purifying respirators, supplied air respirators, or self-contained breathing apparatus. Even giving the patient and the rescuer oxygen via a non rebreather mask will provide a low level of protection from respiratory hazards.

An easy and quick method for isolating the interior of an ambulance is to roll out a section of plastic sheeting (14 feet wide) from bumper to bumper. Then unfold the plastic and tape the edges together to make a long "tube." Then tape the front of the tube closed, forming a sausage shape. Place the entire assembly into the patient compartment very much like an internal condom. Take three 1/2-inch by 8-foot PVC tubing, bend them in an upside down U shape and place them inside of the tube, spacing one in the front of the compartment, one in the middle, and one at the entrance. The PVC will expand and act as a spring, holding the plastic open, similar to a tent inside the unit. Once the patient is placed into the ambulance the support team can fold it shut. This forms an isolated space for patient transport.

This process makes decontamination of the interior portion of the unit easier. However, the space is now confined, and the risk of secondary contamination may be increased. During transport, patient care and monitoring must continue, which may expose the treatment personnel to secondary contamination. By the time the patient is ready for transport, the chemical should be identified and the chemical, physical, and toxic properties known. All equipment that is to be used for patient care (cardiac monitor, pulse oximetry, oxygen cylinder, air monitoring equipment) should be covered with a layer of protective plastic.

Medical providers should use personal protective equipment. Skin protection can be accomplished through the use of lightweight suits made from Tyvek,

supplied air respirator (SAR)

a breathing apparatus that is plumbed into a large reservoir of air

self-contained breathing apparatus (SCBA)

a closed breathing system that contains a limited air supply used in hazardous atmospheres

air purifying respirator (APR)

a mask used to filter outside air

Sarnex, or similar protective material. Respiratory protection may also be required. A **supplied air respirator (SAR)** is the most useful in this environment because it gives the rescuer mobility within the transporting unit and decreases the weight on one's back. The down side is that the medical providers are limited to an area just around the vehicle and air hoses can become easily tangled or hung up on other objects.

If SARs are not available, then a **self-contained breathing apparatus (SCBA)** can be used. An SCBA is a closed breathing system that contains a limited air supply. This system provides a high level of respiratory protection, however the weight of the bottle and the apparatus decreases mobility. The length of air time is limited unless arrangements for a continuous air supply have been made.

A full-face **air purifying respirator (APR)** in which a filtration mask filters outside air may be appropriate if the specific cartridge for the chemical is available. For APRs to afford proper protection, the user must be previously fit tested and the offending chemical identified. Because of these drawbacks, APRs are not often used on emergency scenes and mostly have been phased out of emergency response.

If the radiation method is used, an issue that should be evaluated is that of compliance with confined space regulations (see Chapter 12). A confined space is an area that has a potentially hazardous environment and limited access, and is not intended for human occupancy. Although the patient compartment is designed for human occupancy, the modified configuration must be examined. With this modification the patient compartment has limited access only through the back. The compartment also has the potential for being a hazardous environment. Therefore the air quality should be monitored for oxygen deficiency or enrichment, flammability, and toxicity. Air monitoring should take place as a safety factor. If the environment has the potential of becoming hazardous, then the issue of personnel protection must be addressed.

If the patient was appropriately decontaminated, is it safe to assume that the contaminates have been removed? If the answer is yes, then why go through the time and expense of draping the interior of the truck? These are commonly asked questions that can be answered by stating that total decontamination almost never exists. Decontamination is merely the reduction of contamination on a victim or equipment. Hazards may resurface when the patient sweats, cries, breathes, coughs, or vomits and may recontaminate themselves and the surroundings. In some cases isolation of the patient is the only safe means for handling such a patient.

Organizational Summary

All hazmat response duties must fit together and responders must function as a team to safely mitigate the incident. They may be part of a larger organization called incident management. A hazardous materials release may represent the entire incident overseen by an incident commander or the incident commander

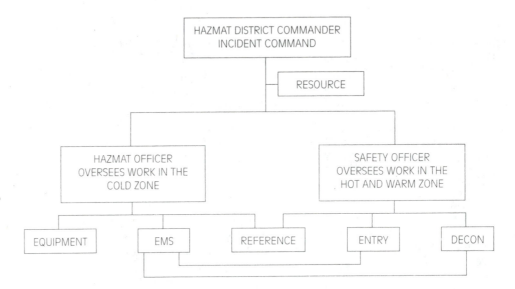

Figure 2-5   *The hazardous materials organizational chart.*

may be overseeing a much larger incident where only one segment is dealing with hazardous materials.

For example, an overturned tractor trailer spilling its contents onto the roadway, into storm drains, and contaminating soil and a nearby waterway represents a large, complex hazardous materials incident. In this case the incident command could be involved and complex but the entire scene is the hazardous materials emergency (see Fig. 2-5). On the other hand, a lumberyard fire with an exposure of a pesticide storage facility receiving damage from heat and water runoff is also a hazardous materials emergency. This incident would involve an incident commander overseeing the entire incident but would necessitate a hazardous materials subdivision with a hazardous materials commander taking charge of the hazmat sector and reporting to the incident commander.

The success of either of these hazardous materials incidents depends on organized and established policies and guidelines. Furthermore, for the policies to work, they must be flexible enough to allow adaptation to meet the emergency at hand.

HAZARDOUS MATERIALS ORGANIZATIONAL STAGES

The hazardous materials team normally progresses through stages as the scene matures. Upon initial arrival of the team certain goals must be met and built upon through termination of the incident. The stages of development of the hazardous materials scene are investigation, planning, preparation, hazard mitigation, clean up, site termination, and critique. Each of these stages will be addressed and the jobs performed in each identified. Many hazardous materials teams use the acronym APIE to help organize their thoughts and actions.

A — Analyze

P — Plan

I — Implement

E — Evaluate

As the stages of the hazardous materials incident are covered, the acronym will be examined to illustrate how the thought process fits into the organization.

Investigation Stage

Investigation starts when the unit is dispatched or when the first information about the incident is received. The hazmat operations officer coordinates any activities through the incident commander. If incident command has not yet been established, then to establish command will be the first step in the framework that will build into the incident command system.

Certain information can be gathered from the EDC while en route to the emergency and placed on the worksheet by the senior officer. Information such as wind speed, wind direction, temperature, preliminary chemical information such as type of chemical, type of problem (spill or leak), and the nature of the site (industry, highway, urban, rural, etc.) can be gathered and recorded.

Reference may begin to research the chemical(s) involved in the emergency. Finding out exactly what chemicals are involved may become a chore if basic information is not available on the scene. Clues from placards, shipping papers, MSDS sheets, or information from preplanning all may be used to identify the chemical. Once a chemical can be identified, then basic information from the NAERG Guidebook can be used for evacuations and precautions. Additional information about the chemical's properties such as health aspects, flammability, reactivity, and environmental damage potential should all be established and verified from three separate sources. The assigned reference personnel should complete a prepared worksheet. Using a worksheet provides a more efficient means of recording research so that the work is not done haphazardly.

The safety officer, with guidance provided by the hazmat operations officer and reference, should establish a hazard zone. The hazard zone includes the hot and warm zones. By establishing this zone, initial evacuation can be identified and predictions about the movement of vapors, warranting future evacuations, can be considered.

Together the hazmat operations officer and the safety officer can address options for an action plan. Documenting such a plan on a worksheet will assist all sectors when the plan is carried out.

ANALYZE—The process of gathering facts and organizing those facts to gain a complete picture of the scene. This is also known as size-up.

Planning Stage

In the planning stage any information obtained is reviewed and an action plan is established. Certain questions must be answered before completing the action plan.

What is the hazard?

What will happen if nothing is done?

What needs to be done?

What is needed to perform the work?

What support is needed?

Should we undertake this task?

If at this point it is determined that the options are beyond the scope or ability of the personnel on the scene, then additional units or mutual aid must be summoned to assist or a decision to terminate your involvement should be made.

The plan must be developed as a group process involving the incident commander, hazardous materials operations officer, and the hazardous materials safety officer (Fig. 2-6). The plan should be recorded and followed. Any significant deviation from the written plan should involve an agreement by the commanding group.

PLAN—The steps taken to mitigate the emergency. The plan should be written and followed. If changes in the plan are needed, then it should be discussed and a new plan formed.

Figure 2-6   *In the planning stage, available options are evaluated to determine if these options are within the scope of the team's ability.*

Preparation Stage

During the preparation stage the plan is approved by the incident commander. Preparation must be made to perform the work. A briefing is held for all hazardous materials personnel. The briefing includes a presentation of the step-by-step plan for operation and identifies each section's responsibility in the plan.

The hazmat operations officer coordinates the activities and keeps command informed of the progress. Hazmat safety oversees the setup of decon and entry, monitors all the areas for safety, maintains the hazard zone security, and ensures that there are no nonessential personnel within the hazmat setup area.

Reference continues to search for additional information that may affect the scene. Reference notifies the hazmat officer of any new information and maintains communication with both the hazmat operations officer and the safety officer, informing them of any changes within the scene that would affect the safety of the operation.

The equipment section gathers and stages equipment needed for entry, decon, and medical. It responds to requests for additional equipment by any other section as rapidly and efficiently as possible.

The entry team checks the suits and equipment for entry. The preentry physicals are conducted and the entry team members are hydrated (Fig. 2-7). Then the entry and backup teams don the equipment and are ready for entry.

The decon area is selected within the identified warm zone. Decon is marked, identifying the entry and exit corridor. If the scene warrants, preentry physi-

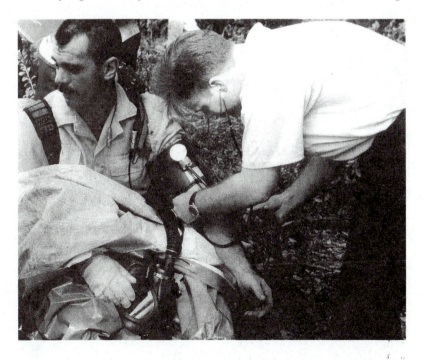

Figure 2-7 *Prior to entry, preentry physicals are conducted and hydration provided.*

cals are completed and the team is hydrated. The decon team dons protective gear.

The medical section establishes a treatment area and ambulance staging area. It also performs the preentry physicals on the entry and decon members. Communications are established with the receiving hospitals, advising them of the types of possible chemical injuries. Medical also communicates with reference to establish the toxicity of the chemicals involved and possible treatments in the event of an exposure. In addition, the medical section prepares for the treatment of an entry team member in case of heat cramps, exhaustion, or stroke. On a high-temperature, high-humidity day, this preparation should be high priority.

> **IMPLEMENTATION—The enactment of the plan. All sections must be ready for the entry to take place and be ready for emergency action if anything goes wrong.**

Hazard Mitigation

Entry   During the hazard mitigation stage of the operation, entry into the scene is made to deal with the problem. The scene may necessitate several entries. The first entry is made to provide reconnaissance of the scene and determine the equipment needs to mitigate the incident. The reconnaissance team may also perform duties within the scene such as closing valves or rescuing victims. Using a protected video camera or a Polaroid to take snapshots provides excellent visual information to the hazardous materials operations officer.

After a reconnaissance is complete, a second entry can be used to move equipment to the scene and start the work. Each entry should have a specific goal. Hazardous materials incidents involve release of a substance from the confines of its container. The ultimate goal of entry is to contain the material or allow it to disperse, making the scene safe again (Fig. 2-8). To accomplish this, several entries may be needed necessitating the rotation of personnel.

> **EVALUATION—The process of assessing the plan's progress and changing the plan to meet changing or new situations.**

Decontamination   Decontamination is also a part of the hazard mitigation stage. After each entry, the team coming out of the hot zone must be decontaminated. Tools carried into the hot zone are usually staged at the tool drop area located at the entry into the decon corridor for decontamination or future use.

Decontamination is not usually thought of as a medical process, but if improperly performed, has many medical ramifications. Improperly decontaminated patients handed over for treatment and transportation can cause harm to medical personnel. For this reason, at least a minimal knowledge of the decontamination process is reviewed in this book. Many teams utilize medical personnel

Figure 2-8   *The goal of entry is to contain, disperse, or neutralize the hazardous material and make the scene safe once again.*

to determine what decontamination process is most appropriate for contaminated patients. Medical personnel may also be ideal for overseeing the decontamination process because of their in-depth knowledge of the body systems and routes of entry of chemicals into the body.

Decontamination is the removal of chemicals or other foreign material from equipment or personnel. Removal of these materials can be accomplished either through a physical (brushing off) or chemical (neutralization) process. There are several types of decontamination:

- Gross decontamination refers to the removal of all clothing from the patient and may include a quick rinsing of the skin.

- Secondary decontamination is the mechanical removal or the diluting of a chemical. Dilution is the most common method used to decontaminate a patient. The solution of choice in the decontamination process is water. For water to efficiently decontaminate a patient, the chemical on the patient must be water soluble. Most decontamination solutions contain water and a mild detergent that will work to break down the non-water soluble chemicals. Harsh detergent solutions are never used as they tear down the protective barriers of the skin and may, in fact, speed absorption through the skin. Occasionally a special solution may be needed to complete the decontamination process.

- Technical decontamination is a method used to decontaminate equipment but is not used for human decontamination. Absorption, degradation, neutralization, and disinfecting are examples of technical decontamination.

Figure 2-9 *The decontamination stations are located in the warm zone. The personnel working in this zone usually wear protective equipment one level below the entry team personnel.*

The decontamination area (sometimes referred to as a *corridor*) must be organized any time patients are contaminated or before entry team members enter the hot zone (Fig. 2-9). The decontamination corridor may be one of the most important sectors within the hazardous materials incident. This corridor "draws" the line between the unsafe area (hot zone) and the safe area (cold zone).

The decon corridor should have a buffer zone to one side that can be utilized by the safety officer. This buffer area is parallel to the decon corridor so the safety officer can see the process without being in high level personal protection. It can also be used for moving clean equipment toward the hot zone. In this area medical equipment can be staged for future use within the decon corridor or once a patient has had primary decontamination (see Figure 1-4, page 19).

Once a patient is brought to the warm zone, the first step is to remove all of the clothing. By doing so, up to 60%–80% of the contaminates may be removed. Once in the corridor, the patient is transferred from station to station using a new backboard at each stage. This transfer can be accomplished using a scoop stretcher to lift the patient from one backboard to the next, washing the back of the victim during each transfer. At each station the individual is washed with bland nonchemically reactive soap and water solutions. Personnel must avoid contact with the removed solution as there is a potential for secondary contamination. If powder is present brush off the powder or vacuum it with a high efficiency particular air (HEPA) vacuum cleaner. Wash the patient in a head to toe fashion, paying particular attention to the top of the head, arm pits, groin, nose, eyes, ears, and nail beds. While washing the patient, avoid high water pressure, which could "inject" the contaminates into the skin. Anywhere there is a high concentration of hair follicles there is a high level of absorption. Folds of the skin should also be identified and cleaned appropriately.

When a patient is completely washed, the skin should be blotted dry and protected from the environment. Most hazmat teams use water that originates in the hydrant system. Even on the warmest days the difference in water temperature and air temperature can result in a drop of core temperature. During the winter months or when the ambient temperature is cool, care should be taken to provide warm water for decontamination. Enclosed decontamination units, although expensive, are gaining in popularity throughout the nation. In these units, conditioned air and temperature-regulated water provide the optimal conditions for decontamination.

Cleanup

During cleanup all the equipment is cleaned, bagged, boxed, or barreled and/or disposed of. Reusable equipment should be cleaned on the scene so that it can be placed back into service. If there is any question about the effectiveness of the decontamination process, the equipment should be sealed in a container and transported to a holding area until expert advice is given on the effectiveness of decontamination or if disposal is necessary.

Depending on the severity of the incident the decon crew may also require decontamination. The decon process should start with the member closest to the hot zone and continue away from the hot zone until all of the decon crew are finished.

During cleanup the medical section performs all postentry physicals, observing entry and decon personnel for signs or symptoms of toxic or heat injury. Medical also informs all the responders and civilians of the signs and symptoms expected from exposure to the chemical(s) involved in the emergency. Depending on the severity of the incident the medical section may also request follow-up care for the team in the form of postincident stress debriefing.

Site Termination

Site termination is the final stage conducted while still on the scene. The goal of this stage is to inform the responsible party of his responsibility for cleanup and replacement of equipment used on the scene. Cleanup can be enforced by informing and working with the local or state environmental departments. Many local departments lack enforcement capabilities when dealing with disposal of contaminated soil or recovered hazardous materials. The state departments of environmental protection or its county-level counterparts have enforcement capabilities and must be informed. Some cities can enforce within storm water and sewer departments and within the fire safety management divisions.

At the very least the following steps should occur during site termination.

1. Advise the responsible party as to the hazards that still may exist.
2. Advise as to the work completed for mitigation.
3. Provide a list of cleanup contractors as necessary.
4. Provide a list of information contacts as appropriate (state and local authorities).
5. Inform responsible party of follow-up checks for compliance.
6. Advise as to potential medical and environmental problems.
7. Advise as to the legal requirements.
8. Answer questions if possible.
9. Decide whether the hazmat responders should remain on the scene during contracted cleanup operations.

Steps 3 through 7 may also be done by state enforcement authorities if they are on the scene. Sector worksheets and any other information should be gathered by the resource section and filed for future reference.

Critiques

A detailed critique should be conducted on each significant incident. The critique should involve all of the sector and commanding officers along with any other interested persons who participated in the event. The goal of the critique is to identify areas of the operation that need improvement or could be handled more safely or efficiently. Specifically, operational and procedural issues should be discussed. Training and preplanning shortcomings should be identified along with any other deficiencies. Not only should weaknesses be discussed but also the strengths of the incident. The critique should never be used to point fingers or discipline individuals. Because the critique is used to actively assess the situation as a whole, the meeting should be conducted informally, with all participants having equal opportunity to express their opinions. The results of the critique should also be recorded and shared at the next scheduled team meeting.

Summary

Organizing the hazardous materials scene takes teamwork and the willing participation of a multitude of personnel and agencies. The scene can only function once the goals are identified and each player understands the part he is to accomplish. This is achieved through proper planning and written operational guidelines. As each stage of an incident is planned for and implemented, it must be evaluated to determine that the goal is being met. The hazardous materials scene is a continual cyclic process of analyzing, planning, implementing, and evaluating.

SCENARIO

You are the officer of an advanced life support (ALS) pumper and have just arrived on the scene of a large structure fire in a vacant warehouse on the riverfront. The warehouse, which has been vacant for at least 20 years, once housed fertilizers and pesticides for the local farming community. Because of a large amusement park built in the area and the booming growth of the tourist industry, the farms have virtually disappeared. In fact just adjacent to the burning structure is an entertainment facility with a gambling riverboat tied up at the dock. The facility is filled with tourists ready to embark on their trip up the river. Many of them are outside watching the efforts of the fire department.

When you refer to the preplan you find that a smaller detached structure contains abandoned tanks of anhydrous ammonia and pallets of ammonium nitrite. You also find that the property was targeted as a hazardous materials site and slated for cleanup during the next year. You request the hazmat team for their expertise in dealing with the additional hazards.

In addition to dealing with the large population of tourists, you now notice the Eye of the Sky helicopter over the fire, fanning the smoke in different directions and the On the Spot News van setting up for a live broadcast just across the street from the fire.

Scenario Questions

1. Who makes up the initial command structure?

2. What functional groups would need to be immediately set up for this incident?

3. Would you order the news agencies out of the area or off the scene? To where and why?

4. When the hazmat team arrives, where in the command structure would they belong?

5. What means can the hazmat team use to identify the hazards present?

6. Once identified, what type of research is necessary to properly control the incident?

7. If the firefighters became grossly contaminated what additional functions would be necessary prior to treatment? Transportation?

8. What would be the main concerns of the hazmat team upon their arrival?

9. Is run-off from this fire a concern?

10. Is a decontamination area necessary? Is this a function to be provided by the hazmat team? Why or why not?

FUNDAMENTALS OF THE HAZARDOUS MATERIALS DISCIPLINE

An emergency worker was once heard saying "When emergency responders arrive on the scene of a hazardous materials incident, isn't it absurd that these responders who know little about chemical processes are the ones rushing in while the chemists are the ones running out?" This observation poses some interesting questions. Why are the scientists the ones running out when they possess vast knowledge about chemical processes? Why are the emergency responders handling such an incident? The answers to these questions have roots in the training of emergency workers. That training is geared toward handling incidents that contain elements of the unknown. Emergency responders are willing to take calculated risks to handle those emergencies. To successfully determine those risks an extensive knowledge about chemistry and toxicology must be achieved.

For many years it was thought that all chemical exposures would affect the human body with disastrous results. Thankfully, for the most part this is not the case. Most exposures, because of concentration, length of time, or individual resistance, do not cause the degree of harm that was once expected. However, this does not imply that one should be complacent. Statistics from research done over the years by the International Association of Firefighters indicate that emergency workers have a life span 10 years less than the general population. Reasons given for this have ranged from poor eating habits and smoking to the lack of physical conditioning. Actually the habits of emergency workers generally are much healthier than most Americans. Some feel that low-level exposures to a wide range of toxins over a career contribute greatly to the shortened life span. Others contend that several substantial acute exposures may be the culprit. Unfortunately, the true effects of these exposures have not been truly established.

There has not been enough research conducted on the human model to positively state what level of chemicals cause toxicity. Chronic effects that are seen in the emergency response arena is high, with few connections between cause and effect. It is thought that the multitude of chemical exposures an emergency responder may have over the course of his/her career has a negative influence on health later in life. These repetitive exposures may, in some cases, lower the immune response, thus allowing opportunistic infections to take place. Others believe these exposures lower the body's tolerance to chemicals in general, stimulating a variety of disease processes in later years. The answers to questions of cause and effect lie in the future of medical surveillance.

The following chapters were included to inform the emergency responder about the possibilities presented during a hazardous materials response. Their main thrust is to aid these responders in developing the skill of product identification. Captain Keith Williams of the Tampa Fire Department encapsulates this idea in one easy phrase, "What hazmat chemistry does for responders is provide them with a means of enemy identification." The purpose of a detailed study of chemistry for emergency responders is to learn to identify the enemy, which requires a knowledge of chemical structure, toxicologic nomenclature, and many abbreviations.

Decisions made on an emergency scene can have an impact on the team, support personnel, and the community at large. These decisions at times are based on limited facts about the incident and an overall understanding of hazardous materials. Responders must take the lead and educate themselves to a high level so that crucial decisions can be made with safety and confidence. To that end a knowledge of chemistry and toxicology is necessary.

This section is designed to give a workable understanding of chemistry and toxicology as it relates to the naming process, toxicologic data, and the referencing of these facts. It is hard to limit the level of difficulty within these disciplines because of the complexity of each. These chapters are not intended to make chemists or toxicologists out of emergency responders, only to provide them with a set of tools to use when faced with a hazardous materials emergency.

Chapter 3

Chemical Behavior

Given a list of chemical reference criteria, you should be able to define, identify, and appropriately describe their importance as they relate to risk assessment, health, and safety surrounding a hazardous materials incident. You should be able to describe the physical, chemical properties, concentrations, strengths, structure, and form of the hazardous material, primarily as they relate to product identification, personnel protection criteria, decontamination procedures, and treatment protocols.

As a hazardous materials medical technician, you should be able to:

■ Describe the relative chemical and physical properties that may present as a health risk to the emergency worker.

Melting point	Expansion ratios	Saturated hydrocarbon
Boiling point	Temperature	Aromatic hydrocarbon
Density	Air reactivity	Specific gravity
Solubility	Chemical reactions	Vapor density
Vapor pressure	Inhibitor	Viscosity
Appearance	Salt	Ignition temperature

Flammable explosive limits	Combustion	Polymerization
	Catalyst	Nonsalt
Flash point	Critical temperature and pressure	Unsaturated hydrocarbon
Reactivity		

- Define the atomic structure and how it relates to risk assessment.
- Describe how bonding can give clues on chemical behavior.
- Identify a chemical and its risk assessment based on the symbols and/or formula.
- Describe the risk assessment of a chemical based on its location on the periodic chart.
- Identify the differences in bonding.
- Describe the principles of bonding and their relevance to the risk vs. gain process.
- Describe the principles of the gas laws and their affect in decision making.
- Describe how the rates of reaction can affect the risk assessment.
- Identify the principles of acid–base reactions in terms of:

Concentration	Strength	pH

- Identify the hazards and risk assessment of the following organic families:

Alcohols	Amines	Esters
Aldehydes	Carboxylic acid	Ketones
Alkyl halides	Ethers	

- When given a scenario, the risk assessment, health effects and possible chemical reactions shall be identified utilizing:

Technical resources	Interpretive data
Material safety data sheets	NAERG handbook
Technical information centers	Reference manuals

Chemistry can be, and is for many, a confusing and bewildering subject. For most individuals, the topic in itself leads to apprehension. As chemistry infringes into our everyday life, the understanding of this exacting science becomes more important and demanding. The emergency response agencies, including fire, EMS, and police, have become the responsible entities when an emergency exists involving chemicals. Therefore, it is the hazmat medical technician's responsibility to understand chemistry in order to mitigate the incident safely.

Chemists and physicists deal with specific quantities of material and work in very controlled environments. Emergency services, on the other hand, must

inorganic chemistry
the study of substances that do not contain carbon (the major components are not carbon based)

organic chemistry
the study of carbon compounds

mitigate incidents that involve uncontrolled situations, many times within a framework of chaos. In most areas of the country, the fire departments provide this service, usually supported by private or public emergency medical services. In many cases police also play an important role in hazardous materials response. Regardless of who is responsible, these emergency professionals understand emergency situations and train for them. The same principles used to manage multiple-casualty incidents or control a large fire can be used on a hazardous materials scene.

Incidents involving small amounts of diluted chemicals make up the majority of situations involving emergency intervention. Because these minor incidents often lead to complacency, safe practices must always be stressed to avoid injury. For example emergency units are dispatched to a patient having difficulty breathing. The dispatcher does not acquire the pertinent information surrounding the incident and therefore is unable to inform the emergency responders that the patient's difficulty was caused by a mixture of chemicals that led to a respiratory injury. More times than not this situation is dispatched as a medical emergency rather than a suspicious chemical emergency.

Chemistry does not have to be boring, confusing, or mystifying—it can be very interesting. A knowledge of chemistry allows responders to predict the way a material reacts under certain environmental conditions. For emergency responders to function safely and efficiently they must understand chemical processes. By no means does this chapter identify all the concepts of chemistry. However, the basic principles of chemistry are discussed in simple terms so that the needed knowledge can be acquired. With this basic knowledge you can predict the behavior of chemicals at the scene of an emergency incident.

Chemistry and the related field of toxicology are relatively new to the emergency services. In the past any study involving chemistry in the fire service had been to gain an understanding of fire behavior. Emergency medical services studied chemistry as it related to human physiology. The next logical step within these emergency services was to incorporate the study of chemistry and physiology into the hazardous materials discipline and apply this knowledge into an understanding of toxins and how they affect the body.

The study of chemistry, in the strict sense, is the study of the natural sciences. It is also the study of the composition of substances, the changes that may occur, and the natural laws that surround them. These laws can be applied under any condition, giving us an understanding of how changes take place. There are exceptions to the basic laws of nature, however, it is beyond the scope of this text to involve ourselves with all the exceptions and the intricacies of that subject. For the purpose of this book, only a general overview of chemistry as it pertains to hazardous materials is presented.

The study of chemistry is divided into two separate but closely related branches: inorganic and organic chemistry. **Inorganic chemistry** is the study of substances that do not contain carbon (the major components are not carbon based), whereas **organic chemistry** is the study of the carbon compounds. Both

biochemistry
the chemistry of
biological processes

have reactions that are distinct and individual to the chemistry under discussion. From these two branches, the study of **biochemistry**, the chemistry of biological processes, or that which involves the human body is derived.

Most of what is discussed in chemistry is theory however, these theories are supported with strong evidence. This intangible, yet observable, part of chemistry provides the anxiety and danger at hazardous materials incidents. Emergency workers and students of chemistry feel this apprehension. Chemistry generally follows very specific laws of nature and has observable results. But these physical laws of nature are the interpretation of man and for the most part are theoretical. The world, and thus, hazardous material incidents follow these basic laws, but occasionally the exceptions become the rules, and it is under these circumstances that the business of hazardous materials response become dangerous.

CHEMICAL AND PHYSICAL PROPERTIES

There are 92 naturally occurring elements and 11 created in the laboratory, making a total of 103 known elements. All of these elements are found on a periodic table. From these base elements, compounds are made.

All elements are found in one of three states: solid, liquid, or gas (Fig. 3-1). Within these states of matter exist tiny discrete particles called *atoms*. All atoms are made up of neutrons, protons, and electrons, which are arranged in a particular configuration. The atom is the smallest unit of an element. Elements comprise all matter at the atomic level and above. Each element can combine in a variety of configurations leading to a diversity of compounds.

A compound is a substance that is made up of two or more elements. Each element and compound also has its own characteristic chemical and physical properties. The properties of these combined elements are different from the properties of the elements separately. Compounds are different from **molecules** (Fig. 3-2).

molecules
a stable configuration
of elements by which
a structural unit is
made that has its own
characteristic chemical
and physical
properties

Each substance has its own set of chemical and physical properties which are useful when describing that chemical. They are the properties that have qualitative numbers (like fingerprints) peculiar to that chemical. For example, a sulfur dioxide molecule in any state always possesses an individual set of chemical and physical properties. So if a sulfur dioxide molecule melts, boils, looks, and freezes like a sulfur dioxide molecule, it must be sulfur dioxide.

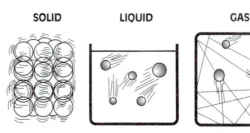

SOLID   LIQUID   GAS

Figure 3-1 *All matter consists of three states or forms.*

In a solid, the molecules are in very close proximity to each other. Although the atoms are vibrating ever so slightly, the movement can not be seen. In the liquid, the molecules are a bit farther apart, slipping by one another. This can be observed when a liquid flows out of its container. Breaking from the liquid environment to the vapor state is easy with enough molecular momentum. In the gaseous state, the molecules are the farthest apart, colliding with other molecules and the container itself. In each state of matter, an increase in temperature increases the molecular movement; likewise, decreasing the temperature decreases the molecular movement.

■ **NOTE**

Standard temperature and pressure (STP) is a condition characterized by a temperature of 32°F (0°C or 273 kelvin) and a pressure of 1 atmosphere (760 torr or 760 mm Hg).

Most of what is seen in the reference material identifies the pressures and temperatures in which reactions take place. These conditions are standard and are called standard temperature and pressure (STP). Technically it is a temperature of 32°F (0°C or 273 kelvin) at a pressure of 1 atmosphere (760 torr or 760 mm Hg) that the reaction has taken place in. Any other set of conditions, as it relates to temperature and pressure will be identified in the literature.

When describing flammable or combustible conditions, not only are temperature and pressure identified, but also what kind of container was used. The two types of containers usually identified are the, o.c. for open cup, and c.c. for closed cup. During open cup testing, the vapors are allowed to evolve above the solution and into the testing chamber. When testing the closed cup, vapors are contained within the cup holding the material.

■ **NOTE**

When describing flammable or combustible conditions, two types of containers are usually identified: o.c. for open cup, and c.c. for closed cup.

Physical Properties

Appearance, melting point, boiling point, density, specific gravity, vapor density, solubility, vapor pressure, viscosity, and freezing point are all examples of physical properties, which are characteristic of a particular element or molecule. They are qualities measured without changing the chemical makeup of the compound or molecule.

Appearance is the form of the chemical. The form may be reported as solid, liquid, or gas. Appearance may also be the size of a particle such as a powder, dust, or fume and can even be the color. The form of the hazardous material dic-

Figure 3-2 *When two or more elements combine to make a compound, the individual elements and the resultant compound have different properties.*

It is sometimes difficult to think in Celsius. Converting to Fahrenheit gives a frame of reference that most of us are familiar with. The following equations can be used to change Fahrenheit to Celsius or Celsius to Fahrenheit.

$$F = 1.8 (C) + 32; \text{ where F is Fahrenheit and C is Celsius}$$

$$C = \frac{F - 32}{1.8}$$

tates the management strategies toward incident stabilization. For example, if the material is a liquid, leak and spill control may be the tactics of choice. When a gas is involved, limiting the release or changing the physical form may be the tactical procedure. If the material is in the solid form, confinement of the material may be the action chosen to minimize exposures.

Melting point is the temperature at which a material changes from a solid to a liquid. From an emergency response point of view, generally speaking, a solid is easier to manage than a liquid. If a material has a low melting point, it may be expected to become a liquid in an emergency situation. Liquid materials maintain the shape of their container but have no form of their own. When they are not in a container, liquids present responders with challenges such as containment or confinement and contamination due to the state the chemical is in.

Boiling point is the temperature at which a liquid's vapor pressure equals atmospheric pressure. At this point the material changes from the liquid into the gaseous state. For example, a pot of water placed on a hot stove to boil in Florida (sea level) will boil more slowly than the same pot of water placed on a hot stove in Denver (mile-high elevation), because the pressure being applied to the surface of the water in Denver is much less than the pressure in Florida. Therefore it

Color of Smoke	Product Possibilities
White	Phosphorus
Grey/White	Benzene Nitrocellulose Sulphur products Nitric acid Hydrochloric acid
Greenish/Yellow	Chlorine
Grey/Brown	Iodine
Dark Brown/Black	Petroleum products

stereochemistry
the spatial arrangements of molecules that affect their chemical and physical properties

Compounds that have the same molecular weight but are structurally different are called isomers. There are different types of isomers. One that we have identified as having a mirror image is called an enantiomer. However, we may have positional isomers, functional group isomers, and chain isomers. Branching of a chained isomer is called a chain isomer. When the difference is in the position of the chain we call these positional isomers, and when the isomer contains a functional group we call it a functional group isomer. However, in identifying and referencing hazardous materials, these isomers present us with some problems. Because the **stereochemistry** can greatly effect the boiling points, we are concerned with identification of isomers.

In general, isomers decrease the boiling point. Two rules can be generally applied:

1. Within the same chemical family, the size is the determining factor. In other words, the molecular weight determines the boiling point. An isomer of a chemical with the same molecular weight will in general have a lower boiling point.

2. Among different families, polarity will affect the boiling point. Having a carboxyl group or hydrogen bonding is the identifier of polarity. If the boiling point is dropping, the vapor pressure is on the rise.

! SAFETY
Hazardous substances that are in a liquid state with a very low boiling point must be kept under pressure or they will boil and change form.

! SAFETY
Materials having high boiling points are usually much safer than those with low boiling points.

takes less time for the vapor pressure to equalize with the atmospheric pressure, so the water boils faster.

This same principle that applies in a kitchen also applies at hazardous materials incidents. Hazardous substances that are in a liquid state with a very low boiling point must be kept under pressure or they will boil and change form. Most of these chemicals present potential fire, reactivity, or health hazards. Conversely, the high-boiling-point liquids have relatively low vapor pressures. These liquids need an active energy source (fire) in order to convert from the liquid state to the vapor state. Materials having high boiling points are usually much safer than those with low boiling points (Fig. 3-3).

Figure 3-3 *As you can see, the lower the boiling point, the higher the vapor pressure. Polarity and hydrogen bonding greatly affect the boiling points.*

$$CH_3-CH_2-O-CH_2-CH_3$$
Ethyl Ether   BP =   94°F
                VP = 440 mm Hg

$$CH_3-CH_2-OH$$
Ethyl Alcohol   BP = 173°F
                  VP = 40 mm Hg

$$HO-CH_2-CH_2-OH$$
Ethylene Glycol   BP = 387°F
                  VP = .05 mm Hg

Density is thought of as how heavy a substance is. Although not a bad way of thinking about density, it is actually a relationship between weight and quantity or volume of a material and can be applied to all three states of matter. Chemically speaking, density is a ratio between mass and volume. For solids and liquids, this ratio is expressed as grams per centimeter cubed and for gases, grams per liter.

Density = Mass/Volume

Both specific gravity and vapor density are applications of this principle for liquids and vapors respectively. They both concern the mass per unit volume. From these calculations the movement of the liquids or vapors can be predicted. For example, the specific gravity of sulfuric acid is 1.84. This is a ratio based on the weight of water at 8.35 pounds per gallon. To calculate the weight of sulfuric acid, multiply the weight of water times 1.84. The calculated weight of sulfuric acid is 15.36 pounds per gallon.

specific gravity

the weight of a solid or liquid as compared to an equal volume of water

Specific gravity is the weight of a liquid or solid compared to an equal volume of water. Water has a value of 1.0 in relation to the compared material. If the tested material has a specific gravity of less than one, the material will float. If it is greater than one, the material will sink. At hazardous materials emergencies it is often imperative to know if the liquid material will float or sink in order to decide the type of hazard control techniques to be used. If the material has a specific gravity of greater than one, then a viable tactic may be to contain vapors by floating water over the top of the material. If the value is less than one and water is used, the material will flow over a larger area, carried on the surface of the water.

vapor density

the weight of a vapor or gas as compared to an equal volume of air

Vapor density is the weight of a vapor or gas as compared to an equal volume of air. Air is assigned the density of one. If the relationship indicates a number greater than one, then the vapor will drop or settle below the air. If the comparison yields a number less than one, then the vapor will rise. The rising gas may or may not dissipate depending on wind, humidity, and other related factors (Fig. 3-4). Vapor density and pressure are two of the properties considered when dealing with plume dispersion models.

■ NOTE

Material with a specific gravity of less than one will float, and with greater than one will sink.

■ NOTE

A vapor density of greater than one indicates a substance that will drop or settle below the air, while a number less than one indicates that the vapor will rise.

Under certain conditions, especially when dealing with the lighter-than-air gases in high humidity environments, the water vapor in the air must also be considered. For example, methane, which is lighter than air, has shown qualities very much like those of liquid petroleum gas (LPG), which is heavier than air. This can be explained if we look at the vapor density of methane and water vapor as a combined effect. Vapor density is equal to the molecular weight of a gas over the molecular weight of air:

$$VD = \frac{\text{Molecular weight of the substance}}{\text{Molecular weight of air (29)}}$$

$$VD = \frac{\text{Methane (16) + Water vapor (18)}}{\text{Molecular weight of air (29)}}$$

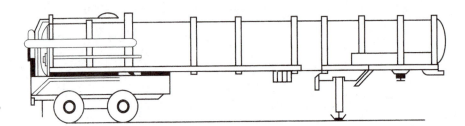

Figure 3-4 *Vapor density, along with the lay of the land, can greatly affect your incident.*

> Heavier-than-air vapors move along the ground, while lighter-than-air vapors tend to dissipate in the air with the wind. Some materials will absorb water, making a lighter-than-air product equal to air density or heavier. Research the chemical and physical properties of all chemicals that are involved.

where 16 is the molecular weight of methane (carbon = 12, hydrogen = $1 \times 4 = 4$, thus $12 + 4 = 16$) and 18 is the molecular weight of water (oxygen = 16, hydrogen = $1 \times 2 = 2$, thus $16 + 2 = 18$). Therefore,

$$VD_{\text{methane in high humid environment}} = 34/29 = 1.17 \text{ times heavier than air}$$

The scenario described in the previous paragraph occurred in North Carolina. A natural gas line was fractured during road construction causing a serious leak. The repair crew attempted to place a jacket over the broken line that was under about 3 feet of water. The gas bubbled up through the water, becoming humidified. When the gas escaped the hole, instead of rising and dissipating it traveled the ground and found an ignition source a distance from the leak and exploded.

Solubility is the ability of a material to blend uniformly within another material. Certain materials are soluble in any proportion, while others are not. One of the dependent factors in solubility is the polarity and concentration of the materials involved and whether the solute (the stuff you are placing into the solution) is a liquid or a solid. The blend is called a *solution*. The material that is in the greatest amount is call the *solvent* whereas the material that is in lesser amount (usually the additive) is called the *solute*. *Miscibility* is often used synonymously with solubility. When a compound is said to be miscible in water, it means that the substance is infinitely dissolvable in water.

solubility
the ability of a material to blend uniformly within another material

Polarity has a lot to do with solubility. Polar substances have a positive end and a negative end and nonpolar substances do not. Generally speaking, like will dissolve in like, polar with polar and nonpolar within nonpolar.

If the solute is a solid the polar/nonpolar dissolvability still holds true, however there is a limited capacity of the solid to dissolve. Each solvent has a saturation point. Above this point the added solute will not dissolve.

Solutions represent a homogeneous mixture where all the parts of the end mixture are composed of the same material. For example, if water is used to dilute a hydrochloric acid spill, then the solution running off the road is a less concentrated mixture of hydrochloric acid and water. Although the solution is less hazardous in terms of concentration, the molecular structure has not changed, thus the strength has not changed.

On the other hand, if a spill of carbon tetrachloride occurred and the intention was to dilute it with water, the outcome would not be the same. Carbon tetrachloride is not soluble in water, therefore the water would only displace the carbon tetrachloride causing it to move around and contaminate other areas. If a patient is contaminated with carbon tetrachloride and is decontaminated only with water, will the procedure be successful? Probably not! But if an alcohol solution were used, the decontamination would be successful.*

Vapor pressure (VP) is the pressure a material exerts against the sides of an enclosed container as it tries to evaporate or boil. Each atom within the liquid material is bouncing about until it reaches enough velocity to escape the liquid. Once this molecule has escaped the liquid form, it has been changed into a molecule traveling through the air space, giving it a gaseous state. This movement of atoms in the gaseous state is measured as vapor pressure. All liquids have a vapor pressure. Vapor pressure is measured in millimeters of mercury (mm Hg).

For example, water has a VP of 21–25 mm Hg. Naphthalene and its derivatives are the only solids that have a vapor pressure (0.08 mm Hg). This is due to a characteristic called *sublimation*, in which the material goes directly to the vapor state from the solid state without appearing to have gone through the liquid state. Under certain conditions vapors can move into the solid state without this intermediate liquid phase.

Those materials having a higher vapor pressure will maintain a higher pressure within a closed container. The same materials that have high vapor pressure also have low boiling points (Fig. 3-5, page 75). These materials have a greater potential for container breach, especially if heated.

When a gas is released into the environment, the wind, ambient temperature, and topography play roles in the chemical's ability to ignite. It must find an available ignition source. If a gas traveling through the air could be measured, three distinct areas of concentration would be noted. First, the area closest to the container from which the material is escaping would be too rich to burn (above the flammable limits). Even if an ignition source were available, the gas would

vapor pressure (VP)
the pressure a material exerts against the sides of an enclosed container as it tries to evaporate or boil

---

*Carbon tetrachloride is soluble in alcohol; insoluble in water.

SAFETY ISSUE

There has been very little research on how vapor pressure affects the biological model (i.e., the inspiration of a vapor laden environment within an organism). However in reviewing the facts, the statement can be made that chemicals that have a high vapor pressure affect the exposed individual at a high rate, whereas chemicals with low vapor pressures have less effect on the individual. There are several reasons for this assumption and the related observations (case studies of individuals affected by small concentrations of high vapor pressure vapor).

1. We know that particle size has its effect within the human body, when inhaled, in terms of depth into the respiratory tree. This mostly has to do with solubility of the chemical, size of the molecule (size, orientation in space, and polarity) and the velocity of the chemical in air (see Chapter 5 for routes of exposure).

2. High vapor pressure relates to the velocity or movement of the molecules in air. Turbulence, air temperature, and barometric pressure can affect the overall movement. However, a higher movement represents a high-vapor-pressure molecule, whereas the chemicals with a lower vapor pressure move relatively more slowly.

3. Testing of mist, vapors, and aerosols have given some insight to the movement of these small molecules. For example, when high-vapor-pressure chemicals are aerosolized, they seem to evaporate completely, whereas the chemicals with low vapor pressure are present as a mist. There seems to be a relationship between the low vapor pressure and the ability to maintain a mist consistency. In other words, when chemicals that have a low vapor pressure are aerosolized, they stay in mist form and don't attach to a droplet.

When testing the mist/vapor air, the vapor is caught and retained within the sample filter if it is a chemical with low vapor pressure. In high-vapor-pressure chemicals the filter caught some of the vapor and none of the aerosol. Because of this observation, vapors and mists can be considered separately, and in fact, are tested together, however analyzed as separate constituents. By adding the rates of absorption between the vapor and mist we see that the effect on the sampling system is not additive. For example, when filters and presamplers (the filter is placed in front of the presampler) are analyzed, the presampler tends to collect the aerosol and none of the vapor. In some studies, the filter caught the chemical with low vapor pressure but not the high-VP molecule.

It is thought that the chemical with high VP evaporates faster than the analysis can occur on the filter. This would seem logical considering that the chemical with high VP has a higher velocity and tends to be low weight, or the size is concentrated as compared to other molecules.

Because of the foregoing facts we can consider chemicals with high vapor pressure as being more harmful than those with low vapor pressure. Although most of this is based on assumptions, the research in this area is very limited, and future testing with biological models will have to be done. However, in the meantime and bordering on the side of safety, we consider the high-vapor-pressure (vapor pressures above 760 mm Hg) substances as having the high potential to cause a severe respiratory injury.

⚠ SAFETY
Flammable limits or ranges—concentrations of vapor to air mixture—differ for each chemical compound.

not burn. The next area holds a mixture that is within the ratio of air to vapor to support combustion. At this point if an ignition source were available, the vapor would ignite (within the flammable range). On the tail end of the vapor cloud the concentration of vapor to air mixture would be too lean. It would not burn even if there were an ignition source available (below the flammable limits). These concentrations are referred to as flammable limits or ranges and are different for each chemical compound.

Figure 3-5 *Most of these trends are seen in the lower weight organic compounds. These properties are in relation to the size of the molecule, hydrogen bonding and placement of functional groups. When the properties on the right are low, the properties on the left will be relatively higher for small compounds, and vice versa.*

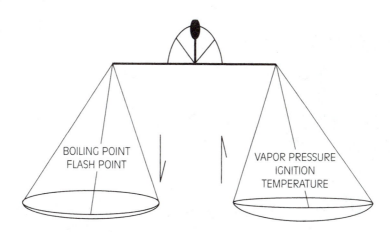

viscosity
a measure of flow

flash point
the minimum temperature at which a liquid will give off vapors to form an ignitable mixture in air, but not enough vapors to sustain combustion

flammable limits
the range between the lower and upper limit in which the concentration of the mixture of vapor in the air is favorable for ignition which is directly related to liquid materials' boiling point, and thus, to vapor pressure

ignition temperature
the minimum temperature to which a material must be raised to be ignited (by an outside source) and sustain combustion

Viscosity is a measure of flow. It is a determination of the thickness of a liquid or how well it flows. A low viscous liquid will flow like water. The lower the viscosity, the higher the tendency for the liquid to spread. On the other hand, high viscosity liquids flow more slowly (like molasses).

When dealing with combustible and flammable hydrocarbons, viscosity can be related to the production of static electricity. Because hydrocarbons have polarity they are very much like magnets. As these tiny magnets move over each other they create a small charge of static electricity. Once this electricity finds a ground, a spark can be created and if the conditions are right, vapors ignite.

Freezing point and melting point can be thought of as synonyms, depending on the context of the chemical reaction. If, for example, the chemical is moving from a solid to a liquid state, then the term *melting point* is used. If the product is going from a liquid to a solid state, then the term *freezing point* is used. The amount of heat that is required to move the chemical from a solid to a liquid, or the amount of heat that must be removed to move the liquid to a solid is dependent on the chemical itself. Each chemical and the state of matter that it is being contained in will have an impact on these two properties.

Chemical Properties

All chemicals try to reach equilibrium. In their quest for equilibrium, chemicals undergo changes. These changes give chemicals their properties which include the heat of combustion, reactivity, and flash point.

Flash point, a component of vapor pressure, is that minimum temperature under which a liquid will give off vapors to form an ignitable mixture in air, but not enough vapors to sustain combustion. This is provided that the **flammable limits** are within favorable parameters.

Ignition temperature is the minimum temperature to which a material must be raised to be ignited (by an outside source) and sustain combustion. Materials

Figure 3-6 *When the flash point is lower than the ambient temperature, vapors are produced. A lower boiling point equals a low flash point and a higher possibility of toxic vapors.*

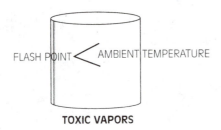

FLASH POINT < AMBIENT TEMPERATURE

TOXIC VAPORS

autoignition temperature

the minimum temperature required for a material to spontaneously ignite and maintain combustion

heat of combustion

the heat evolved when a definite quantity of a product is completely oxidized or complete combustion takes place

> ! **SAFETY**
> When dealing with combustible and flammable hydrocarbons, viscosity can be related to the production of static electricity.

that have a low ignition temperature will have a relatively higher flash point (Fig. 3-6). This concept is most important when dealing with organic compounds.

Autoignition temperature is the minimum temperature required for a material to spontaneously ignite and maintain combustion.

Heat of combustion is the heat evolved when a definite quantity of a product is completely oxidized or complete combustion takes place.

All chemicals are subject to change, each exhibiting its own chemical and physical properties. They interact to produce a chemical conversion. The transformation into a stable compound is the result of a chemical reaction. These reactions takes place because of the specific architecture of the elements or compounds within the surrounding environment.

ATOMIC STRUCTURE

Protons, Neutrons, and Electrons

Within an atom a precise assembling of nature occurs. In the center of the atom is a nucleus containing two types of particles, positively charged particles called *protons* and uncharged particles called *neutrons*, with the mass close to that of a proton. The relative proportions of these particles make up an element. Revolving around the nucleus are particles called *electrons* (Fig. 3-7). They are in "orbit" about the center portion of the atom. The rotation of this particle is so fast that if we could see such a particle rotating, it would look like a cloud about the center nucleus. Electrons carry a negative charge to balance the positive charge of the proton. The mass of an electron is approximately 1,800 times smaller than that of a proton.

The electron is the lightest of all the foundation particles (there are smaller particles, called subatomic, however the particles that we are discussing are the fundamental particles). All elements have electrons. The sharing or the transfer of electrons is the basis of chemical reactions and the bonding of elements. Hydrogen, for example, has only one electron, helium, on the other hand, has two. Electrons rotate around the nucleus in orbits called *shells*. The outer orbit

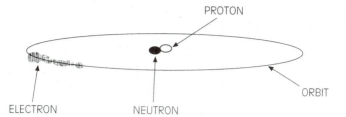

Figure 3-7 *Atoms contain three fundamental structural units, the neutron, proton, and the electron.*

PROTON

ELECTRON   NEUTRON

ORBIT

can never have more than eight electrons and the innermost orbit never has more than two. For an element to be perfectly stable, it must have eight electrons in its outer shell, or two electrons if it is hydrogen.

All atomic nuclei contain an equal number of protons in relation to the number of electrons that are in orbit. The number of protons dictate the number of electrons within a neutral atom. If the atom has fourteen protons, then there will be fourteen electrons in orbit around the atom. The number of protons determines the atomic number. This atomic number is used to "catalog" in order, the elements.

Isotopes

isotopes

one or more forms of an element that have the same number of protons but differ by the number of neutrons

Isotopes are one or more forms of an element that have the same number of protons but differ by the number of neutrons. For example, hydrogen regularly does not have a neutron in the nucleus but can have one (deuterium) or two (tritium) neutrons. Therefore, hydrogen is said to have three isotopes (Fig. 3-8). Most elements that exist in nature only have one isotope, however, some of the elements contain several coexisting isotopes.

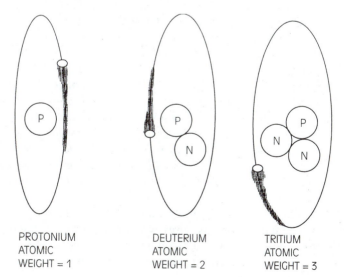

Figure 3-8 *The atomic weight of hydrogen is 1.0080, which accounts for the three isotopes.*

PROTONIUM
ATOMIC
WEIGHT = 1

DEUTERIUM
ATOMIC
WEIGHT = 2

TRITIUM
ATOMIC
WEIGHT = 3

Weight

The mass number is the total number of protons, neutrons, and electrons within the atom. For simplicity an electron's weight is so minute that it is usually dropped when calculating atomic weight. The mass number is the "weight" of an atom, or the average weight of all isotopes. The atomic weight seen on the periodic chart is actually an average of all the isotopes and their presence in nature. By knowing the number of protons and neutrons in the nucleus, the atomic weight of an element can be calculated. In fact the atomic weight of a substance could be calculated by adding the number of all of the protons and neutrons of the substances within an element. These calculations would be close, but still off, because in any given volume a number of isotopes would naturally occur.

THE PERIODIC CHART

The periodic chart (Fig. 3-9) provides a general framework of the elements to give emergency responders a look at chemistry as a whole. This chart arranges the chemicals according to the atomic number. Looking at the chart, chemicals that are arranged in vertical columns are called groups. Within each group, elements that have similar properties are found. Unfortunately, some charts are labeled across the groups differently. For example, the first column is called alkali metals, or Group IA. On the newer charts this group has been renamed Group One. In this section we will attempt to give the most common naming variations.

Looking at each group as it appears on the periodic chart gives insight to the general properties of the elements. For example, by looking at the chart, the number of electrons in the outer shell can be determined by identifying which group the element falls under. Group IV, which starts with carbon, has four electrons in the outer shell. Fluorine is within Group VII, a group in which each element has seven electrons in the outer shell. The number of electrons in the outer shell determines the reactivity and bonding potential of each element. Those chemicals on the far left column of the periodic chart are the most reactive and those in the far right column are the most stable. Various degrees of reactivity are found in between these outermost columns.

Once introduced into the human body, all chemicals go through a process called *metabolism*. When dealing with chemicals that may pose a health hazard, this process of metabolism changes chemicals and is termed *biotransformation*. The biotransformation of chemicals is, in part, due to enzyme reaction in the body. This reaction is the body's attempt to utilize or excrete the chemical that has entered the body.

Most naturally occurring chemicals are considered nontoxic and do not pose a health hazard or risk. However, once these chemicals are manipulated in industry or synthetically produced, they are changed and many then posses toxic qualities. These chemicals are discussed in examples listed below.

Group 1	Group 2	Group 3	Group 4	Group 5	Group 6	Group 7	Group 8	Group 9	Group 10	Group 11	Group 12	Group 13	Group 14	Group 15	Group 16	Group 17	Group 18
1 **H** Hydrogen																	2 **He** Helium
3 **Li** Lithium	4 **Be** Beryllium											5 **B** Boron	6 **C** Carbon	7 **N** Nitrogen	8 **O** Oxygen	9 **F** Fluorine	10 **Ne** Neon
11 **Na** Sodium	12 **Mg** Magnesium											13 **Al** Aluminum	14 **Si** Silicon	15 **P** Phosphorus	16 **S** Sulfur	17 **Cl** Chlorine	18 **Ar** Argon
19 **K** Potassium	20 **Ca** Calcium	21 **Sc** Scandium	22 **Ti** Titanium	23 **V** Vanadium	24 **Cr** Chromium	25 **Mn** Manganese	26 **Fe** Iron	27 **Co** Cobalt	28 **Ni** Nickel	29 **Cu** Copper	30 **Zn** Zinc	31 **Ga** Gallium	32 **Ge** Germanium	33 **As** Arsenic	34 **Se** Selenium	35 **Br** Bromine	36 **Kr** Krypton
37 **Rb** Rubidium	38 **Sr** Strontium	39 **Y** Yttrium	40 **Zr** Zirconium	41 **Nb** Niobium	42 **Mo** Molybdenum	43 **Tc** Technetium	44 **Ru** Ruthenium	45 **Rh** Rhodium	46 **Pd** Palladium	47 **Ag** Silver	48 **Cd** Cadmium	49 **In** Indium	50 **Sn** Tin	51 **Sb** Antimony	52 **Te** Tellurium	53 **I** Iodine	54 **Xe** Xenon
55 **Cs** Cesium	56 **Ba** Barium	57–71* Lanthanides	72 **Hf** Hafnium	73 **Ta** Tantalum	74 **W** Tungsten	75 **Re** Rhenium	76 **Os** Osmium	77 **Ir** Iridium	78 **Pt** Platinum	79 **Au** Gold	80 **Hg** Mercury	81 **Tl** Thallium	82 **Pb** Lead	83 **Bi** Bismuth	84 **Po** Polonium	85 **At** Astatine	86 **Rn** Radon
87 **Fr** Francium	88 **Ra** Radium	89–103** Actinides															
		Lanthanides	57 **La** Lanthanum	58 **Ce** Cerium	59 **Pr** Praseo-dymium	60 **Nd** Neodymium	61 **Pm** Promethium	62 **Sm** Samarium	63 **Eu** Europium	64 **Gd** Gadolinium	65 **Tb** Terbium	66 **Dy** Dysprosium	67 **Ho** Holmium	68 **Er** Erbium	69 **Tm** Thulium	70 **Yb** Ytterbium	71 **Lu** Lutetium
		Actinides	89 **Ac** Actinium	90 **Th** Thorium	91 **Pa** Protactinium	92 **U** Uranium	93 **Np** Neptunium	94 **Pu** Plutonium	95 **Am** Americium	96 **Cm** Curium	97 **Bk** Berkelium	98 **Cf** Californium	99 **Es** Einsteinium	100 **Fm** Fermium	101 **Md** Mendelevium	102 **No** Nobelium	103 **Lr** Lawrencium

*Lanthanides–the rare earth elements.

**Actinides–mostly synthetic, radioactive elements.

Figure 3-9 *The periodic table organizes the elements in order of atomic number. Horizontal rows are called periods and the vertical columns are called groups or families.*

Table 3-1   *Group 1 (or Group IA) alkali metals.*

Chemical Symbol	Density (g/cm^3)	Melting Point	Boiling Point	First Ionization Potential (ev)
Li	0.534	180.54° C	1347° C	5.392
Na	0.971	97.81° C	883° C	5.139
K	0.862	63.65° C	774° C	4.341
Rb	1.532	38.89° C	688° C	4.177
Cs	1.873	28.40° C	678° C	3.894
Fr	?	27.00° C	677° C	?

Alkali Metals—Group 1 (GROUP IA)

The first column to the left when viewing the periodic chart lists the alkali metals as Group 1 or Group IA. This group contains lithium (Li), sodium (Na), potassium (K), radibium (Rb), cesium (Cs), and francium (Fr), listed from top to bottom
(see also Table 3-1). Highly reactive in the pure form, the elements in this group
give up an electron easily because there is only one electron in the outer orbital
shell. Therefore it is rare that any of this group are found in pure form in nature.
      The reactions of these elements are so violent that the commercial use is
limited to sodium and potassium (see Table 3-2). Although the primarily used

Table 3-2   *Some common compounds containing sodium or potassium
and their uses in industry.*

Na_2CO_3	Sodium carbonate	Soda ash
NaCl	Sodium chloride	Table salt, rock salt
$NaHCO_3$	Sodium bicarbonate	Bicarbonate
NaClO	Sodium hypochlorite	Bleach
NaOH	Sodium hydroxide	Lye
KOH	Potassium hydroxide	Potash
KNO_3	Potassium nitrate	Saltpeter

Radiation is the energy produced during ionization. It is spontaneously emitted by materials that decay. This decay emits a naturally occurring energy and is called ionization. One must remember that this energy is given off by the material without any outside energy sources. This ionization usually occurs when the isotopes have an atomic number greater than 83. This ionization is termed radiation and exists in two basic forms: one as a particle and the other as an electromagnetic wave of energy. The particle forms have mass and density, thus travel is dependent on the initial escape velocity of the particle. The distance from the source is significantly lower than the electromagnetic energy wave.

The half-life of a radioactive substance is the amount of time that is required to decrease the ionization by half. It is the rate at which the substance decays. Some examples are:

Calcium 45: 164 days

Carbon 14: 5,600 years

Cobalt 60: 5.3 years

Iodine 131: 8 days

Plutonium 239: 24,000 years

Strontium 89: 52 days

Strontium 90: 28 years

Tritium: 12 years

Uranium 235: 700 million years

physics half-life
the amount of time it takes a radioactive isotope to lose one-half of its radioactive intensity

chemicals in this group, sodium and potassium have their own health hazards. They produce strong corrosive alkalis that can result in severe burns in the eyes, skin, and respiratory system.

Francium, a radioactive element, is not a concern, as it has only been created in the laboratory. Francium has a **physics half-life** (physics half-life is the amount of time it takes a radioactive isotope to lose one-half of its radioactive intensity) of 22 minutes.

The general hazards of this group are:

- Highly water reactive
- Solutions can become extremely caustic
- Some will produce hydrogen gas on contact with water

Alkaline Metals—Group 2 (Group IIA)

The second group is the alkaline metals, Group 2 or Group IIA, sometimes referred to as alkali earth metals. This group contains in order beryllium (Be), magnesium (Mg), calcium (Ca), strontium (Sr), barium (Ba), and radium (Ra) (see also Table 3-3). These elements appear to be gray with a metallic brilliance. The most common commercial uses of these elements relate to their radioactive qualities (see Table 3-4).

Table 3-3 *Group 2 (or Group IIA) alkaline metals.*

Chemical Symbol	Density (g/cm³)	Melting Point	Boiling Point	First Ionization Potential (ev)
Be	1.848	1279.0° C	2472° C	9.322
Mg	1.738	648.8° C	1090° C	7.646
Ca	1.550	839.0° C	1484° C	6.113
Sr	2.540	769.0° C	1384° C	5.695
Ba	3.500	725.0° C	1898° C	5.212
Ra	5.000	700.0° C	1536° C	5.279

⚠ SAFETY

The general hazards of Group 2 elements are that they are water reactive, their solutions may be caustic, and hydrogen gas may evolve if in contact with water.

Radium, a strong radioactive element gradually degenerates losing atomic weight. Even in a degenerated form radium poses a strong radiological hazard. Every precaution should be exercised around this element. Radium has the potential to emit alpha and beta particles, and gamma radiation, and has a half-life of 1,620 years.

As with Group 1, this group is very reactive. Group 2 elements have two electrons in the outer shell and freely release them for bonding. As with Group 1 elements, when reacting with water, these elements release hydrogen gas that is extremely flammable.

An example of chemical hazards found in this group is CaO, calcium oxide or quicklime which can produce an exothermic reaction with water, while producing calcium hydroxide $Ca(OH)_2$, a strong base. If calcium hydroxide contacts moist skin, it produces serious thermal and chemical burns.

Table 3-4 *Some commercial uses of elements from Group 2.*

$MgSO_4$	Magnesium sulfate	Epsom salts
$(MgO)_3MgCl_2$	Magnesium oxychloride	Cement
$C_a(OH)_2$	Calcium hydroxide*	Slacked lime
$CaSO_4$	Calcium sulfate	Gypsum
C_aC_2	Calcium carbide	Used to make acetylene

* Calcium hydroxide is the hydrated form of lime C_aO.

Since 1989, firefighters have had to deal with a new breed of fires, the high temperature accelerant fires (HTA). These fires have produced such intensity that steel has melted and flowed toward low points, silica in concrete has turned to glass, all without any chemical residue for fire investigators to gather as evidence. This chemical cocktail is unparalleled by any other type of event in terms of firefighter safety.

It is speculated that the chemical contents of these cocktails have their chemical bases within the alkali and alkaline metals. Some have witnessed white plumes of smoke with shooting sparks, highly indicative of phosphorous, while other reports have indicated green, blue, and hot white fire within the building, all signs of transition and Group 13 elements.

The concern as stated is the safety of fire crews if presented with this type of incident. Temperatures have been noted to increase to beyond several thousand degrees within a matter of seconds. This can and has been disastrous. With most fire departments responding more quickly, with aggressive interior attacks, firefighters are at high risk when presented with this type of hazardous materials incident.

So far the profile of these fires has been abandoned, or non-occupied buildings, with extremely fast fire spread. Fire colors are similar to fireworks, with abnormal smoke production. Water has been ineffective. Temperatures have been so great that molecular bonds in water have broken, resulting in oxygen as a oxidizer and hydrogen as a fuel.

■ NOTE

Elements in groups 3-12 contain health hazards and are frequently monitored for medical consequences.

The general hazards of this group are the same as the alkali metals:

- Water reactive
- Solutions may be caustic
- Hydrogen gas may evolve if in contact with water

Transition Elements—Groups 3–12 (Groups IIIB, IVB, VB, VIB, VIIB, VIII, VIII, VIII, IB, IIB)

The next ten groups, Group 3 through Group 12, are called the transition elements. These groups contains metals that are stable compounds. The stability is so great that these elements are sometimes referred to as the noble metals.

! SAFETY

The general hazards of group 3-12 elements are heavy metal poisoning and other varied effects, dependent on the element combination.

Most of the elements in this area of the periodic chart have to do with the true metals and their properties. The properties of the metals and their respective alloys are a whole topic in itself. However, of importance in this text are the medical concerns of some of these common elements. This area of the chart contain elements referred to as the *heavy metals*. These are elements that contain health hazards and are frequently monitored for medical consequences. (See Medical Surveillance, Chapter 7.)

Heavy Metals   *Mercury (Hg).* Mercury is the only metal that is in a liquid state at room temperature. Mercury in the elemental form is poisonous, especially when it vaporizes. The mercury vapor penetrates the pulmonary system, where it is absorbed into general circulation. It is oxidized in the red blood cells where it is biotransformed into mercury (mercuric ion) with a +2 charge. Mercuric ion bonds with sulfur (sulfhydryl) within the cell, deactivating enzymes required for cellular metabolism (pyruvic acid interruption). Accumulation also occurs in the nerve tissue and kidneys. Symptoms may include, but are not limited to, convulsions, numbness in the extremities, difficulty speaking, tunnel vision, and renal compromise. Once a toxic level is reached, the capacity of the central nervous system (CNS) is severely damaged producing a variety of CNS disorders.

The toxicity of this element has been misinterpreted for years. For example, if little Johnny decides to break the thermometer between his newly formed teeth, the effects from ingestion of this amount are limited to a silvery diaper present a few hours later and maybe some lacerations around the gums from the broken glass. However, if mercury is allowed to form vapors that are inhaled, severe health effects may result. This consequence has been reported among the American Haitian population where elemental mercury is sprinkled around the rooms of the house to ward off evil demons.

In the early 1800s this exact effect of mercury was noted among those individuals that made hats (hatters). Mercury was used in the final stages of the hat production. Although the wearer of the hat did not show any ill effects of mercury poisoning, the hatters had a variety of nervous system dysfunction and hyperactivity. This event was due to the use of fuming mercury that occurred during the curing process of the hat. The hatter was directly exposed to the mercury fumes which in turn damaged the nervous system. This is where the phrase "mad as a hatter" originated.

Chromium (Cr). Chromium in the oxidative states of +3, +2, and 0 has minimal toxic effects. In the +6 oxidative state it is extremely toxic (chromic acid, chromates) and is a known carcinogen. When in the powder or fume state, it is often inhaled. Solubility of chromium is dependent on the oxidative state. When exposure is dermal, the chemical acts as an irritant and is also known as a strong sensitizer. Either of these reactions can be severe.

If an inhalation injury occurs, the pulmonary tree becomes irritated causing pulmonary edema 24–72 hours after exposure. Cancer can form in the lung tissue months to years after the exposure. Some research indicates that low-level exposure leads to adult respiratory distress syndrome.

Both chromium and mercury are considered toxic heavy metals. These two elements are commonly encountered at hazardous materials waste sites, but there these elements do not produce the above symptomology because they are monitored very closely. In the emergency services, these elements should be taken seriously and biological monitoring performed on a regular basis.

Within the transition elements are Titanium (Ti), Zirconium (Zr), and Zinc (Zn). Under the right conditions these elements pose significant fire threat. Magnesium (Mg) and Aluminum (Al) also possess the same basic qualities. Generally these chemicals in the dust form are considered nuisance dusts.

Table 3-5 *Examples of groups 3-12, transition elements.*

Chemical Symbol	Density (g/cm³)	Melting Point	Boiling Point	First Ionization Potential (ev)
As	5.72	808° C	603° C$_{sublimes}$	9.810
Be	1.85	1287° C	2472° C	9.322
Cd	8.65	321° C	767° C	8.993
Cr	7.19	1857° C	2672° C	6.766
Pb	12.00	327° C	1750° C	7.416
Hg	13.53	−39° C	357° C	10.437

Titanium, for example, is an upper airway irritant that has in a few cases caused chemically induced pneumonia. However it is unclear whether these symptoms are attributable to the titanium ion or the halogen attached to it. It is known that these metals in the dust form can exacerbate preexisting pulmonary conditions with symptoms that range from flu-type symptoms to severe respiratory distress.

Cadmium (Cd). Cadmium is mostly found around zinc ores and is produced through isolation. Other metals of isolation are zinc, copper, and lead. The primary use for cadmium is during the production of silver solder, picture tubes, and nickel-cadmium (ni-cad) batteries. It is extremely toxic and care should be taken when around this element.

When medical surveillance programs are outlined for hazmat response teams, biological monitoring should include blood levels of arsenic (As), beryllium (Be), cadmium (Cd), chromium (Cr), lead (Pb), and mercury (Hg) (see also Table 3-5).

General hazards of these groups are heavy metal poisoning and other varied effects, dependent on the element combination.

Group 13 (Group IIIB or IIA)

This group contains one nonmetal, boron (B). The remainder of the group, aluminum (Al), gallium (Ga), indium (In), and thallium (Tl) are metals (see also Table 3-6). The physical state is solid for the most part (except gallium, which is a solid below 30°C or 86°F).

Boron is used in flares and some of the pyrotechnic rockets; it gives fireworks their distinctive green color. Although not a potent poison, if ingested it may result in chronic toxic effects. Indium is primarily used for making low melting alloys and is toxic through inhalation. Thallium is a toxic substance: The compound thallium sulfate was mostly used in rodenticides during the 1970s

Table 3-6   *Group 13 (or Group IIIB or IIA) elements.*

Chemical Symbol	Density (g/cm³)	Melting Point	Boiling Point	First Ionization Potential (ev)
B	2.34	1806° C	4002° C	8.298
Al	2.70	660° C	2519° C	5.986
Ga	5.91	29° C	2205° C	5.999
In	7.31	156° C	2073° C	5.786
Tl	11.85	304° C	1473° C	6.108

Table 3-7   *Some common uses of Group 13 elements.*

H_3BO_3	Boric acid	Boric eye solution
Al_2O_3	Aluminum oxide	Abrasives
$Al(OH)_3$	Aluminum hydroxide	Indigestion relief tablets

(see Pesticides in Chapter 6). Because of its highly toxic qualities, household availability is prohibited (see also Table 3-7).

General hazards of this group are poisoning and other varied effects dependent on the element combination.

Group 14 (Group IVB or IVA)

SAFETY

The general hazards of Group 14 elements are heavy metal poisoning and other varied effects, dependent on the element combination.

Group 14 has a mixture of nonmetals and metals. These elements are carbon (C), silicon (Si), germanium (Ge), tin (Sn), and lead (Pb) (see also Table 3-8). In color these elements range from gray/black to silvery white. They are all solids and have high densities. Germanium initially did not have many industrial chemical applications. However, with the introduction of computers and the communication optic fiber for the information superhighway, this element has found many useful applications. It also has been found to be advantageous in fighting specific bacteria, which makes it highly useful in the medical field. Fortunately Germanium's toxicity is low, nevertheless it may present long-term effects if the exposure is chronic.

Lead is a toxic heavy metal that, much like mercury (although the mechanism is different), can cause a variety of CNS disease. The toxic effects of lead are

Table 3-8   *Group 14 (Group IVB or IVA) elements.*

Chemical Symbol	Density (g/cm³)	Melting Point	Boiling Point	First Ionization Potential (ev)
C	1.8 –3.15	3550° C	4197° C	11.260
Si	2.33	1412° C	3267° C	8.151
Ge	5.32	937° C	2834° C	7.899
Sn	7.30	232° C	2603° C	7.344
Pb	12.00	327° C	1750° C	7.416

long term, which is one reason why lead is no longer used in plumbing and wall coverings.

Carbon and oxygen sometimes combine (during incomplete combustion as an example) to produce carbon monoxide (CO). Once attached to the hemoglobin molecule, this molecule creates a very stable compound called *carboxyhemoglobin* and this molecule is so stable that oxygen can no longer attach to the red blood cell, causing chemical asphyxiation.

General hazards of these groups are heavy metal poisoning and other varied effects dependent on the element combination.

Group 15 (Group VB or VA)

SAFETY
The general hazards of Group 15 elements are heavy metal poisoning and other varied effects, dependent on the element combination.

Group 15 contains the elements of nitrogen (N), phosphorus (P), arsenic (As), antimony (Sb), and bismuth (Bi) (see also Table 3-9). Nitrogen is the only element

Table 3-9   *Group 15 (Group VB or VA) elements.*

Chemical Symbol	Density (g/cm³)	Melting Point	Boiling Point	First Ionization Potential (ev)
N	0.81	–210° C	–196° C	14.534
P	1.82–2.69	44° C	280° C	10.486
As	1.97 / 5.73	808° C	603° C	9.810
Sb	6.68	631° C	1587° C	7.344
Bi	9.75	271° C	1564° C	7.289

Table 3-10 *Some common uses of Group 15 chemicals.*

N_2O	Nitrous oxide	Medical analgesic
HNO_3	Nitric acid	Explosives, fertilizers
P	Phosphorous	Poison, incendiaries, fertilizers
$Ca(H_2PO_4)_2$	Calcium acid phosphate	Fertilizer
$Pb_3(AsO_4)_2$	Lead arsenate	Insecticide

within this group that is a gas; the rest are in solid form. Nitrogen is most often used in the production of ammonia for refrigeration. Its diversity makes it one of the most used commercial elements. It is used in the production of fertilizers, explosives, and medical applications (see Table 3-10). A 1986 survey of a journal citing EPA estimations disclosed that the production in the United States alone was 12 million tons annually. Some authorities consider this figure to be extremely conservative.

Because of its prevalence in nature, nitrogen is used often in industry, often to form oxides. These oxides, called *nitrogen oxides*, are a very toxic gas. If inhaled, the oxides are strong irritants that cause pulmonary edema. Oxides cause a chemical oxidation in the lung tissue injuring the alveolar sacs and stimulating bronchiole fibrosis. The result of this type of injury is pulmonary edema and chronic lung disease.

Nitrogen or, to be exact, nitrates and nitrites used in fertilizers affect the body by causing a twofold injury. The cardiovascular system responds with temporary vasodilatation and hypotention. This property is used medicinally for the treatment of angina. Nitroglycerin, a nitrate, causes blood vessel dilatation, which relieves angina chest pain. The second effect, which is longer lasting and more dangerous is the change of hemoglobin to methemoglobin, a nonoxygen carrying compound (see nitrates and nitrites in Chapter 6).

Phosphorous has three forms, each having a different set of physical and chemical properties. For example, the specific gravities vary among the three types:

White phosphorous	1.82 g/cm³
Red phosphorous	2.20 g/cm³
Black phosphorous	2.25–2.69 g/cm³

Phosphorous is a very reactive element. It exists in several forms (four to be chemically correct). White (yellow) phosphorous, red and black (this is actually a dark violet color). The white phosphorous has two types, one called alpha and the other called beta.

White phosphorous is very unstable. As it stands in a water environment it will start to turn yellow. White phosphorous is trying to convert in order to attain red phosphorous-type qualities. If the material is then exposed to air, a violent reaction can take place. If the white phosphorous is exposed to sunlight or if it is heated, it will convert to red phosphorous. Red phosphorous is more stable and will not oxidize as readily in air. Because of this quality white phosphorous is sometimes transported by encasing it with red phosphorous and placing it in a container of water or in a nitrogen-enriched atmosphere (sometimes stored in fuel oil).

Red phosphorous is relatively safe as compared to white. However, as with all chemicals, caution should always be taken. Although this variation of phosphorous does not spontaneously burn in air as does the white variety, if an ignition source is available burning will result.

In both variations the vapors that are released from this chemical are highly toxic. One should avoid all exposure to the vapors. Phosphorous and halides can form to produce compounds called *phosphorous halides*. These too should be avoided. If in solid form, such as powders or granules, the material will irritate the skin and/or mucus membranes. If in a vapor state, avoidance should be the number one priority. In these cases, pulmonary edema is observed due to the tissue damage in the lungs.

Solid-state computer component industry, agricultural poisons, and insecticides are all examples of arsenic uses. All compounds of arsenic are considered to be poisonous. (See Pesticides in Chapter 6).

General hazards of these groups are heavy metal poisoning and other varied effects dependent on the element combination.

Group 16 (Group VIB or VIA)

Group 16 contains such elements as oxygen (O), sulfur (S), selenium (Se), tellurium (Te), and polonium (Po) (see also Table 3-11). The first two elements in this group are of the most concern. As has been mentioned, grouping can give us a clue as to what to expect during a chemical reaction.

SAFETY

The general hazards of Group 16 elements are poisoning and other varied effects, dependent on the element combination.

Table 3-11 *Group 16 (Group VIB or VIA) elements.*

Chemical Symbol	Density (g/cm³)	Melting Point	Boiling Point	First Ionization Potential (ev)
O	1.43 g/l	−218° C	−183° C	13.618
S	1.96	115° C	445° C	10.360
Se	4.79	217° C	685° C	9.752
Te	6.24	449° C	990° C	9.009
Po	9.32	254° C	962° C	8.420

It is known that living organisms, including humans, need oxygen to survive. Humans breathe in approximately 21% oxygen and expire roughly 16% (see Chapter 12 for exact O_2 values). In order for us to remain in the conscious state, a level of at least 12–16% must be maintained. However, as precious as this element may be, it has dangerous forms. When diatomic oxygen (O_2) becomes exposed to a strong electrical charge, it changes to an active three atom molecule. This newly formed compound is called ozone (O_3). This product is highly toxic, and all exposure to this gas should be avoided. However, as dangerous as it may be to humans, this compound forms a protective layer that encompasses the earth above our atmosphere. Oxygen is very reactive and will combine in a variety of configurations.

Sulfur is used commercially in a variety of compounds and also occurs naturally (see Table 3-12). A large percentage of sulfur is used to make sulfuric acid. Sulfur is also used in explosives, in gunpowder, as a fumigant (mercaptan), as a component to the production of rubber, in fertilizers, in bleaching of fruits, as a fungicide, and in electrical insulators to name a few.

Sulfur is an essential part to the life cycle in humans (and other organisms). Some sulfur compounds are extremely toxic and found not only in industry but are also naturally occurring, leading to problems in confined spaces. Hydrogen sulfide is an example of such compounds.

Selenium, although not often encountered, is a potentially toxic element. The chemical properties are similar to sulfur, although it is primarily used in industry because it can convert light into electricity. This single property makes it beneficial to us for photoelectric cells, solar cells, light meters, and solid state electronics.

Tellurium is usually found in combination with other metals. It is used to machine stainless steel and copper. It can also be used to decrease the potency of sulfuric acid. It is often used in blasting caps and ceramics. An interesting side effect is that when individuals are exposed to this element they develop "garlic breath." Although toxic levels are in debate, it does possess chronic toxic qualities.

Table 3-12 *Examples of sulfur compounds.*

H_2S	Hydrogen sulfide	A toxic gas from the decay of organic compounds. It is very irritating to the mucosa and can cause chemical asphyxiation. The characteristic smell is that of rotten eggs.
SO_2	Sulfur dioxide	A refrigerant in old systems. Air pollutant, which can cause irritation and pulmonary edema in compromised patients. It is also used in wastewater operations.
H_2SO_4	Sulfuric acid	Currently it is estimated that more than 20 million tons are in use. Widespread in chemical processes, often seen in clandestine laboratories

Halogens are used in combination with methane or ethane, resulting in a nonflammable, colorless gas that can be used as an extinguishing agent. The numerical naming of these halogenated hydrocarbons has to do with the number of carbon fluorine, chlorine, bromine, and iodine (in that order) that are within the molecule:

		Carbon Atoms	Fluorine Atoms	Chlorine Atoms	Bromine Atoms
Halon 104	Carbon tetrachloride	1	0	4	0
Halon 122	Dichlorodifluoromethane	1	2	2	0
Halon 1011	Bromochloromethane	1	0	1	1
Halon 1202	Dibromodifluoromethane	1	2	0	2
Halon 1211	Bromochlorodifiluoromethane	1	2	1	1
Halon 1301	Bromotrifluoromethane	1	3	0	1
Halon 2402	Dibromotetrafluoroethane	2	4	0	2

Pound-for-pound, halons are the most effective extinguishing agent. The main medical concern is the displacement of oxygen within the room or area of discharge.

The halons are in the process of being replaced with more environmentally safe extinguishing agents. For example, Halon 1301 is replaced by CEA 410 (CEA is the abbreviation for the new halon replacements. It stands for Clean Extinguishing Agents) and Halon 1211 replaced with CEA 614. CEA 410 is perfluorobutane (C_4F_{10}), a gas with a boiling point of 28.4°F. CEA 614 is perfluorohexane (C_6F_{14}), a liquid with a boiling point of 132°F.

! SAFETY
The general hazards of Group 17 elements are that they are the most reactive, cause extreme tissue irritation, and are highly toxic, dependent on the element combination.

Polonium is a radioactive element with a half-life of 103 years (^{209}P). This element is very dangerous in small quantities. All handling of this material should be avoided. This element is primarily used to remove dust from photographic films.

General hazards of these groups are poisoning and other varied effects dependent on the element combination.

Halogens—Group 17 (Group VIIB or VIIA)

The halogens—fluorine (F), chlorine (Cl), bromine (Br), iodine (I), and astatine (At)—comprise Group 17 (see also Table 3-13). This group of elements is one of

Table 3-13   *Group 17 (Group VIIB or VIIA), the halogens.*

Chemical Symbol	Density (g/cm³)	Melting Point	Boiling Point	First Ionization Potential (ev)
F	1.7 @ 0°C	–220°C	–188°C	17.422
Cl	3.2 @ –33°C	–101°C	–35°C	12.967
Br	7.6 @ 0°C	–7°C	59°C	11.814
I	11.3 @ 20°C	113°C	184°C	10.451
At	?	302°C	337°C	?

the most widely used group of chemicals. They are all toxic and produce a variety of serious health effects. The group is utilized in the production of the halons. In this associated state with methane or ethane, the resulting compound is used for extinguishment. In the free state (in nature the halogens are in combination with another element), the halogens are corrosive, irritating and are strong oxidizers. Great care should be exercised during a hazardous materials operation dealing with these elements.

Within this group, the outermost shell has seven electrons. Because of this fact, all have a strong desire to combine with chemicals that will "share" the needed electron. This makes this group very reactive. They all exist as a diatomic molecule. (Most elements that are gases exist in this state naturally; this configuration provides a certain degree of stability.) The halogens are used extensively in organic chemistry, which is one reason they are so often found in clandestine drug laboratories. All laboratory processes should be investigated for the use of this chemical group.

Fluorine is the strongest electronegative element; it is also the most reactive. Its uses vary from rocket propellant to fluoridation of drinking water. Fluorine is highly toxic and a corrosive gas (Hydrofluoric Acid in Chapter 6). In the form of hydrofluoric acid, it is used to etch glass. Hydrofluorosilicic acid is one of the forms used for water fluoridation. All acids (or products) that contain fluorine have a high potential for the production of a halogen acid gas.

Chlorine as well as fluorine, are used in the purification of drinking water. Chlorine is used in organic chemistry as a chemical substitute and oxidizing agent. Chlorine is extremely irritating to the skin, mucosa, and the respiratory tree. A few breaths of chlorine gas can be fatal. In WWI, chlorine gas was used as a chemical warfare agent, one of the first applications of chemicals during war.

Chlorine and its by-products are extensively used as disinfectants for swimming pools, hot tubs, and jacuzzis. The two most common are chlorinated isocyanurates and calcium hypochlorite. These two chemicals are incompatible with each other and with other pool treatments. If they come in contact with gasoline, ammonia, kerosene, solvents, or reducing agents, the heat produced can start a reaction that can result in spontaneous combustion.

In reviewing case studies on chlorine and chlorinated products, the most common emergency is caused from placing incompatible agents, usually chlorinated isocyanurates and calcium hypochlorite, together. If the products come in contact with each other, a heavy dense white toxic smoke is generated. If emergency response crews are to enter such a scene, high-level personal protection is warranted.

Remember, this chemical group is full of strong oxidizers that can saturate the clothing and structural firefighting gear, allowing it to burst into flames at the next contact with flame. At the time of the incident, decontamination may not seem important, however, all gear should be immediately decontaminated and inspected for further use (NFPA 1500; 5-1.8).

Iodine has several radioactive isotopes. ^{131}I has a half-life of 8 days and is used to bombard the thyroid gland as a therapeutic procedure for a variety of disease processes. As with the other halogens, iodine gas is extremely toxic and irritating to the mucosa, skin, and respiratory tree.

Astatine is a radioactive element with a half-life of eight hours (^{210}At). When reviewing the literature, we found only one use for astatine. Some physics laboratories use astatine to measure the subatomic particles. These particles are thought to be the components of the electrons, neutrons, and protons. As far as a hazardous materials response goes, there is no documentation indicating a likelihood of an incident occurring where this element is involved. However as science moves forward, we are not sure where this element may present itself. Astatine's properties are similar to iodine with the hazards of the halogens.

These gases as well as other toxic vapors are unpredictable, however, evidence supports that the halogen gases may cause a variety of pulmonary diseases, which have been identified as obstructive processes. Bronchiolitis obliterans, peribronchial fibrosis, and bronchitis have all been noted following accidental releases of the halogen gases.

General hazards of these groups are that they are the most reactive, cause extreme tissue irritation, and are highly toxic, dependent on the element combination.

Noble Gases—Group 18 (Group VIII or VIIIA)

The noble gases consists of helium (He), neon (Ne), argon (Ar), krypton (Kr), xenon (Xe), and radon (Rn) (see also Table 3-14). They are all stable, without the introduction of energy (fire). The reason for stability is due to the electron configuration in the outside shell. All of these elements contain eight electrons in the outer shell, making them extremely stable. Use of this group is focused on the extreme stability of its members. They are used to purge pipelines between com-

Table 3-14 *Group 18 (Group VIII or VIIIA), the noble gases.*

Chemical Symbol	Density (g/cm³)	Melting Point	Boiling Point	First Ionization Potential (ev)
He	0.179	−272	−269	24.587
Ne	0.901	−249	−246	21.564
Ar	1.784	−190	−186	15.759
Kr	3.740	−157	−154	13.999
Xe	5.890	−112	−108	12.130
Rn	9.91	−71	−62	10.748

! SAFETY
The general hazard of Group 18 elements are that they are asphyxiants.

modities because of their lack of reactivity. Some are used in fluorescent light bulbs for color. The greatest danger of these elements is their ability to displace oxygen and cause asphyxiation.

The general hazard of this group is that the elements are asphyxiants.

BONDING PRINCIPLES

chemical bond
the attachment between elements to form molecules

The basic structural unit of the compound are the elements. For compounds to be formed elements must "attach" themselves together and form a molecule. The attachment is an attraction that contains a relatively moderate to strong force, holding the molecule together. This force is called a **chemical bond**, the attachment between elements to form molecules.

Each molecule has a "structural" plan that nature follows. This system has its own geometry that provides a variety of molecules. Matter, as we know it, must be electrically neutral. Thus, the negatives must equal the positives. When ions come together to create a compound (cations and anions in equal numbers), a transfer (or sharing) of electrons takes place. Once the electrical neutrality is satisfied, a compound is generated. Ionic compounds are three-dimensional lattice structures, extremely strong, and continuous. Thus most are solids at normal room temperatures. In order to break this bond, a tremendous amount of energy must be placed into the compound.

In some cases, elements have their outermost electron stripped from its orbit. Elements that have one or two fewer electrons are called *cations*, and possess a greater positive charge. On the other hand, some elements gain one or two electrons making them negatively charged. These are called *anions* (Fig. 3-10). Collectively this group of atoms is called *ions*.

Figure 3-10 *When an atom from the left side of the periodic chart loses an electron it becomes positively charged (cation), whereas the elements from the right side of the chart become negatively charged (anion).*

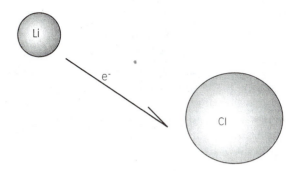

This group of compounds has specific naming nomenclature. Without a knowledge of chemistry, an emergency responder will not be able to determine the elements present or identify the hazards presented by them. Here they are presented as a source of reference (and an easy list to memorize). When bonding is presented, the chemical concepts involving that nomenclature is discussed.

There are several monatomic (a single atomic ion) anions that are named by adding the suffix -ide to the corresponding term. Each of these has electrical charges representing the number of additional electrons picked up to satisfy the octet rule (filling the outermost shell with eight electrons or two in the case of hydrogen) These are addressed in more detail in the section on metal salts.

–1 anions	*–2 anions*	*–3 anions*
H Hydr*ide*	O Ox*ide*	N Nitr*ide*
F Fluor*ide*	S Sulf*ide*	
Cl Chlor*ide*	Se Selen*ide*	
Br Brom*ide*		
I Iod*ide*		

If the placement of these elements on the periodic chart is examined it can be seen that the elements are from groups 15 (VA), 16 (VIA), and 17 (VIIA).

When there exists a relatively small number of oxygen atoms within the anion, the suffix *-ite* is used. The oxygenated inorganic compounds are identified here.

–1 anions	*–2 anions*
NO_2 Nitr*ite*	SO_3 Sulf*ite*
ClO_2 Chlor*ite*	

With a relatively large number of oxygen atoms within the anion, the suffix *-ate* is used. This is considered the normal state.

−1 anions	−2 anions	−3 anions
NO_3 Nit*rate*	SO_4 Sulf*ate*	PO_4 Phosph*ate*
ClO_3 Chlo*rate*	CO_3 Carbon*ate*	BO_3 Bor*ate*
	CrO_4 Chrom*ate*	

Chlorine (as well as other oxygenated ions or radicals) provides a special case with the variety of anions. When a relatively higher number of oxygen atoms are present, they are denoted with the prefix *per-* and the suffix *-ate*. If *-ite* is the suffix then the ion has one less oxygen than the normal state. If *hypo-* is a prefix and *-ite* is a suffix, then the ion has two oxygen atoms less than the normal state.

For example Chlorate, ClO_3, is in the normal oxygen state in this combination of three oxygen and one chlorine atom. The overall electrical charge (valence) of this ion is −1. If the chlorate ion had one more oxygen than the oxygenated state, one above normal, it is thus called a perchlorate, ClO_4 ion. If an oxygen was taken from the original chlorate ion, then the ion would have one less oxygen, giving rise to a change in nomenclature as a chlorite, ClO_2 ion. If two oxygen were taken from the original ion then hypochlorite, ClO would be the name given. Within this nomenclature, the oxygenated state of the base ion dictates the naming process.

	−1 anion	−2 anion	−3 anion
	ClO_4 *Per*chlor*ate*	CO_4 *Per*carbon*ate*	PO_5 *Per*phosph*ate*
Normal State →	ClO_3 Chlor*ate* →	CO_3 Carbon*ate* →	PO_4 Phosph*ate*
	ClO_2 Chlor*ite*	CO_2 Carbon*ite*	PO_3 Phosph*ite*
	ClO *Hypo*chlor*ite*	CO *Hypo*carbon*ite*	PO_2 *Hypo*phosph*ite*

Over the last few years we have seen an increase in terrorist activity within the United States. The two that have received high profile media coverage have been the World Trade Center in New York City in 1993 and the Federal Building in Oklahoma City in 1995. Each presented with the "homemade" version of an explosive device.

In the World Trade Center bombing, nitrourea was used in conjunction with hydrogen. Nitrourea is a white powder that has high explosive potential. In relationship with its standard of explosive power (TNT and picric acid), nitrourea has approximately 34% more power than TNT, kilogram-for-kilogram.

The Oklahoma City bomb was a mixture of ammonium nitrate and fuel oil. Over-the-counter ammonium nitrate is mixed with ammonium sulfate or calcium carbonate, which decreases the explosive potential of the ammonium nitrate. However, during the decomposition reaction ammonium nitrate will melt, especially at high temperatures releasing nitric acid vapor and water vapor. Within a closed container the result is quite explosive.

With monatomic cations, the ions names are the same as seen on the periodic chart. For example Na+ is sodium, both the element and the cation.

However, some metals have more than one cation. In this case, the Latin names are used. With the suffixes *-ous* to represent the lower charged cation, and *-ic* to represent the higher charged cation. This is usually found in the naming of common chemicals.

+1 cations	*+2 cations*	*+3 cations*
Hg Mercur*ous*	Hg Mercur*ic*	
	Fe Ferr*ous*	Fe Ferr*ic*
	Cu Cupr*ous*	Cu Cupr*ic*
	Sn Stann*ous*	Sn Stann*ic*
	Mn Mangan*ous*	Mn Mangan*ic*

IONIC BONDING

Normally, an atom of high electronegativity (an atom that has a stronger negative quality than positive) and low ionization (ionization is the ability to lose or gain an electron) will react to form an ionic bond. This occurs when elements from the left side of periodic chart bond with elements from the right side of the chart. This is better expressed by stating that the bond is between a metal (left side) with a nonmetal (right side) (Fig. 3-11). Electronegativity is an element's ability to attract electrons. These bonded materials are called salts.

$$Metal + Nonmetal = Salt$$

So let us look at ordinary table salt for an example:

$$Na^+ + Cl^- \longrightarrow NaCl$$

By transferring the electrons, each attains the state it is comfortable with. It is said that they have achieved a noble gas state, remembering that the noble gases all have eight electrons in their outermost shell and will not react.

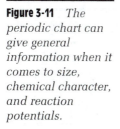

Figure 3-11 *The periodic chart can give general information when it comes to size, chemical character, and reaction potentials.*

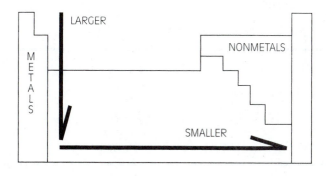

Down the columns on the periodic chart the atoms tend to get larger and heavier. They increase in electropositive character. As we travel across the chart from left to right the atoms tend to become smaller and more compact, thus giving them a strong electronegative character.

These compounds can exist in water solutions or in a solid form. When in a solid form, water can cause reactions to occur. In general they are usually solids, water soluble, and have a high melting point when found in nature. Fortunately, they do not burn under normal conditions. Salts are comprised of several subgroups. Each of these subfamilies are ionically bonded and possess their own health and reactivity hazards.

Metal Salts

The first types of salts to be examined are when a metal and a nonmetal combine with each other and form a metal salt. Here a metal cation combines with a nonmetal anion. If the overall negative charge is one, then the respective positive charge for combination is one. However, in the case of calcium nitride, calcium has an overall plus 2 charge, and nitrogen has a negative 3. When these two elements combine, it takes three calcium atoms to combine with two diatomic molecules of nitrogen to make calcium nitride.

$$3Ca^{+2} + 2N_2^{-3} \longrightarrow Ca_3(N_2)_2 \text{ Calcium nitride}$$

! SAFETY
When in contact with water, carbide produces acetylene gas, hydrogen evolves hydrogen gas, nitride gives off ammonia gas, and phosphide produces phosphine gas.

The -ide identifies the compound as a metal/nonmetal salt. We name the compound with the metal first and the nonmetal second, ending it with ide. The hazards are general, depending on the metal of attachment. Four are water reactive and extreme care should be taken. They are: carbide, which produces acetylene gas when water comes in contact with it; hydride, which evolves hydrogen gas; nitride, which gives off ammonia gas, and phosphide, which produces phosphine gas. Some examples of the metal salts are shown in Table 3-15.

Table 3-15 *Examples of metal salts.*

Potassium sulfide	K_2S	Flammable; may spontaneously ignite; explosive in dust form
Aluminum chloride	$AlCl_3$	Violent reaction with water, hydrogen chloride gas is evolved
Cupric chloride	$CuCl_2$	Toxic by ingestion or inhalation
Calcium nitride	Ca_3N_2	Evolution of ammonia gas when in contact with water

Table 3-16   *Examples and uses of metal oxides.*

Aluminum oxide	Al_2O_3	Used for electrical insulators, heat resistant fibers; dust is toxic
Arsenic oxide	As_2O_5	Used in insecticides, herbicides, and colored glass
Beryllium oxide	BeO	Toxic by inhalation; used in reactor systems, electron tubes
Calcium oxide	CaO	Once exposed to water evolves heat

Metal Oxides

The next group has a metal and oxygen in attachment. The name ends with -oxide. These are called the *metal oxides* and, they tend to be water reactive, which in turn produces heat once the reaction takes place. The resulting solutions are corrosive, which is dependent on the base metal that was attached to the oxygen. They are moderately hazardous. If the oxygen were attached to an alkali metal, the corrosivity would increase (see also Table 3-16).

Inorganic Peroxides

Inorganic peroxides (see also Table 3-17) are similar to metal oxides, except that one more oxygen is attached in the molecule. They are a metal in combination with a peroxide radical (O_2). The *per* in peroxide means that there is one more oxygen than the normal state of one. They tend to be strong oxidizers, which generate heat upon reaction. This can ignite ordinary combustibles if in the general area. Most of them as a family are water reactive. Corrosivity depends on the base metal.

Hydroxides

Hydroxides (see also Table 3-18) are metals that combine with the hydroxide radical (-OH). The name ends with hydroxide. Sodium hydroxide, potassium

Table 3-17   *Examples of inorganic peroxides.*

Barium peroxide	BaO_2	Oxidizer; fire upon contact with organic materials
Calcium peroxide	CaO_2	Oxidizer; fire upon contact with organic materials
Cesium peroxide	Cs_2O_4	Oxidizer; fire upon contact with organic materials
Hydrogen peroxide	H_2O_2	Oxidizer; severe fire hazard

Table 3-18 *Examples and uses of hydroxides.*

Aluminum hydroxide	$Al(OH)_3$	Used in flame retardants, mattress batting, and cosmetics
Beryllium hydroxide	$Be(OH)_2$	Decomposes to an oxide; extremely toxic
Cesium hydroxide	CsOH	Extremely toxic
Potassium hydroxide	KOH	Toxic and corrosive

! SAFETY
● **The corrosivity of hydroxides is dependent on the base metal of attachment.**

hydroxide, and calcium hydroxide are all examples of this type of salt. All are caustic in the solid state. If mixed with water, the result will be a caustic solution. All are classified as corrosives and produce heat during reaction. Again the level of corrosivity is dependent on the base metal of attachment.

Oxygenated Salts

The last group in our discussion are the oxygenated salts (see also Table 3-19). Here the terms *Per-* and *Hypo-* aids in the designation of the level of oxygen that the compound contains. By definition the oxygenated salt is a metal in combination with an oxygenated radical. Nomenclature is based on the normal state of the oxygenated radical. These are termed as normal state and the name starts with the metal and ends with the oxygenated compound as *-ate*. If the oxygenated compound contains one more oxygen than the normal state, the metal and the oxygenated compound is identified. In this case the oxygenated compound has the prefix of *Per* and an ending of *-ate*. If one oxygen less than normal, the compound ends with *-ite*. If two less oxygen than normal, the metal again is identified and the oxygenated compound radical has a prefix of *Hypo-* and a suffix of *-ite*.

For example, the chlorine molecule as an oxygenated molecule has three oxygens attached to the chlorine. This is the normal state. One more oxygen would give a Per—ate state. One less, a -ite state, and two less a Hypo—ite state.

Table 3-19 *Examples of oxygenated salts.*

Sodium perchlorate	$NaClO_4$	Oxidizer; fire upon contact with organic materials
Sodium chlorate	$NaClO_3$	Oxidizer; fire upon contact with organic materials
Sodium chlorite	$NaClO_2$	Oxidizer; flammable; irritant to tissue
Sodium hypochlorite	NaClO	Oxidizer; fire upon contact with organic materials; irritant

! SAFETY
Oxygenated salts
are extremely strong
oxidizers; heat that can
ignite combustibles is
generated when they
become wet; and, if
mixed with fuels, the
resulting compound
becomes violently
explosive.

So in our example of chlorine the radicals would look like this:

$-ClO_4$ *Perchlorate*

$-ClO_3$ Chlor*ate*

$-ClO_2$ Chlor*ite*

$-ClO$ *Hypochlorite*

The metal that is attached would be placed in front of the oxygenated radical. As may be seen these are extremely strong oxidizers. If they become wet, heat is generated which in turn can ignite combustibles. If mixed with fuels, the resulting compound becomes violently explosive (Fig. 3-12).

Figure 3-12 *When an element from the right side of the periodic chart combines with an element from the left side, an ionic compound is formed. In general these ionic compounds are solids, dissolve in water, conduct electricity, and do not burn. However some may have hazardous qualities due to chemical reactions with other materials (such as water). The above categories show the hazards and naming of each group. Recognize these groups and their hazards!*

Salts
-ide
General Hazards
Toxic gas with water contact with following:

Carbides	=	Acetylene
Hydrides	=	Hydrogen
Nitrides	=	Ammonia
Phosphides	=	Phosphine

Metal Oxides
-oxide
Hydroxides when contact with water
All water reactive
Heat generation and caustic solutions

Hydroxides
Hydroxide
Solids and solutions are corrosive

Inorganic Peroxides
Peroxide
Water reactive
Strong oxidizers
Heat production with high ignition potential

Oxygenated Inorganic Compounds
Per -ate; ate; ite; Hypo -ite
Normal oxygenated states

-1	-2	-3
FO_3	CO_3	PO_4
ClO_3	SO_4	BO_3
BrO_3		
IO_3		
NO_3		

More than 1	O :	Per -ate
Normal	O :	-ate
Less than 1	O :	-ite
Less than 2	O :	Hypo -ite

Figure 3-13 *In the ionic bond, the electron clouds are close to transfer the electron(s), where in the covalent bond, the overlap is an equal sharing of electrons.*

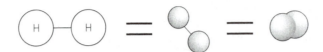

COVALENT BONDING

covalent bond

electrons that are shared between atoms to make a molecule

Not all compounds are held together by ionic bonds. A **covalent bond** is the chemical bond created by the sharing of electrons between atoms (Fig. 3-13). Here the bond involves a sharing of electrons in order to form the molecule. The sharing requires that two or more elements get close enough together to allow the outer shells to share a pair of electrons in the orbit. In this case, the bond creates a stable molecular configuration. The more stability a compound possesses, the more energy is required to break the compound apart.

The concept of covalent bonding is easier to explain if two hydrogen ions are examined. Each of the ions has one electron circling the center structure. For hydrogen to feel satisfied it must have two electrons in the shell. As the two atoms move closer together, the electron cloud about each atom overlaps so much that a sharing of two electrons takes place, each filling the outer orbital shell with two electrons (the maximum electron capacity for the first shell). Thus these two atoms are satisfied and somewhat stable because the shell is saturated.

Unfortunately, this example is the simplest form of covalent bonding. When other covalent bondings are examined the picture becomes increasingly more complex. As the compound becomes larger, the configuration of the electron cloud changes. To simplify, it looks similar to a tetrahedron. As was discussed earlier, the number of electrons increases with the size of the atom. If the atom has 14 protons, then there will be 14 electrons about the atom.

We have shells or orbits much like the planets that revolve about the sun. The shells in the atomic world interact with orbits of other compounds' orbits to create spaces for the electrons to travel. So for example, our atom that has 14 electrons has two shells full, with a third shell partially filled. It is this third, or outer, shell that loses or gains electrons.

The point here is that there are relative energy levels in which electrons fall. Each energy level has a finite number of sublevels or orbits. As the atomic number increases, so does the number of energy levels. Because each atom inherently possesses a certain degree of energy, when the covalent bond is broken, there is a corresponding release of energy. For this reason, covalent bonds can be relatively

■ NOTE

When a covalent bond is broken, there is a corresponding release of energy.

unstable. If these bonds are broken, they may produce a violent reaction. For example if acetylene, a triple-bonded hydrocarbon, is shocked, the bonds between the carbon-to-carbon link are broken, resulting in a chain reaction—an explosion.

In the covalently bonded molecule, one or more of the atoms within the compound may possess a relatively stronger electronegative pull. This pull shifts the density of the electron cloud toward the proximity of the stronger electronegative atom. The shift creates a positive end and a negative end of the molecule. A tiny magnet is created. **Polarity**, the possession of two opposing tendencies, is created. An nonsymmetrical electron cloud distribution is termed *being polar*.

polarity
the possession of two opposing tendencies

If two of these molecules "rub" together, a small amount of electromagnetic energy is produced. How does this relate to a hazardous materials incident? A quick look at a fuel spill will provide an example of this principle. The spill involves a hydrocarbon fuel with a low flash point. When the molecules of fuel flow through piping, a static electric charge is generated and a spark may be produced. If that fuel is gasoline, JET A, or a derivative of low flash point hydrocarbon, the end result could be a fire. If this situation occurred near a large fuel source, such as a fixed-storage facility, the result could be devastating.

nonpolar bond
a nondiscernable attraction between atoms

If the electronegative shift does not take place the bond type is described as nonpolar. **Nonpolar bonds** (nondiscernible attraction between atoms) are found whenever the two atoms that are bonding are identical. The diatomic (two atoms of the same element in one molecule) molecule of oxygen is a good example of this. As stated earlier, when the atomic number increases so do the number of electrons, and thus the number of orbits.

! SAFETY
Multiple bonds also tend to break during metabolism.

In the world of covalent bonding, for the molecule to reach a stable energy level, sometimes a double or triple bond is required. Remember, as the complexity of the molecule increases so does the complexity of the bonding framework. Under certain conditions, double and triple bonds are more effective than the single bond which are easier to break. When these multiple bonds are broken, a great deal of energy is released. This energy is tremendously greater than if they were single bonds. Multiple bonds also tend to break during metabolism.

To simplify matters, only the outermost shell of the elements will be considered. This outermost orbital either transfers electrons or shares electrons. This orbit is stable enough that the transfer or sharing will not take place without applying some external force to the atoms in question.

All of the elements, once engaged in a reaction would like to have their outermost shell filled with eight electrons. (Hydrogen and helium are an exception to this rule as their shell is only capable of two electrons. Hydrogen has only one and is anxious to gain a second. Helium has two electrons making it very stable.) Covalently bonded molecules are able to satisfy the eight electron rule by borrowing electrons from another element. This configuration is satisfying the octet rule. For example, carbon has only four electrons in its outer shell. For carbon to be satisfied it must have four more electrons within the outermost shell. Carbon can either "give-up" four electrons or it can "gain" four electrons. For this example, if four hydrogen atoms come in close proximity to this carbon atom, four covalent bonds will take place, Satisfying the octet rule for the carbon and the duet rule for the hydrogen, stabilizing both. This creates a stable molecule. Covalently bonded molecules are stable electrically, however they are potentially unstable if outside energy is available. In this example, four hydrogen atoms covalently bonded to one carbon atom form the molecule methane, a flammable gas. If methane comes in close proximity to a heat source, it will burn.

IONIZATION ENERGY

Ionization potential (IP)
the minimum energy required to release an electron or a photon from a molecule

Ionization energy is the amount of energy that is required to remove an electron from the outermost shell. Energy is introduced into the atom to pull the electron away from the orbit but stay within the realm of the outer shell. This minimum energy will succeed in doing so, giving either a release of the electron or light energy generation once the electron bounces back into its normal path. This is termed the *ionization potential* (IP). This term is discussed again in Chapter 11.

The size of the atom plays a strong role when discussing ionization. For example, if the atom under discussion has a large diameter (this is actually measured in atomic radius: a point from the middle of the nucleus to the outer portion of the electron cloud), it would lose electrons easily because the ionization energy is small. So what does all this mean? Large atoms with electrons far from the nucleus and low ionization potential are fairly reactive in terms of giving up their electrons. Conversely, if the atom has a small radius, placing the electrons close to the nucleus, the ionization potential is great and there is less of a desire to give up electrons.

A look at the periodic chart can give clues to the ionization potential of the elements. As a line is followed from left to right, there is an increase in ionization potential. Following a line from the top of the chart to the bottom, a decrease in ionization potential is seen. These changes in ionization potential are because the radius of the atoms decreases moving to the right of the chart. Moving from top to bottom, the radius increases. Understanding atomic placement will give clues to the reactivity of the compound(s) in question.

PROPERTIES OF IONIC AND COVALENT COMPOUNDS

Polar Ionic Compounds

In polar ionic compounds electrostatic forces create an attachment between all the molecules. These forces are so strong that in order to melt a polar ionic compound, a great amount of thermal energy must be supplied. The energy that is thus supplied breaks the intermolecular attachment. The energy must be in proportion to the magnitude that the solid will become a liquid. Because of this intermolecular attraction, polar ionic bonded compounds have high melting points, high boiling points, are hard, and are very soluble in polar solvents.

Figure 3-14 *Van der Waals force is a magnet type attraction. If this compound is a liquid, this material can be volatile. A very strong type of dipole interaction is hydrogen bonding. Van der Waals and dipole forces are electrostatic.*

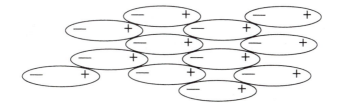

Nonpolar Ionic Compounds

Nonpolar ionic bonds are held together by Van der Waals forces (Fig. 3-14). These forces are weak and are basically caused by the movement of the electron about the nucleus and its relative position to the proton. Sometimes the nonpolar compound has an area that is negative one second and positive the next. This constant moving of positive and negative fields creates the "magnetism" needed to hold the compound together. Because they are weak (due to the slight movement of the electron in relation to the proton), these compounds are extremely soft and have low melting points. If the compound is a liquid, the material is extremely volatile.

Covalent Properties

Covalent bonds are somewhere between the polar ionic and the nonpolar ionic bond. The covalently bonded molecules have high melting and boiling points that actually are dependent on the size of the molecule (Fig. 3-15).

Basically each molecule acts as a unit within a chemical reaction. Just as was seen with individual atoms, molecules also strive to reach a comfortable state (octet satisfaction), in order to become stable.

GAS LAWS

In order to continue the discussion of chemistry, the concept of kinetic molecular theory must be discussed. Understanding this subject is important to gaining a better comprehension of general chemistry and its principles (not to mention the processes that can occur at our hazardous materials or EMS incident). Beyond that it will aid in understanding the laws that surround the behavior of gases.

Figure 3-15 *The bond moment is a direction of negative or positive charge. It is an orientation in space that gives the chemical configuration a positive and negative end, or the moments will cancel to give an overall zero effect. In this example, ammonia has a negative moment. This is due to the unshared pair of electrons.*

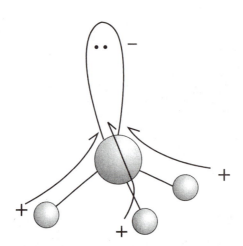

Kinetic Molecular Theory

kinetic molecular theory
the theory that states that all molecules are in constant motion

The **kinetic molecular theory** states that all molecules are in constant motion. This movement may be as simple as slight vibration within a solid, or rapid constant motion as in a vapor or gas. The theory states that if a material, either solid, liquid, or gas is brought down to a particular temperature, at this temperature all motion within the atom, and thus the molecule, will stop. This temperature is referred to as the absolute temperature. Absolute temperature or Kelvin is –459.67° F. This number is referred to as the temperature of absolute zero (0 Kelvin). Remember the definition of standard temperature and pressure (STP). It is a temperature of 32°F (0°C or 273 K) at a pressure of 1 atmosphere (760 torr or 760 mm Hg) that a reaction is tested or occurs.

To demonstrate this theory, the molecules of hydrogen sulfide gas placed in a closed container at normal pressure (sea level 14.7 psi) and temperature will rapidly strike the container's wall. If the temperature is increased, the molecular movement will increase, striking the walls with more force or pressure. If the gas is heated even more, the force and pressure will increase and eventually lead to container failure.

Gradually lowering the temperature of the same gas will slow the molecules within the container. As the temperature of –77°F (–76.72°F is the boiling point of H_2S) is approached, the molecules will slow to a point that the gas now turns to a liquid. If the temperature continues to decrease, freezing of H_2S will occur at roughly –122°F and the liquid hydrogen sulfide gas changes into a solid state. The molecular movement has been reduced. At –459.69°F, our once gas would be a solid in which all molecular movement has stopped.

Torricelli, an Italian scientist, recognized the concept of pressure in the 1600s. He measured the pressure of the atmosphere by utilizing a tube in which mercury was placed. He then measured the difference of levels of the mercury within the tube. This measurement has come to be known as millimeters of mercury (mm Hg at 32°F) or torr. So the pressure that is exerted by the atmosphere onto a column of mercury is 760 mm Hg high and is called one atmosphere.

14.7 psi. = 760 mm Hg = 1 Atm. = 760 torr.

As was previously discussed certain chemicals in the liquid or gaseous state will release a pressure in an enclosed container, which is referred to as the vapor pressure of the substance. Using the gas laws it will be seen why this occurs.

Boyle's Law

> **! SAFETY**
> **Boyle's law demonstrates that volume decreases as the pressure increases.**

Boyle's law is a relationship between the pressure of a substance and the volume this gas occupies. Here the temperature remains fixed, usually at 77°F (the temperature that a chemical should be noted when referencing chemical and physical properties). This law demonstrates that as the pressure increases the volume decreases.

$$V = 1/P$$

where 1 can be represented by a constant known as K but is equal to volume times the pressure. So that algebraically arranging the formula we see:

$$V = K/P$$
$$V_1 \times P_1 = K \text{ or } P_1V_1 = K$$
$$\text{where } K = PV$$
$$P_1V_1 = P_2V_2$$

So we can see that the original volume and pressure of the gas are equal to the resultant pressure and volume.

Charles's Law

> **! SAFETY**
> **Charle's law demonstrates that volume and temperature are directly proportional.**

Charles's law deals with a gas volume as related to temperature but the pressure remains constant, usually one atmosphere. As was seen with Boyle's law, a relationship is expressed. Here the volume and the temperature are directly proportional. We can express this mathematically as followings:

$$V = T$$
$$V = KT \text{ where } K \text{ is equal to } V/T$$
$$V/T = K \quad \text{or} \quad V_1/T_1 = V_2/T_2$$

This expression denotes that the volume of a gas increases one to one as the temperature increases. J.A.C. Charles and Joseph Gay-Lussac were interested in hot air balloon flight. By observing the above law they were able to ascend into the sky in the late 1700s and early 1800s.

We know that in the real world, especially at the scene of a hazardous materials incident, attempts are made to control temperature and pressure. Water streams are sprayed onto hot pressurized containers, however the pressure may still increase, until the gas cools. Before the kinetic activity of the gas is reduced, the container may have a failure. After the gas within the container has cooled from the application of water, the incidence of container failure is also reduced.

If the laws are combined it can be seen that temperature, volume, and pressure all play a critical role when discussing gas emergencies. The ability to control them may lie in a general understanding of these laws.

$$P_1 V_1 = P_2 V_2 + V_1/T_1 = V_2/T_2$$

can be expressed as

$$P_1 V_1/T_1 = P_2 V_2/T_2$$

Dalton's Law

Dalton's law demonstrates what happens when several gases are together in a closed container. Dalton's law simply states that the pressure within a container holding several gases is the sum of all the pressures of each individual gas. In a mathematical form:

$$P_{Total} = P_1 + P_2 + P_3 + \ldots P_n$$

From a medical standpoint this concept can be utilized once all of the pressures are added. The resulting pressure can aid in deciding between Level A or Level B encapsulation. If the resulting pressure* is well above the atmospheric pressure, then the atmosphere is potentially hazardous and has a high risk of severe respiratory hazard. If the pressures are below the atmospheric pressure, the atmosphere may still be toxic however, with a lower potential for a respiratory hazard.

This is all dependent on the vapor pressures of the material. If you have a substance that has a very low vapor pressure and the partial pressure all add up to below the atmospheric pressure, then the possibility of saturation of the vapor within the ambient air is low. This does not mean that the environment is clear of toxic gases; it only means that the atmosphere can be entered given the appropriate level of personnel protection.

However, if the vapor pressure is high and the resulting partial pressures are also high, then the potential for a severe respiratory hazard exists. When the thoracic cavity expands, the atmospheric pressure within the chest decreases. If

---

*Vapor pressure.

the chemical already has high vapor and partial pressure, then the chances for a severe respiratory injury exists.

Henry's Law

Henry's law states that the amount of gas absorbed by a particular volume of liquid and a particular temperature is directly proportional to the pressure of the gas.

Henry's law addresses what happens when a gas is placed into solution. When a soda bottle is opened, tiny bubbles come out of solution. This example demonstrates the extreme of this phenomenon. Gas can be placed into solution through the use of pressure, however, two factors govern this type of reaction. First the pressure of the gas that surrounds the solution, and second the solubility of that gas in the fluid. Temperature also plays a part. Doubling the partial pressures of a gas while maintaining certain critical temperatures can double the gas in the liquid state, thus doubling the concentration in the gaseous state.

Prior to the discussion of gas laws, it was stated that some chemicals need a certain pressure and temperature before they proceed to the next state of matter. This temperature and pressure were termed the critical temperature and critical pressure. Now that the concepts of these laws have been reviewed, an application into the hazardous materials emergency can be examined.

With the addition of pressure, gas can be forced into a liquid state. For some gases, pressure and a reduction of temperature may be needed to force a gas into a liquid. Most of the liquefied gases seen by emergency responders have been placed in that state through the use of pressure only. Others, less common, are those that require a combination of pressure and reduced temperature.

All gases can be forced into a liquid or a solid state providing that the critical pressure and temperature are attained. The critical pressure is that pressure required to place enough restriction on the moving molecules to change the materials state from a gas to a liquid. When temperatures are below the critical temperature, the gas can be liquefied. Above this temperature, a gas cannot be changed into a liquid. The lower the temperature below the critical temperature, the less pressure is required for the conversion of state.

In other words, in order to liquefy a gas, pressure and temperature must be at respective finite points for the change in state to occur. Pressure alone may not place enough restriction on the molecules. The molecules still contain enough kinetic energy to maintain the gas's state. Once the temperature is dropped to the critical level with the associated pressure, the material then moves from a gas to a liquid.

cryogenics
gases that are turned into liquids by cooling to very low temperatures and associated pressurization

Whenever a gas is liquefied under the influence of temperature and pressure, a liquid-to-gas conversion will have a volume-to-volume ratio of expansion, once this chemical is released. (For one volume of liquid, how many volumes of gas will be produced?) Reference sector must be prepared to investigate this ratio (see also Table 3-20) and how it may affect the scene. Health hazards and fire potential are realized when the expansion ratios are known. As a rule of thumb, liquefied gases without temperature decrease (noncryogenics) will have expansion ratios between 1 to 300. **Cryogenic** liquids have expansion ratios in the 1 to

Table 3-20   *Examples of gases.*

Product Name	Boiling Point	Critical Temperature	Critical Pressure	Expansion Ratio
Propane	−43.8 °F	206.3 °F	617 psi	1 to 270
Methane	−258.9 °F	−115.8 °F	672 psi	1 to 637
Hydrogen	−421.6 °F	−390 °F	294 psi	1 to 848
Nitrogen	−319.9 °F	−230.8 °F	485 psi	1 to 694
Oxygen	−297 °F	−180 °F	735 psi	1 to 857

! SAFETY

As a rule of thumb, liquefied gases without temperature decrease (noncryogenics) will have expansion ratios between 1 to 300, while cryogenic liquids have expansion ratios in the 1 to 600 to 1000+ range.

! SAFETY

The typical injuries seen in connection with dealing with cryogenics are traumatic and hypothermic injuries; the sudden release of cryogenic liquids can cause traumatic lacerations, in association to the chemical exposure (absorption, injection).

600 to 1000+ range (694 for nitrogen to 1445 for neon). The technical definition of a cryogenic is applied to those gases that have a boiling point lower than −130° F at one atmosphere.

When we are dealing with cryogenics it is important to remember that these are materials that have been forced into a liquid state. Once released, this liquid will revert back to its natural state, that of a gas. Medical problems arise when an individual is within the area of this state change. The typical injuries that are seen are traumatic and hypothermic injuries. Obviously the cold nature of the cryogenic will produce frostbite type injuries or freezing of the tissue itself. The sudden release can cause traumatic lacerations, in association to the chemical exposure (absorption, injection). Once this liquid heats up to revert back to its gaseous state, the volume of the chemical has multiplied. The increase of volume is, of course, dependent upon its physical properties, for example, the boiling point. However, the rapid expansion from the liquid state to the vapor state will cause an increase in vapor pressure, which further increases medical considerations of respiratory injury.

Cryogenics are the gases that have boiling points below −130° F. Three basic hazards are associated with cryogenics: (1) The ability to liquefy or solidify other products, (2) hypothermic and traumatic injuries upon contact and (3) extremely high expansion ratios once released.

With liquid oxygen and liquid fluorine, if a spill should occur, the vapor is a fast-moving expanding oxidizer. These products are rapidly absorbed into surrounding material. If for example, the cryogen is absorbed into asphalt, the resulting product is macadam, a shock-sensitive material. The shear weight of an individual could result in a detonation.

HEAT DYNAMICS

Heat dynamics has its origin in the need to know about workable energy, that is, the amount of energy required to produce work within the industrial process or the energy lost or the work or energy gained during a mechanical operation. The beginnings of such a study had a practical premise in the beginnings of the industrial revolution.

Without becoming too complicated, the basics in this area of chemistry have a relationship to the surrounding environment and the chemical reaction itself. In the firefighters' work environment, extinguishing fires is of concern. For emergency medical personnel the concern is somewhat less and actually is not a consideration in terms of most medical treatment. However, for hazardous materials response personnel who deal with chemically exposed individuals and the incident itself, a general understanding of this area of chemistry is helpful.

Emergency workers must consider the relationship between the surroundings and chemical reactions. More specifically, the movement of heat should be examined. Conduction (the movement of heat from one material to another), convection (the movement of heated gases or liquids), and radiant heat (the linear movement of thermal energy, energy in the form of electromagnetic energy or energy packets) all indicate the flow of heat energy to and through materials. In terms of conduction, this flow of heat energy through a material produces heat dispersion away from the seat of the fire. Convection, is the movement of the hot gases and radiant heat is the particles, packages, or electromagnetic waves of heat energy that can be felt that move linearly through space.

The first law of heat dynamics states that energy can be changed but never created or destroyed. When a house burns, it is the chemicals within the house that are burning. The chemicals are changing, releasing energy (fire in terms of heat) and giving off smoke (products of combustion). Energy is not created; it changed through the burning of the house material. The energy that was inside each molecule within the house's contents is now released in the heat of the fire, the light produced, and products of combustion. The same set of rules occur in the human body in terms of metabolism and thermoregulation. You ingest food (fuel), which the body converts into water and energy. This energy is used to maintain bodily functions and the activities of the day.

calorie
the amount of heat required to raise the temperature of one gram of water one degree Centigrade

A **calorie**, the amount of heat that is required to raise one gram of water one degree Centigrade at one atmosphere, is used to measure the energy of heat that is lost or gained. Materials can give off heat, an exothermic reaction, or they can absorb heat, an endothermic reaction. Emergency responders are most familiar with exothermic reactions, which give off heat, such as in fire. Oxidation and combustion are examples of this reaction.

Another application concerning heat dynamics is the concept of outage. In a storage tank, liquids will expand as the temperature increases. Vapor will also be produced. The outage, space provided for expansion, is calculated based on the percent of volume expansion expected during increased temperature. If the outage is not calculated, or ignored, the container would overflow or rupture when the volume increased.

RATES OF REACTIONS

So far this chapter has covered chemical theories explaining why things occur. It began with a study of the individual elements and their representative chemical and physical characteristics. Then the discussion reviewed gas laws and how they may affect a hazardous materials scene. However to present a complete discussion of why a reaction takes place, an overview of the reaction sequence is necessary. During a career in the fire service, most firefighters experience that a fire progresses at an unexpectedly fast rate or an unexpectedly slow one. Hazardous materials incidents have the capability of displaying the same type of unusual rates. Generally the rate irregularities are because of the reactions taking place during the event.

    Some reactions occur very fast—dangerous reactions. Others occur slowly, sometimes so slowly that the reaction is useless. Why this diversity of reactions? If all the molecules act (or react) in the same basic way, then all reactions should also occur at the same rate. But we know that this statement is not always true.

    During the discussion of the gas laws we saw that as the temperature increased so did the pressure. The pressure increase was attributed to the frequency that molecules were released from the liquid state to the gas state. Actually with an increase of 10°C the rate of reaction doubles. The increase of temperature caused the molecules to move faster and faster until they were able to leave the liquid and gain the state of a gas. This increased movement increased the frequency of actual molecular collisions. This collision energy in combination with the concentration of the reactants creates the reaction.

❗ SAFETY
Rate of reaction doubles with an increase of 10°C.

Energy through Collision

Consider the normal reaction without the aid of temperature to increase the frequency of collisions. Looking at a normal reaction, it can be seen that two molecules colliding can either strike each other head-on, from either side, or from behind. If the molecule strikes the molecule from behind the energy is displaced and could push the second molecule further. A side strike causes a situation that displaces the movement at right angles to each other. However, if the moving molecules strike each other head-on, then an effective collision results.

Effective Collision

Why are effective collisions important? Bonds must be broken in order for new bonds to be created. Breaking of bonds takes enormous energy, which is gained only from head-on collisions while the molecules are moving extremely fast.

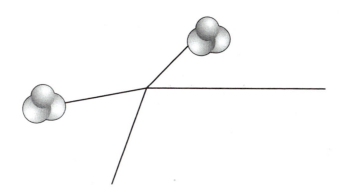

Figure 3-16 *The orientation, speed, and collision environment must be in harmony for the bond to be broken and another one established.*

When the bond is broken, it is actually an absorption of energy (endothermic); the creation of a bond is giving off energy (exothermic). This exothermic reaction is what we interpret as fire or explosion. From this example it can be seen why a low energy collision may not break a bond and form a new one.

On the other hand, if we have two molecules colliding and their speed is too high, again a reaction may not take place. The velocity was too high to translate the energy into a broken bond and create another one. Just like in the story Goldilocks and the three bears, the porridge temperature had to be just right, so must the collision be just right.

Another aspect of high speed (or moderate speed) molecular collisions is the molecule's spatial orientation. Some molecules are crowded, some are not. If the molecule is bombarded from an orientation that attacks the crowded molecule or the crowded end of the molecule, then the reaction cannot take place. However, if the collision is of the appropriate speed and the orientation is correct, then a reaction will take place (Fig. 3-16).

Reactions are like hills that must be crossed to reach the other side. If the hill is low, then the energy that is required is small, and the reaction will progress with relative ease. If the hill is large, then the reaction will need a lot of energy (Fig. 3-17). This reaction will need to be forced through with a supply of energy. Generally speaking, ionic bonds are broken and created more easily than covalent bonds.

activation energy

the minimum amount of energy needed for a molecular collision to be an effective one

catalysts

chemicals that lower the activation energy

inhibitors

chemicals that increase the activation energy

Activation Energy

This concept of a "hill" of energy is called the **activation energy**. This energy is the minimum amount of energy needed for the collision to be effective. **Catalysts** are chemicals (or reactions themselves) that lower the activation energy (make the hill lower) so that the main reaction can proceed faster. **Inhibitors** are chemicals that increase the activation energy (make the hill higher), thus slowing the main reaction down. Organic peroxides are a good example of activation energy.

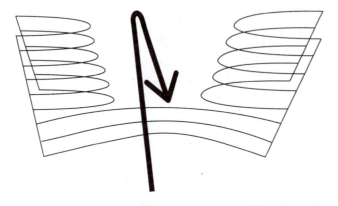

Figure 3-17 *Rate of reaction is like going over a mountain pass or hill. The lower the pass, the faster the reaction and the higher the pass, the slower the reaction.*

A wide mountain pass means that the number of molecular collisions are high, with appropriate molecular orientation and speed. A narrow pass means that the molecular conditions must be perfect. A wide and low pass tells us that the reaction will go fast.

Organic peroxides and the hydroperoxides are chemicals that produce free radicals. This characteristic is very useful in the production of plastics. Here the radicals polymerize (a reaction that allows one or more smaller molecules to combine into larger ones), creating long chains of the desired product—plastic. However, the problem is that these compounds are very unstable before the polymerization reaction. Normally, they have very low activation energies. An inhibitor is placed in the solution to increase the activation energy, thus slowing the reaction down. Once organic peroxide reactions start, if they are not within a controlled environment, they will decompose to completion. They are shock and heat sensitive therefore, very unstable.

A look at each molecule reveals four potential areas of energy:

1. Energy of electronic excitation
2. Energy of rotation
3. Energy of vibration
4. Energy of translation.

These energy sources, as minute as they may seem, explain why some reactions take place and others do not. It also explains the heat dynamics of a reaction that would normally take place in nature, providing that there is some external energy (heat).

Energy of Electronic Excitation   Here, the electrons within each shell can become excited, or in chemical terms, move from a lower energy level to a higher energy

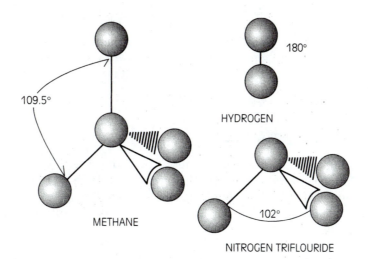

Figure 3-18 *The structure of the bonding can have a variety of geometric shapes. These configurations lead to the polarity, reactivity, solubility, and toxicity of the chemical compound.*

109.5°

180°

HYDROGEN

METHANE

102°

NITROGEN TRIFLOURIDE

level. For example, when a neon atom is bombarded with electrons, the outer shell of the neon atom becomes excited and the electrons move from a lower energy level to a higher one. When the electron is allowed to move back to the original ground state, energy is released. This reaction is viewed as light emitting from the neon fixture.

Energy of Rotation   When atoms are bonded together, the bond itself can move to a certain degree. The atoms that are attached can also move. The movement of these two atoms is the energy of rotation. Ethane for example, can be represented as a dumbbell. On this dumbbell each carbon atom can rotate around the center axis. This is termed the *energy of rotation.*

Energy of Vibration   In the two previous types of energy potentials, the electrons move at higher energy levels, while at the same time the molecule rotates on a center axis. Both are occurring at the same time. This movement causes some inward and outward movement and is called the *energy of vibration.* This type of motion is seen when two or more atoms of different sizes make up the molecule.

Energy of Translation   This energy potential is a combination of all that has been discussed, with the addition of full chains rotating around one another. Each of these atoms are rotating upon their individual connecting bonds. This rotation of all the atoms within the molecule is termed the *energy of translation* (Fig. 3-18).

These laws govern whether a chemical reaction takes place. This whole concept explains and gives insight into molecular collisions. In order for a reaction to take place the appropriate "atmosphere" for each molecule must be provided. Included within this atmosphere is the concentration of the materials. Not only will the orientation in space affect the reaction, but so will the concentration of the reactant. As discussed in Chapter 4, rates of reaction, orientation, and concentration of the toxin have an effect on the organism exposed to a hazardous material.

ACID–BASE REACTIONS

Years ago chemists placed substances into categories based on the acidity of that particular compound. The testing procedure used at the time was to taste the chemical. This procedure did not last very long, especially when tissue damage or pain resulted. This way of testing chemicals within the discipline of chemistry ended 150 years ago. In the mid-twentieth century when fire departments became involved with hazardous materials (chemicals), ironically the same obsolete identification procedure was born again. Because of the resulting injuries, our increase in chemical knowledge, and the awareness of the potential hazards, this testing procedure once again has been put to rest. However, looking at this technique gives an understanding cognitively of how we relate to acids and bases. If a certain substance tasted sour or bitter, it was classed as an acid. Bases had a variety of tastes (aromatics when we get to organic chemistry were classed under this heading for the fragrances a substance possessed), but mostly the bases created a soapy film. Although this film was created from damaged tissue, this was the classification method used at the time.

Acid and Base Definition

Acids and bases can be defined in terms of qualitative numbers. These numbers can be used to decide what form of protective gear an entry team member will don. The numbers are calculated from the chemical's ability to produce the hydronium ion, H_3O^+, or the production of a hydroxide ion, OH^-.

The acid is a compound that is capable of transferring a hydrogen ion in solution or it can donate a proton. The base is a group of atoms that contain one or more hydroxyl groups. The hydrogen is replaced by the acid. The base is referred to as the proton acceptor. This acceptance and donation produces the acid–base reaction. Acids and bases exist on their ability to dissociate. An equilibrium between the hydronium ion and the hydroxide ion creates a neutral solution.

Classification of Acids and Bases

Acids and bases (see also Table 3-21) are classified according to their particular ionization within water. This ionization, or degree of cation or anion state, we term *strength*. The scale to measure acidity and alkalinity has a range from 0 to

Table 3-21   *Examples of acids and bases.*

Chemical	Ionization Constants	pH
Acids		
Sulfuric acid	1.2×10^{-2}	0.3
Acetic acid	1.8×10^{-5}	2.4
Hydrocyanic	6.2×10^{-10}	5.1
Bases		
Ammonia	1.8×10^{-5}	11.6
Calcium hydroxide	3.74×10^{-3}	12.4

14. Zero to 6.9 is termed *acidic*, while 7.1 to fourteen is *basic*. Seven is considered neutral. This scale is used to measure the relationship of cations and anions within solution. Generally, compounds that form acids are covalently bonded. Compounds that form bases are ionically bonded.

The reference sources may use the scale referred to as the pH scale (pH is an abbreviation for positive hydronium ions) (Fig. 3-19). Some of the reference material uses a numerical system based on the number of hydronium ions in comparison to the number of hydroxide ions, but the format is different. When dealing with reactions, the reference material may cite total reaction as expressed and measured in K, where K is the ionization constant (usually measured at 25°C). This expression gives us the true relationship between acids and bases. K will have a subscript of either a or b. These subscripts identify the constant for an acid or a base, respectively (K_a, K_b).

The hydronium ion and the hydroxide ion are in equilibrium when a solution is neutral. However, when the solution contains a greater concentration of

Figure 3-19   *The pH scale is a numerical representation of the ionization within a solution. The strength of a solution is the percent of this ionization, while the concentration is the amount of a substance within a solution.*

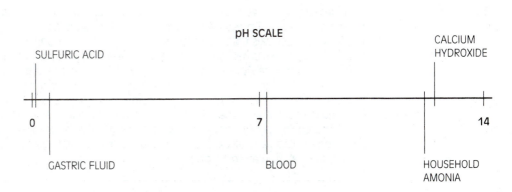

The term *strength* of an acid is the ratio of ionization that occurs in water. Inorganic acids and organic bases have different ionization percentages. With inorganic acids low concentrations are extremely hazardous. This has to do with the degree of ionization. Inorganic acids have an almost 100% ionization within water. Organic acids are relatively weaker because their degree of ionization is lower than the inorganic acid. The degree of ionization is typically less than 1%.

Do not determine the level of concern based on strength but rather the concentration of the acid (or base). Before making an assessment, understand the strength, concentration, and the family the corrosive solution belongs to.

hydronium ion with respect to the hydroxide ion, the solution is said to be acidic. Conversely, if the hydroxide ion is greater than the hydronium ion, the solution is said to be basic. In general, the larger the K value, the stronger the acid or the base.

! SAFETY

In general, the larger the K value, the stronger the acid or the base.

Extremely Strong	K value greater than 10^3
Strong	K value between 10^3 and 10^{-2}
Weak	K value between 10^{-2} and 10^{-7}
Extremely Weak	K value less than 10^{-7}

This whole concept is a relationship between two ideas: the concentration and strength of a solution, and, in this particular case, the discussion of a corrosive solution. An analogy to demonstrate this process involves a pool table. The green felt represents water. If ten cue balls were placed on the table all moving about the table with equal speed and duration of velocity, the solution would have the strength of 100%. If ten eight balls were added also moving with the same velocity about the pool table, the concentration is now 50%, however the strength of either cue balls or eight balls is still respectively the same. Strength

Hydrofluorosilicic acid is utilized in the paint manufacturing industry, as a cement hardener, a wood preservative, and water fluoridation agent. Its name makes it one of those chemicals that are seldom heard of but commonly used. It is an inorganic acid that has extremely strong ionization when dissolved in water. The medical concerns are threefold: (1) the corrosivity of this acid may present the emergency worker with minor to severe chemical burns; (2) the inhalation injuries are present if the liquid state of the material is moving into a gas; (3) once the chemical has come in contact with the skin or mucosa (including lungs), the fluorine molecule will search for calcium or magnesium causing further damage. This chemical is extremely toxic and all care for evacuation, emergency response protection, and decontamination are high priorities.

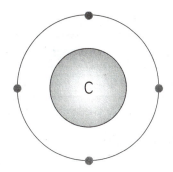

Figure 3-20 *These four bonding electrons give carbon the unique ability to form strong covalent bonds with other carbons and/or other molecules.*

and concentration are not interchangeable concepts: concentration is the relative content of the material within the solution, and the strength is the degree of influence or the velocity of the molecules.

ORGANIC CHEMISTRY

Organic chemistry, the largest branch of chemistry, deals with the compounds that contain carbon. Most of the chemicals interacted with on a daily basis are organic compounds. A majority of the products dealt with by emergency responders fall within the realm of organic chemistry.

In this section, the unique properties of carbon-based compounds are discussed. What makes carbon compounds unique is that the outer shell of carbon has four electrons (Fig. 3-20). This means that bonding occurs in a systematic fashion.

Hydrocarbons

The main constituent of the organic chemistry family is the group of compounds called the hydrocarbons (see also Table 3-22). Here, a carbon-to-carbon bond exists with hydrogens attached about the center carbon or along the chain of carbons.

If we have a chain of carbons that have a single bond between the carbons we call these compounds **alkanes**, or saturated hydrocarbons. They are called saturated hydrocarbons because every available point where a hydrogen bond

alkanes

saturated hydrocarbons with single bonds

Alkanes	Single bonds	Saturated	C_nH_{2n+2}
Alkenes	Double bonds	Unsaturated	C_nH_{2n}
Alkynes	Triple bonds	Unsaturated	C_nH_{2n-2}
Aromatics	Resonant bonds	Saturated	C_nH_n

Table 3-22 *Examples of hydrocarbons.*

		Boiling Point	Flash Point	Synonyms
Alkanes C_nH_{2n+2}				
Methane	CH_4	−161.6°C	−188°C	Marsh gas
Ethane	C_2H_6	−88.6°C	−135°C	Methylmethane
Propane	C_3H_8	−42.5°C	−105°C	Dimethylmethane
Butane	C_4H_{10}	−0.5 °C	−60 °C	n-Butane
Pentane	C_5H_{12}	36.1 °C	−40 °C	Amyl hydride
Alkenes C_nH_{2n}				
Ethylene	C_2H_4	−103.9 °C	−135 °C	Ethene
Propylene	C_3H_6	−47.7 °C	−162 °C	Propene
Butylene[a]	C_4H_8			
1-Butene		−6.3 °C	−79 °C	Ethylethylene
2-Butene		3.7 °C	−72 °C	Dimethylethylene
Alkynes C_nH_{2n-2}				
Ethyne	C_2H_2	−84 °C	−17.7 °C c.c.	Acetylene
Propyne	C_3H_4	−23.1 °C	N/A	Methyl acetylene
Aromatics C_nH_n				
Benzene	C_6H_6	80.1 °C	−11 °C c.c.	Phenyl hydride
Toluene	$C_6H_5CH_3$	110.7 °C	4.4 °C c.c.	Methyl benzene
Xylene	$C_6H_4(CH_3)_2$	144.4–138.3 °C	17.2–27.2 °C	Dimethylbenzene

[a] The numbers in front of each butylene identify the placement of the double bond. Looking at the molecule, the double bond can be at each end (1-butene), or in the center (2-butene). See discussion on IUPAC nomenclature.

can form is a carbon-to-hydrogen bond, or a saturation of hydrogen. There are cases when not all of the points of attachment involves a hydrogen molecule. So to be technically correct it is the number of carbons multiplied times two, plus two, which gives the number of hydrogens that can attach.

At times a double bond exists within the carbon-to-carbon chain. These compounds are called **alkenes**, or unsaturated hydrocarbons. In other words, a full saturation of hydrogen is not possible with these compounds. If there exists a triple bond within the carbon-to-carbon chain these are called **alkyne**. Again, this is an unsaturated hydrocarbon.

Therefore, if a chemical has the suffix of -*ane*, we know it is a single-bonded hydrocarbon. If the suffix is -*ene* it contains at least one double bond, and a suffix of -*yne*, indicates a triple bond. In the nomenclature of organic chemistry the number of carbon-to-carbon bonds is also important. Each carbon has a name when in a chain. This name or prefix designates the number of carbons within the compound:

alkenes
unsaturated hydrocarbons with double bonds

alkynes
unsaturated hydrocarbons with triple bonds

One carbon	Meth(yl)- (form)	
Two carbons	Eth(yl)- (acet)	{vinyl}
Three carbons	Prop-	{acryl}
Four carbons	But-	
Five carbons	Pent-	
Six carbons	Hex-	
Seven carbons	Hept-	
Eight carbons	Oct-	
Nine carbons	Non-	
Ten carbons	Dec-	
Eleven carbons	Undec-	
Twelve carbons	Dodec-	

For every attached carbon, a different prefix is denoted. The suffix indicates whether it contains a single, double, or triple bond, whereas the prefix denotes the number of carbons in the chain (Fig. 3-21).

Figure 3-21 *Multiple bonds do not occur within the same geometric plane. Although stable they have tremendous potential energy.*

$$CH_3 — CH_3$$

$$CH_2 = CH_2$$

$$CH \equiv CH$$

For example, C_2H_6 is ethane; C_2H_4 is ethene or ethylene; C_2H_2 it is ethyne (ethylyne), better known as acetylene. If the ethane has a chlorine atom attached, then we call the compound ethyl chloride. These are general nomenclature rules that are not always applied because the common name is better understood.

IUPAC Naming

As new chemicals were produced, a system for naming was needed. The International Union of Pure and Applied Chemistry (IUPAC), has established rules for naming the compounds in order to have uniformity of nomenclature so that all can understand the basic structure of the compound and the chemistry can be immediately identified. These rules follow a logical progression of application:

1. The molecule is looked at from the respect of the longest chain. If the chain is found to have a continuous single bond, then the suffix of *-ane* is used. When the chain has a double bond or multiple double bonds, the carbon chain is referred to as a *-kene*. Likewise if a triple bond is seen in the chain it is named as an *-yne.*

CH_3-CH_2-CH_2-CH_2-CH_2-CH_2-CH_3 Heptane

CH_3-CH_2-CH_2-CH_2-CH_2-CH=CH_2 Heptene (see next section for naming further)

CH_3-CH_2-CH_2-CH_2-CH_2-C≡CH Heptyne (see next section for naming further)

1a. If there is a halogen coming off the branch, we name the compound utilizing the foregoing rule and the number of the carbon to which the halogen is attached.

$$1 \quad 2 \quad 3 \quad 4 \quad 5 \quad 6 \quad 7$$
$$CH_3CH_2CHCH_2CH_2CH_2CH_3 \qquad \text{3-Chloroheptane}$$
$$| $$
$$Cl$$

2. The number of carbons are then derived by counting the carbon-to-carbon bonds. When a double or triple bond is noted then the carbon at the right or left of the double or triple bond is denoted as the number one carbon. The appropriate suffix and prefix are applied to derive the name, and the number is incorporated into the name in order to identify the double or triple bond.

LPG, as it is commonly called, stands for liquefied petroleum gas. It is a mixture of several gases, such as propane, butane, ethane, ethene, propene, butene, isobutane, isobutene, and isopentene. LPG is an odorless, colorless, heavier than air gas, usually "marked" with a sulphur compound called mercaptan. When propane and butane (the main constituents of LPG) are placed under a moderate amount of pressure the gases are liquefied. Mercaptan is a sulfur-based product and is heavier than air. The medical problems seen with sulfur are also observed with mercaptan.

$$CH_3\text{-}CH_2\text{-}CH_2\text{-}CH_2\text{-}CH_2\text{-}CH=CH_2$$

7   6   5   4   3   2   1

Heptene or 1-Heptene (1 is used to identify the placement of the double bond

$$CH_3\text{-}CH_2\text{-}CH_2\text{-}CH_2\text{-}CH=CH\text{-}CH_3$$

7   6   5   4   3   2   1

2-Heptene or Hep-2-ene

$$CH_3\text{-}C\equiv C\text{-}CH_2\text{-}CH_3$$

1   2 3 4   5

2-Pentyne or Pent-2-yne

$$CH_2\text{-}CH=CH\text{-}CH=CH\text{-}CH_3$$

6   5   4   3   2   1

2,4-Hexadiene or Hexa-2,4-diene (diene for two double bonds)

$$CH_3\text{-}CH_2\text{-}CH_2\text{-}C\equiv C\text{-}CH_3$$

6   5   4   3 2 1

2- Hexyne or Hex-2-yne

3. When an alkyl (these are chained groups which will receive their name as independent groups) group comes off the chain, the appropriate group name is applied and incorporated into the naming process. The carbon closest to the end of the chain with respect to the alkyl group is how the carbon chain is derived or named. For example:

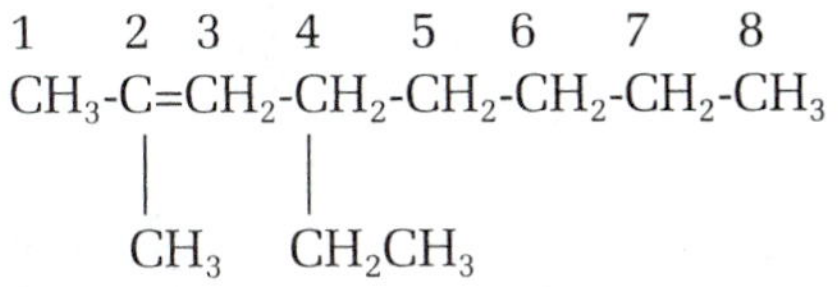

1   2 3   4   5   6   7   8

$$CH_3\text{-}C=CH_2\text{-}CH_2\text{-}CH_2\text{-}CH_2\text{-}CH_2\text{-}CH_3$$

        | |

      CH_3 CH_2CH_3

The chain is octane.

We see a methyl group and an ethyl group off the carbon number 2 and 4 respectively. So the compound as pictured is 2-methyl-4-ethyloctane.

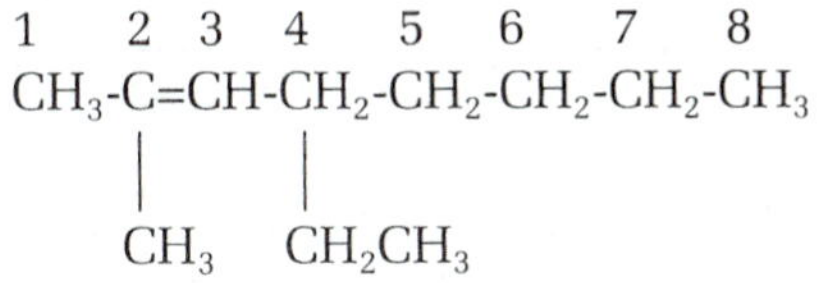

1   2 3   4   5   6   7   8

$$CH_3\text{-}C=CH\text{-}CH_2\text{-}CH_2\text{-}CH_2\text{-}CH_2\text{-}CH_3$$

        | |

      CH_3 CH_2CH_3

2-methyl-4-ethyl-2-octene

The chain is octane.

Many cities are utilizing alternative fuels for fleet vehicles, mass transit, and private use. In general these fuels are classified under the following acronyms: CNG, compressed natural gas; LPG, liquefied petroleum gas; and LNG, liquefied natural gas. In addition to these products, methanol is also utilized. The difference between the methanol and the just-mentioned products is that methanol is a liquid at ambient temperature and pressure. The CNG, LPG, and LNG all have expansion ratios due to the pressure (and in some cases, temperatures) they are under in order to remain in the liquid state within their respective containers.

In organic chemistry, because of the magnitude of different compounds within the same basic chemical family, each compound has associated chemical and physical properties. However, at times we see like compounds with different properties. When this occurs we have an *isomer* of the compound. Because of this isomerism we see a huge variety of organic compounds. In each "class" we can have several different molecular structures, all of which have different chemical and physical properties, yet have the same number of atoms within the structure.

Consider, for example, dichloroethlene, $C_2H_2Cl_2$. Here we see three isomers, all having a different set of properties yet they look basically the same. The only difference between each is the chlorine atom orientation to the double-bonded carbon.

1,1-Dichloroethlene	Cis-1,2-Dichloroethlene	trans-1,2-Dichloroethlene
(1)	(2)	(3)

In dichloroethlene #1 we have the chlorine atoms opposite each other at one end of the molecule. With dichloroethlene #2 the chlorine is on the top (if on the bottom it would be the same compound. Likewise the #1 if the chlorine both on the right as opposed to the left, it would be the same compound). With dichloroethlene #3 we see the opposite ends of chlorine placement. In order to denote the differences of each compound we call #1 as dichloroethlene, #2 as *cis*-Dichloroethlene, and #3 as *trans*-Dichloroethlene. The *cis* and *trans* designations describe the orientation of the added groups on the parent chemical structure.

	Boiling Point	Melting Point	TLV	IUPAC Name
#1 Dichloroethlene	31.7° C@ 760 torr	−122.5°C	?	1,1-Dichloroethene
#2 cis-Dichloroethlene	59.6° C@ 745 torr	−81.5°C	200 ppm	1,2- Dichloroethene
#3 trans-Dichloroethlene	47.2° C@ 745 torr	−49.4°C	200 ppm	1,2- Dichloroethene

The alkanes, alkenes, and alkynes are all related chemically. This general group is often referred to as the *aliphatic hydrocarbons* (nonaromatic). When there is a benzene ring present we term these as the aromatics. Derivatives of the aromatics include toluene and xylene.

In organic chemistry, it is extremely prevalent that a compound needs to be stable. Electronic stability is what all compounds are trying to achieve. Because of this we see that six carbon chains are sometimes incorporated into a circle structure. We call this circle structure a *benzene ring*. In the benzene ring, a high level of stability is achieved by virtue of the ring structure. Because the six-

Table 3-23   *Examples of aromatic derivatives.*

Name		Synonym	Formula	Properties	Uses
Aniline	NH$_2$	Phenylamine	$C_6H_5NH_2$	BP 184.4 °C FP 70 °C c.c TLV[a] 2ppm	Aniline is used primarily in dyes and the drug industry.
Phenol	OH	Carbolic Acid	C_6H_5OH	BP 182 °C FP 78 °C c.c. TLV 5ppm	Phenol is used in the drug industry, in dyes, and in some plastics.
Styrene	CH=CH$_2$	Phenylethylene	$C_6H_5CHCH_2$	BP145.2 °C FP 31.1 °C TLV 50ppm	Styrene is used in making synthetic rubber, polystrene, plastics, and resins.

[a] TLV = threshold limit value, a term used for airborne concentration. See Chapter 4.

carbon ring is more stable and stronger than the straight chain, six-carbon structures will sometimes fall on to itself to create the more stable ring structure. This
benzene ring (hexagonal) has a chemistry unto itself. When the electrons are
arranged in a stabilized fashion above and below the ring it is called an *aromatic*.
A special class of organic compounds, aromatics are the benzene ring and all the
related structures as examples (see Table 3-23).

Before proceeding into some chemistry that may hit a little closer to home,
let us discuss the polarity of organic compounds. The structure in organic chemistry is very precise. The orientation in space within the molecule is important
when talking about reactions. As we will see, when we become contaminated
with a chemical, the body's metabolism greatly affects this orientation. In some,
it will make a completely benign chemical, in others an extremely toxic compound. The exact orientation and the movement of the molecule within itself
will affect the reaction. However, here we are going to look at the orientation to
light waves and the nomenclature that surrounds this phenomena. Isomers of
organic compound can sometimes have a mirror image of themselves. Here we
call these materials *enantiomers* (Fig. 3-22).

Under certain optically active conditions, the molecule will move to the
right or to the left. If a molecule moves to the right or in a clockwise direction, we
call it (+), if the rotation is to the left it is termed counterclockwise and is denoted
to be (–). If we have a mixture of both the right (+) and the left (–) enantiomer we
call the mixture a *racemic* mixture (±). Epinephrine is sometimes denoted as such.

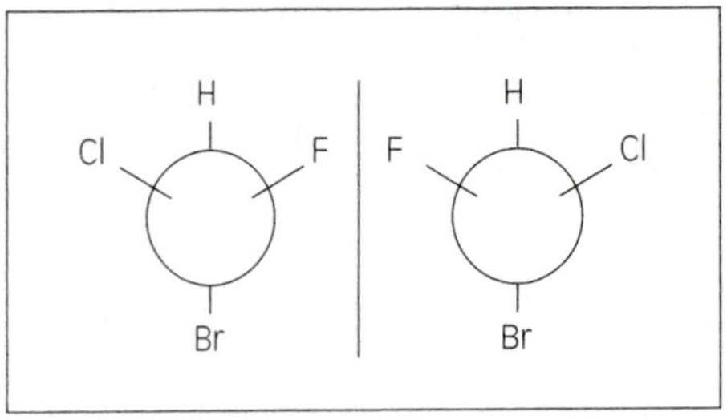

Figure 3-22

Enantiomers are mirror images of each other. Each can have different boiling points, melting points, densities, and overall chemical and physical properties.

This discussion is not to give you the ability to identify these materials. But in reality, if such a chemical were on fire or spewing fumes, the end result health-wise will be roughly the same, even if you did not know what these designations were. However, when you are researching a chemical that has such a designation in front of the name, you will now know what you are looking for in the reference manuals. It is extremely important that the exact chemical is referenced. Each of these naming designations must be viewed as necessary components of the referencing process.

In organic chemistry (as in other discussions of chemistry), we have abbreviations that denote a typical or definite structure. The formulas are sometimes very complicated and for the most part do not become a part of the chemical reaction. For this reason we can indicate the structure of this group as an "R" group. By doing this we are only addressing the functional group. The following is a brief discussion of such functional groups and how they impose a medical threat to us and our patients.

Alcohols   R-OH

The R here denotes that the carbon-to-carbon chain can be two or three or ten carbons long (the length does not matter). What is involved in the reaction is the –OH group. This is not to say that as the chain becomes longer, the chemistry of the compound will not change. On the contrary it will.

For the most part, the boiling points are higher than seen in the parent hydrocarbon, yet lower than most other molecules. Lower chained alcohols are very soluble in water, because of the hydroxyl group (–OH), and insoluble in the higher chained molecules. The –OH group affects the boiling point, melting point, and solubility. It is the –OH placement on the carbon chain, and the length of the chain with respect to the –OH placement that affects BP, MP, and solubility.

Alcohols (see also Table 3-24) react with the same compounds that water reacts with. Reactions can produce alkenes, or ethers, and under controlled conditions, form aldehydes, esters, alkyl halides, ketones, or acids. Because of the

Table 3-24   *Examples of alcohols.*

Name	Boiling Point	Vapor Pressure	Synonym	TLV
Methyl alcohol	147°F	92 mm Hg	Methanol	200 ppm
Ethyl alcohol	172 °F	43 mm Hg	Ethanol	1000 ppm
Ethylene glycol	387 °F	0.05 mm Hg	1,2-ethanediol	50 ppm

! **SAFETY**
● **Because the vapor pressures in the alcohols group is on the high side, toxic fumes are a common problem.**

wide variety of compounds that can be generated from alcohols, its use is frequent in clandestine drug laboratories.

The vapor pressures in this group run on the high side, making the possibility of toxic fumes a common problem. If these fumes find an ignition source, fire is yet another hazard to consider.

Two associated families of the alcohols are the glycols and the glycerins. The chemical definition of each is that the glycols have two –OH groups, whereas the glycerins have three or more –OH groups. Two examples of this subgrouping of the alcohols are ethylene glycol commonly referred to as antifreeze, and nitroglycerin.

The toxic values of these products are limited, however. Ethylene glycol poisoning occurs fairly often, usually during the siphoning of a radiator. The dose is fairly small to elicit a toxic reaction, usually a mouthful. The reaction occurs in three phases, each bringing the patient closer to death.

The first phase, called the *glycol phase*, lasts approximately 30 minutes to 12 hours. In this phase, the glycol is metabolized in very much the same way drinking alcohol would be. However, here the principal metabolite is glycolic acid. Signs and symptoms are similar to ethanol intoxication. In 4 to 24 hours the toxic metabolite peaks to indicate the second phase called the *glycolate phase*. Here noncardiogenic pulmonary edema, acidemia, and neurological toxicity manifests. Once renal damage is detected, the process has moved into the third phase. This occurs in 24 to 72 hours and is called the *nephropathy phase*. Renal edema and tubular necrosis are observed from the buildup of oxalate crystals in the form of calcium oxalate, thus causing hypocalcemia. Because the production from the

The Food, Drug, and Cosmetics Act of 1937 was assisted by an incident that occurred when a medication that was placed in diethylene glycol and consumed. This act required accountability of the manufacturer before public use. The management and metabolism that occurs with diethylene glycol is similar to ethylene glycol.

metabolic process produces an acidoic state, ECG "T" waves are pronounced, which is consistent with hyperkalemia. Renal failure soon follows. In some of the literature it is suggested that a second neurologic phase occurs. At this point the oxalate crystals have entered the tissues in the brain, causing further damage.

Currently the treatment of choice for the first phase patient is high loading doses of ethanol. Because the metabolism of ethylene glycol utilizes the enzyme alcohol dehydrogenase, ethanol is used to compete for this enzyme. This competition reduces and in some cases prevents the toxic metabolites. Once the patient has moved into later stages of the toxic event, other antidotes must be used. For example once in the glycolate phase, thiamine and pyridoxine are used to reduce the oxalic acid that is combining with the calcium. Leucovorin, folic acid, and 4-methylpyrazole are also used in ethylene glycol poisoning.

Naming the alcohols is usually done with the common nomenclature as seen in Table 3-24. Here the carbon-to-carbon chain is identified and the prefix utilized. The word alcohol is then placed to identify the complete structure.

In the IUPAC naming system the prefix is used to identify the carbon-to-carbon structure and the suffix of *-ol* to identify the alcohol. For example methyl alcohol is methanol, and ethyl alcohol is ethanol. The suffix *-diol* is used to identify the glycol and glycerol for the glycerins

Phenols   OH

Phenols are closely related to alcohols. The main difference is that the carbon chain in phenols are in a circular pattern, the benzene ring. The simplest member of this family is referred to as phenol (Fig. 3.23), but is also known as carbolic acid. It has a destructive action on animal tissue.

In this classification the boiling points are higher than the parent hydrocarbon. Solubility in water is slight with true solubility in alcohol or organic sol-

Figure 3-23 *The ortho, meta, and para is used in nomenclature to designate the carbon group of attachment respective to the main attachment. In this case, the functional group at the top of the ring is carbon number 1. The ortho, meta, and para are carbon 2, 3, and 4 respectively.*

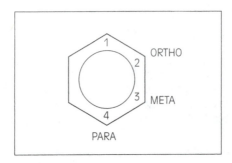

Table 3-25 *Examples of phenols.*

Name	Boiling Point	Synonym
Benzophenol	359.6 °F	Phenol
p-Chlorophenol	422.6°F	4-chloro-1-hydroxybenzene

vents. The boiling points range from 181° C for phenol to 217° C for nitrophenol. The acidity of this family is high in comparison to the alcohol cousin. See examples of phenols in Table 3-25.

Carboxylic Acid

$$\begin{array}{c} O \\ \| \\ -C- \end{array}$$

If attachment occurs at the oxygen an acid results. If attachment occurs at the carbon a base results.

Carboxylic acids have high boiling points, with solubility in water only in the lower molecular weighted compounds. Boiling points are characteristically higher than seen in the alcohols. The reason for the higher boiling points is the molecule's ability to hold and maintain two hydrogen bonds. The lower chained acids are very irritating and can have toxic potentials.

Basically this functional group makes either a ketone or an aldehyde. Some common carboxylic acids, sometimes referred to as the organic acids:

Common Name	*IUPAC name*
Formic acid	Methanoic acid
Acetic acid	Ethanoic acid
Propionic acid	Propanoic acid
Butyric acid	Butanoic acid
Valeric acid	Pentanoic acid
Caproic acid	Hexanoic acid
Caprylic acid	Octanoic acid
Capric acid	Decanoic acid

Ketones

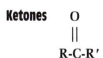

$$\begin{array}{c} O \\ \| \\ R\text{-}C\text{-}R' \end{array}$$

Ketones (see also Table 3-26) have similar properties to those of aldehydes and make up a large class of solvents. However, unlike the aldehydes, they are not

Table 3-26 *Examples of ketones.*

Name	Boiling Point	Synonym	TLV
Methyl ethyl ketone	175.2°F	MEK	200 ppm
Butyl ethyl ketone	298.4°F	BEK	50 ppm
Dipropyl ketone	290.7°F	Butyrone	50 ppm

SAFETY Ketones are volatile, and their narcotic effects make them extremely hazardous.

affected by oxidizing agents used in synthesis. Ketones may form peroxides and polymerize. They are volatile, and their narcotic effects make them extremely hazardous. This compound is also used in clandestine operations.

The lower carbon-weighted ketones are soluble in water, as well as in the usual organic solvents. Vapors are typical resulting in toxic atmospheres, so this classification has a high inhalation hazard. Avoid all skin contact. Sensory and motor neuropathy is often seen in this group.

Acetone is a common solvent that is rapidly absorbed through the respiratory and gastrointestinal tracts. Toxicity manifests itself in effects similar to ethyl alcohol ingestion, with stronger anesthetic effects. With inhalation injuries, the depression of respiratory status is most common. There is no known antidote and therapy is governed by symptomology. Ingestion of acetone usually present without any problems with the exception of vomiting, however because of the prolonged elimination with associated metabolic effects, hospitalization may be required for observation for 30 to 48 hours.

The common nomenclature identifies the carbon-to-carbon chains exclusive of the carboxyl group, naming the R groups as independents. The IUPAC system identifies the ketone with the suffix *-one* and the carbon of the C=O as a part of the chain.

$CH_3CCH_2CH_3$ || O	Methyl ethyl ketone	O || $CH_3CCH_2CH_3$ 1 2 3 4 2-butanone
$CH_3CH_2CH_2CH_2CCH_2CH_3$ || O	Butyl ethyl ketone	O || $CH_3CH_2CH_2CH_2CCH_2CH_3$ 1 2 3 4 5 6 7 3-heptanone
$CH_3CH_2CH_2CCH_2CH_2CH_3$ || O	Dipropyl ketone	O || $CH_3CH_2CH_2CCH_2CH_2CH_3$ 1 2 3 4 5 6 7 4-heptanone

Aldehydes

$$\underset{\text{R-C-H}}{\overset{\overset{\textstyle O}{\|}}{}}$$

Aldehydes (see also Table 3-27) and ketones are very closely related, therefore the properties of each overlap. Both have the potential for forming peroxides. Although there are considerable likenesses, we discuss each as a separate reaction and compound.

Aldehydes have low boiling points just under the alcohols that have comparable length. However the boiling points are above the ethers. The solubility is a mix of that of the alcohols and ethers. For the most part, the lower carbon chains give solubility in water whereas the higher chained aldehydes are insoluble in water. Aldehydes are so reactive that they can be converted to acids (carboxylic).

From the medical standpoint, they account for a majority of the irritant and sensitivity reactions that are observed. Chemical bronchitis, pneumonitis, pulmonary edema, and immunological effects can be caused from the exposure of aldehydes.

As an example of this group, acetaldehyde is found in varnishes and photographic chemicals. It has severe respiratory injury potential leading to bronchitis and pulmonary edema in quantities as low as 134 ppm. Narcoticlike effects are noted, however, antagonists are not suggested. There is no known antidote, and treatment should be directed toward the symptoms. In an inhalation injury, high flow oxygen, along with a nebulized bronchodilator for the bronchospasm may be used. Endotracheal intubation may be necessary early on. Be prepared for fast-acting bronchospasm.

Naming these compounds under the common nomenclature requires a count of all carbons in the chain, inclusive of the carbon within the aldehyde functional group. After counting the carbons, the chain name is used to identify the number of carbons. The only exception to this is the first two. They are : One carbon, Meth(yl)- (form) and two carbons, Eth(yl)- (acet). For example CH_3-CHO is *Acet*aldehyde. CH_2O is *Form*aldehyde. If the compound is an aldehyde and it

SAFETY

Aldehydes are water reactive and they can be converted to acids (carboxylic).

Table 3-27 *Examples of aldehydes.*

Name	Boiling Point	Vapor Pressure	Synonym	TLV
Formaldehyde	2.2°F	less than 1mm Hg	Methanal	1 ppm
Acetaldehyde	68.4°F	740 mm Hg	Ethanal	100 ppm
Acrylaldehyde	126.9°F	210 mm Hg	2-Propenal	0.1 ppm

has a double bond, such as the case in acrolein $CH_2=CH-CHO$, then the prefix of *acryl* is utilized and the name of acrylaldehyde.

In the IUPAC naming system, the identification as an aldehyde is done by the suffix *-al*. Formaldehyde is methan*al*, acetaldehyde is ethan*al*, and acrylaldehyde becomes 2-propen*al*.

Esters

$$\begin{array}{c} O \\ \parallel \\ R\text{-}C\text{-}OR' \end{array}$$

Boiling points and solubility are seen to be analogous to the same molecular weighted compounds as seen in the aldehydes and ketone family. Here the R groups can be the same or different lengths. All are toxic at relatively low levels and total avoidance should be maintained.

The toxicity of esters are dependent on the parent reactants. Esters (see also Table 3-28) are derived from a combination of carboxlic acids and a variety of alcohols. The toxic event usually occurs from a reaction that occurs within the stomach once ingested. The process that occurs in the stomach separates the parent alcohol from the carboxyl group. Not only is there a toxic event from the ester but also from the attached components. For example, if methyl carboxylate was ingested, methyl alcohol and the ester would be of toxic concern. Intoxication from the methyl alcohol would ensue fairly rapidly. Symptomology should guide treatment, with the addition of local protocols for cathartics. Seizure activity can be managed by the administration of diazepam or phenobarbital. Inhalation is rare, however, the respiratory injury is best managed by high flow oxygen. In high concentrations the central nervous system is depressed.

The common naming system identifies the carbon groups on either side of the compound. The carbon group that is attached to the oxygen is named first and is derived from the number of carbons in the chain. The second group is derived from counting the carbon chain that is attached to the carbon. The carbon chain and the carboxyl carbon is used in the nomenclature. The second carbon chain has the suffix of *-ate*.

Table 3-28 *Examples of esters.*

Name	Boiling Point	Vapor Pressure	Synonym	TLV
Ethyl acetate	170.6°F	73 mm Hg	Acetic ether	400 ppm
Ethyl acrylate	210.9°F	29 mm Hg	Ethyl propenoate	5 ppm
Isopropyl acetate	192.9°F	40 mm Hg	2-Acetoxypropane	250 ppm

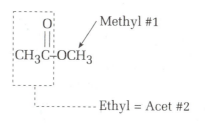

Methyl #1

Ethyl = Acet #2

Methyl Acetate

The IUPAC naming system is the same as the common nomenclature.

Alkyl Halides   R-X (Halogenated Hydrocarbons)

In this classification the R is the carbon-to-carbon chain as we have seen before, and the X is the functional group. In this case the X is a halogen.

Here we see higher boiling points than the parent alkane. As a general rule the boiling point increases with the higher weighted halogen that is attached to the carbon chain. They are insoluble in water, which makes them soluble in organic solutions. They are very useful in preparing a variety of compounds (see also Table 3-29), thus very beneficial in the synthesis of other compounds. Again we see a way to produce compounds through fairly simple reactions. For this reason clandestine labs utilize the alkyl halide in their synthesis of drugs.

The toxic vapors affect the central nervous system, which is seen as a narcotic-induced effect that can lead to convulsions and respiratory depression. This "narcotic"-observed effect is just that—an outward sign. Technically the narcotic receptor sites are not bound. Literature suggests that antagonists such as Narcan should not be used. It is unknown whether the antagonist may potentiate or create a synergistic effect.

Epinephrine is also contraindicated in the alkyl halide patient. Alkyl halides stimulate the central nervous system, therefore additional epinephrine would not be useful and could overstimulate the CNS. This reaction has been seen as sudden death due to lethal arrhythmia from the high levels of catechola-

Table 3-29   *Examples of alkyl halides.*

Name	Boiling Point	Vapor Pressure	Synonym	TLV
Dichloromethane	104.2°F	350 mm Hg	Methylene chloride	50 ppm
Trichloromethane	142.2°F	160 mm Hg	Chloroform	10 ppm
1,1,2-Trichloroethyane	236.7°F	16.7 mm Hg	Vinyl tichloride	10 ppm
1,1,1-Trichloroethyane	167.0°F	100 mm Hg	Methyl chloroform	350 ppm

mines. It is actually the myocardium's response to the high level of cate-cholamines within an hypoxic environment within the body that leads to the arrhythmia. Even in incidents where victims of an exposure are in cardiac arrest, epinephrine should be avoided.

Most of the halogenated hydrocarbons have vapor densities that are greater than one, making them an asphyxiant. Prolonged exposure to this classification leads to kidney and liver damage.

The common naming system for the alkyl halides is fairly straightforward. Here the carbon-to-carbon chain is identified and the halogen is placed before or after the alkyl group. If more than one alkyl group exists, then the prefixes *di-*, *tri-* and *tetra-* are used, for example, dichloromethane CH_2Cl_2 (or methyldichloride) and trichloromethane $CHCl_3$ (or methyltrichloride).

With the IUPAC system, the carbon-to-carbon chain is identified and the placement of the halogens are numbered according to the lowest carbon (see the discussion on IUPAC). This enables the reader to identify which carbon the halogen is attached to, for example, 1,1,2-trichloroethyane and 1,1,1-trichloroethyane.

1,1,2-Trichloroethyane

$$
\begin{array}{ccc}
\text{Cl} & \text{H} & \\
| & | & \\
\text{H}-\text{C} - \text{C} & -\text{H} \\
| & | & \\
\text{Cl} & \text{Cl} &
\end{array}
$$

1,1,1-Trichloroethyane

$$
\begin{array}{ccc}
\text{Cl} & \text{H} & \\
| & | & \\
\text{Cl}-\text{C} - \text{C} & -\text{H} \\
| & | & \\
\text{Cl} & \text{H} &
\end{array}
$$

Amines RNH_x

The R group is a carbon chain, wherein the functional group is the nitrogen group. Here we can have derivatives of ammonia, in which the hydrogen has been replaced: RNH_2, R_2NH, or R_3N all are considered lower amines. These compounds are polar, which give them high boiling points comparative to the nonpolar compounds of the same molecular weight. Overall they have boiling points lower than seen in the alcohols.

⚠ SAFETY

Amines are extremely toxic and classified are corrosives with the same potential as a corrosive.

This group of compounds (see also Table 3-30) have the characteristic smell of fish and are considered extremely toxic. In general they are classified as corrosives with the same potential as a corrosive. This is especially true if the compound is in the vapor state. Rapid absorption can occur anywhere (i.e., the skin and mucus membranes), however, if inhaled the result can be death. However this reaction is very chemical specific. The reason for the toxicity in general is amines raise the pH of the exposed tissue, which causes tissue necrosis. If the amine becomes systemic, the kidneys and liver can become necrotic. Lung tissue can bleed with pulmonary edema.

Although the literature states that the amines are extremely toxic only fifty cases of severe exposure have been found. In all of these exposures that are cited, the exposure took weeks to months before symptoms appeared. Most of the described symptomology was dermatitis and skin discoloration. This would be

Table 3-30   *Examples of amines.*

Name	Boiling Point	Water Solubility	Synonym	TLV
Methylamine	19.8 °F	Very soluble	Monoethylamine	10 ppm
Ethylamine	61.9 °F	Very soluble	Monoethylamine	10 ppm
Dimethylamine	40.7 °F	Very soluble	DMA	10 ppm
Triethylamine	193.5 °F	Soluble		10 ppm

consistent with the corrosive qualities that amines possess. The only acute case involves the chemical acrylamine where neurological effects were noted (hallucinations, slurred speech, and weakness) along with the dermatitis. The largest hazard that is found with the amines is their tendency burn.

The common name for amines is found by naming the attached hydrocarbon chain(s):

CH_3NH_2	Methylamine
$CH_3CH_2NH_2$	Ethylamine
$(CH_3)_2NH$	Dimethylamine

The IUPAC system is basically the same as the common naming system. The only difference occurs when there are double bonds or isomers of the carbon chain. Then the numbers are used to identify the carbon groups or the double bonding:

$CH_3(CH_3)_2NH_2$	2-methyl-2-propanamine
$CH_2=CHCH_2NH_2$	2-propenamine

Ethers   R-O-R'

In this family, the R groups can be the same or different carbon chain lengths. The members in this family have low boiling points (see Table 3-31). The weak bonding that occurs between the hydrogen and water tends to give this group extremely slight solubility factors, but only for the low carbon weighted ethers. As we saw with the alcohols, the higher the carbon chains, the higher the potential for insolubility. The chemical dilemma we face with this group is the possibility of oxidation into organic peroxides.

Most of the ether solvents have slow absorption dermally. However, inhalation and gastric absorption is high. The outward symptoms are fast and short lived and are similar to the effects of ethanol ingestion. The most common route of entry is through inhalation. Irritation, cough, vomiting, dizziness, euphoria, and local necrosis are the usual symptoms.

! SAFETY

Ethers possess the possibility of oxidation into organic peroxides.

Table 3-31 *Examples of ethers.*

Name	Boiling Point	Synonym
Dimethyl ether	12.1°F	Methyl ether
Methylphenyl ether	311 °F	Anisole
Vinyl methyl ether	42.8°F	MVE

Polar Compounds				
Alcohol	**Ester**	**Aldehyde**	**Ketone**	**Organic Acids**
R-OH	R-C-OO-R'	R-C-O-H	R-COR'	R-COOH
-ol	-ate	aldehyde	ketone	acid
Flammable and toxic qualities	Flammable and polymerization	Flammable, toxic polymerization	Flammable and CNS effects	Toxic and corrosive

Figure 3-24 *Polar compounds.*

The glycol ethers have low vapor pressures with high dermal absorption rates. The pathophysiology is similar to the glycols with the addition of extremely high absorption through inhalation and dermal contact. Rescue workers can expect symptoms as in the foregoing list.

The common IUPAC nomenclature is to name the R groups, situating them in the name according to size (Fig. 3-24).

MISCELLANEOUS COMPOUNDS

Water-Reactive Metallic and Nonmetallic Compounds

SAFETY

Some water-reactive metallic and nonmetallic compounds can be found in the anhydrous (without water) state, and water vapor in the air can start a reaction with them.

Many inorganic compounds have the same basic naming process, however, they do not fall into a classification that we have identified thus far. The following compounds are examples of this group and belong to the general classification of water-reactive metallic and nonmetallic compounds (sometimes called *binary nonsalts*).

They have a principal component of a nonmetal that will gain electrons from another nonmetal. The name is identified by *-ide*.

Phosphorous trichloride	PCl_3
Aluminum trichloride	$AlCl_3$
Antimony Pentachloride	$SbCl_5$

Most all of these compounds have violent water-reactive qualities. Some can be found in the anhydrous (meaning without water) state. In these compounds the water vapor in the air can start the reaction.

Nonmetal Oxides

The next group of inorganic compounds are the nonmetal oxides. Most of these gases are found in the gases of combustion and are referred to as the "fire gases." They contain a nonmetal with oxygen, and the name starts with the principle atom (not oxygen) with the suffix of *oxide*.

Sulfur Dioxide	SO_2
Sulfur Trioxide	SO_3
Dinitrogen Tetroxide	N_2O_4

Oxygenated Inorganic Compounds

As we observed in the oxygenated compounds, a group of acids can also be formed by the oxygenated inorganic compounds. As in the oxygenated compounds, here the ion complex is the base for the name. However, Per- -ic; -ic; -ous; and Hypo- -ous are the prefixes and suffixes used ending in the word acid.

*Per*chlor*ic* acid	$HClO_4$
Chlor*ic* acid	$HClO_3$
Chlor*ous* acid	$HClO_2$
*Hypo*chlor*ous* acid	$HClO$

These compounds are both strong oxidizers and corrosives.

Inorganic Peroxides

The inorganic peroxides are called the *metal peroxides* and have diatomic oxygen attached to a metal. The naming is the metal followed by peroxide. They are strong oxidizers as a group and are water reactive.

Potassium peroxide	K_2O_2
Mangenese peroxide	MnO_2

ORGANIC COMPOUNDS

Cyanide Salts

The cyanide salts are placed in the organic grouping because, like the functional groups studied earlier, a carbon chain can be represented as R. This chain can be one carbon or several carbons long. This group in general is used in the manufacturing of plastics and as solvents. Naming this compound can either be by naming the carbon chain then adding the term cyanide or by using the carbon of the cyanide ion as a basis of the name (nitrile):

Methyl cyanide	CH_3CN	Acetonitrile
Propenenitrile	C_2H_3CN	Acrylonitrile (vinyl cyanide)
Benzonitrile	C_6H_5CN	Phenyl cyanide

The R group may be a metal:

Ammonium thiocyanate	NH_4SCN	Ammonium sulfocyanide
Barium cyanoplatinite	$BaPt(CN)_4$	Barium platinum cyanide

Nitrogen Group

The nitrogen groups, sometimes referred to as the *nitros*, are commonly used as explosives but have other uses. The naming of these compounds utilizes the hydrocarbon chain (tolune as an example) beginning with the word nitro, (trinitro, so in this example trinitrotolune) or IUPAC rules for naming.

Nitroglycerin	$CH_2NO_3CHNO_3CH_2NO_3$	Glyceryl trinitrate
Nitrocellulose	Nitro group attached to cotton	Cellulose nitrate
Trinitrotoluene	$C_6H_5CH_3(NO_2)_3$	TNT
Trinitrophenol	$C_6H_5OH(NO_2)_3$	Picric acid
Cyclotrimethylenetrinitramine*	$N(NO_2)CH_2N(NO_2)CH_2N(NO_2)CH_2$	Cyclonite

Sulfur Group

Sulfur groups, counterparts of the alcohols, phenols, and ethers, are a combination of carbon chained compounds and functional groups. Sometimes referred to as organosulfur compounds, they too have a specific naming system. They are identified as follows:

Thiols R-SH (Mercaptan)   Thioethers R-S-R′   Disulfides R-S-S-R′

$$Sulfoxides\ R\text{-}\overset{\displaystyle O}{\overset{\|}{S}}\text{-}R'$$

$$Sulfones\ R\text{-}\overset{\displaystyle O}{\underset{\underset{O}{\|}}{\overset{\|}{S}}}\text{-}R'$$

$$Sulfinic\ acids\ R\text{-}\overset{\displaystyle O}{\overset{\|}{S}}\text{-}OH$$

$$Sulfonic\ acids\ R\text{-}\overset{\displaystyle O}{\underset{\underset{O}{\|}}{\overset{\|}{S}}}\text{-}OH$$

Organic Peroxides

The peroxides are usually identified by recognizing the name as "peroxide" or "peroxy-", without the inclusion of a metal group, as we saw in the inorganic peroxides. Here the peroxide is of a carbon chain origin.

Methyl ethyl peroxide	$CH_3\text{-}O\text{-}O\text{-}CH_2CH_3$
Dimethyl peroxide	$CH_3\text{-}O\text{-}O\text{-}CH_3$

The concept to remember with the organic peroxides is that they are man-made and very unstable. They normally have inhibitors placed within them to slow down the reaction. If the temperature increases or the reaction starts, a self-accelerating decomposition reaction will occur which goes to completion. They are potent oxidizers that are highly flammable.

! SAFETY
Organic peroxides are man-made and very unstable.

*This is also considered as a derivative of an amine.

Summary

There is a specific need to acquire and maintain the knowledge of chemistry. Reference, entry, decontamination and treatment all depend on this knowledge. Once faced with pressing hazardous conditions, the answers do not come easily. As emergency responders, we must be able to logically place the facts in such an order as to arrive at appropriate decisions. The facts and the laws of chemistry should support that decision-making process.

We saw how the effects of pressure, temperature, and chemical and physical properties can affect us in each aspect of the hazardous materials event. By placing a material under pressure we have the molecules in very close contact. This is not the normal environment for such a material. Vapor pressure and temperature can make the material react, and in some cases, react faster.

The concentration of a material can increase the probability of having each molecule come in close proximity with another material, thus causing a reaction. Each chemical group and classification posed its own inherent potential fire, reactivity, and health concerns. By knowing the general principles of nature, you, the emergency responder can make honest, well-informed decisions.

Know the chemicals that you are dealing with. Obtain reliable source information on all chemicals. Document all information that was gleaned from the incident. This will aid in the critique of the incident and the documentation that is required of all emergency responders.

Figures 3-25 and 3-26 are tactical reference sheets. These reference plans enable the science sector to assemble a variety of technical information onto an easily accessible guide.

MEDICAL and CHEMICAL REFERENCE TACTICAL WORKSHEET

Chemical Name:_____ Synonyms:_____ Chemical Formula::_____ DOT:_____ CAS:_____

State of Matter:_____ State Found:_____ Water Reactive: Y N Oxider: Y N Other Hazards:_____

Wind Direction:_____ Wind Speed:_____ Temp.:_____ Humidity:_____ Adjusted Temp.:_____ Dew Point:_____

Manufacture:_____ Contact Person:_____ Phone:_____ Fax:_____

INORGANIC COMPOUNDS	COMPOUND STRUCTURE and HAZARDS	ORGANIC COMPOUNDS

INORGANIC COMPOUNDS

SALTS

-ide

General Hazards; Toxic gas with water contact with following:

Carbides = Acetylene
Hydrides = Hydrogen
Nitrides = Ammonia
Phosphides = Phosphine

METAL OXIDES

-oxide

Hydroxides when contact with water; All water reactive
Heat generation and caustic solutions

HYDROXIDES

Hydroxide
Solids and solutions are corrosive

INORGANIC PEROXIDES

Perodixe
Water reactive; Strong oxidizers
Heat production with high ignition potential

OXYGENATED INORGANIC COMPOUNDS

Per -ate; ate; ite; Hypo -ite
Normal oxygenated states

–1	–2	–3
FO_3	CO_3	PO_4
ClO_3	SO_4	BO_3
BrO_3		
IO_3		
NO_3		

More than 1 O : Per -ate
Normal 0 : -ate
Less than 1 O : -ite
Less than 2 O : Hypo -ite

COMPOUND STRUCTURE and HAZARDS

BP: _____ IT: _____

FP: _____ LEL: _____

VP: _____ UEL: _____

Hosp.

Notified: _____

Amb. Dressed: _____

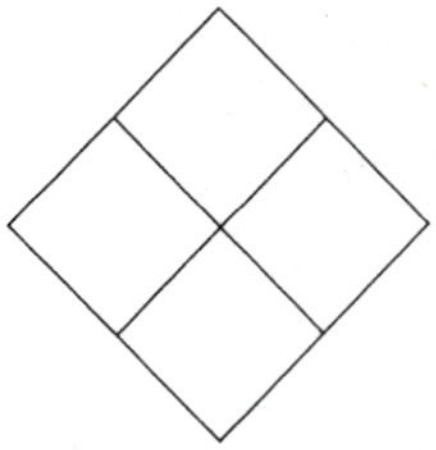

TLV: _____

PEL: _____

STEL: _____

SUIT COMPATIBILITY:_____

ORGANIC COMPOUNDS

NON-POLAR

Compounds

ALKYL HALIDES

R-X

R radical + halogen
Toxic and Flammable

NITROGEN GROUP

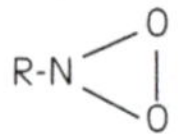

R-N (ring with O)

R radical +Nitrogen Group
Explosive

AMINE GROUP

R_xH_x

R radical +H or other ion
Toxic and Flammable

ETHERS

R-O-R

R and R' may be same or different radicals
Anesthetic and Flammable

ORGANIC PEROXIDES

R-O-O-R'

R and R' may be same or different radicals
Explosive and Oxidizer Potential

		POLAR COMPOUNDS				
Entry / Decon		**ALCOHOL**	**ESTER**	**ALDEHYDE**	**KETONE**	**ORGANIC ACIDS**
Level		R-OH	R-C-OO-R'	R-C-O-H	R-COR'	R-COOH
A	A	-ol	-ate	aldehyde	ketone	acid
B	B					
C	C	Flammable and toxic qualities	Flammable and polymerization	Flammable, toxic polymerization	Flammable and CNS effects	Toxic and corrosive
D	D					pH:_____

Figure 3-25 *Reference sheet for a single chemical.*

MEDICAL and CHEMICAL REFERENCE TACTICAL WORKSHEET

Chemical Name:_____ Synonyms:_____ Chemical Formula::_____ DOT:_____ CAS:_____
State of Matter:_____ State Found:_____ Water Reactive: Y N Oxider: Y N Other Hazards:_____
Wind Direction:_____ Wind Speed:_____ Temp.:_____ Humidity:_____ Adjusted Temp.:_____ Dew Point:_____
Manufacture:_____ Contact Person:_____ Phone:_____ Fax:_____

Chemical Name:_____ Synonyms:_____ Chemical Formula::_____ DOT:_____ CAS:_____
State of Matter:_____ State Found:_____ Water Reactive: Y N Oxider: Y N Other Hazards:_____
Wind Direction:_____ Wind Speed:_____ Temp.:_____ Humidity:_____ Adjusted Temp.:_____ Dew Point:_____
Manufacture:_____ Contact Person:_____ Phone:_____ Fax:_____

Chemical Name:_____ Synonyms:_____ Chemical Formula::_____ DOT:_____ CAS:_____
State of Matter:_____ State Found:_____ Water Reactive: Y N Oxider: Y N Other Hazards:_____
Wind Direction:_____ Wind Speed:_____ Temp.:_____ Humidity:_____ Adjusted Temp.:_____ Dew Point:_____
Manufacture:_____ Contact Person:_____ Phone:_____ Fax:_____

INORGANIC COMPOUNDS	COMPOUND STRUCTURE and HAZARDS	ORGANIC COMPOUNDS

INORGANIC COMPOUNDS

SALTS

-ide

General Hazards; Toxic gas with water contact with following:

Carbides = Acetylene
Hydrides = Hydrogen
Nitrides = Ammonia
Phosphides = Phosphine

METAL OXIDES

-oxide

Hydroxides when contact with water; All water reactive

Heat generation and caustic solutions

HYDROXIDES

Hydroxide

Solids and solutions are corrosive

INORGANIC PEROXIDES

Perodixe

Water reactive; Strong oxidizers

Heat production with high ignition potential

OXYGENATED INORGANIC COMPOUNDS

Per -ate; ate; ite; Hypo -ite

Normal oxygenated states

–1	–2	–3
FO_3	CO_3	PO_4
ClO_3	SO_4	BO_3
BrO_3		
IO_3	More than 1 O : Per -ate	
NO_3	Normal O : -ate	
	Less than 1 O : -ite	
	Less than 2 O : Hypo -ite	

COMPOUND STRUCTURE and HAZARDS

BP: _____ IT: _____
FP: _____ LEL: _____
VP: _____ UEL: _____

TLV: _____
PEL: _____ STEL: _____

BP: _____ IT: _____
FP: _____ LEL: _____
VP: _____ UEL: _____

TLV: _____
PEL: _____ STEL: _____

BP: _____ IT: _____
FP: _____ LEL: _____
VP: _____ UEL: _____

TLV: _____
PEL: _____ STEL: _____

SUIT COMPATIBILITY:_____

ORGANIC COMPOUNDS

NON-POLAR

Compounds

ALKYL HALIDES

R-X

R radical + halogen

Toxic and Flammable

HYDROXIDES

R-N (with O groups)

R radical +Nitrogen Group

Explosive

AMINE GROUP

R_xH_x

R radical +H or other ion

Toxic and Flammable

ETHERS

R-O-R

R and R' may be same or different radicals

Anesthetic and Flammable

ORGANIC PEROXIDES

R-O-O-R'

R and R' may be same or different radicals

Explosive and Oxidizer Potential

Entry / Decon Level		POLAR COMPOUNDS				
A A		**ALCOHOL**	**ESTER**	**ALDEHYDE**	**KETONE**	**ORGANIC ACIDS**
B B		R-OH	R-C-OO-R'	R-C-O-H	R-COR'	R-COOH
C C		-ol	-ate	aldehyde	ketone	acid
D D		Flammable and toxic qualities	Flammable and polymerization	Flammable, toxic polymerization	Flammable and CNS effects	Toxic and corrosive pH:____

Figure 3-26 *Reference sheet for multiple chemicals.*

SCENARIO

During a company survey of an occupancy, your engine company finds the storage of some chemicals. As a part of department protocol, the hazmat team is called to the facility to research some of the found chemicals. Because of another incident occurring in the south part of town, the hazmat team is riding one firefighter short. The hazmat captain knows that you just completed a course on Hazardous Materials Medical Considerations and asks you to perform the research. You have available to you the MSDS, commonly found reference texts, and computer access through a local emergency planning agency.

Scenario Questions

1. What chemical and physical properties should you initially research?

2. Knowing the chemical and physical properties, how would the temperature affect a variety of inorganic, organic, ionic, or covalently bonded chemicals?

3. What would water do to your inorganic chemicals, and what category should you be concerned with?

4. If a cryogenic was involved, what are the considerations of the above discussions?

5. Discuss the inorganic compounds in terms of fire potential and health hazards.

6. Discuss the organic compounds in terms of fire potential and health hazards.

7. What other categories of compounds should you identify?

Chapter 4

Essentials of Toxicology

Objectives

When presented with a possible toxic exposure, the student should be able to work through the risk assessment as it relates to the acute toxic event. In addition, the numbers that are referenced as levels of possible concern should be understood for their scientific values rather than their absolute virtues.

As a hazardous materials medical technician, you should be able to:

- Explain the differences between toxicity and toxic.
- Discuss the role metabolism has within the exposure event.
- Discuss the four basic factors of exposure.
- Explain the principles of toxicology and how they relate to the exposure individual:

Graded response	Phase I and II reactions
Quantal response	Variables affecting toxic levels
Dose response	Acute exposure
Effective concentration	Subchronic exposure
Routes of exposure	Chronic exposure
Combined effects	

- Define the following toxicological terms:

TLV	IDLH	LC_{50}
TLV-TWA	PEL	LD_{50}
TLV-STEL	CL	LCT_{50}
TLV-EL	PK	MAC
TLV-s	LClo	RD_{50}
TLV-c	LDlo	

- Discuss the one basic truth about threshold limit values.
- Discuss the margin of safety.

It would be foolish to think that we could discuss all the information that falls under the heading of toxicology in one short chapter. Until recently toxicology has been included in the discipline of pharmacology. Ancient writings describe "compounds" as being used for medicinal purposes, poisoning animals for their easy capture, and as an assassin's tool. Even the great scholar Socrates fell prey to the ingested poison hemlock (which, incidentally, was the Grecian government's "standard" poison).

However, it wasn't until the sixteenth century that a development within the discipline of pharmacology (toxicology) was seen. A Swiss alchemist and physician, Phillipus Paracelsus, was the first to recognize the dose-response relationship—that it is the quantity of a poison that is harmful. He stated that "it is the dosage that makes it either a poison or a remedy." Paracelsus realized the difference between the therapeutic levels of a substance and the toxic levels of the same substance is a fine line over time.

For years the subjects of toxicology and pharmacology followed the same curriculum. Both were taught at the same time, with pharmacology receiving the highest concentration of time and educational intellect. Toxicology became the offshoot of pharmacology. Only in 1961, with the formation of the Society of Toxicology, did the science of toxicology begin to develop. One reason for this change in focus was the idea of poison detection, rather than analyzing possible antidotes. It is interesting to note that while this philosophy was changing, another important development was taking place: A new discipline, emergency medical services, was being born.

The establishment of emergency medical systems across the country also provided a reduction in the time between the poisoning and the response of medically trained technicians. This response of medical technicians lowered the mortality and morbidity in some cases, while in other situations secondary exposure occurred, with the medical technicians becoming patients and some suffering long-term effects.

Today there is a vast diversity of disciplines under the heading of toxicology—environmental toxicology, industrial toxicology and industrial hygiene, clinical toxicology, biological toxicology, to name a few—examining chemical effects from food, drugs, the workplace, and the environment. All provide one goal: understanding how chemicals effect or affect our environment and the living organisms within their surroundings.

There can be more than one response produced by a drug or chemical. The extent of injury may be one target organ or it may be several. The organism as an individual directs the outcome of the toxicity of a chemical or chemical family. There are no single solutions to the technical problem of health risk assessment, rather the solution of any analysis is weighing all the information then gathering viable solutions. This solution process is dynamic and is not terminal, however it must have a fundamental basis in ethics and morality (see referencing techniques in appendix A).

Several interrelated topics are discussed as they relate to the chemically injured patient. Each topic is discussed in order to gain a further understanding of the material presented. As in any discipline, the terminology of the science helps to establish building blocks for clarification. Toxicology is no different.

■ NOTE

A chemical compound foreign to the human body reacts with the biometabolism and causes an effect that may range from very slight to catastrophic.

TOXICOLOGY

The term *toxin* will be used in the general sense. Specifically, it is a *hazardous* substance that has the potential to be toxic if it effects/affects the organism, usually in a negative context. For example, some compounds are reported to have toxic properties in the pure or diluted form (Fig. 4-1). Combined with another "activating" compound or impurity, the combination can be lethal.

In all of chemistry, toxicology (for our purposes) must be thought of as a biochemical reaction. In other words, a chemical compound that is foreign to the

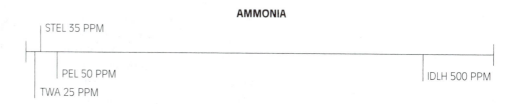

AMMONIA

STEL 35 PPM

PEL 50 PPM

TWA 25 PPM

IDLH 500 PPM

MEK

PEL 200 PPM

TWA 200 PPM

STEL 300 PPM

IDLH 3000 PPM

Figure 4-1 *We can see that chemicals have "levels" that are considered to be hazardous.*

human body is reacting with the biometabolism and causing an effect. On occasion that effect may be very slight or it may be catastrophic.

Thousands and thousands of chemicals are present in our environment, millions of chemicals within our society. Some are naturally occurring (organic) and some are man-made (synthetic). Because of this vast quantity of chemicals, it is sometimes hard to link the cause and effect to an exposure. Did the exposure (the cause) lead to the medical problem (the effect). The fundamental reason this relationship is so hard to establish is that it is difficult to understand the human (and animal) physiological functioning on the biochemical (cellular/tissue) level. Our overall understanding of biochemical reactions and functions are very limited in terms of other influencing factors. It is true that the exposure can be tracked, given the chemical compound and the route of entry. However, the intricacy of the human body is deeper than the function or structure of tissue. In many instances, chemical reactions between the toxin and the organism take place at the cellular level. So the real problem is a simple fact of whether we can detect an adverse reaction. Is it truly a reaction to the toxin, or is this reaction from a defensive mechanism? In either case, it is difficult, maybe impossible, to determine if the reaction will affect the organism in any way.

Toxicity

What is an exposure? What is contamination? How do they relate to a toxic event? Some simple terms must first be defined. Toxin and toxicity are often used interchangeably. **Toxicity** is the ability of a chemical or substance to cause an injury or harm. In simple terms it can be thought of as the exposure. For example if an emergency responder enters an atmosphere without the appropriate level of protection and is exposed, the environment has toxic potential, or its toxicity (the ability to cause harm) is present. A **toxin** is the reaction caused by a specific substance that causes the harm.

Contamination

Contamination is contact with the toxin. This agent can now cause potential harm. It is combined with exposure to indicate the event of contact. How do these ideas interrelate? The classical definition of exposure is to leave one unprotected. **Toxic exposure** can be defined as the concentration of a dose that causes a response. If the exposure leads to a contaminated patient with entrance into the body, then the individual has had a toxic exposure from contaminates.

Contamination can result in the introduction of a chemical into an organism. To be contaminated is to be exposed, but the chemical may not have entered the body to cause harm. Or conversely, one may be exposed or injured but never truly contaminated. A simple analogy is walking in a cow pasture. Once you climb over the fence, you are now within an environment that has a potential to cause harm (exposure). You could walk several feet within the pasture and never step onto a cow patty. You are still exposed to the environment. Once you take that one step onto a patty you have now become contaminated.

toxicity
the ability of a chemical or substance to cause an injury or harm

toxin
the reaction caused by a specific substance that causes the harm

contamination
contact with a toxin

toxic exposure
the concentration of a dose that causes a response

Factors Affecting Toxicity

Most often the route of entry determines the level of harm (toxic exposure) that is received. The entrance of a chemical can cause immediate or delayed signs and symptoms. Factors such as temperature, humidity, trauma, previous exposure, and type of personal protection all affect the toxic event.

So when exposure and dose relationships are discussed, what is actually being discussed are the biochemical reactions, the risk of influence, and the individual's body response to the chemical stimuli. This stimuli is described as a biochemical (pharmokinetic) reaction and is expressed by a toxicological numerical term.

The normal assumption is that we think in terms of "how safe chemicals are." For the most part they are safe in the typical surroundings. From a responder's standpoint, important questions must be asked: What chemical(s) are involved? What is the amount of the chemical(s) and what are its inherent harmful levels? These answers will usually lead to the appropriate action and mitigation objectives. However, when an individual is contaminated, the exposure may lead to chronic effects. The reality is that chemicals are not truly safe. Chemicals can and will cause harm at some level. The real question is, How will the human body metabolize the chemical if exposed? Supporting evidence has shown that some individuals that are subjected to a single chemical (or like chemical families) over time remain unaffected by the repetitive exposure. The body has the wonderful ability to become acclimated to some chemical exposures dealing with the levels it expects day after day. This principle allows cigarette smokers to be exposed to the chemicals in cigarette smoke day after day with little or no *acute* effect. But if a person who normally does not use cigarettes were to smoke twenty-five or thirty cigarettes in a day, the results would be massive acute and possible critical symptoms. This is not to say either person (the smoker and the non-smoker) will not have chronic effects. This is known. However, the body can become acclimated. We see with smoking and repetitive smoke exposure from structure fires, a variety of long-term effects.

Within the emergency response field, the problem is not a single exposure that one may have but the repetitive exposures to a variety of chemicals. Some of the literature supports the theory that with single exposures, the body has the ability to compensate for the chemical insult. Others state that the immune system has in some instances the ability to maintain homeostasis. However, the fact still remains that as emergency responders become more involved with chemical accidents, the variety of chemical and biological insults that one may have within a career is well beyond the body's ability to create an immunity or even compensate. The question that must be asked is, Will there be a measurable and qualifying influencing effect once a toxic exposure has occurred? (See Chapter 7.)

We can look back and see that low-dose exposures to chemicals have produced disease processes across the job environment. Why do firefighters have a significantly lower life expectancy than most Americans? Will EMS see the same lower life span in the future due to the biological insults and chemicals that they are exposed to? In law enforcement, there is an increasing incidence of liver, kidney, and bone cancers. What will the future hold?

Laws have been propagated from a public outcry to control chemicals within our society. We, as a civilization, have been conditioned to an industrial chemical dependency. In every household, products of convenience exist—fertilizers, insecticides, and cleaning solutions are found along with many more. In turn we, as a society, must also understand and become responsible about the dangers of these same chemicals that have contributed to making our everyday life enjoyable.

This reasoning must be employed when dealing with a hazardous material. Chemical materials are dangerous, possessing the potential for toxic effects. Emergency responders must become fully aware of the health hazards these materials pose. As stated in the introduction to this section, the idea is to learn enemy identification. The nomenclature and testing procedures used to define the toxicity of chemicals must be known before the referenced materials at an incident are understood. It is up to emergency workers to understand and apply this information so that decisions based on a risk versus benefit analysis can be accurately made.

METABOLISM

To fully understand the concepts of toxicology, dose response, and antidotal treatment, the concept of metabolism must be reviewed. The outcome of an exposure, or better yet, the dose response is predicated on our understanding of this concept.

■ NOTE
Dose response is predicated on understanding metabolism.

General Metabolism

General metabolism is the body's (or organism's) ability to utilize foods for fuel. These foods undergo chemical changes to provide for the energy requirements of the body. For example, the body must produce enough energy to maintain daily muscular function, the nervous system's electrical potentials, body temperature, glandular secretions, and general maintenance of cellular activity to name a few. It is the general "combustion" of fuels that creates energy referred to as metabolism. When food is ingested, the body is taking in carbohydrates, proteins, and fat. Each respond and are manipulated chemically to provide the organism with a usable fuel source.

carbohydrates
a major class of food formulated by plants

Carbohydrates, a major class of food formulated by plants, are made up of chemical compounds that we call glucose. This "sugar" molecule is broken down (*catabolized*) and placed into two chemical processes. These processes convert the fuel into carbon dioxide and energy. Initially the glucose molecule is placed into the first reaction called glycolysis. In this reaction, oxygen is not required, yet energy is produced along with pyruvic acid.

Once the conversion takes place, the pyruvic acid enters into the secondary reaction, which is oxygen dependent, creating carbon dioxide, heat, water, and energy. The energy is utilized by the organism to maintain cellular activity. Without a constant delivery of water and food to the body, the organism will cease to exist.

proteins

naturally occurring complex molecules that are assimilated by animals

Proteins, naturally occurring complex molecules that are assimilated by animals, are also utilized by the organism. However the proteins (amino acids) are built upon to create (*anabolize*) complex compounds. These compounds are utilized by the cells to establish enzymes and cell structure. In a healthy individual, the amount of protein that is anabolized is extremely small. Normally, this type of metabolism is used to build the individual needs of the cell. In an unhealthy individual, one that may not have any fat reservoirs, the protein stores will become metabolized. This metabolism of proteins is available for the use of energy. However, the "food" is being taken away from the muscles, enzymes, and structures of the cell. If this is allowed to continue, the organism will eventually die.

fats

stored energy-rich food that may be assimilated in the future by the animal

Fats, stored energy-rich food that may be assimilated in the future by the animal, can also be utilized by the body to produce energy. In a normal healthy individual, fats are metabolized by the body when the carbohydrate stores have been depleted. For example, a responder eating a small breakfast in the morning will produce enough carbohydrate to function for several hours. The day becomes busy and dinner is delayed until nine o'clock that evening. Once the body utilizes all the carbohydrates that were ingested in the morning, then metabolism shifts and the body starts to catabolize fats as the day progresses. If, in the same example, the responder had the opportunity to eat lunch, the food is converted into fat and stored in the adipose tissue for future use. Stored fat tissue has the ability to hold fat-soluble chemicals. When these fat cells are later catabolized, the chemicals can again be released into the body and cause harm a long time after the exposure took place.

BIOCHEMISTRY

The liver, kidneys, and lungs are involved with the detoxification of a chemical. Two basic scenarios can occur that lead to injury. The first is a nontoxic substance that gains access into the body and during metabolism is converted into a toxic substance. The second situation involves a toxic substance gaining access and being metabolized into different toxic substances.* In both situations the issue of solubility plays a role in long-term toxicity.

Liver Metabolism

Liver metabolism is a complex and detailed process. In general, the liver, bile, gastrointestinal tract, and blood constituents are responsible for the filtration of chemicals (the kidneys also play a role). Compounds that are ingested are absorbed from the small intestine and go straight to the liver. Within the liver, chemicals that are unwanted by the body are metabolized or chemically converted. This metabolism may result in several outcomes. It may produce a nontoxic substance from a toxic one; a higher degree of toxicity; a lower degree of toxicity; or it may not engage in a reaction at all. In all cases, the metabolite being produced enters the general circulation and is transported to other areas of the organism.

---

*Nontoxic can remain nontoxic; or toxic can become nontoxic.

> Nontoxic can become highly toxic, moderately toxic, mildly toxic, or remain nontoxic. Toxic compounds can remain highly toxic, moderately toxic, mildly toxic, or remain at their entrance level of toxicity. Intermediates can also go through the same process.

There are two types of toxic substances, one that enters the body as a toxin and one that is converted into a toxin during the metabolic process (protoxic). Chemicals also have the ability to enhance, cancel, support, and give an additive response to toxicity. The primary reason for these outcomes is due to what is called Phase I and Phase II reactions.

Phase I and II Reactions

In discussing a Phase I reaction, we must revert back to the discussion on the functional groups and remember that the chemistry of these groups depends on the surrounding available compounds. An assortment of outcomes can be created out of a single compound because of the other available compounds that will combine with it. These possibilities are multiplied once a chemical enters the body, not only because of the chemical conversions that take place there, but also because of the prevalence of other chemicals.

For the most part the phase I reaction takes the highly active polar chemicals and converts them into lipophilic toxic chemicals (in some reactions). Once this chemical becomes modified, the product is a water soluble compound. However, the solubility is relative to the original compound. In other words, if the compound was slightly soluble then the product will, more than likely, be water soluble. On the other hand, if the compound was not soluble in water at all, then although the product compound may have water soluble qualities, it may not be as soluble as is necessary for the excretory process to handle.

Examples of Phase I Reduction Reactions

$$R-\overset{\overset{O}{\|}}{C}-H \qquad \text{Aldehyde} \quad R\text{-CHO} \longrightarrow RCH_2OH \qquad R-\overset{\overset{H}{|}}{\underset{\underset{H}{|}}{C}}-OH$$

$$R-\overset{\overset{O}{\|}}{C}-R' \qquad \text{Ketone} \quad R\text{-CO-R'} \longrightarrow RCH(OH)R \qquad R-\overset{\overset{OH}{|}}{\underset{\underset{H}{|}}{C}}-R$$

$$\overset{\diagdown}{\diagup}C=C\overset{\diagup}{\diagdown} \qquad \text{Alkene} \quad C=C \longrightarrow \text{-CHCH(OH)} \qquad \overset{H\ H}{\underset{}{-}}\overset{|\ |}{\underset{}{C}}-\overset{|\ |}{C}-OH$$

$$R-NO_2 \qquad \text{Nitro groups} \quad R-NO_2 \longrightarrow \begin{array}{l} R\text{-NO; } RNH_2; \\ RNH\text{-OH} \end{array} \quad R\text{-NO; } R-N\overset{\diagup H}{\diagdown H} \text{ ; } R-N\overset{\diagup OH}{\diagdown H}$$

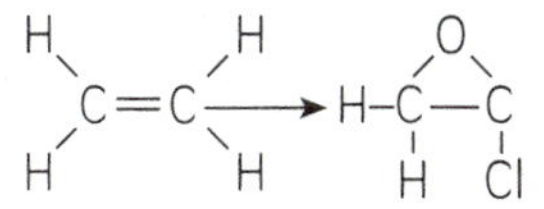

Figure 4-2 *The vinyl chloride molecule is an easily inhaled material. Once within the body, it attacks the CNS, liver, blood, and lymphatic system.*

Once in the liver, vinyl chloride molecules are converted to an epoxide through oxidation. The epoxide binds with the DNA within the cells and causes cancer. OSHA 1ppm.

If the substance becomes not as soluble as it should be for the kidneys to take over, then an "active site" is placed on the compound so that the substance may be eliminated through the kidneys. This attachment of an active site is termed the *Phase II* reaction.

It must be remembered that these reactions are the body's attempt to detoxify an intruding chemical (Fig. 4-2). In some cases, the chemical intermediate is a highly toxic compound, one that can actively destroy cells and tissue. In other cases, the cells are hindered from accepting oxygen and nutrients or from releasing carbon dioxide. The intermediates may have a synergistic, additive, potentiative, or antagonistic quality. These qualities are discussed later.

Most of the antidotal treatment is based on the organism's ability to excrete the end product. There are limited amounts of antidotes as compared to the vast variety of chemicals that one may become exposed to. Antidotes that are not discussed in this book but that require mentioning are the **chelating agents**, chemicals that are used to combine with the toxic complex for biochemical elimination. Chelating agents are used to form a chemical complex between the insulting chemical and the agent. Either another chemical is introduced to eliminate the newly formed complex or the chelating agent combines with the insulting chemical and is eliminated. With these particular agents, the attachment of the antidote to the toxic chemical makes it "water soluble," allowing it to be filtered through the kidneys.

All substances excreted by the kidneys are water soluble. For an introduced substance to be excreted from the body via the kidney, the substance must be water soluble and polar (Fig. 4-3). Remember, the polarity determines the solubility of a substance. With some chemicals the body "looks" for a way to convert the nonpolar (thus nonsoluble) substance into a polar molecule. This must occur in order for the excretory process to handle the metabolite. Unfortunately, the body may take a nontoxic, nonpolar substance and convert it to a polar substance for excretion. During this process a toxic polar intermediate may be formed that can damage tissue.

NOTE

A chemical intermediate may have a synergistic, additive, potentiative, or antagonistic quality.

chelating agents

chemicals that are used to combine with a toxic complex for biochemical elimination

Figure 4-3 *Common functional groups that are metabolized into water soluble products, allowing the individual to eliminate the toxin.*

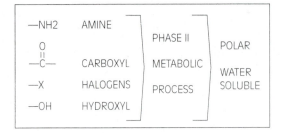

These reactions for the most part occur at the enzyme protein level. Actually, the alteration of the metabolism of these chemicals (enzymes) may interrupt the body's ability to recognize particular unwanted chemicals. If this cycle of reactions continues (repeated exposure), the body is unable to return to the preexposure state. If the exposure to the chemical is maintained at a minimum, specific enzymes within the body act on the toxic metabolite. The chemical reactions continue to take place until the unwanted substance is excreted from the body (Fig. 4-4).

It should be remembered that enzymes are chemically specific compounds, that is, they have very precise chemical and physical properties, which give them the ability to carry out the very precise, specific roles for which they were intended. Enzymes are responsible for changing nonpolar and nontoxic substances into the desired substances, if for example, the body may produce an enzyme to counteract a particular substance. If the intermediate or the original chemical itself hinders the production of this particular enzyme, a toxic state is produced.

Detoxification by the Lungs

The lungs may also play a major role in the detoxification process by removing carbon dioxide along with other products via the respiratory pathway. As we saw in

Figure 4-4 *The molecular orientation dictates the combination between the enzyme and the substrate. Once this occurs, the complex can enter into the necessary metabolic process. Some chemicals affect this combination by binding an "attachment" site.*

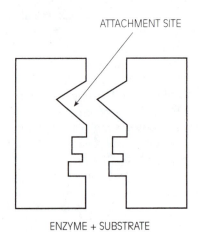

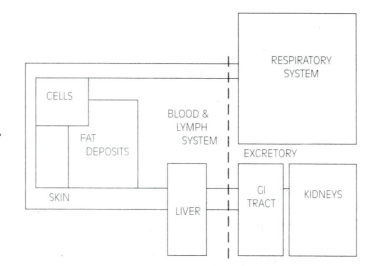

Figure 4-5 *The routes of exposure are important when discussing toxicology. How a chemical enters the body can affect the entire organism.*

> Once in the organism, the metabolic pathway a chemical takes has to do with the chemistry of the organism and the chemical itself. The interdependence between the systems can give a chemical several paths. Metabolism can change the chemical into a higher toxic chemical or one with a lower toxicity level, nonpolar to polar or polar to nonpolar, excreted or retained within the organism.

normal metabolism, carbon dioxide is produced from the combustion of oxygen and fuel to create energy, water, and heat. It is the lungs' responsibility to eliminate the carbon dioxide while providing oxygen needed by the cells in order for metabolism cycles to take place (Fig. 4-5).

Chemical Qualities

The chemical qualities of a compound influence the toxic character that a particular compound possesses. As discussed in Chapter 3, its orientation in space will greatly influence the characteristics that a chemical may have (Fig. 4-6). This same concept holds true when the toxic qualities of a compound are discussed. It is the shape of a molecule that the body recognizes. This recognition places the molecule into the appropriate chemical reaction. Remember that activation energies that drive reactions are like a hill or a mountain—the lower the required energy, the faster and more probable that the reaction will continue. The opposite is also true; the higher the activation energy, the lower the probability of the reaction to finish.

This arrangement in space identifies one structure as polar and another as nonpolar (remember, polarity is another way of saying how soluble a substance

Figure 4-6 *The orientation of the attached chains or functional groups have an effect on the toxicity of the chemical.*

OH

ORTHO

META

PARA

■ **NOTE**

During metabolism, a fat-soluble chemical may become stored within the adipose tissue.

■ **NOTE**

Through metabolism or by their own chemical shape, some chemicals may "look" like an enzyme and thus occupy receptor sites meant for certain enzymes.

may be). As an example, the polar substances are soluble in water (thus excreted out through the kidneys) whereas nonpolar substances are usually soluble in fat. This provides yet another problem when dealing with toxic chemicals. A chemical that is soluble in fat may, through the metabolism process, become stored within the adipose tissue, where it may be released in small quantities over and over again, whenever the fat is utilized in a metabolic cycle. This concept explains why an individual can ingest a poison and not suffer any discernible effect, yet some time later during the process of losing weight may become affected.

The shape of a molecule may also lead to toxicity within the body. Enzymes utilize their shape to bond at receptor sites (Fig. 4-7). Some chemicals can, through metabolism or by their own chemical shape, "look" like the enzyme and thus occupy the receptor sites meant for certain enzymes. This receptor site bonding can lead to both acute and chronic symptoms.

So far, several problems associated with chemical exposure as it relates to the metabolic process have been discussed. Many processes contribute to the toxic effects of a chemical but four factors determine the overall response of a chemical exposure (Fig. 4-8).

1. The amount and concentration of the chemical
2. Rate of absorption, which has to do with the shape and polarity of the chemical (absorption and distribution)
3. Rate of detoxification, which is dependent on the organism's metabolism
4. Rate of excretion, which is conditional to the end result of the metabolic pathway

Figure 4-7 *Some chemicals, through metabolism or due to their own shape, can "look" like the enzyme that normally occupies that receptor site. In this figure, the required chemical (RC) is competing for the site with the toxic chemical (T).*

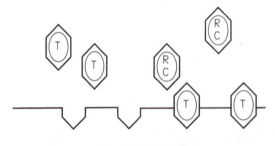

RECEPTOR SITES

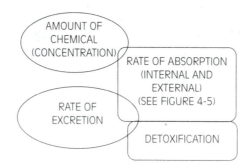

Figure 4-8   *It is up to the body to detoxify and excrete the toxin. Antidotes help in this process.*

THE DOSE RESPONSE

Two principles are used when describing the toxic levels of a chemical. Each, in itself, explains the pharmacological response to the chemical in question. The first is the concentration–response relationship and the second is the dose–response relationship.

When a chemical is viewed at the physiological level, the concentration of that chemical plays a part. The concentration determines the ability of the chemical to bind to a receptor site. Chemicals bind to normal receptor sites in order to inhibit, excite, or control the site. The chemical must "fit" into the site as if it belongs there for one of these responses to take place. The fit in turn causes a response. The chemical may simply replace a different chemical that would normally fit into that receptor site. Or a reaction can result at the enzyme production phase. The change in the molecular "signal" from the cellular level is what is observed as the signs and symptoms (or the demise of signs and symptoms in the case of medicine), all depending on the receptor sites affected. Logically this reaction must occur more than several times and often enough to produce an effect.

■ NOTE

Chemicals bind to normal receptor sites in order to inhibit, excite, or control the site.

Graded Response

So the true question may be, How much of a chemical will produce this observation? What concentration of the chemical will bind enough receptor sites, or disrupt enough enzyme pathways, to cause an effect? Two types of responses are seen in terms of concentration. The first is a *graded response*, which is represented by a gradual increasing of receptor sites binding with the chemical. It is a gradually more pronounced response that is seen as the concentration increases. The greater the concentration, the greater the number of sites that are occupied. The increase of signs and symptoms displayed by the individual is proportional to the bound receptor sites.

Quantal Response

In the *quantal response*, an all or none response is observed. In other words, the increase in concentration does not necessarily mean an increase in observable

■ **NOTE**

EC_{50} means that 50% of
a tested population
responded with an
observable effect at
this concentration.

effects. It may take a large concentration to produce an effect. Up until that time no effect may be noted. The concentration producing the effect is thought to have a direct relationship on body weight and metabolism. The point at which the chemical produces an effect is called the *effective concentration*, abbreviated as EC. If a number is associated with the EC like, EC_{50}, this means that 50% of the tested population responded with an observable effect at this concentration.

Many models have been developed to understand the effects of drugs and chemicals. The body of any organism is a complex machine in which complicated biochemical reactions take place. To understand all that occurs, the concept of what is thought to take place must be examined first.

Most models look at the body as having five groups:

1. Lung group
2. Vessel rich group
3. Blood rich compartment
4. Muscle group
5. Fat group

The lung group consists of the respiratory tree, partial pressures within the respiratory system, and how these pressures relate to the other four groups. When the effects of vapors or gases are discussed, the lung group is isolated for injury study. The vessel-rich group looks at the site in which metabolism may occur, such as in the liver, kidneys, heart, and the gastrointestinal tract. Some scientists place the hormonal activity within this group. It is observed that this area is where most of the damage from a variety of chemicals occurs. The blood-rich compartment is the brain. An inability of the brain to metabolize some chemicals is observed as a central nervous system disorder. The muscle group is composed of muscle and skin. When studies are done with topical agents, this group is isolated for the experiment. The fifth and last group is the fat group, which is composed of the adipose tissue and bone marrow. In this group, the adipose tissue may respond by absorbing the material, and toxic events can take place when the body chooses to metabolize fat.

All of the compartments have one thing in common—blood perfusion. It is assumed that as an individual assimilates the chemical, the said chemical is equally distributed throughout the bloodstream. Through this equal distribution, the chemical's physical and chemical properties target a particular organ. For example, a fat soluble chemical will target the fat group, hiding in the adipose tissue until metabolism starts to utilize that particular tissue. Because the liver and the kidneys are very well perfused (this is also where phase one and two reactions take place), we will see a higher degree of organ damage within these tissues. With some chemicals once a reaction takes place (the intermediate or end product of the reaction process), the chemical precipitates out of solution. Usually it is within the brain and kidney that this occurs (see ethylene glycol in Chapter 3). Most of the chemicals, at least to some degree, target the liver, lungs, and kidneys.

Dose Response

The dose response is an overall observable reaction to a chemical. It takes into account the difference in tissue function by looking for a response as it relates to the particular compartment. The problem here is that we have a wide variety of responses that a particular chemical may elicit. When looking at dose response, the total group of responses must be evaluated. Not only are the issues of time versus dose examined but also the effect experienced by the total animal population.

For example, two theoretical chemicals will be examined: one that affects a small percentage of the population and one that insults a large percentage. These chemicals will be referred to as Chemical A and Chemical B. When chemical A is given to a population of 100 individuals, an effect within a small population base is observed. With chemical B, a large population of 100 was affected. Figure 4-9 is the graphic representation of this experiment.

From these statistical curves, we could assume that a small concentration of chemical A is relatively safer than a small concentration of chemical B. At a hazardous material event where two chemicals are involved, reference may indicate that one is not as toxic as the other but may not be able to give the exact toxicity of a particular chemical, highlighting the fact that there may not be a safe level of exposure. For chemical A we could say that a one-tenth concentration is safer than the one-tenth concentration of chemical B, but that is all. The large range of individuals affected by chemical B shows us that a larger safety factor would be needed. It is for this reason that we as emergency responders should not predicate our decisions based solely upon the LC_{50} or LD_{50}.

Toxicity is not a tangible, measurable quantity like chemical and physical properties, which can be measured and reproduced. When chemical properties are measured, the same points should be reached each time given that the tests were performed under the same temperature and pressure (Fig. 4-10). When measuring toxicity, this is not the case. Too many factors (metabolism, specie, size, and weight, to name a few) propagate the quality of toxicity.

Figure 4-9 *Chemical B has a wide range of effect while chemical A has a narrow range of effect. In low concentrations, chemical A is relatively safer than chemical B.*

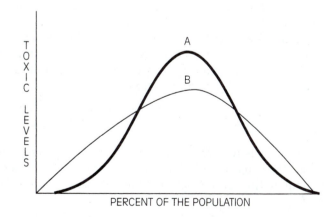

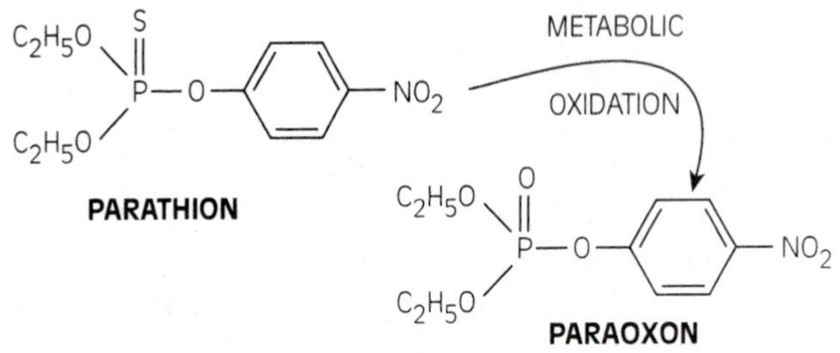

Figure 4-10   *When parathion is absorbed it is converted to paraoxon through an oxidative process. Both are cholinesterase inhibitors, however, paraoxon is the higher toxin of the two. The potent effects of parathion is from this conversion. Parathion BP = 157°, insoluble in water, LD_{50} 13mg/kg. Paraoxon BP = 169°, soluble in water, LD_{50} 1.8 mg/kg.*

The toxic event can be displayed as a mathematical representation of what is observed. It is a graph that represents the observations pictorially. The numbers cited in the reference material are a single point on this graph. This single point only makes reference to a particular point in time with a very specific exposure and animal study group or in some cases groups of animals. The numbers that are found in the reference material do not give the standard deviation of the graph. Did the testing procedure account for 95% of the population (two standard deviations) and if so what about the other 5%? As shown in the foregoing example, the differences between chemical A and chemical B had to do with the width of the graph, a greater or smaller standard deviation. These particular facts are not given in most of the referencing literature.

Before we move on, let's look at one more general concept, that of flammable limits. In Chapter 3, flammable limits were defined as a range of concentration of a chemical that would burn, expressed in a percent mixed in air. When discussing toxicity these ranges also apply. Generally speaking, the toxic levels of a chemical lie just below the lower flammable limit (lower explosive limit, LEL). Just like these flammable limits, the toxic limits or levels have a graduation of toxic effects. For this reason percentages are taken for "safe" levels when we are dealing with confined space (usually lower than 15% of our LEL). In general, under the LEL we shall say is a graduated toxic atmosphere.

ROUTES OF EXPOSURE

Exposure also deals with the way that a chemical enters the body, termed *routes of exposure*. We can generalize exposure routes into four basic modes: absorption, ingestion, inhalation, and injection. Many of these routes have components of the others, for example inhalation will have a certain degree of absorption quality. All of these routes are covered in great detail in Chapter 5.

COMBINED EFFECTS

So far several concepts have been discussed to explain the degree of the toxic exposure. Each exposure entails certain degrees of toxification. It must be realized that most often the exposure is to more than one chemical. At a clandestine lab, for example, exposure would include a number of chemicals, each possessing its own inherent physical and chemical properties. At what level does the chemical in question harm us? Does the chemical become enhanced or does it cancel out the effects of toxicity? We describe these chain reactions as synergism, antagonism, additive, and potentiation (Fig. 4-11).

- Synergism—Occurs when the combined effects are more severe than the individual chemicals. In this case, there exists a chemical that by itself is moderately toxic, however in combination with another, the toxic qualities are enhanced (1 + 1 = 3).

- Antagonism—Occurs when the combined effects cancel the effects of each other, decreasing the toxic event. What actually happens is that one of the chemicals acts to decrease the effects of the other chemical (1 + 0 = 0.5).

- Additive—Occurs when some chemicals that are different in chemical structure (shape and polarity) have the same physiological response in the organism. Thus the effect is a twofold (or more) enhancement (1 + 1 = 2).

- Potentiation—Occurs when a chemically inactive species acts upon another chemical, which enhances the chemically active one (1 + 0 = 2).

Figure 4-11 *The dose response and combined effects.*

THE DOSE RESPONSE AND COMBINED EFFECTS

- SYNERGISTIC
  - 1 + 1 = 3
- ANTAGONISM
  - 1 + 0 = 0.5

- ADDITIVE
  - 1 + 1 = 2
- POTENTIATION
  - 1 + 0 = 2

VARIABLES THAT AFFECT TOXIC LEVELS

You may have heard that the physical health of an individual can determine the outcome of a catastrophic disease process. This idea of holistic health is a primary goal in many medical philosophies. One must have a good emotional and physical health status in order to ward off the disease. This same idea is also present when it comes to exposure to a chemical. One's state of health both emotionally and physically must be intact.

Diet

It is easily seen that part of this concept is diet dependent. From the discussion on metabolism, carbohydrates were identified as providing the fuel to produce energy. Beyond this the body will find other substances, for example, fats and proteins, in order to maintain the energy output required of the body. This same concept of diet having an influence on toxicity has been demonstrated in laboratory animals. During the performance of toxicity testing in several documented cases, the caloric qualities were limited prior to the exposure. This change in diet alone made drastic changes in the outcome of the experiments.

During metabolism, energy is produced for the body to complete many tasks that it must perform. These tasks include interior functions such as digestion and cellular function and outside tasks such as carrying heavy objects or running. The energy is dependent on the foods placed in the body. It is easy to see that the diet consumed by a person has a direct effect on one's physical health. The other aspect of physical health is the current state of health one maintains. Take for example a healthy individual applying additional table salt to food. This would probably not trigger any dramatic effect in this person. On the other hand, giving the same amount of salt to a person with reduced cardiac output because of a previous infarction could be devastating. This particular victim could suffer from pulmonary edema, increased blood pressure, and an additional workload on the kidneys, which, in effect, could be interpreted as a toxic event. It is reasonable to expect less of a toxic response from a healthy individual than would be seen with an unhealthy one.

When laboratory animals are tested for chemical toxicity, one factor that is controlled in the experiments is the animal's diet. This concept of metabolism versus diet is further enhanced when experiments on cancer-causing agents are performed. In one experiment, two sets of animals were fed two different diets, one a nutritionally correct diet and one that was less than nutritious but tasted good (we don't know how they established that the food tasted good to a bunch of rats). The nutrient-enriched diet animals produced less than the expected response (cancer), while the nutrient-starved animals produced a greater frequency of tumors as an associated disease response.

General Health

To a point we can control the physical "body temple" with exercise, limited alcohol ingestion, no smoking, and a well-balanced diet. However, the emotional side of our health may not be controllable. It can be hard to maintain or control the emotional ups and downs everyone experiences in life. Yet several studies have indicated that the emotional well-being of an individual can greatly influence the state of health.

The medical community is also not sure why some individuals are more susceptible to a chemical (hypersensitivity) and others are less susceptible (hyposensitivity). There have been several experiments wherein rats were used as the test animal against **organophosphates**, phosphorous-containing pesticides that inhibit synaptic response. It was found during these experiments that the female is more sensitive to the chemical than the male. During the research for this book, several studies that looked at the human exposure to organophosphates identified that the female can be more sensitive to the **acetylcholinesterase** inhibitors than the male. The male was identified as being hyposensitive to this category of poisons. However, research further identified that males are more sensitive than females to the chemicals that affect the liver. Within this same experiment, the effected systematic (average) exposure for the female was not appreciably affected. In other words, on average, the female did not show a sensitivity to the chemical.

organophosphates
phosphorus-containing pesticides that inhibit synaptic response

acetylcholinesterase
a chemical enzyme found at the ending of the neuron that prevents the accumulation of acetylcholine

Previous Exposure

In some individuals, a single previous exposure can potentiate the effects of a later chemical exposure. In recent years there have been many exposures that correspond to this single event exposure and a hypersensitive state. To explain these events it is hypothesized that the genetic makeup of the individual may cause this single-dose response.

One group of individuals that seem to have a building hypersensitivity are those that have had repetitive small dose encounters with the organophosphates and **carbamates** (synthetic organic insecticides). This is especially true if the same individuals were exposed to DDT in the early 1950s to the middle 1960s. This idea of hypersensitivity is further confused by the allergic reaction one may have toward a chemical. In this case it is unknown whether this was a hypersensitivity state, or it is the hypersensitive state analogous to an allergic reaction.

The contrary to this account is also true—a chronic exposure to a chemical has produced an "adaptation" to the chemical and a hyposensitive state exists. This case is rare for most chemicals, yet is often seen in low-level dose exposure over very long periods of time. Examples of this would be alcoholics (low level to ethyl alcohol) and COPD (low level exposure to carbon dioxide/monoxide). The time element between each event is long enough for the chemical to detoxify and a possible genetic rearrangement has taken place (Fig. 4-12).

carbamates
synthetic organic insecticides

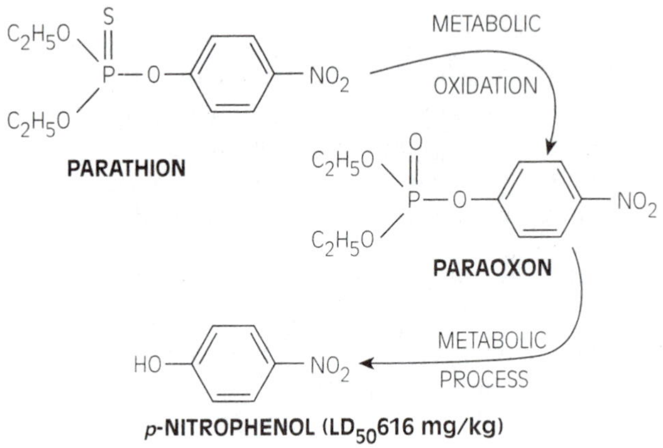

Figure 4-12 *Through metabolism, the original molecule is reduced into a "nontoxic" compound, which is then eliminated through the kidneys.*

Age

Usually when we refer to age there is an assumption that the discussion is about middle-aged individuals or those who are in an advanced stage of life. For the most part we have very little information on these two age groups. Most of the information pertains to the newborn and children. This is not to say the geriatric patient may not suffer under the same circumstances, but we do not have the supportive documentation to see what happens to the advanced aged individuals.

When trauma is taught in EMT and paramedic school, it is taught that two high-risk groups exist when an injury occurs—the young and the old. The explanation to this statement is that the ability to compensate during shock was either very well developed or unable to compensate respectively.

This same idea holds true for poisonings in that the young do not have the body mass to compensate for the ingestion of a chemical. The older individual may be unable to compensate, possibly due to a disease process that is already present or other drugs that they take for other illnesses. The elderly for the most part do not have the health status conducive to ward off the effects of an exposure. In either case the chemical can invade the body and start to manipulate metabolism easier in these high risk individuals than it can in the middle age population.

It has been seen in the laboratory that newborn rats are extremely sensitive to organophosphates, but the effect of exposure to DDT was nondiscernible. However, with the adult rat, the effects of both chemicals were apparent. This shows that the concept that young individuals are more sensitive to a chemical than an adult is not a honest statement.

Take for example an elderly individual who has angina. The angina is controlled through the use of nitroglycerin PRN. Other than an occasional bout with chest pain, this individual leads an active "normal" life. One day this person runs the car in the garage to warm it up before heading to the store on a cold winter day. The garage is left closed to keep out the cold and a dangerous level of carbon monoxide builds up. Within just a few breaths the individual starts to experience crushing chest pain that is not relieved with nitroglycerin. The reduced oxygen carrying capacity of his blood has lead to a serious event and may cause an myocardial ischemia. If a young healthy individual had experienced the same levels of carbon monoxide poisoning, the outcome would of been much different

Sex

We can easily see that there are physical and physiological differences between a male and a female. In the past, the male was predominant in the workforce therefore, most exposures occurred to the male. In recent years the female has entered the same workforce exposing her to the same consequences as that of the male. The impact of this has not yet been fully addressed, yet could have a profound influence on genetic changes caused from exposure.

Today, women are working around the same chemicals as men, but women become pregnant. In some cases women are working as long as they can during pregnancy. This exposure of chemicals to the fetus is an area that has had limited research. The ramifications and ethical implications of this issue are well beyond the scope of this text. Nevertheless, the influence on the reproductive functions, chemical exposure, and genetic modifications are all factors that influence the toxic event.

■ NOTE

Enzyme production, lung capacities, and muscle mass to body frame ratios can all influence a toxic event.

Enzyme production, lung capacities, and muscle mass to body frame ratios can all influence the toxic event. For example, consider lung capacity. In the male, the average air volume movement is .4–.5 m^3/hr, with the female, lung volume movement is an average of .3–.4 m^3/hr. Does the male have a higher frequency of respiratory injuries? We don't know. Body weight is less in the female as compared to the male on an average, not to mention the heart rate and enzyme production. However, fat stores are higher in the female as compared to the male. All may influence the effects of exposure of a chemical on the male and the female.

We saw earlier the difference in dose response between the male and female, especially when dealing with chemicals that affect the liver as the target organ. Some research has implied that this is because males are more apt to have that "single" beer with the guys after work. Repetitive consumption of alcohol may depress the liver's ability to create the enzymes needed to detoxify chemicals. It will be interesting to see if females in the future become just as susceptible to these chemicals as the males. Now that females are in the workforce, they too are going out with the gals for that beer. Could this also have an effect on these individuals? Only time will tell.

Genetics

The discussion of mutations on the genetic material is most definitely beyond the scope of this text, but a simple discussion of this subject is included in order to give the reader a complete picture of the chemical problem.

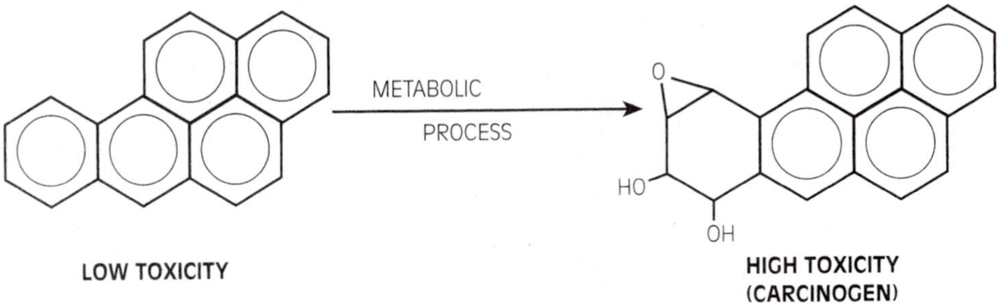

Figure 4-13 *Some chemicals will produce a carcinogen through the metabolic process. This process is common for chemicals that can form epioxides.*

LOW TOXICITY   HIGH TOXICITY (CARCINOGEN)

We are not sure if an exposure will mutate the genes so as to cause a more "acclimated" individual. We do know that some chemicals will affect and change the DNA within the cell. Rearrangement and even chromosomal breaks have been studied. Ethylene oxide, hydrazine, and ionizing radiation all have histories toward genetic mutations. These mutations could have a role in the carcinogenic or teratogenic responses, that are noted in some lab experiments (Fig. 4-13).

But let us discuss a more common everyday type of problem. As firefighters for example, the one general chemical group we are all exposed to is products of combustion, smoke. During this process of combustion, a variety of compounds are produced. For many years, we all have seen firefighters entering a building charged with smoke to fight the fire and perform salvage and overhaul. Once finished, the individual puts her/his gear back on the truck, only to remove the gear later and wear it again and again. When the firefighter puts the gear back on and begins to sweat, all of the chemicals found on the gear are over and over again contaminating this firefighter. Since the introduction of NFPA 1500, this problem has decreased. It is now a strongly recommended standard to wash gear periodically (NFPA 1581 and NFPA 1500). Frequent cleaning is an engineering control in order to lessen the occurrence of the chronic exposure. (See Toxicity of Smoke and Combustion Gases in Chapter 5.)

Sleep

There have been several studies indicating the effects of sleep on the toxic event. The lack of sleep has been indicated to produce a greater effect during a toxic event. This follows the concept that emotional well-being also contributes to a greater effect. What most of the studies cite is that as a person becomes tired, the enzymes needed to metabolize the chemical may not be produced at a high enough rate, leaving the body's defenses down. In other reports, the tired individual does not exercise safe precautions resulting in an exposure. In either case, lack of sleep does affect toxic exposures.

Little is truly known about the foregoing seven factors. What we do understand is that as diverse as humans may be, there are some factors that can be addressed to all of our patients. However, in some instances the understanding of the exposure and why it effects one individual and not another is still a mystery.

For this reason the concept of medical surveillance to monitor the health of emergency responders and those who are in constant contact with chemicals was established. This one concept is important in terms of prevention and treatment. It may be years and thousands of exposures before there is a true understanding of the metabolism of chemicals within the human organism. In the meantime the chemicals to which emergency responders are exposed must be identified.

TERMINOLOGY OF TOXICOLOGY

Threshold Limit Value

threshold limit value
the level of exposure that starts to produce an effect

In the United States, the recommended **threshold limit values** (TLV) are established by the American Conference of Governmental Industrial Hygienists (ACGIH). The TLV is that level of exposure that starts to produce an effect. Under this level if a worker is exposed to a chemical repeatedly there will not be any discernible effects. Most of these values refer to airborne concentrations and are based on a standard work week. When OSHA enumerates this principle, it is called it the permissible exposure limit (PEL), which is the same as the TLV-time weighted average (TWA). For NIOSH it is called the recommend exposure limit (REL). ACGIH denotes it as TLV, however, some articles have also referred to this level as the published (personal) exposure limit (PEL).

In Europe and across the Soviet block nations, different standards exist. One standard that comes out of Germany is high as compared with the American criteria. The German Research Society (GRS) has a maximum allowable concentration value (MAC or MAK), which is analogous to the ACGIH standard.

■ NOTE

The maximum allowable concentration value (MAC or MAK) of the German Research Society (GRS) is analogous to the ACGIH standard.

The numbers that each organization use are generally the same. In some instances the testing procedures may vary. For the most part the ACGIH standards are customarily used. Their threshold limit values (TLV) are the industry standard and a trademark of the ACGIH. The committee that introduced these levels has stated that the TLVs are not intended for the following uses:

- The TLV should not be used to describe the level of hazard as it relates to the toxicity for a chemical or group of chemicals.
- The TLV should not be used for the evaluation of air pollutants.
- The TLV should not be used for deciding on an extended work period within the said environment or when estimating the toxic capacity of a chemical.
- The TLV cannot be used to describe the cause and effect of a chemical substance.
- The TLV cannot be used if the working conditions are greatly different than those within the United States work environment.

Their only intended purpose was to establish safe working conditions within the United States, for those individuals who may become exposed on a daily basis. They were not intended to be used by emergency response personnel at the scene of a hazardous materials accident! However, by utilizing these values associated with other toxic parameters and our understanding of chemistry (here is where most of the chemistry comes in), we can plan our "safe level" and manage the incident with one objective—to reduce the possibility of exposure to response personnel.

We can use these numbers with associated information because most of these numbers were established for safe working conditions without the need for personal protective equipment (PPE). At an incident we almost always wear some level of PPE, thereby increasing the margin of safety. However, all the values must be respected for the testing parameters under which they were derived.

Workweek Application of TLV

When the discussion is from the ACGIH it is based on an 8-hour day, 5 days a week to establish a 40-hour work week. OSHA also considers this the exposure standard. In both cases they refer to this level as the TLV-TWA, where TWA represents a time-weighted average. It is the average concentration of a chemical that most workers will be exposed to during the normal 40-hour work week, at 8 hours a day. If NIOSH denotes a TLV-TWA it is based on a 10-hour work day 4 days a week, to give the 40-hour work week.

It was found that during the 1960s a variety of work schedules started to emerge. Ten-hour days, 4 days a week; two 16-hour days on week ends to be paid for as 40 hours; 4 hours repetitively for 10 days and 12 hours for 6 days with a week off with pay, set at 40 hours a week. What ever the schedule, the TWA did not apply. When the concept of the TWA was initially conceived, it was under the assumption that schedules are traditionally based on the 40-hour work week with either an 8-hour day or a 4-day period with 10-hour work days. These values were designed to allow an individual to be exposed to a chemical for the length of a day, provided that the maximum TLV was not reached. It was thought that the off time would allow the body to eliminate any stored chemicals. As you can see, the TLV has some limitations. For this reason, in 1976, the committee that designed the TLV began to assign short-term limit values or what is called short-term exposure limits (STEL).

Short-Term Exposure Limit

The STEL is an exposure that only occurs for 15 minutes and is not repeated for more than four times a day. Each 15-minute exposure event is interrupted by a 60-minute nonexposure environment, so that in the course of an 8-hour day, the individual can only be exposed to a chemical for 15 minutes with an hour break in between exposures not to exceed four times within a day. The STEL is sometimes referred to as the *emergency exposure limit* (EEL). The ACGIH has recently recommended the use of *excursion limits* (EL). In contrast, the ELs are more realistic than the STEL. The EL is a weighted average with the exposure time not to exceed five times the published 8-hour TWA. This can only occur for 30 minutes within any one work day that is 8 hours in duration.

Other Limiting Values

The ACGIH has other limit values, these are TLV-s and TLV-c. The threshold limit value -s denotes the TLV for skin. S identifies that the material is absorbed through the skin. The TLV-c, where c denotes ceiling levels, indicates that exposure at the published level insures death. Ceiling levels should NEVER be exceeded. This level is similar to the NIOSH's IDLH (Immediately Dangerous to Life and Health).

ACGIH Terminology

ACGIH TLV is that level of exposure that starts to produce an observable effect. It refers to airborne contaminants.

TLV-TWA—An 8-hour day 40-hour work week with repeated exposure without any adverse effects.

TLV-STEL—Fifteen-minute excursion in which the worker is exposed to the chemical continuously. Must not have any of the following effects:

1. Any irritation.

2. Chronic tissue damage.

3. The impairment of a self-rescue

TLV-EL—An average exposure not to exceed 5 times the published 8-hour TWA. This will not occur for more than 30 minutes on any work day.

TLV-s—Identifies a material that is absorbed through the skin.

TLV-c—A ceiling level.

OSHA Terminology

The Occupational Health and Safety Administration (OSHA), which is a branch of the Department of Labor, has for the most part adopted the ACGIH TLV's as their own PEL (Permissible Exposure Limit). OSHA and NIOSH have given the following as a definition of Immediately Dangerous to Life and Health (IDLH), as the maximum airborne concentration that an individual could escape within 30 minutes without sustaining any adverse effects. The IDLH (and TLV-c) give us a level at which the only form of protection is that of full encapsulation with self-contained breathing apparatus. In some cases, the skin absorption factor must also be evaluated. Here for the level of protection that would afford no health effects is that of full encapsulation

As we will see later LC-lo (lethal concentration low) can be used to approximate the IDLH (or TLV-c). It is recommended that IDLH atmospheres are not to be entered unless properly trained and the appropriate protective garments are donned. Individuals who are caught and are unconscious within this environment are more than likely dead by the time the hazmat team can make entry.

The problems are many when dealing with IDLH, TLV-c, or LC-lo should be obvious from the foregoing discussion. When using IDLH (TLV-c, LC-lo), one must realize that IDLHs are based on three factors:

1. The IDLH is only for the healthy young active male working population. It does not account for the possibility of hypersensitive or hyposensitive individuals. IDLH does not take into consideration the sensitivity of females to a variety of organic compounds.

2. IDLH are based on a total of 30-minutes exposure time. This time frame may or may not be realistic in an emergency situation. As one becomes excited, respiratory rates increase. We are not sure how this affects the 30-minute time frame.

3. NIOSH does not have values for all IDLH toxic chemicals. TLV-c and LC-lo can only be used to approximate the level of concern. Because we are dealing with a variety of unknowns, TLV-c and LC-lo should only be used as a guideline and not considered absolute levels.

OSHA has two other levels of concern. TLV-c or CL (ceiling level) may be used. Both are displaying the same objective, a ceiling level. PK sometimes is seen in the literature and is described as the peak value.

PEL—Has the same meaning as TLV-TWA.

IDLH—The maximum airborne contamination that an individual could escape from within thirty minutes without any side effects.

CL—This is similar to the TLV-c.

PK—Is different depending on the testing agency. However, it is the apogee of the daily allowable limit.

NIOSH Terminology

REL—This is the same as PEL and TLV-TWA.

Lethal Concentrations and Lethal Doses

We have already looked briefly at the following toxicological terms. Here the terminology that we are going to discuss comes from the dose response of drugs. In the beginning of this chapter we discussed to the concept of graded response, quantal response, and effective concentration. Each by their own quality showed the exact moment in time for which the population shows an adverse reaction. At this point we call that the threshold limit, effective concentration, or dose.

The qualifier here is that 50% of the total population under study died. This is based on the understanding that there will be some individuals within the population that are hyposensitive and some that are hypersensitive to the chemical agent. This qualifier refers to a finite time frame (usually 14 days), in which a total of 50% of the population died. If the chemical in question is airborne and is primarily a inhalation hazard, it is termed as the *lethal concentration* (LC). If the chemical poses a threat other than an inhalation injury it is termed the *lethal dose* (LD). In both cases it is the percentage of the population that is being described. A 50% death rate after the introduction of the chemical is denoted as lethal concentration fifty (LC_{50}) and is the calculation of an expected 50% death toll after the exposure of the chemical to the test animals as an inhalation hazard. The lethal dose fifty (LD_{50}) is the calculation of an expected 50% death toll after an exposure to a chemical, which may or may not include an inhalation injury.

For example, an LC_{25} or LD_{25} denotes a 25% death toll within the tested population. The number that is given with the associated LC or LD is relative. In general the smaller the number, the more toxic the chemical. Likewise the greater the number, the less toxic it is. For the most part LD_{50} and LC_{50} are an average of lethal doses (ALD). These mean values are an observation and calculation based on researched suicides or homicides.

One must remember that for the most part these are values observed in the laboratory, under controlled conditions. These tests are not done on human beings. However there are a few cases in which the human experience has been documented. In these cases the reference texts usually denote this by placing the chemical exposure animal in parenthesis (Fig. 4-14). (The TLVs during the 1960s were largely based upon the human exposure.) At times this value is referred to as the (TDL) toxic dose low or (TLC) the toxic concentration low. As we noted before,

■ NOTE

Lethal concentration fifty (LC_{50}) is the calculation of an expected 50% death toll after test animals are exposed to a chemical inhalation hazard; lethal dose fifty (LD_{50}) is the calculation of an expected 50% death toll after exposure to a chemical, which may or may not include an inhalation injury.

Figure 4-14
Abbreviations commonly found in the reference material.

ABBREVIATIONS

ROUTE OF EXPOSURE		ANIMAL	
IHL –	INHALATION	CAT –	CAT
IMS –	INTRAMUSCULAR	FRG –	FROG
IPR –	INTRAPERITIONAEAL	HMN –	HUMAN
IVN –	INTRAVENOUS	MAN –	MAN
ORL –	ORAL	MUS –	MOUSE
SCU –	SUBCUTANEOUS	RBT –	RABBIT
SKN –	SKIN	DOG –	DOG
UNK –	UNREPORTED	GPG –	GUINEA PIG
		RAT –	RAT
		MAM –	MAMMAL

the concentration enumerates the inhalation injury and the dose is a liquid/solid state of matter. This could be absorbed either through the skin or by ingestion.

The Environmental Protection Agency (EPA) utilizes the values as presented from ACGIH, NIOSH, and OSHA. They are mostly concerned with the impact that the chemical may have on the environment and the organisms that make up that environment.

The reference books that enumerate the LD and LC for humans are actually statistical extrapolations. They are derived from the mean lethal dose or the average lethal dose or concentration. From there they are calculated to the human experience.

LClo—The lowest concentration of airborne contaminates that caused injury.

LDlo—The lowest dose (solid/liquid) that caused an injury.

LC_{50}—50% of the test population died from the introduction of this airborne contaminate.

LD_{50}—50% of the tested population died from the introduction of this chemical, which may be a solid, liquid, or gas.

LCT_{50}—A statistically derived LC_{50} (LDT_{50} statistically derived lethal dose).

MAC—This is the maximum allowable concentration (European).

RD_{50}—A 50% calculated concentration of respiratory depression (RD = respiratory depression of 50% of the observed population), toward an irritant, over a 10–15 minute time frame.

ANIMAL STUDIES

There are several shortcomings to the testing designs used within the toxicology field (Fig. 4-15). Despite these problems the testing procedures employed are the most accurate. The main problem is so many body systems interact, along with the chemistry involved that it is very difficult to set testing parameters. However, for emergency responders to make the decisions that they will be faced with in the field, several problems with the testing procedure must be mentioned. The reference material unfortunately does not identify any of these possible problems. These tests only reference a single point in time, which under testing conditions, was used as an effect parameter. We have seen that the effect a chemical has on an organism has to do with the rate of absorption, detoxification, and excretion (the amount of the chemical is also a factor, however, some of the testing procedures give such a high dose that any animal would show an adverse reaction). If we want to consider the possibility of intermediates having a profound effect on the organ-

ANIMAL TESTING

PROBLEMS WITH ANIMAL TESTING ARE:

- TOO SMALL A STUDY GROUP
- STATISTICAL METHOD IS OFTEN BIASED
- DOSES ARE HIGH
- ANIMAL SENSITIVITY
- COMBINED EFFECTS

Figure 4-15   *The problems with animal testing.*

ism, then we have to add another parameter within the testing procedure. Hypersensitivity, hyposensitivity, synergistic, additive, potentiation, and antagonist reactions are not truly considered.

Statistically these factors are deliberately looked at from a mathematical model standpoint. In other words, a certain percent of a population will be hypersensitive and a certain percent will be hyposensitive. By utilizing statistics it is assumed that randomness in the population will equal itself out, using the above two factors as an example. However, humans are not statistical points on a graph. Each point observed on such a graph represents a life. As we will see in the acute and in the subchronic testing, the chance of having a statistically biased test is quite high. This occurs when the testing group is too small or limited to only a few species (Fig. 4-16).

Some testing laboratories have been accused on the basis of dose rates that have been utilized. In some of the test procedures, the dose rates are so high that any animal, whether sensitive to the chemical or not, will produce an effect. If the group did not include sensitive individuals then the results are skewed, giving an unrealistic result.

■ NOTE

Gender differences, race differences, age, and physical activity are not usually considered in toxicologic testing.

Gender differences, race differences, age, and physical activity are not usually considered. In terms of race, chemicals are not tested on humans. To that end we are not sure how the genetic makeup will be affected from one race to another.

In the testing process, some of the standard tests that are applied are administered to "genetically pure" groups. These groups have been fed a certain diet, bred to limit genetic anomalies, and controlled to a certain "clean" environment. This does not occur in the human existence: We humans do not live in a totally "clean" environment. We do not breed in order to have the genetically perfect offspring. Besides, how many people for the entire length of a life span, eat food that is nutritionally healthy?

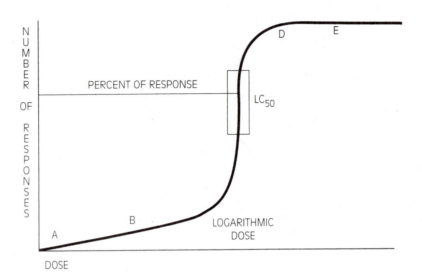

Figure 4-16 *From the hazmat perspective we are not sure where on the curve an individual may have a response. Unless the individual has a history of (A-B) hypersensitivity, or (D-E) hyposensitivity, we are not sure of the acute or chronic responses.*

However blunt this may sound, the reality is that the scientists are trying to block out as many factors as possible that may affect the toxic event. We do not have a better practice in order to evaluate the toxicity of chemicals. When the health effects are being considered at the scene of a hazardous materials incident, all the concepts that we have mentioned must be considered in the decision-making process.

Acute Exposure

■ NOTE
The term <u>acute</u> is used to describe a sudden onset or exclusive episode.

The term *acute* is used to describe a sudden onset or exclusive episode. It relates to the hazardous materials exposure as a single event that causes an injury. This single exposure is usually a short duration type of exposure. It is sometimes classified as nonpredictable. This single dose occurs within a 24-hour period, or it could be a constant exposure for 24 hours or less. In certain cases it could also be multiple exposures within the 24-hour time frame.

There is a testing routine that for the most part is a standard. However, anyone that is testing a chemical may, on their own, come up with what they may think is a scientifically sound testing procedure. Generally when a chemical is about to be tested, the first step in the procedure is to reference all the information that is currently known about that chemical. The testing technique is then based on the historical reference of the material. A new procedure may be done, however, and the results compared to what is already known about the particular chemical.

In terms of difficulty, the oral toxicity tests are relatively easy. The scientist uses rats or mice to study the chemical or drug in question. These tests are done initially to rate the chemical as to the toxic levels. In other words, they give the test animals an ever-increasing amount of the chemical until a lethal dose is attained. From there they can evaluate the chemical in relation to other chemicals in terms of toxicity.

From there they take a group of animals and place them into four or five (sometimes six) groups. In group one the rats and/or mice are given a set amount of the chemical that may or may not be at the threshold limit value. It is in general, a no-death dose. All the animals are anticipated to live. However there may be an observable toxicity event. From group one, increasing to group four, five, or six (depending on the chemical and the testing technique), the animals are given increasing dosages. The last group is given a dose that is known to completely destroy that group. From this observation a further subranking of the chemical is done. They observe for the next 14 days the death rate, sickness, fur loss, and other effects. The dose that produced a 50% death toll is determined to be the LD_{50}.

Each group produces a statistical picture of the chemical dose versus the death rate, fur loss, sickness, and so forth. This is compared to effective doses that are known for the chemical or like chemicals. Both graphs are statistically "flattened" out into a straight line or what is called the *dose response curve.*

Through statistical modeling the effective dose and the lethal dose are calculated and the published LD_{50} is now established. Both oral toxicity and inhalation toxicity are measured in this way. The oral toxicity levels are usually a one-time dose. Where the inhalation exposure is an exposure that occurs for 1 hour to each test group, each test group receives the appropriate dose. In each case the groups of animals are observed for a period of 14 days. At the end of the 14 day cycle the 50% death rate must be observed. If the percent is higher or lower, then the doses are reevaluated until the end result is 50%.

With dermal exposure the test animals are usually pigs, however rats and mice are also used. The pig's skin represents a very close facsimile to human skin. The animal is exposed for a period of 24 hours on the bare skin (if hair is present it is shaved off). Only 10% of the body surface area is used. The animals are again observed and the reporting is matched with the known level of that chemical or chemical family.

If the chemical needs further study or the results are such that the human element needs to be established, then the procedure is repeated using different species. In general if during the testing process all the test animal species respond in a similar manner and the statistical slope of the dose response is steep, then it is considered to be an accurate LD_{50} or LC_{50}. To the contrary, if the testing battery showed that there was a diversity of slopes across the animal species spectra, while shallow slopes were observed statistically, then the accuracy of the outcome is questioned. The problem is that the numbers that we see are some point in time in which a toxic dose was received. Not only does it only reflect a small window into the dose response but it does not tell us the angle of the testing slope. Was the slope steep, thus, a good correlation to the human event, or was it a shallow slope, which does not tell us the true toxicity of the chemical in man.

Lethal concentration or dose is not performed on man. If in the literature a document states that this was the level that caused death in man, this information was achieved through an autopsy from suicide, homicide, or accidental release.

In general the smaller the number that the lethal concentration or dose projects, the more toxic the chemical. If, for example, the number is large, then the toxicity of the chemical is quite low relatively. It is hypothesized that it is the variation of exposure that gives the true toxic picture. In simpler terms, if the chemical in question has been demonstrated to be toxic in most all plant and animal life without the documentation of a human experience, then it is considered to be a health hazard to humans. If the literature states that most plant and animal life is evenly destroyed when an a exposure exists associated with a human event, then again it is a human health risk.

Lethal concentrations and doses are only an *estimate* of the potential health problems that may exist. They are by no means an all-or-none limit, as with the TLVs, PELs, and RELs. They should not be taken as a set limit that will identify an exposure as a health effect or as a limit that no exposure will occur.

Most chemicals are weight dependent. In other words, the more physical mass of the animal in relation to the quantity of chemical exposure, the higher the likelihood of the animal combating the exposure. On the converse, the lighter the weight of the animal, the lower the potential of combating the exposure, or more accurately, the toxic event.

Looking at the LD_{50} of some chemicals, in order to estimate the lethal dose we could take the weight of the person and multiply it by the quantity of the material. For example, in a small child weighing 40 pounds (18.18 kg), the LD_{50} is multiplied by 18.18 kg. So if this child ingested Sevin (a moderately toxic pesticide, LD_{50} of 500 mg/kg), we would multiply 18.18 kg by 500 mg/kg or 9090 mg, or on a more convenient measuring system 9.090 grams or .02 pounds. If our patient was a 220-pound man for example (220/2.2 = 100 kg; 100 × 500 = 50,000 mg or 50 grams which is .11 pounds) it would take a little over a tenth of a pound to reach the LD_{50}. Remember that metabolism, sensitivity, humidity, temperature, and vapor pressure are a few of our influencing factors that are not considered here. We are only discussing body weight as it relates to the LD_{50} of the chemical.

Generally there are more LDs than there are LCs, because most chemicals do not change states of matter. Most are in the solid or liquid state. Very few become a vapor problem (this is of course if we are not experiencing flame impingement on our product). There are not many chemicals, relatively speaking, that are an airborne contaminate. However this brings up yet another problem: the testing procedures that these chemicals are experimented with. We as emergency responders have to deal with the health, flammability, and reactivity issues that a chemical may possesses. Under fire conditions what will this chemical represent? As stated before, what are the synergistic, additive, antagonistic, or potentiative reactions? The problems of how this chemical may react, combust, or jeopardize our health integrity are all significant management problems.

■ NOTE

Lethal concentrations and doses are only an estimate of potential health hazards that may exist and should not be taken as a set limit that will identify an exposure as a health effect or as a limit that no exposure will occur.

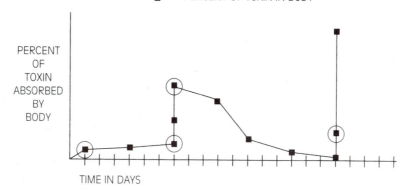

TIME IN DAYS

Figure 4-17 *Once a chemical is absorbed, it takes time for the body to metabolize it and eliminate the product.*

With some chemicals the first exposure may take several weeks to several months. If during that time frame, a second exposure occurs, the body's reaction is as if the initial dose was higher. It will take longer for the body to recover. A third dose during this elimination process can result in extremely toxic levels. For hypersensitive individuals the graph in Fig. 4-17 would escalate logarithmically.

Subchronic Exposure

Subchronic and subacute are two terms that at times are used interchangeably. Both have been used to describe the same type of exposure in printed literature. However, in order to be correct, the appropriate term is subchronic. This type of exposure involves an acute exposure that is repetitive. It is by definition a recurring event. In total it is an exposure that happened during approximately 10% of the organism's life span.

The testing procedures are based upon the LD_{50} or LC_{50}, which established death in 50% of the test group and are conducted on two species (two species as a minimum) with each having a control group. Three to four groups are tested simultaneously, as we saw in the acute testing procedure. At the top end, the dose that is given is under the LD or LC. The dose is high enough to show signs of injury, however, not death. At the low end of the dose range, the treatment is such that there should not be any noticeable effects. Depending on the test chemical, the middle point is chosen. If the curve in the acute testing was shallow, then the two or more middle points are picked. If the curve was steep, one point is selected.

The exposure is given for a period of 90 days. At the end of the time frame all the animals are autopsied for any histological evidence of effect. It is this effect

■ NOTE

A subchronic exposure is one that happened during approximately 10% of an organism's life span.

observed during autopsy that is compared to the control groups and histological evidence documented and reported in the literature.

Chronic Exposure

Chronic exposure is a long-term effect a chemical may have on an organism. Technically speaking it is the length of time that the animal was subjected to an exposure of a chemical, usually 80% of the total life span. Chronic effects are much harder to establish than the acute effects. Many factors come into play when discussing the chronic effects of chemicals. As we saw in the acute section and the animal sensitivity discussion, there are limits to the testing procedure. Our understanding of toxicological responses and the knowledge of biochemistry is limited. Chronic exposure is deeper than just an exposure that occurs for 80% of a life span. It also can influence some, any, or all the offspring. Accumulation, or acclamation, may lead to hyposensitivity, or hypersensitivity.

As in the acute testing process, a certain degree of inaccuracy is also seen in the chronic exposure figures. Most of what we see and read about is an after-the-fact chronic event. In other words, someone or a group of individuals have become ill or have died. What we have to continue to consider is that we are not going to see a cause and effect with all chemical exposures.

Nonetheless, it is the oral acute study that we start to investigate for the chronic toxicity of most chemicals. There is good reason for this particular starting point. The first studies were to isolate those chemicals that could potentially become a problem, if introduced into the food, for example preservatives and production additives. Basically, this method of exposure study is easy to do. Only in recent years has this chronic exposure study been expanded to include chronic inhalation and topical studies.

The experiment is similar to the subchronic (subacute) method of evaluation. The experiment starts out by finding a toxicity range of the chemical in question. Most of the information is gathered from the acute studies. The chemicals are introduced into the animal by placing them in the food. This process lasts for 90 days. During this time, a variety of dose ranges is used. This establishes the high, medium, and low toxic ranges. In the high toxic range the effect is limited. In the low no desirable effect is noted. In the moderate range the effect is mild. In addition to these three groups under study there is also a control group.

■ NOTE

Chronic exposure is deeper than just an exposure that occurs for 80% of a life span.

Typically many animals are used. There is also a two or more different species utilization. The oral ingestion is started from birth and continued to approximately 80% of the animal's life expectancy. Every day the animals are observed for negative effects and weekly a battery of tests is performed. At the end of the experiment all animals are autopsied and histological tests are analyzed. From the data, a statistical model is used to give the dose effect response.

As you see, this type of testing can become extremely expensive. Also, the length of time that is required to observe some effect are years in the making. If new concepts were introduced within the experiment during the testing period, the results would be skewed.

Since the Gulf War, approximately 45,000 American soldiers have manifested health problems, from environmentally induced asthma to recurring pneumonia to patterns of severe birth defects in their children. These health problems may have originated from the chemicals these soldiers were exposed to during their tour of duty. The following is a brief description of one chemical, soman (a military nerve gas) and its combined effects:

1. During the Gulf War crisis, the use of chemical warfare was a very real possibility. If one was paying attention during the news bulletins while the initial attack was taking place, you would have seen the deployment of antidotes to military as well as the civilian staff at the front lines. Antidotal kits were given to these individuals. Each kit contained two prepackaged syringes. One containing Atropine and the other 2-Pam (Pralidoxime), antidotes for nerve agents. In an attempt to reduce the effectiveness of the nerve agents, a pretreatment pyridostigmine bromide (PB) can be given. Alone PB has a mild effect, binding approximately 25–30% of the acetylcholinesterase. The theory is that with a small amount of the acetylcholinesterase bound, when the nerve agent attacks the body, the rest will be bound. Here the PB half life is respectively shorter than that of the nerve agent and will release the PB bound acetylcholinesterase. This, associated with the antidotal therapy, frees and reactivates acetylcholinesterase to be used by the body for normal nerve ending response. However, because PB is basically a Carbamate it can have an additive or potentative effect, exacerbating the effects of the nerve gas or organophosphate.

2. After the bombing of strategic Iraqi factories, gases and plumes of smoke were elevated over the terrain. These particles landed miles away in the form of soot. It is speculated that these factories were the storage areas for their chemical arsenal. At the time, several chemical alarms went off, indicating the presence of a nerve gas, however, in low level concentrations.

Margin of Safety

■ **NOTE**

The FDA has established a "margin of safety" by arbitrarily assigning a separation between hyposensitive and hypersensitive populations of 100-fold differences.

We have seen that there is a difference between the hyposentive and the hypersensitive populations. The FDA has established a "margin of safety" by arbitrarily assigning a separation between these two distinct groups. The convention establishes 100-fold difference. Two assumptions are made:

1. Human beings are more sensitive than the animal population.

2. The hyposensitive are even more sensitive than the normal human population.

Each assumption represents a degree of sensitivity of 10. Ten times ten gives the 100-fold margin of safety. As an example, if a chemical showed no adverse effect until 10 ppm, then the maximum concentration that would be considered acceptable for human exposure is .1.

Although this is an attempt to lower the level in a somewhat logical fashion, who is to say that 100 is the key number? In some circumstances 1000-fold may

TOXICOLOGICAL AND MEDICAL REFERENCE TACTICAL WORKSHEET

#1
Chemical Name:_____ Synonyms:_____

Wind Direction:____ Wind Speed:____ Temp.:____ Humidity:_____ Adjusted Temp.:_____ Dew Point:_____

#2
Chemical Name:_____ Synonyms:_____

Wind Direction:____ Wind Speed:____ Temp.:____ Humidity:_____ Adjusted Temp.:_____ Dew Point:_____

#3
Chemical Name:_____ Synonyms:_____

Wind Direction:____ Wind Speed:____ Temp.:____ Humidity:_____ Adjusted Temp.:_____ Dew Point:_____

#4
Chemical Name:_____ Synonyms:_____

Wind Direction:____ Wind Speed:____ Temp.:____ Humidity:_____ Adjusted Temp.:_____ Dew Point:_____

ASSESSMENT CRITERIA

#1	#2	#3	#4
BP: _____ IT: _____	BP: _____ IT: _____	BP: _____ IT: _____	BP: _____ IT: _____
FP: _____ LEL: _____	FP: _____ LEL: _____	FP: _____ LEL: _____	FP: _____ LEL: _____
VP: _____ UEL: _____	VP: _____ UEL: _____	VP: _____ UEL: _____	VP: _____ UEL: _____
SG: _____ Solub: _____	SG: _____ Solub: _____	SG: _____ Solub: _____	SG: _____ Solub: _____
IDHL/TLV: _____	IDHL/TLV: _____	IDHL/TLV: _____	IDHL/TLV: _____
PEL: _____	PEL: _____	PEL: _____	PEL: _____
STEL: _____	STEL: _____	STEL: _____	STEL: _____
Medical Monitoring:	Medical Monitoring:	Medical Monitoring:	Medical Monitoring:
Signs and Symptoms:	Signs and Symptoms:	Signs and Symptoms:	Signs and Symptoms:
Antidote:	Antidote:	Antidote:	Antidote:
Treatment Facility:	Treatment Facility:	Treatment Facility:	Treatment Facility:
Patient Load: _____	Patient Load: _____	Patient Load: _____	Patient Load: _____
Transported By:_____	Transported By:_____	Transported By:_____	Transported By:_____
Decontamination:_____	Decontamination:_____	Decontamination:_____	Decontamination:_____
Reference Source:	Reference Source:	Reference Source:	Reference Source:
Book:_____pp:__	Book:_____pp:__	Book:_____pp:__	Book:_____pp:__
Book:_____pp:__	Book:_____pp:__	Book:_____pp:__	Book:_____pp:__
Book:_____pp:__	Book:_____pp:__	Book:_____pp:__	Book:_____pp:__

Figure 4-18 *A worksheet can organize the magnitude of information required at the scene of a hazardous materials incident.*

be the appropriate margin of safety. Is more really better? Or could ten-, twenty-, or fiftyfold be the safe number? By just reducing these numbers we feel that we are protected. As you have seen, chemicals and their effects are not this simple. One chemical may in fact require a 1000-fold reduction in permissible exposure whereas another chemical may only need tenfold. The amount of information that is required at the hazardous materials incident can become overwhelming. In order to simplify our resource gathering capabilities, a tactical worksheet can provide the necessary organization of information while at the same time enabling us to identify all key issues associated with the chemical in question. We cannot always remember all the important facts to gather, however, by having a worksheet available, all important scene assessment criteria can be referenced and the facts gathered (Fig. 4-18).

Table 4-1 *Chemical and physical properties of selected poisons.*

Product	Hazard Class	Color	Odor	Physical State	Vapor Density	Specific Gravity	Flash Point
Acetaldehyde	Flammable liquid	Colorless	Fruity	Liquid below 69°F	1.50	0.80	−38°F
Acrolein	Poison Flam. liquid	Colorless	Disagreeable	Liquid	1.90	0.80	−15°F
Aniline	Poison	Colorless	Aromatic, amine like	Liquid	3.20	1.0+	158°F
Arsine	Poison gas Flam. gas	Colorless	Garlic	Gas	2.70	1.69	
Benzidine	Poison	White or slightly reddish		Powder crystal	6.36	1.25	
Benzotrifluoride	Flammable liquid	White	Aromatic	Liquid	5.00	1.20	54°F
Bromine	Corrosive Poison	Red-brown gas; liquid is yellow	Choking, irritating	Liquid	5.50	3.12	
Bromoacetone	Poison Flam. liquid	Purple in water		Liquid		1.63	
Chlorine trifluoride	Poison gas Oxidizer	Greenish-yellow. almost colorless	Sweet, irritating	Gas	3.14	1.88	
Chloropicrin	Poison	Colorless	Intense, penetrating	Liquid	5.70	1.65	
Cyanogen	Poison gas Flam. gas	Colorless	Pungent, penetrating, bitter almonds	Gas	1.80	.95	
Dimethyl sulfate	Poison Corrosive	Colorless	Faint onionlike	Liquid	4.40	1.30	182°F
Ethylene dichloride	Poison Flam. liquid	Clear	Chloroform	Liquid	3.40	1.30	56°F
Fluorine	Poison gas Oxidizer	Yellow	Pungent	Gas	1.30	1.50	
Hydrogen cyanide	Poison Flam. liquid	Colorless	Bitter almonds	Liquid below 79°F	0.90	0.70	0°F

Ignition Temp.	Flammable Range	Threshold Limit Value	Immediately Dangerous to Life & Health	Median Lethal Conc. or Dose	Life Hazard
347°F		200ppm	10,000 ppm	20,000 ppm	Eye, skin, and lung irritant, narcotic effect from inhalation
428°F	2.8–31	0.5ppm			Small amounts are highly poisonous
1139°F	1.3–11	2ppm			Skin absorption or inhalation causes anoxia due to the formation of methemoglobin
extreme flammability		.05ppm (TWA)	6 ppm		Immediately dangerous to life and health
won't readily burn					Human carcinogen, exposure not permitted
					Moderately toxic in high concentrations
		0.1ppm	10 ppm		Both liquid and vapor cause severe burns, highly toxic
					Extremely toxic, warfare agent
		0.1ppm	20 ppm		Extremely toxic and corrosive
		0.1ppm	4 ppm		Very toxic. Short exposures may cause fatal lung damage
	6.6–32	10ppm	50 mg/m³	16 ppm	Highly toxic when heated or when in contact with acids, water, or steam
370°F		0.1ppm	10 ppm		Toxic by inhalation, skin contact, or ingestion. Extremely irritating
775°F	6.2–16	100ppm			Toxic by inhalation, skin contact, or ingestion
		0.1ppm		60 ppm	Highly toxic gas causes severe burns to eyes, skin, and respiratory tract
1000°F	5.6–40	10ppm			A few breaths can cause death. Can be absorbed through the skin.

Table 4-1 *(Continued)*

Product	Hazard Class	Color	Odor	Physical State	Vapor Density	Specific Gravity	Flash Point
Hydrogen fluoride (Hydrofluoric acid)	Corrosive Poison	Colorless		Liquid below 67°F	0.70	1.00	
Hydrogen sulfide	Poison gas Flam. gas	Colorless	Rotten eggs	Gas	1.20	1.54	
Methyl bromide	Poison gas Flam. gas	Colorless	Chloroform-like vapor	Liquid or gas	3.30	1.70	
Methyl isocyanate	Poison Flam. liquid	Colorless	Sharp, tear causing odor	Liquid			19°F
Nitrogen peroxide	Poison gas Oxidizer	Colorless solid; yellow liquid; brown	Irritating	Liquid below 70°F; solid below 15°F	1.58	1.45	
Parathion	Poison	Yellow to dark	Garlic	Liquid		1.27	
Phenol	Poison	Colorless to white		Crystals	3.20	1.10	175°F
Phosgene	Poison gas Corrosive	Colorless	Sweet, hay-like	Gas	3.40	1.40	
Silver nitrate	Oxidizer	Colorless crystals	Odorless	Crystals melt at 414°F		4.33	

Ignition Temp.	Flammable Range	Threshold Limit Value	Immediately Dangerous to Life & Health	Median Lethal Conc. or Dose	Life Hazard
		3ppm	20 ppm		Both liquid and gas states are irritating to eyes, skin, and respiratory tract
500°F	4.0–44	20ppm	300 ppm		Eye, skin, and respiratory tract irritant
990°F	13.5– 14.5	20ppm	200 ppm		Toxic by inhalation, ingestion, or repeated exposure. Can cause severe burns.
994°F	5.3–26	0.02 ppm	20 ppm		Highly toxic, intense irritant to eyes, skin, and mucous membranes
may ignite other material		3ppm	50 ppm		Very toxic. Eye, skin, and respiratory tract irritant
		0.11mg/m^3	20mg/m^3		Very toxic and can be fatal by inhalation, skin contact, or ingestion
1319°F	1.8–8.6	5ppm		.	Severe tissue burns. Skin absorption or inhalation can cause death
		0.3ppm	201 ppm		Death or delayed lung injury can result from inhalation
		0.01mg/m^3			Absorbed through respiratory and GI tract. Silver accumulates in elastic tissue and nervous system

Summary

Most of our knowledge on toxicity stems from the insight gained in the acute toxicity studies. Even though there are problems built into these studies, they enable us to understand the concept of healthy thresholds. Most of the limit values are done on unprotected organisms. Most of the time our entry within these environments are in some level of protection. If the protection remains intact and an effective decontamination is done, the exposure has been managed. However, if the incident produces victims that are unprotected, is this equal to the testing population? It is essential that appropriate levels of decontamination take place prior to patient treatment such that medical response personnel are not affected by the toxic qualities of the chemical. It is also equally essential to manage our patients realistically. Some of the chemicals that we deal with are extremely toxic. In these cases we are looking at body recovery and not rescue as there is no reason to subject ourselves to that type of environment based on what we know and what we can assume from the chemistry and toxicology we have discussed.

Both acute and chronic exposures should be limited for the emergency crews. The understanding of biochemical reactions is still in its infancy. We are not sure about the single hyposensitive or hypersensitive event. The point where the damage begins and the observable effect occurs is the true basic problem with toxicity levels. The chronic exposure must occur for 80% of the organism's life span, but how do we know that the first exposure may not produce a negative effect?

For example, present-day liver tests are a current test for the surveillance of known diseases that are produced from chemicals. Liver tests of years past were not as accurate as the liver tests that are performed today. The tests of today are probably not as accurate as the tests of the next decade will be. Therefore, liver problems that are detected today may have already existed, but because of better testing can now be detected.

Acute and chronic effects can be better controlled. This control can only be accomplished by reducing the incidence of exposure. Recognition of the health effects of chemicals combined with the appropriate management of an incident both will limit exposures.

■ **NOTE**

Both acute and chronic exposures should be limited for emergency crews

Scenario

You have been called by a neighboring hazardous materials team that is on the scene of several spilled chemicals. Because of reduced manpower, they are overwhelmed in reference sector. They have called you to provide the necessary medical research of the hazardous materials scene.

As you are recording the information, you recognize a few of the chemicals. One in particular has a very shallow toxicological curve, which occurs over large animal populations. You further assess that most of the chemicals are metabolized during phase one into highly polar and soluble products.

During the physical description of the incident, your cellular phone connection becomes full of static, and your connection is lost. Your only piece of solid information is that the incident is occurring in a populated area of town. You know the members of the other team and feel confident that they will call or utilize another method of communication shortly. In an attempt to ready yourself for the research task, you start by writing down a set of questions. With the addition of information as it becomes available you will base your research on these questions.

Scenario Questions

1. What form of matter is the material in?
   A. How will temperature affect the chemical and physical properties?
   B. What are the potential routes of exposure?
   C. What are the potential target organs that could be involved?
2. If several chemicals are involved, how could it affect the potential exposure?
3. You remember that one of the hazmat team members has had a previous exposure. How might this affect the operation if this individual becomes exposed?

4. List the terminology that you will look for once the exact chemicals are known.
5. What general chemical classification could these phase one chemicals come from?
6. Discuss the differences between the acute, subchronic, and chronic testing procedures.
7. How does the curve of the toxicological data affect the decision-making process?
8. By utilizing the toxicological worksheet provided at the end of this chapter, what are the considerations?

Section

3

HAZARDOUS MATERIALS PHYSIOLOGY AND TREATMENT

Care providers on an emergency incident have a duty to the citizens they serve. This commitment makes them responsible for having the knowledge necessary to safely and efficiently control emergency incidents. In hazmat response, that control may be as simple as plugging a minor gas leak or as complex as treating a toxic patient. However, in the realm of emergency care the medical responder must have the tools to adequately treat patients. In the hazmat field, this is not as simple as it sounds.

It was once said that a monkey could be taught to give a drug when a certain symptom is noticed. Even in the early days of paramedicine, emergency responders were instructed to give the drugs contained in the "blue box" or the "yellow box." The profession has grown by leaps and bounds since that time. Paramedics' knowledge of physiology is vast and growing by the day. Schools require more each year just to earn that sought after piece of paper that states "Paramedic" or "EMT." Chapters 5 and 6 give the medical provider the knowledge to recognize symptoms and then, with an understanding of the events taking place, administer appropriate care. This subject is not easy. These chapters are full of information that a medical provider can use, having met with a toxic patient. The satisfaction of knowing that you possess the knowledge to adequately assess and treat a complex patient whose injuries are not typical is gratifying.

But the emergency responder's responsibility doesn't stop at understanding and knowledge; he or she must be able to apply this knowledge in a reasonable and efficient manner. The use of antidotal treatments are but one facet to the entire response. Emergency medical personnel must be on the lookout for the "other" problems that are subtle at the hazardous materials incident. For example, the common house fire is a situation that has the same elements of a hazmat response. The controls that have been discussed can apply to each incident type. The assumption is that an emergency provider, given a new set of tools, can apply them during the common everyday incident.

Issues such as fire toxicity, heat stress, antipersonnel weapons, and terrorist activity all, unfortunately, have a place within the emergency response. We, as a profession, must become aware of the issues that confront us and understand the duties that we must perform in the field. During the performance of duties we must understand our limitations, while delivering above average health care under adverse conditions. Today's communities expect the emergency service to be aware and resourceful during emergency response. We must apply this knowledge and plan for the technical incidents of the future.

Recent history has shown the devastating effects that chemical and technological incidents can have on a community. No one will be truly 100% prepared, however, by educating ourselves within this discipline, understanding the limitations of our field work, and training the medical facilities to engage in hazardous materials operations, the simple calls will come naturally. If and when the larger incident occurs, then maybe we will have enough systems educated so that we all can assist in the mitigation.

Chapter 5

Body Systems and the Environment

Objectives

Given a hazardous materials incident you should be able to recognize and identify the routes of entry of a chemical and predict the possible injury incurred by the patient based on the properties of the material. You should also possess an in-depth knowledge of the pathophysiology involved with chemical exposures to the skin, respiratory system, cardiovascular system, and eyes as well as the effects of heat, cold, or smoke exposure on the body systems.

As a hazardous materials medical technician, you should be able to:

■ Recognize potential hazards to medical providers, transporters, and hospital employees and take the necessary action to lessen these hazards.

■ Understand the routes of entry a chemical can take to gain access into the body.

    Inhalation   Ingestion

    Absorption by skin and eyes   Injection

■ Describe the anatomy and physiology of the skin and how injury to the skin takes place.

■ Describe the anatomy and physiology of the eye and how injury to the eye takes place and how to rate the severity of that injury.

- Demonstrate the use of specialized eye medications and equipment.

     Alcaine, ponticaine, etc.

     Use of nasal cannula for irrigation.

     Use of Morgan Lens.
- Describe the anatomy and physiology of the respiratory system and how an injury to that system takes place.
- Identify specialized respiratory equipment that may be useful during a respiratory system chemical injury.

     BVM with PEEP valve.

     Pulse oximetry, uses and predictable effects from chemical exposures.

     Field ventilators.
- Describe the physiologic effects suffered by the cardiovascular system during chemical exposures.
- List the common signs and symptoms associated with and the BLS and ALS treatment for the following:

     Corrosives

     Irritants, both respiratory and topical.

     Asphyxiants, both simple and chemical.
- Identify the signs and symptoms related to heat exposure as it pertains to encapsulated entry team members.
- Identify the physiology associated with an exposure to a hot, high humidity environment.
- Identify the physiology associated with acute exposure to cryogenic materials and the treatment for this injury.
- List the common chemicals produced during the combustion process and identify what fuels produce them.

Body systems and the environment were grouped together in this chapter so the student could identify some of the most common means of injury to both the victims and responders of hazardous materials accidents. By no means have all of the possibilities been reviewed here, although the vast majority of injuries encountered during these incidents are discussed. The chapter follows a logical format of identifying body systems as they relate to acute exposures then moves on to heat and cold related injuries, and injuries related to inhalation of products of combustion. The physiology associated with each injury is presented followed by a discription and rationale for each treatment.

ROUTES OF ENTRY AFTER CONTAMINATION/EXPOSURE

When exposure or contamination takes place, a chemical gains an opportunity to injure the body. This effort to injure the body takes place through one or more different means, which are referred to as *routes of entry* into the body. Toxicology books list many routes of entry including intravenous, inhalation, intraperitoneal, intramuscular injection, subcutaneous injection, oral, and cutaneous exposure. This book only uses examples of routes commonly found in hazmat emergencies. Each section deals with the anatomy pertaining to the exposure and explains in detail the physiology involved in that exposure.

The general routes of entry into the body experienced by those exposed to hazardous materials (chemical exposure) are inhalation, absorption (including skin and eyes), ingestion, and, to a much lesser degree, injection. Biohazard exposure through injection accounts for more reported injuries than any other route. Some would state that regardless of how the body contacts a chemical all routes into the body involve absorption. In fact some authorities list routes of entry as "routes of absorption." In the basic sense of the phrase, this statement is true. But, for reasons of simplicity absorption is addressed as an exposure to the skin and eyes.

Once a toxin is exposed to tissue it may cause damage to that tissue. Other times the toxin may utilize that tissue as a means of gaining access into the body where it seeks out an organ system completely dissociated from the point of entry. The organ system affected by an exposure is then referred to as the target organ. Those chemicals that specifically target the nervous system are called **neurotoxins**. For example, organophosphates target their effects on the **parasympathetic nervous system**. **Solvents**, another neurotoxin, target the central nervous system. Almost all organ systems can be affected by some type of toxin. The most common target organs are the nervous system, the cardiovascular system, and the liver and kidneys.

The effects from exposure may be felt almost immediately, causing acute signs and symptoms. Probably more often the effects are noticed days, weeks, or years later, lasting for long periods of time. These chronic symptoms appear because of damage caused to an organ system by the invading toxin. Some types of cancer can be related to toxic exposures many years earlier. Asbestosis, a form of chronic lung disease, is linked to the inhalation of asbestos, often 10 to 20 years earlier.

Inhalation

Inhalation exposure is the taking in through the respiratory system of a chemical that causes harm to the body. Some chemicals will cause considerable damage to the respiratory system during an inhalation exposure; others use the respiratory system as an opportunity to gain access into the body to damage other organ systems. Examples of chemicals that harm the respiratory system are strong irritants such as chlorine or ammonia. These two commonly used chemicals inflict their damage to the

neurotoxins
toxins that affect the nervous system as the target organ

parasympathetic nervous system
the side of the autonomic nervous system that is mediated by acetylcholine release

solvent
the liquid in which a substance is dissolved

inhalation exposure
the taking in through the respiratory system of a chemical that causes harm to the body

tissue, making up the surface of the external respiratory system (that part of the respiratory system open to the outside environment, above the alveoli).

Other chemicals use the respiratory system as a means of transportation, to gain access into the body. Once the chemical is within the bloodstream or tissues, it seeks out and harms other organ systems. Examples of this type of inhalation exposure is organic nitrogen compounds or halogenated hydrocarbons. An organic nitrogen compound, once inhaled and absorbed into the bloodstream, causes profuse **vasodilatation** and shock. Furthermore, it chemically changes hemoglobin, interfering with its normal function. Its target organ therefore, is the cardiovascular system. Halogenated hydrocarbons may not cause injury to the respiratory system or the circulatory system but it targets the central nervous system (CNS), evidenced by muscle fasisculations and seizures. Neither of these chemicals causes significant damage to the lungs, which was the organ of entry.

Still other chemicals will cause injury to both the respiratory system and other target organs during an inhalation exposure. Hydrogen sulfide gas, a respiratory irritant that causes significant damage to the airways of the lungs, causes this type of twofold injury. Once it passes through the alveolar membrane and into the bloodstream, the injury continues. At a cellular level it binds with an enzyme necessary for utilization of oxygen, resulting in **cellular hypoxia**.

Inhalation exposure is one of the most common types of exposures to hazardous materials. The fact that so many hazardous chemicals have no odor, color, or irritating qualities, adds to the danger. Probably the most common inhalation injury seen by a medical responder is the inhalation of products associated with smoke. Fortunately, smoke inhalation by professional firefighters has been greatly reduced because of mandatory department regulations enforcing the use of self-contained breathing apparatus.

Absorption, Skin and Eyes

Absorption of chemicals through the skin and eyes presents the most common and most urgent emergencies in the field. The properties of a chemical will determine how rapidly it is absorbed. Lipid soluble chemicals seem to have a more rapid absorption quality. Moist skin is ten times more permeable by chemicals than dry. Many other factors also affect the absorption rate of the chemical. The temperature of a chemical, the surrounding air, and the body all affect the absorption rate. The condition of the tissue in direct contact with the chemical also determines the absorption rate. Injury, inflammation, and sensitivity increases the absorption quality of a chemical.

Much like an inhalation injury, the surface absorbing the chemical may or may not be injured by the chemical itself. The skin and eyes have the ability to act as a transport medium for chemicals that affect other organ systems. Nitroglycerin paste is a pharmaceutical most medical persons are familiar with. Once the paste is placed on the skin there is no effect at the site. The effects caused by the chemical moving through the skin and transported throughout the body is a wide-

---

vasodilatation
the dilating of the blood vessels

cellular hypoxia
a lack of oxygen within the cell

absorption
the process that will take up and hold a gas, liquid, or dissolved substances on the surface of another substance

spread change in the systemic vascular tone. On the other hand, if hydrochloric acid were placed against the skin, the results would be a chemical burn at the site of exposure, rather than a systemic injury.

Some chemicals have the ability to cause a twofold effect, one effect found at the site of contact and a different effect somewhere in a disassociated body system. This type of injury is the most complex to diagnose and treat. Hydrofluoric acid is an excellent example of a chemical that evokes this type of injury. Once against the skin, hydrofluoric acid causes a devastating chemical burn. The fluoride within the acid then penetrates below the surface, where it binds with calcium and magnesium. This results in injury to the nerves and bones, and has the potential of causing hypocalcemia within the circulating bloodstream, resulting in EKG abnormalities and possibly cardiac arrest.

Ingestion

ingestion
the incorporation of a material into the gastrointestinal tract

Ingestion injuries make up only a small percentage of hazmat contamination emergencies. Interestingly, ingestion makes up the majority of poisonings throughout the United States. For an industrial worker or hazmat responder to suffer from an ingestion injury, the chemical must enter the mouth. In the industrial setting this takes place most often when a worker inadvertently contaminates a food or drink with a chemical. Ingestion poisonings for the hazmat responder take place when proper techniques are not taken during decontamination or rehabilitation. Probably less common but still worth mentioning is the ingestion of a chemical, postinhalation exposure. Many times significant amounts of a chemical can be trapped by mucous in the respiratory system only to be coughed up into the pharynx where it is swallowed and absorbed in the gastrointestinal system.

Once the chemical is ingested, absorption within the gastrointestinal system is dependent on many factors. Chemicals that are water soluble (hydrophilic) are not absorbed as rapidly as those that are fat soluble (lipophilic). Others are altered by the chemical makeup of the contents of the stomach. The stomach fluids are made up of strong acidic gastric juices, enzymes, and deeper in the gastrointestinal tract, bacteria. Any of these substances can alter the makeup of the ingested chemical and cause a wide variety of effects. In many cases it is impossible to predict which symptoms to expect. When ingestion of a chemical is suspected, consult your local poison control center for advice on dilution, adsorption, stimulating emesis, or actuating a rapid movement through the bowel. Always monitor these patients for systemic symptoms commonly found with ingestion injuries.

Injection

injection
the forced introduction of a substance into underlying body tissue

Although **injection** is not a common route for accidental hazardous materials exposure, it remains a possibility on an emergency scene. Injection may take place in two ways, both possible on varying hazmat scenes.

The first way involves injection of a chemical by a pointed or sharp item that is contaminated. When the sharp item (wood, metal, etc.) punctures the skin and enters the subcutaneous tissue, the chemical is left behind causing an injury to the interior of the body. Although the amount of chemical would be minute, absorption is guaranteed, possibly leading to a chronic site-specific injury.

The second way becomes possible when a chemical (solid, liquid, or gas) is released under pressure. The pressurized chemical, moving at a high rate of speed, is propelled through the epidermis (injected) and into underlying subcutaneous tissue. Either of these mechanisms can inject toxic chemicals causing a chemical exposure and injury to underlying tissues and structures.

Emergency medical responders often report injection injuries from contaminated needles. These injuries alone number more exposures than any other route of entry but are confined to needle sticks or other types of medical sharps.

SKIN EXPOSURE AND INJURIES

Overview

■ NOTE

Because the skin is the first line of defense against chemical exposure, dermal exposures make up the majority of work-related injuries identified with hazardous materials.

Since the skin offers the first line of defense against chemical exposure, it stands to reason that dermal exposures make up the majority of work-related injuries identified with hazardous materials. These injuries occur in the workplace, but a large percentage of exposures to chemicals also take place in the home. Take a look at many of the chemicals used in the home for cleaning. Oven cleaners contain sodium hydroxide, a strong alkali; paint removers and floor stripers contain hydrofluoric acid; disinfectants may contain high percentages of phenol and phenol-related compounds. The list goes on and on. It is important for the medical responder to be familiar with the effects common to these types of chemicals.

Anatomy and Physiology

The skin is one of the largest organs of the body. Its surface area on an average adult is approximately 3,000 square inches (roughly the playing surface of a pool table). The thickness of the skin varies from about 0.02 to 0.12 inches thick and makes up about 10% of the total body weight. The skin is generally thicker on the dorsal (back) side and thinner on the ventral (front) side. It is also thicker on the palms of the hands and the plantar surface of the foot. It is thinner on the eye lids and scrotum. These thinner areas allow chemicals to gain easy access to underlying tissues.

Function   Skin provides the external covering for the body. It protects the underlying structures from invasion of parasites and bacteria, assists in temperature regulation, limits dehydration, excretes salts and other organic materials, is a reservoir for food, senses cutaneous stimulation, is a source for vitamin D, and protects the body from external injuries. It also provides surprisingly good protection from the invasion of chemicals.

Structure   It is important for us to examine the structure of the skin for many reasons. By understanding the structure one can sensibly estimate the amount of injury and provide appropriate treatment for a victim exposed to a hazardous chemical. Understanding the structure of the skin will also guide us in decontamination efforts.

The skin consists of two principle parts, each made up of sublayers (Fig. 5-1). The epidermis makes up the top layer, and in comparison, is the thinner layer of the skin. It is cemented to the thicker part called the dermis. The epidermis varies in thickness depending on the protection or durability needed for the differing areas.

The epidermis is generally made up of four layers of tissue. Where friction is great, such as on the palms of the hands and the soles of the feet, the layers of tissue are five deep. The top layer of the epidermis is the stratum corneum, or horny layer. It provides the first line of defense against chemical exposures because of the chemical resilience offered by it. The stratum corneum is made up of 25 to 30 rows of flat dead cells. The dead cells of this layer are the remains of cells from lower levels of the epidermis that rise to the surface as they die. These cells are brushed, scraped, or flake off on a continuous basis. Since they are

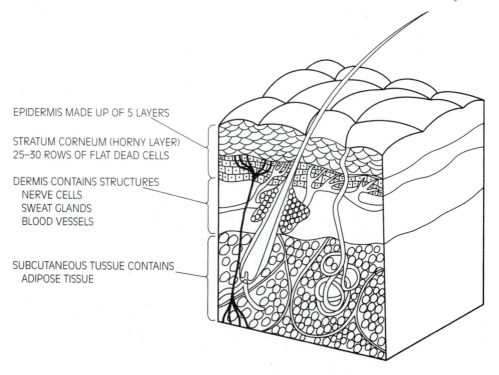

EPIDERMIS MADE UP OF 5 LAYERS

STRATUM CORNEUM (HORNY LAYER)
25–30 ROWS OF FLAT DEAD CELLS

DERMIS CONTAINS STRUCTURES
   NERVE CELLS
   SWEAT GLANDS
   BLOOD VESSELS

SUBCUTANEOUS TUSSUE CONTAINS
   ADIPOSE TISSUE

Figure 5-1   *The skin has two major layers, the dermis and the epidermis. Overlying the epidermis is a layer of flat dead cells called the horny layer.*

already dead, these cells are not injured by direct contact with a chemical. Because this layer is so resilient to chemical invasion, it is important that rigorous brushing of the skin not take place during decontamination. Brisk rubbing of the skin disrupts this layer and allows chemicals an opportunity to easily invade the underlying living cells.

The other three layers, stratum granulosum, stratum spinosum, and the stratum basale, are all living cells. Where friction is expected to be the greatest, a fifth layer, called the stratum lucidum, exists. This layer is like the stratum corneum in the fact that it is made of flat dead cells that are expected to take the brunt of external injury or chemical exposure.

● CAUTION

Facials and chemical peels increase skin permeability, thus creating an opportunistic means for chemical invasion.

Men and women having facials performed, including chemical peels, are exposed to an intensive cleansing of the face to bring out brighter, smoother, younger looking skin. This cleansing is nothing more than a controlled chemical burn, many times done with phenol (carbolic acid), to remove the rough dead skin associated with the stratum corneum. What is left of the stratum corneum is then moisturized with a lipid base moisturizer, essentially increasing permeability greater than ten times. In these cases the initial protective feature of the skin is removed, effectively creating an opportunistic means for chemical invasion through the surface.

The second principal section of the skin is the dermis. The dermis, which covers a layer of subcutaneous tissue, is anchored in place with fibers from the dermis extending into the subcutaneous layer. It is thicker than the epidermis and is the thickest on the palms and soles. It is thinnest on the eyelids, scrotum, and penis. It is important to recognize these thin areas as they are easily missed during decontamination efforts. Embedded in the dermal layer are blood vessels, nerves and nerve endings, glands, and hair follicles. This layer is made up of connective tissue containing collagenous and elastic fibers that allow the skin to stretch without tearing. Adipose tissue (fat cells) are found dispersed throughout the dermis.

Injuries

The skin can be injured by a variety of means such as exposure to hot or cold, and by mechanical injuries such as abrasions, lacerations, and contusions. These typical skin injuries disrupt the protective features of the skin, allowing a chemical easy access for absorption. Traumatic injuries themselves are not addressed here, but injuries resulting from chemical exposures are discussed, followed by the appropriate treatments for these specific injuries.

Irritation   Irritation of the skin is caused by an exposure to chemicals that are not extremely toxic to the skin. Typically these include diluted forms of acids, alkalis, and solvents. An exposure of this type causes an inflammation of the skin in the area of contact. If the area is decontaminated within a reasonable amount of time, the injury is limited to localized redness, minor swelling, and occasionally blistering. The symptoms may develop immediately or be delayed, and can last from hours to days. In most cases the injury is self-limiting and heals without adverse long-term effects or scarring.

An example of this type of injury experienced by many people while operating a pump at the gas station is the splattering of gasoline onto the back side of a hand. Gasoline, being a strong solvent, causes dehydration of the skin cells, stimulating an injury that is usually temporary. Within a few minutes, a burning or stinging sensation develops and continues until the gasoline is washed from the skin. In some cases the burning is the only symptom; other times, contact dermatitis appears in the form of reddened, raised spots. In either case the injury subsides once the product is washed or wiped from the skin.

Burns   Burns are caused from chemicals found in a higher concentration. Most commonly acids and alkalis are the assaulters. They have significantly different physiologic effects on the skin and underlying tissues. Because acids are the most produced chemical in the United States, it only stands to reason that injuries due to acid exposures make up the majority of hazardous material injuries.

• Acid burns are one of the most common chemical injuries to the skin. Acid burns can be devastating but most are self-limiting due to the effect acids have on the proteins within the skin. Proteins exposed to acids are precipitated, forming a thick impenetrable layer that protects underlying structures. The coagulum formed within the skin as a result of an acid exposure is very similar to the effect of dropping a raw egg on a hot grill. A hot grill turns the clear protein of a raw egg white into a white and rubbery coagulum. This same type of coagulum forms in the skin as a result of an acid exposure. This coagulum is tough and, for the most part, impenetrable by the chemical intruder. Stronger acids, such as those with a pH of less than 2, do have the ability to penetrate further and cause extensive soft tissue damage. These acids are an exception to the rule and finding a patient exposed to them would be a rare occurrence.

Treatment of acid injuries involves removing the chemical from the surface of the body. Most of the time, water is the physiologic decontaminating substance, although certain acids require special decontaminating agents. Phenol (carbolic acid) is only partially soluble in water. It has been found that decontamination only with water will not completely remove the chemical. Utilizing another agent that is fat soluble, such as alcohol, or another non-water-soluble agent, such as olive oil, mineral oil, or vegetable oil works well as a final decontaminating agent. Care must be taken in selecting these specialized decontamination agents. Some chemicals when mixed with a fat soluble agent may increase the absorption rate through the skin.

• Alkali burns cause a change in the fatty substances in the skin. An alkali chemical causes a liquefaction of fatty substances, allowing the chemical to gain access deep into the tissue. The liquefaction process caused by alkalis is called liquefaction necrosis. The chemical term used to describe this process is *saponification* (the conversion of a fat by a strong alkali into soap).

This process was used when making lye soap. Lye (sodium hydroxide) was added to animal fat forming a slick, wet paste. The paste was then pressed into bars and left to dry.

The skin injury caused by alkalis is devastating. An alkali burns deep into the tissue and continues until the body neutralizes it. Rapid decontamination is useful and may reduce the amount of damage to the tissues by removing any alkali still found on the surface, although the alkali that had already gained access below the surface will continue to burn until it is neutralized by the body tissues. Because of this, strong alkalis have the ability to penetrate deep underlying structures, causing devastating wounds.

The injury site closely resembles what was described as lye soap. The surface of the skin appears wet and is slick or soapy to the touch. These wounds usually linger for a long period of time and heal slowly. Occasionally these wounds need multiple grafts to properly close and sometimes leave disfiguring scars similar to deep thermal burns.

An excellent way to demonstrate the type of injury expected from acids and alkalis is to place the egg white of two eggs into two different bowls (clear glass). By applying a few drops of muratic acid (approx. 30% HCl) on the egg white, an encapsulation of the acid is produced by the coagulation necrosis. This is similar to the reaction experienced by the skin because the egg white is made up of proteins and fats, similar to skin cells. In the second bowl apply a few drops of sodium hydroxide (drain cleaner). The liquification process caused by the alkali does not make an immediate change like the acid did, but within a few minutes a milky slurry is present. Within 30 minutes the acid bowl will still contain an encapsulated drop of acid but the alkali bowl will show difuse liquification necrosis.

Other chemicals can cause injuries to the skin but these chemicals account for only a small percentage of the overall injuries related to exposure. Phosphorus and alkali metals are two such chemicals, and if the medical responder is unfamiliar with their properties, the injury can become much worse.

• Phosphorus is a nonmetallic element that has a variety of uses in industry. It is produced in white (yellow), red, and black forms. There is presently no industrial use for black phosphorus so you would not expect to find it in industry. Phosphorus is also used in production of methamphetamine (crack) in clandestine drug labs. These drug labs may provide a source for this and other toxic injury types.

• White phosphorus, also referred to as yellow phosphorus, is a crystalline, waxlike, transparent solid. It may be used in rodenticides, analytical chemistry, and munitions. White phosphorus is dangerous when in contact with the skin because it self ignites at temperatures greater than 86°F (30°C). If the skin is contaminated, the excess chemical should be brushed away followed by submersion in cool water to prevent ignition.

During World War II, munitions specialists handled white phosphorus during loading procedures. White phosphorus was called "willie p" by the specialists who handled it often. These specialists suffered from the chronic effects of this chemical, which included a breakdown of the bone material in the jaw. This condition was termed *phossey jaw*.

• Red phosphorus is not as serious a danger to the skin but will spontaneously ignite if exposed to oxidizing material. It reacts with oxygen and water vapor to form poisonous phosphine gas. Red phosphorus is used to produce semiconductors, incendiaries, pyrotechnics, and safety matches. Under certain conditions, heating red phosphorus may form white phosphorus, which can self ignite at temperatures greater than 86°F. More information about this process is discussed in Chapter 10.

• Alkali metals are metals that, in the pure form, will react violently with water. The elements sodium, lithium, magnesium, and calcium are the most commonly found alkali metals. Once in contact with the skin they must not be washed away with water. The reaction with water forms hydrogen gas, heat, corrosive material, and eventually ignition. Excess chemical should be brushed away, then the affected area covered with cooking oil to prevent a reaction with air. The metal itself is a strong, caustic irritant to tissue and when exposed to water forms a strong alkali (hence the name alkali metal).

It is used for the production of alkali liquids and solids, in nonglare highway lighting, and as a heat transfer agent for solar-powered electric generators.

EYE EXPOSURE AND INJURIES

The eye is a very complex, very delicate structure that allows a person to process visual stimuli into thought projected images. Of surface chemical injuries, eye exposures pose the greatest emergency. Rapid, appropriate action by the rescuer can make the difference between the loss of sight or complete recovery of a victim injured by a chemical exposure. This section reviews how different chemicals affect the structure of the eye, and identifies means for emergency treatment of those exposures.

Anatomy

The eyeball or globe is about 1 inch in diameter. Of the total surface area only the anterior one-sixth of the globe is exposed to the outside environment. The remainder is well protected within the socket or orbit into which it fits. When exploring toxic exposures only the anterior one-sixth of the globe and its underlying structures are important and are discussed in this section.

The ocular surface is made up of several layers consisting of the conjunctiva or *epitheleal tissue, cornea, sclera*, and the *tear film* (Fig. 5-2). This ocular surface provides the first line of defense against a chemical attacker. Each level that

OCULAR SURFACE LAYERS

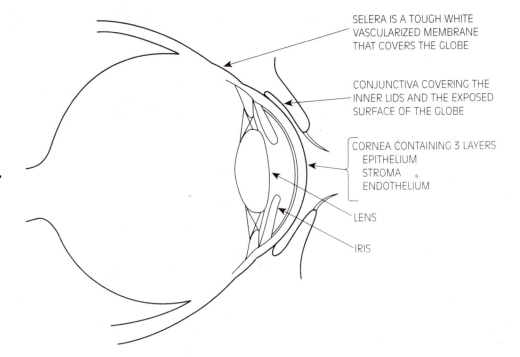

SELERA IS A TOUGH WHITE VASCULARIZED MEMBRANE THAT COVERS THE GLOBE

CONJUNCTIVA COVERING THE INNER LIDS AND THE EXPOSED SURFACE OF THE GLOBE

CORNEA CONTAINING 3 LAYERS
EPITHELIUM
STROMA
ENDOTHELIUM

LENS

IRIS

Figure 5-2 *An eye represents a great emergency when exposed. Its structure provides an excellent means for absorption into the body or it can, itself, be permanently injured.*

is affected by the chemical represents a differing degree of injury. The means of grading the injuries are identified in the *Assessment of Eye Injuries* portion of this chapter.

The tear film is a clear fluid layer that covers the open surface of the globe. Its purpose is to provide lubrication for the lid to slide easily over the globe. It also traps any invasive particles so that the lid can sweep them into the lower canthus for movement out of the eye. It also allows the surface tissues to stay moist and transparent to allow light movement into the inner structures of the eye.

Tear film is made up of *mucin* covered with a lipid layer. Mucin is excreted by the goblet cells located in the epithelium (surface tissue layer) of the conjunctiva. Combined with the saline solution excreted by the lacrimal glands, mucin decreases the surface tension of the aqueous tear, allowing the tear film to coat the globe completely. The tear film is covered with a lipid layer to decrease evaporation between blinks.

The outer layer of the eye and surrounding tissue consists of the *conjunctiva*, which covers the interior segments of the upper and lower lids, the sclera, and provides an epitheleal layer over the clear cornea. The corneal covering is three clear layers of tissue, allowing light to pass through it.

The top layer of epitheleal tissue is composed of lipoproteins that readily allow the passage of lipid soluble chemicals. Once the epithelium is injured, chemicals gain easy access to the second layer, known as the *stroma*. The stroma makes up 90% of the corneal thickness and is permeable to water-soluble chemicals. It is also the tough membrane that provides rigidity and puncture resistance. Once a chemical injury has reached the stroma, scarring, opacification, and irregularities result.

The endothelium makes up the third level and is the basement membrane of the cornea. It consists of a single layer of cells that lack the ability to regenerate in adults. If the endothelium is affected by chemicals the result is usually a loss of sight or a total loss of the globe.

When chemical exposure occurs, the epithelium protects the underlying tissue by taking the brunt of the injury then sloughing off, carrying with it much of the chemical contaminant. After the injury, new epithelial tissue is generated to replace damaged sloughed tissue. During the healing process a patient is usually treated with an antibiotic ointment or solution to prevent infections of exposed surfaces.

The sclera is a tough, vascularized, white, fibrous membrane that covers the globe and extends from the cornea to the optic nerve. It is made of fibrous material that protects and gives shape to the globe. The sclera is referred to as the white of the eye, although upon close examination, the color is actually off white, and it is interlaced with numerous blood vessels. If during evaluation of an injured eye, it is noted that the sclera is truly white or porcelain in nature, then there is reason to suspect that circulation to that area has been compromised. Because the eye lacks the ability to form scar tissue to replace ischemic areas, this condition will almost certainly lead to necrosis, perforation, and loss of the globe.

All of the tissue of the eye requires a supply of nutrients and oxygen to sustain life. This supply comes from a circulatory system that provides an uninterrupted flow of blood for these tissues. The system is called *perilimbal circulation* and is evidenced by the vessels seen on close exam within the sclera (see Fig. 5-3).

Assessment of Eye Injuries

There are many signs for determining how badly an eye is injured. Usually, the crucial signs are related to opacification of the stroma and loss of circulation. Both of these can be examined in the field and relayed to the hospital prior to or during transport. The key to a rapid assessment is knowing what to look for and realizing what is seen. An opacification of the epithelial layer, from an acid, is not necessarily a critical sign. Many times the epithelial tissue is damaged protecting the underlying structures. Then the tissue sloughs off exposing a virtually undamaged cornea stroma beneath it. For simplicity reasons assessment of burned eyes can be broken down into mild, moderate, and severe injuries.

• Mild burns are distinguishable by the several signs noted on a rapid assessment. They are characterized by sloughing of epithelial tissue over the cornea and

Figure 5-3 *A healthy eye has a clear cornea with well-defined pupil and easily seen iris detail.*

surrounding tissue. The sclera and associated tissue remains well circulated, evidenced by intact blood vessels in the sclera and reddened or pink tissues surrounding the eye. The iris is clear and the iris detail (the fine lines within the iris) is easily seen.

• Moderate burns are characterized by the signs and symptoms noted on a mild burn but include haziness of the cornea indicating that the chemical has reached the stroma. Circulation of the eye and surrounding tissue remains intact. Depending on the chemical and the amount of time exposed, the prognosis for saving the eye is good. There will probably be some loss in visual acuity but not total blindness. This injury is usually associated with acids or solvents.

• Severe burns are usually found when there is an exposure to alkalis or very strong acids. The cornea and sclera usually have total epithelial loss. The cornea appears very hazy or opaque masking the iris detail. The sclera may be white and porcelainized indicating ischemia to that area. The prognosis of a severely burned eye is very poor. The ischemia, as mentioned earlier, will lead to necrosis, perforation, and loss of the globe. Concrete dust causes severe burns that have a slow onset but can result in complete loss of vision and globe (Fig. 5-4).

Surface Toxins

Identification of the toxin is not immediately imperative for initial treatment to begin, although, eventually, identification is important for definitive care. Straightforward first aid such as irrigation is relatively simple and safe. Irrigation should begin immediately upon determining scene safety. Questions about the chemical offender should be asked only after initial irrigation has started. Irrigation can be accomplished utilizing any nonirritating water base solution. It can include tap water, hydrant water, lactated ringers, or preferably normal saline. Many refer-

Figure 5-4 *One of the most common causes of permanent eye injury is the exposure to concrete dust or Portland cement. The dust's subtle action cause a painless caustic injury that discolors the corneal stroma and endothelium.*

ences state that irrigation of the eye, after an acid exposure, can be stopped once the pH of tears reaches 7. It is our opinion that irrigation of an exposed eye, regardless of the offender, continue until the patient reaches the hospital and a physician evaluates the injury. It can not be stressed enough that immediate irrigation can make the difference between maintaining sight or losing the globe.

Exposures to toxic substances may bring on many diverse symptoms such as altered color and blurred vision, decrease in visual acuity, photophobia, excessive tearing, and blindness. Complete destruction of the eye globe may result from an overwhelming exposure to a toxic material, although anything less should receive immediate irrigation and transportation to the hospital.

Corrosives represent the most common type of eye exposure found at hazardous materials accidents. Acids and alkalis are not only found in industry but are also common chemicals found around the home. As mentioned earlier in the chapter, acids are the most commonly produced chemical in the United States. As would be expected then, acids comprise the most common corrosive injury. Alkalis cause the most devastating injuries to the eyes.

Alkali Alkali injuries are particularly devastating to the corneal structures of the eye. The process in which alkalis produce damage is through a liquefaction of the fatty substances within the tissue of the eyes. The liquefaction process allows the alkali to penetrate deeper into the underlying structures, causing a much more devastating injury as compared to acidic chemicals. So much more devastating, that it has been said that alkalis have a twenty times greater dose response upon the eye than do acids. Chemicals with a pH of greater than 11.5 cause particularly devastating results. The more severe the penetration and damage accrued by the alkali, the less pain will be experienced by the victim. Much like third degree thermal burns, deep alkali burns destroy the corneal nerves, desensitizing the tissue as it damages it. Penetration is also related to the toxicity of the chemical involved. For example, ammonium hydroxide penetrates the fastest, followed by sodium hydroxide, potassium hydroxide, and calcium hydroxide respectively.

Damage to the circulatory system or underlying structures of the eye may lead to necrosis of tissue and loss of the globe. Opacification of the cornea and related underlying structures may lead to blindness without loss of the globe. In either case, alkalis have the ability to evoke serious injury to the eye and must be treated aggressively by the first responder to arrive with the patient.

Treatment involves immediate irrigation of the eyes. This irrigation should continue for a prolonged period. It is suggested that once irrigation is established in the field it should be continued throughout transportation to the hospital and until stopped by an ophthalmologist. Alkalis continue to burn even after the surface chemical is washed away because of their ability to penetrate the upper layers of the tissue through a liquefaction of the fats (soponification). Unlike acids, when alkalis are the offender, stabilization of the surface pH is not a good indication for irrigation to stop.

Neutralization of an acid or alkali should never be considered on the surface of the eye, due to the generation of heat during the reaction, furthering the injury. The one and only treatment for this type of injury in the field should be rapid and immediate irrigation and transportation. Once irrigation has started and the eye is washed of excessive chemical, an irrigation lens, such as the Morgan Lens, can be placed to make the irrigation process more efficient. The placement of a therapeutic lens is discussed in the section Specialized Eye Equipment.

Acids   Acid burns to the eye usually are not as devastating as alkali injuries. The epithelial tissue reacts with acids forming a tough coagulum. Acids act on the proteins found within the tissue causing a process called *coagulation necrosis*. The tissue becomes thickened and rubbery either stopping or slowing the absorption of acids below the surface. The process actually acts as a natural protective barrier against acid penetration. Highly concentrated acids, those with a pH of less than 2, have the ability to overwhelm the protective barrier and cause deeper more severe injury, similar to those caused by alkalis.

Again, treatment consists of rapid irrigation of the eye by the first responder. Because the physiology involved in the acid burn is different than that evoked by an alkali, irrigation can be halted in the field. To determine if enough irrigation has been done, test the pH of the fluid found in the pocket of the lower lid. A medical provider can stop irrigation if the pH is found to be at 7. It is suggested this test be done only after irrigation has continued for 30 to 60 minutes. In most cases it is more feasible to irrigate until the patient reaches the hospital, then allow the emergency department physician to make the decision to stop irrigation.

! SAFETY
Neutralization of an acid or alkali should never be considered on the surface of the eye because the generation of heat during the reaction will add to the injury.

Using an irrigation lens, such as the Morgan, will allow the medical provider to irrigate the victim's eyes for longer periods of time, without having both hands involved in the operation. The lens can be placed after rapid irrigation has already started, then irrigation can continue throughout transport to the hospital.

Solvents   Solvents such as gasoline, alcohols, toluene, and acetone are commonly used in industry. These chemicals are fat soluble, meaning they mix with and disrupt fats found in the tissues of the eyes. Unlike alkalis that cause death to the cells by liquefying the fats, solvents disrupt the fats causing normally transparent tissue to become opaque and dehydrated. The effects from the injury include epithelial sloughing and pain. These symptoms generally persist several days and most heal without lasting damage. In industry most solvents are heated to accentuate their effect. Workers injured with heated solvents are suffering not only from the chemical effects of the solvent, but also from the thermal effects.

Treatment in the field is, once again, continuous ocular irrigation and rapid transportation to the hospital for follow-up care.

Surfactants and Detergents   Surfactants and detergents are used in industry to promote wetting. They also disperse and dissolve fatty substances and decrease foaming. Short-term exposure to such substances that are rapidly rinsed from the eyes will only cause epithelial sloughing and pain but usually not long-term injury. Everyday soaps and shampoos are an example of this class of chemicals. Most people have experienced the discomfort of soap in the eyes. Much stronger versions are found in the industrial uses of these chemicals. When splashed in the eye, severe discomfort, epithelial cellular injury, and nerve irritation results. Rapid irrigation will minimize injury and if done within a reasonable period, recovery is usually complete.

Lacrimatory Agents   Lacrimatory agents stimulate corneal nerve endings, causing reflex lacrimation and stinging pain. This effect is usually self-limiting and many times can be dealt with by the use of short-term irrigation followed by a topical anesthetic. In higher doses corneal and conjunctival inflammation may result but this injury is most often self-limiting, not requiring follow-up care.

Tear gas is the most common lacrimator associated with ocular injuries. The police departments today use a spray that contains capsicum, a chemical found in hot peppers. Capsicum (pepper gas) is a strong lacrimator usually with no lasting effects. There have been reports of severe reactions to capsicum but the reports are rare and to this date none have been founded. If treating a patient for capsicum in the eyes, usually a topical anesthetic, like ponticaine, will take care of the burning and lacrimation. By the time the anesthetic effect wears off the chemical is no longer effective in the eye and the symptoms do not reappear. More serious injury to the eye is usually the result of mechanical damage received from the propellant being sprayed into the eyes rather than an injury resulting from the chemical.

Metals   Many metallic salts can cause damage once the ocular surface is exposed. Injury can be as minor as a mild irritation or as severe as tissue necrosis. Many of these metals bind with the proteins and form metallic complexes, which may result in the formation of permanent granular deposits within the ocular tissue.

Solubility within the cornea determines the toxicity of the metallic salt. Mercury has the highest solubility followed by tin, silver, copper, zinc, and lead. Iron, on the other hand, causes little damage to the ocular tissue but, as a foreign body, causes a staining of surrounding tissue, referred to as a "rust ring."

Treatment

Because toxic exposure of the eye continues until the toxin is removed, time is the most important factor. Irrigation must begin immediately and continue until the injury stops. The most accessible bland solution, preferably water or saline should be used. NEVER attempt to neutralize an acid or alkaline irritant in the eye. After a chemical insult to the eye a spasm of the upper and lower lids occurs (**blepharospasm**) causing the eye to involuntarily close tightly. This occurs as a natural protective feature of the eye. The victim should be provided with all available assistance to open the eye and provide irrigation. Many different means for irrigation are available to emergency responders, but none are efficient unless the eye can be opened. Digital opening of the eye must be done to insure proper irrigation.

The lids provide a watertight, airtight seal when tightly closed. Irrigation without opening the lids is useless so care must be taken by the medical provider to open the lids without placing excessive pressure on the globe. As long as the blepharospasm continues, the patient will be unable to hold his own eyes open, therefore, an attendant must continue to assist with the eye opening during irrigation.

Utilizing a nasal cannula placed on the bridge of the nose and connected to an IV solution provides a means for applying irrigating solution at the eyes. This apparatus frees a caregiver's hands to open the lids and ensure that the irrigation solution reaches the globe surface.

Utilizing an anesthetic solution, such as ponticaine, alcaine, or opthalmacaine will minimize or relieve the blepharospasm. The use of the anesthetic alone allows irrigation to be accomplished more easily. It also minimizes the pain suffered by the patient and reduces anxiety. If Morgan lenses are available, the anesthetic agents are necessary prior to placing the lens in the eye. As a point of interest, most of the optical anesthetic solutions are heat sensitive, so if they are carried in an emergency vehicle for field use, they must be kept in a cool environment.

Irrigation should be continued for 30 to 60 minutes or until the physician stops the solution. A pH reading of 7 may indicate that irrigation can stop in an acid injury but not in an alkali injury. pH readings can be done utilizing litmus paper or a urine dipstick by placing the reactive surface in the tear found in the pocket formed by the lower lid of the affected eye.

blepharospasm
involuntary closing of the eye that occurs when the eye is irritated or injured

■ NOTE
Eyelids provide a watertight, airtight seal when tightly closed, so irrigation without opening the lids is useless; however, care must be taken to open the lids without placing excessive pressure on the globe.

SPECIALIZED EYE EQUIPMENT

The Morgan Lens

The Morgan therapeutic lens (Fig. 5-5) was developed to ease the process of irrigation when an eye is exposed to chemicals. It provides the eye with a constant flow of physiologic saline evenly distributed over the exposed globe. If properly used, it is placed with no additional trauma. This is not true of other irrigation techniques.

Once a caustic substance reaches the eye and an injury starts, the lids are forced closed in a natural response to the injury. This reaction disallows further exposure of an eye to the chemical. The medical provider arriving on the scene is faced with two problems in irrigating the eye. First, holding the eye open in an attempt to irrigate the globe properly is extremely difficult or at times impossible without specialized equipment. This is not because the patient is uncooperative, but because the patient is unable to open the eye due to the spasm. Second, there is no efficient way one person can pour the solution into the eye once the lids are opened digitally.

The Morgan lens is designed as a contact-style lens (Fig. 5-6). It is to be used in conjunction with an anesthetic solution such as ponticaine or alcaine. Once anesthesia has taken place, the lens is moistened with a steady flow of solution.

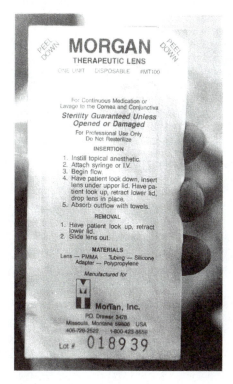

Figure 5-5 *The Morgan therapeutic lens is a prepackaged device for the irrigation of eyes injured by a chemical. It is easy to install and provides excellent hands-free irrigation to an injured eye.*

Figure 5-6 *The Morgan lens is a contact-style lens that is held in place by the lids. A constant flow of irrigation fluid is provided from the IV bag into the lens.*

The solution of choice is normal saline (0.9 sodium chloride) and is attached via an IV administration set to the lens. The technique is to first place the lens into the lower lid, while instructing the patient to look up (Fig. 5-7). Then place the lens under the upper lid, against the globe while instructing the patient to look down. The drip is adjusted to provide a continuous flow over the cornea and under the lids. Just like intravenous lines established in the field, great care should be used to attach the tubing to the patient. So often when moving patients the solution bag goes in one direction while the patient goes in another. Since the Morgan Lens is

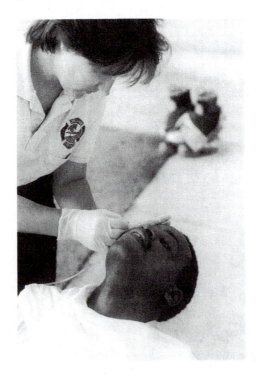

Figure 5-7 *Once the eye is numbed, the lens is placed under the upper lid then under the lower lid. The lens tubing must be held firmly in place to avoid accidentally pulling the lens out of the eye.*

Figure 5-8 *Once inserted, the Morgan lens provides excellent, continuous hands-free irrigation that is comfortable to the patient.*

shaped much like a suction cup, it appears that mechanical damage would probably result from the sudden, violent removal of the lens from the eye. Securing can be accomplished by looping the tubing on the forehead and taping it there. Removal of the lens is done by following the procedures for placement in reverse. Room should be left on the lens side of the tubing for periodic removal. If irrigation is continued for long periods, then the lens should be removed and the eye remedicated with the anesthetic agent.

The inside cup of the lens is manufactured smooth so no mechanical injury can be caused over the injured eye. The Morgan Lens and other contact style lenses are comfortable to the patient and provide a better irrigation to an injured eye than a medical provider struggling to hold open an injured eye while providing a steady flow of solution over the globe (Fig. 5-8). If care is taken with installation and removal, little or no trauma should result from this irrigation technique.

Nasal Cannula

Using a nasal cannula over the bridge of the nose is an optional technique available for those providers who do not have access to irrigation lenses. By placing the nasal cannula and supplying it with a steady flow of irrigation solution via an IV set, a provider can wash the eyes from the medial canthus to the lateral canthus, a practice preferred for this type of irrigation. Interestingly, we have found that the Morgan Lens also fits to the prongs of a nasal cannula and can be used to irrigate two lenses with one IV set (Fig. 5-9). This also solves the problem of fastening the lens in place to avoid accidental removal. The nasal cannula is attached to the patient in the same manner to provide oxygen, but instead of the prongs being inserted into the nose they are placed across the bridge of the nose at the eyes. The lens's tubing can then be attached to the prongs.

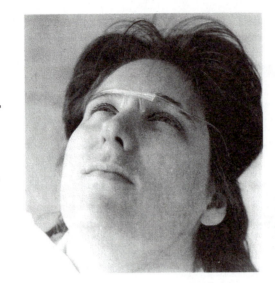

Figure 5-9   *A nasal cannula can be used for irrigating the eyes. This device is not hands-free, as the care provider must ensure that the victim's eyes remain open during irrigation.*

Providing initial irrigation in the most rapid means is of vital importance. Once initial irrigation is established, then a provider can take the time to set up a more elaborate means of irrigating the eyes. Another important point is that the runoff solution may be of sufficient concentration to further injure the patient or the health care provider. Great care should be taken to control and/or contain runoff to reduce the chance of secondary contamination of both the provider and the victim.

RESPIRATORY EXPOSURE AND INJURIES

Overview

"Breathing" is a popular term used to describe respiration, but breathing simply means inhaling and exhaling air. The respiratory system's function is to draw into the body fresh air containing oxygen to be utilized by the cells in the production of energy. This process of moving, circulating, and utilizing oxygen, then disposing of carbon dioxide as a waste product is called *respiration*.

Respiratory Anatomy and Physiology

Respiration is subdivided into external and internal respiration. External respiration is the movement of air into, and out of, the body through the passageways to the alveoli. Internal respiration involves the transportation of gases through the bloodstream to the cells. Furthermore, internal respiration also includes the uptake and utilization of oxygen by the cells and the disposal of the waste product, carbon dioxide. The dividing barrier between internal and external respirations is the alveolar membrane. To understand the effects that toxic inhalation has on the respiratory system, it is necessary to review the respiratory system as a whole.

During the inhalation phase the atmospheric air is moved from the surroundings into the body through a process of pressure changes within the thoracic cavity. While the air is moving, en route to the alveoli, it is warmed or cooled, moistened or dried, and, to a point, purified, as it progresses through the gradually decreasing passageways, to the terminal ends, the alveoli. The normal adult inhalation involves the movement of approximately 3,600 ml of air. After exhalation there remains approximately 2,400 ml. The sum of both volumes is the total lung capacity, or about 6,000 ml in the average adult.

When air is inhaled through the nose, the air moves through passages that are lined with coarse hairs meant to filter out large dust particles. Once past the hairs, air reaches the turbinates, which are irregularly shaped surfaces within the nose that swirl the air as it passes. The swirling facilitates rapid movements of foreign objects suspended in the air, causing them to brush against the sides. Most of the smaller foreign objects are then trapped along the mucous-lined passageways.

The passageways below the trachea leading to the alveoli are made up of cells that excrete mucous for the intent of trapping impurities and water soluble toxins carried in during inhalation. Other cells contain small hair-like structures, called *cilia*, that sweep the mucous toward the upper respiratory area for expectoration or swallowing.

As the air moves from the nose and mouth deeper into the lungs, the passageways continually divide into smaller and smaller passages. Even though the passageways become smaller, they divide thousands of times, significantly increasing the surface area that the inhaled air is exposed to. The cross-sectional surface area from the bronchioles to the alveoli represents an increase of two thousand times, in area. The passageways are continually adjusting to accommodate air flow and are changed depending on physical or emotional stress.

The normal lung contains approximately 300 million alveoli and makes up the largest surface area within the lungs (Fig. 5-10). In fact the surface area of the alveoli, in an adult, is about the same size as found on the playing surface of a tennis court or 1,000 square feet.

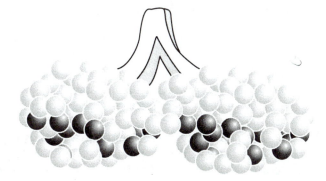

Figure 5-10   *There are approximately 300 million alveoli in the adult lung, helping to form a surface larger than any other organ.*

Alveoli are made up of millions of tiny sacs. Gas exchange, through a process of diffusion (the movement from a higher concentration to a lower concentration through a semipermeable membrane) takes place at these terminal ends. The alveolar wall is made up of delicate cells that allow only specific gasses to move freely between them.

Once the fresh air reaches the alveolar wall (semipermeable membrane) the higher concentration of oxygen in the alveoli versus the lower concentration of oxygen in the venous bloodstream stimulates diffusion to take place. The opposite is true for carbon dioxide (lower concentration inside the alveoli and higher in the bloodstream).

hemoglobin

a molecule found in the blood that is responsible for the transportation of oxygen from the lungs to the cells

oxyhemoglobin

hemoglobin bound with oxygen

The circulating **hemoglobin**, within the blood contains an iron atom (Fe) that readily accepts oxygen for transportation through the arterial system. The bonding of oxygen to the hemoglobin molecule forms **oxyhemoglobin**. This joining of oxygen and hemoglobin forms a weak bond that can easily be separated. Once the oxyhemoglobin reaches the cell membrane the bond is broken so that the oxygen can move into, and be utilized by, the cell.

Circulating oxygen can be measured in several ways. Probably the most reliable test performed is a blood gas study. Blood is drawn from an artery of the patient and, along with other chemical studies, the partial pressure of oxygen is measured. At the present time this is not a practical test for the field. The other common measuring tool is the pulse oximeter. This device is not invasive; it only requires the medical provider to clip the sensor to a well-circulating appendage and turn on the machine. A light source within the clip shines into the tissue and, through a process of measuring the color and opacity of the blood, can determine oxygenation. The pulse oximetry unit will be examined more closely in the section on specialized respirator equipment.

The waste product of cellular respiration (a series of reactions where glucose is oxidized to form carbon dioxide, water, and energy), carbon dioxide, is taken into solution in the blood and is exhaled through a process of diffusion by the external respiratory system. The level of carbon dioxide in the blood is partially responsible for the stimulation of respiration. Higher levels of carbon dioxide within the blood stimulate sensors, that in turn send messages to the respiratory center of the brain. The brain then signals for faster, deeper breathing. Lower levels, in turn stimulate a slowing response resulting in slower, shallower breathing.

Exposure Types

Exposure of the respiratory system to hazardous materials can cause a group of predictable and, for the most part, treatable maladies. These chemical exposures can affect any and all parts of the respiratory system, depending on the properties of the product involved.

The two most common types of respiratory system exposures are to those chemicals considered to be asphyxiants and irritants. Other chemical exposures can affect the respiratory system by hindering the central nervous system or directly affecting the respiratory centers of the brain. These particular chemicals will be addressed in Chapter 6. Other chemicals that enter through the respiratory system may affect the body, but the target organ is not the lungs or respiratory system.

To determine the type of respiratory injury to expect from a chemical, three factors about the involved chemical must be examined: the physical nature, concentration and duration, and the solubility of the chemical.

Physical Nature of the Chemical   Inhalation injuries to the respiratory system result from the ability of the chemical to gain access into the respiratory system. For the injury to take place, the chemical must be in a form that allows access into the respiratory system. Gases, vapors, solid particles, and liquid aerosols are all easily inhaled.

- A gas is a chemical with a boiling point below normal room temperature. A vapor, which is similar to a gas or fume, is the molecular dispersion of a solid or liquid chemical within the air. Toxic gases and vapors readily enter the respiratory system because they easily mix with the atmospheric air, and depending on the concentration and physical properties, can injure all levels of the respiratory system.

- Solid particles represent small particles of a solid, floating within the surrounding air. When entering the respiratory system (Fig. 5-11), the larger particles are filtered by the hairs in the nostrils. Others are trapped in the upper respiratory system and are rapidly swept out by the cilia. Very small particles (smaller than a micron) suspended in the air may have the ability to fall deeper in the smaller passageways and even the alveoli. If the particle is larger than one micron but smaller than three microns, it can reach the fine bronchioles but not penetrate the alveoli. Some may be blown out during the exhalation phase. The particles that may become mixed with secretions of the passageways can break down, form solutions, and be absorbed through the lining and into the bloodstream. Others may become trapped forever, leading to pathologies years later. Pneumoconiosis is a disease of the lung that develops due to inhalation of dust particles. More specifically, asbestosis and silicosis are two such occupational diseases caused from the inhalation of asbestos or silica.

- Aerosols or mists are fine droplets of a liquid suspended within the surrounding air. Depending on their solubility in water they are capable of injuring all levels of the respiratory system.

SAFETY

A child breathing toxic air receives a much higher dose per pound of body weight because children breathe several times more air, in a given period, than do adults.

Concentration and Duration of the Exposure   The amount of air drawn into the lungs with each breath and the concentration of the chemical within the air both determine the dose of the toxin received. A person breathing rapidly and deeply during exercise receives a higher dose of chemical than a sedentary person breathing lightly, regardless of the concentration in the air. An example of this principle is a child breathing toxic air. The child receives a much higher dose per pound of body weight, because children breathe several times more air, in a given period, than do adults.

With brief encounters at lower concentrations, a chemical may be trapped and evacuated by the respiratory system's natural functions, not causing any

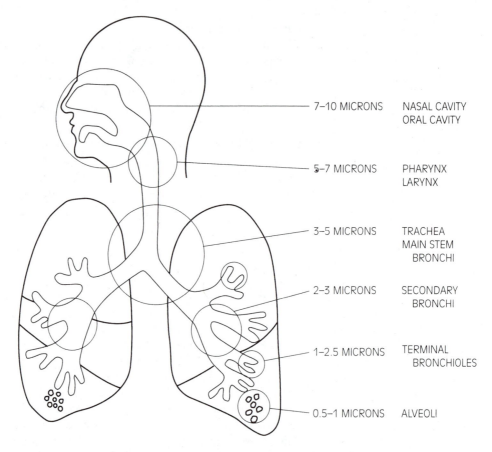

7–10 MICRONS   NASAL CAVITY
                ORAL CAVITY

5–7 MICRONS   PHARYNX
                LARYNX

3–5 MICRONS   TRACHEA
                MAIN STEM
                  BRONCHI

2–3 MICRONS   SECONDARY
                  BRONCHI

1–2.5 MICRONS   TERMINAL
                  BRONCHIOLES

0.5–1 MICRONS   ALVEOLI

Figure 5-11 *Solid particles can only penetrate to the depth allowed by the size of the opening. This diagram indicates the predictable penetration from differing particle sizes.*

injury at all. At higher concentrations the protective functions of the respiratory system may be overwhelmed and certain injury would occur. Low concentrations over a long period of time create a similar devastating injury. One reason this occurs is that eventually the natural protective features of the respiratory system are overwhelmed, allowing a toxin access to deeper areas of the system.

The relationship between the concentration and duration during a toxic exposure is referred to as the dose–time relationship. To express the dose–time relationship in other terms, an exposure to a high concentration for a short time may be similar to that of a low concentration over a long time.

Solubility of the Chemical   Evidence has demonstrated that the solubility or insolubility of a chemical in water has a bearing on where the effect may occur. Since the mucous of the upper airways is a water-based material, water soluble chemicals are easily mixed with the mucous. These water soluble chemicals are either absorbed at the site of mixture or cause damage to the underlying tissues. An irritant gas, such as chlorine, is soluble in water. When inhaled, it mixes with the water-based mucous and causes irritation at the site. The irritation can lead to edema and/or tissue sloughing. These water soluble chemicals inhaled in high concentrations or over long periods of time not only cause injury to the upper airways but also reach the alveoli and cause damage there.

Non-water-soluble chemicals when inhaled usually bypass the upper airways and cause their damage deeper, in the fine bronchioles and alveoli. The water-based mucous in these upper airways actually acts to repel non water soluble chemicals, propelling them deep into the respiratory system.

The alveoli contain a substance called *surfactant* (dipalmitoyl phosphatidyl-choline). Surfactant causes an increase in surface tension that functions to keep the alveoli open. Surfactant is a lipid based chemical, therefore non-water-soluble chemicals are usually soluble in surfactant. Irritants such as phosgene, a non-water-soluble gas, are easily inhaled to the alveoli. Once in the alveoli, phosgene displaces surfactant and damages the tissues that make up the alveolar membrane. The damaged tissue allows fluid from the blood to flow into the lungs. This condition is called *chemically induced* or *noncardiogenic pulmonary edema*. The phosgene also disrupts the surfactant allowing the alveoli to collapse.

Type of Gases That Harm the Respiratory System

The two most common types of respiratory exposures are those involving asphyxiants and irritants. Other chemicals enter the body through the respiratory system but affect other organ systems. There may be little or no effect on the respiratory system because it is not the target organ system for that chemical.

Asphyxiants   Asphyxiants can be classified in two ways: those that displace oxygen (simple asphyxiants) and those that interfere with transportation or uptake of oxygen by the cells (chemical asphyxiants).

1. Simple Asphyxiants. Simple asphyxiants cause a decrease in the amount of oxygen entering the external respiratory system. These asphyxiants displace oxygen, resulting in lower concentrations reaching the alveoli and ultimately supplying the cells. The end consequence is cellular hypoxia, acidosis, and eventually death. Examples of oxygen displacing asphyxiants are carbon dioxide, nitrogen, and halogenated extinguishing systems.

On August 21, 1986, an African village on the shore of Lake Nyos was struck by a freak of nature that took the lives of 1,700 residents of that community. It happened during the evening hours when a change in underground geographic conditions caused release of a large cloud of carbon dioxide gas. Because carbon dioxide is heavier than air, the cloud of approximately one billion cubic meters followed river beds a distance of several miles. All of the oxygen in these low-lying areas was displaced, rapidly causing death to all living creatures. Witnesses to the area after the incident stated that not even an insect was found alive. Although this incident represents an extremely unusual circumstance, it indicates the danger associated with simple asphyxiants.

2. Chemical Asphyxiants. The chemical asphyxiant enters the respiratory system, and due to some chemical means, disallows oxygen from entering the cells. Carbon monoxide, one type of chemical asphyxiant, combines with the hemoglobin, disallowing it from carrying oxygen. This in essence starves the cells of needed oxygen, slowing cellular respiration and energy production. Therefore, carbon monoxide is an example of a chemical that interferes with the transportation of oxygen to the cells. Cyanide, on the other hand, combines with an enzyme found inside of the cells. This enzyme is necessary for the uptake and utilization of oxygen. In this case there is plenty of oxygen circulating in the bloodstream but the cell is unable to utilize it, and therefore, cellular respiration slows and becomes very inefficient. Cyanide is classified as a chemical asphyxiant because it interferes with cellular respiration.

The treatment of exposures to asphyxiants is to always give 100% oxygen by nonrebreather (NRB) mask if the patient is conscious and is respiring adequately. If breathing is inadequate or the patient is unconscious, treatment should consist of intubation and positive pressure ventilation, utilizing the PEEP (Positive End Expiratory Pressure) valve. See Specialized Respiratory Equipment section in this chapter. In the case of carbon monoxide or cyanide poisoning, arrangements to transport the patient to a facility that can provide **hyperbaric oxygen therapy** (HBO) should be considered. Further antidotal treatment of specific asphyxiant poisonings is discussed in detail in Chapter 6.

hyperbaric oxygen therapy (HBO)
placing a patient for medicinal treatment in an oxygen environment that is higher than atmospheric pressure

Irritants   Irritants can cause several and sometimes severe symptoms involving the external respiratory system. Bronchospasms, laryngospasms, tissue sloughing, localized edema, and pulmonary edema are the most common. Irritants have the ability to injure all levels of the external respiratory system. Chemicals that are water soluble such as chlorine and ammonia, unless in very high concentrations, affect the upper areas of the respiratory system. Non-water-soluble chemicals gain access deeper into the respiratory system, causing their damage at the fine bronchioles and alveoli, leading to pulmonary edema.

Predictable Areas of Injury Related to Irritant Inhalation

Injuries to the upper respiratory areas are usually a result of exposure to water soluble chemicals that readily dissolve into the moisture-coated airways. These exposures result in localized chemical burns and inflammation. Laryngeal edema and laryngeal spasms should be expected and treated aggressively with 100% oxygen and updraft treatments. If breathing is compromised, intubation and positive pressure ventilations are indicated. When auscultating breath sounds on the victim of a water soluble irritant, expect to hear upper airway rhonchi and wheezing.

Injuries to the lower regions of the respiratory tract usually result when the chemical inhaled is not water soluble, was in a high concentration, or was inhaled over an extended period of time. This deeper injury causes swelling of the finer bronchioles, sloughing of damaged tissue, and damage to the alveoli. Cilia that may be damaged are unable to rid the fine bronchioles of the sloughed cells and increased mucous production caused by the damaged airways. This condition adds to the complexity of the injury. When auscultating breath sounds of victims exposed to non-water-soluble chemicals, most often lower airway ronchi and rales (crackling) will be heard.

Noncardiogenic Pulmonary Edema, aka Chemically Induced Pulmonary Edema   Noncardiogenic pulmonary edema results when a chemical irritant reaches the alveoli, causing damage. The damage interferes with the alveoli's natural ability to keep fluids out of the alveolar space. This results in fluids filling the injured alveoli and fine bronchioles. In the most serious cases, up to 2 liters of fluid can be displaced from the blood into the external respiratory system.

The healthy alveolar wall is made up of a single layer of semipermeable cells. This semipermeable membrane allows certain gases and some fluids to permeate into, and out of, the bloodstream from the alveolar space. The fluids that do permeate into the alveolar space are carried back into the bloodstream through a cellular pump action, keeping excessive fluid from building up inside of the alveoli.

The alveoli also produce surfactant, which increases the surface tension, which assists in keeping the alveoli open. If the surfactant is disrupted, either by an inhaled chemical or by the invasion of fluid from the blood, the alveoli closes (atalectasis), hindering the transfer of gases across the membrane.

Because the production of chemically induced pulmonary edema is noncardiogenic, the normal treatment of diuretics may be of limited use. That is not to say that diuretics should not be used, only that the expected effects seen on cardiogenic pulmonary edema may not be witnessed in the case of chemically induced pulmonary edema. It is virtually impossible to determine if the cause of pulmonary edema is the result of a respiratory chemical exposure or of cardiac origin. History may be the only clue to rely on.

If the alveoli and fine bronchioles are involved in the injury, then the lung sounds will mock that of cardiogenic pulmonary edema. Typically, rales will be heard on auscultation indicating fluid invasion in the alveoli. Wheezing may also be heard due to the narrowing of the airways from edema caused from the irritant passing through and injuring the upper airway tissues.

Treatment of Chemically Induced Pulmonary Edema   The treatment of choice is to initiate oxygen therapy with 100% oxygen and intubate with an endotracheal tube if necessary. Utilizing a bag valve device equipped with a PEEP valve to ventilate has several favorable effects. First, if the surfactant has been altered or damaged, positive pressure ventilations will assist to keep alveoli open. Next, some research has shown that the utilization of positive pressure will assist in the reabsorption of fluid from the alveoli back into the vascular space.

The onset of the symptoms may be immediate, or delayed for as long as six or more hours. These patients will typically complain of chest tightness, pain, and report a history of toxic inhalation. A productive cough may be present showing signs of blood stains in the sputum. Asthma symptoms may also be present with diffuse bronchospasms, wheezing, and increased expiration times.

Bronchospasms   Bronchospasms may occur when a patient is exposed to an irritant, but, more typically, they happen when the victim is sensitized to a chemical. Asthma symptoms may happen without a previous history of such a sensitivity. Because of previous or long-term exposures, some individuals may experience these symptoms at very low doses when others in the same area, and even at higher concentration, may not experience any ill effects at all. This condition is known as a sensitivity. Some chemicals are more prone than others to stimulate a condition of sensitivity.

Hydrogen sulfide is known to cause this type of sensitivity reaction. In many reported cases, workers who have been periodically exposed to low dose hydrogen sulfide over extended periods of time, suddenly develop an acute sensitivity reaction to the chemical, causing bronchospasms, along with other allergic type reactions.

Treatment of Bronchospasms   Generally, the symptoms can be reversed with the standard means of treatment. The use of updrafts of alupent or albuterol are commonly used field bronchodilators (Fig. 5-12). These dilators work well whether the spasms are due to a sensitivity or chronic asthma. If the bronchospasms continue after updrafts, subcutaneous brethine or epinephrine can be used. Caution must be observed in the use of epinephrine with certain chemical exposures. Cardiac stimulants such as halogenated hydrocarbons can accentuate the cardiac effects of epinephrine causing ventricular tachycardia or fibrillation. Humidified oxygen, in high concentrations, is also important in the treatment. Oximetry, if available, should be measured from the initial assessment, through transportation to the hospital.

Figure 5-12
Bronchodilators such as albuterol, alupent, in updraft form, or brethine as a subcutaneous injection all offer good results. Care must be taken to cause no additional stimulation to an irritable heart.

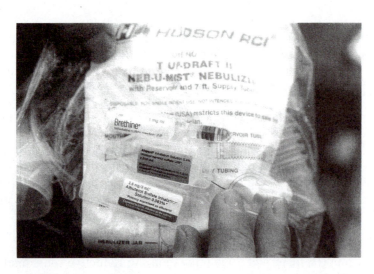

TOXICITY OF SMOKE AND COMBUSTION GASES

History

Both fire and EMS responders frequently encounter victims of smoke inhalation. Conditions found in smoke represent probably the oldest hazardous materials emergency and one of the most studied and analyzed. Since every fire condition is different, the toxicity of each fire is different. In more recent years the use of synthetic materials in construction and furnishings has greatly increased the variety of chemicals found in smoke.

The lethal effects of smoke have been recognized and utilized as far back as the first century. Romans used the smoke of green wood to execute prisoners. The United States Fire Administration states that, on an average, more than 5,700 people die as a result of fire each year and approximately 28,000 injuries result from fire and the toxicity of smoke (Fire in the U.S. 1983–1990, FEMA). Many dramatic events in our recent history have illustrated the seriousness of the problem related to smoke inhalation. In November of 1980, a fire at the MGM Grand Hotel in Las Vegas claimed the lives of eighty-four visitors and workers at the hotel. At least sixty-eight of these victims died as a result of the inhalation of toxic gases. In January 1990, a fire in Zaragoza, Spain, claimed the lives of forty-three who died from an apparent exposure to toxic gases generated during the fire. These incidents are only examples of the devastation that can be caused by the inhalation of smoke gases generated during a fire. On a smaller scale, virtually every community has experienced deaths resulting from fires.

No profession is more aware of the effects of smoke than firefighters. Prior to 1940 most of the fires encountered by these professionals contained typical organic products used in construction. Exotic plastics were not yet invented and synthetic materials were not widely used during this era and therefore not a concern. Firefighters believed that the smoke was not toxic, a belief that we now know is not true. For example, burning Douglas fir wood gives off in excess of seventy-five harmful chemicals. Other wood and cotton products give off harmful levels of carbon monoxide and carbon dioxide. Wools and silks when burning produce large amounts of cyanide gas and hydrocyanic acid. Unfortunately, this belief lead to the early deaths of most of the workers who participated in this profession over a long period.

Today, firefighters use equipment that protects them from many of the products of combustion that are found in typical structure fires. Self-contained breathing apparatus provide excellent protection while in the atmospheres containing the toxins. Unfortunately, compliance throughout the fire fighting process including, attack, knockdown, and overhaul is still low. Therefore, fire fighting remains one of the most hazardous occupations in the United States. Today's structures contain a wide variety of exotic materials that when exposed to heat and fire, release a myriad of chemicals that not only enter the respiratory system, but also can penetrate the firefighters' protective clothing, making entrance into the body through the skin.

The challenges that the modern and future firefighter have are many when it comes to typical fires. Wood, which has been the usual material used for construction, has become an expensive commodity. To make housing more affordable, new materials have been introduced into the construction industry. In recent years, aluminum wiring, laminated trusses, and lightweight roofs utilizing plastic materials have become more the standard. In each of these construction techniques, money, affordability, and ease of construction are the motivating factors. PVC (polyvinyl chloride), which has virtually replaced steel pipes and is used for other household items, has been identified to release fifty-five products of combustion when involved in fire. Polystyrene, used in plastic polymer blends to make coffee makers, hair dryers, power tools, pump housings, and so forth, and polyamide, a high-temperature plastic, both release more than thirty different products when exposed to fire.

Recent statistics on the life expectancy of a career firefighter range from 10 years less than the average population to an incredible 58.5 years old. This, at least, is in part due to the continued low-level exposures to chemicals in their workplace. Decontamination and medical surveillance are discussed in this book with the hope that better education and compliance concerning the use of protective gear will change past trends.

■ **NOTE**

Recent statistics on the life expectancy of a career firefighter range from 10 years less than the average population to 58.5 years old, due in part, at least, to continued low-level exposures to chemicals in the workplace.

Fire Toxicology

Fire is an unpredictable rapid chemical reaction and, depending on the fuel, amount of oxygen present, and the heat generated, an unstable reaction. Therefore, studies relating to this subject have been difficult. Many studies, usually utilizing a firefighter, involve the collection of the gases produced during structure fires. By these studies, scientists are able to determine what chemicals are typically found in fires of this type. Unfortunately, the amount of synthetic materials and the variety of these materials used within structures are changing almost daily. The study of the gases produced during fires involving modern structures look like a hazmat practitioner's worse nightmare.

Most Common Toxicants Found in the Fire Environment

Carbon Monoxide	Ammonia	Nitrogen Oxides
Carbon Dioxide	Hydrogen Cyanide	Hydrogen Chloride
Sulfur Dioxide	Isocyanates	Halogenated Acid Gases
Hydrofluoric Acid		

Other Toxic Gases

Acrolein	Methane	Ethane
Ethylene	Acetaldehyde	Benzene
Toluene	Chromium	

When examining the acute effects that lead to injury or death, these chemicals can be divided into two main groupings: asphyxiants and irritants. Although more chronic in nature, carcinogens and those chemicals that cause long-term effects on organ systems are also discussed.

Asphyxiants   Asphyxiants include simple asphyxiants or those chemicals that displace oxygen. The most common of these found in a smoke environment are carbon dioxide and nitrogen. Chemical asphyxiants that interfere with the transportation or use of oxygen within the body include carbon monoxide, cyanide, and nitrates and nitrites. Asphyxiants incapacitate victims by causing sensory deficits and diminishing the victims ability to reason. If a victim is inside a building involved in fire, these symptoms, combined with decreased visibility and respiratory irritation, make escape almost impossible. Many victims never receive a burn injury but instead die from the respiratory exposure to toxic gases.

Irritants   Irritants are found in both solid particulate matter and gaseous form. The size of the particulate matter determines if the irritant affects the upper airways, lower airways, or alveoli. Particles less than 3 microns can gain entry into the lower airways, while particles at 1 micron or smaller can penetrate the alveolar space. Smoke itself is made up of carbon aerosol and soot which is less than 1 micron and can coat the linings of the alveolar surface. Respiratory irritants found in smoke include acrolein, ammonia, chlorine, hydrochloric acid, nitrogen oxides, phosgene, sulfur dioxide, and formaldehyde. Irritants that reach the respiratory system injure the underlying tissues causing irritation, inflammation, and tissue destruction. If the irritants reach the alveoli, surfactant can be destroyed or displaced. This condition, combined with tissue damage to the alveolar capillary membrane, allows fluid from the intervascular space to easily penetrate the respiratory passageways and further hinder gas exchange.

> ■ **NOTE**
> Particles less than 3 microns can gain entry into the lower airways, while particles at 1 micron or smaller can penetrate the alveolar space.

The signs and symptoms associated with victims of smoke inhalation are related to either injury of the airways or hypoxia caused by some interruption of oxygen transport to the cells. The signs and symptoms associated with irritation include coughing, shortness of breath, tachypnea, bronchial and/or laryngeal spasm, rales, chest pain, and tightness. Those signs and symptoms caused by hypoxia include headache, confusion, dizziness, and coma. Other signs include facial burning, singed nasal hairs, gross soot in sputum, and poor pulse oximetry readings.

> ● **CAUTION**
> Hot steam has 4,000 times the heat capacity of superheated air and can easily produce thermal injuries all the way to the alveoli.

Interestingly, singed nasal hairs and other evidence of superheated air inhalation does not mean that the lower airways are injured. As stated in the respiratory section, the lungs are very efficient when cooling hot air. In reality, unless the air was charged with superheated steam, the lower airways are usually untouched by the hot atmosphere. It must be emphasized that although a lower airway injury may not result from heat exposure, the most common cause of death during the early phases of burn treatment is an upper airway occlusion, secondary to an upper airway burn injury. Hot steam, to the contrary, has 4,000 times the heat capacity of superheated air and can easily produce thermal injuries all of the way to the alveoli.

Victims of smoke inhalation should be held and monitored at the hospital for at least 24 hours, even if no symptoms originally exist. Changes in pulmonary tissue permeability can take place up to 24 hours after an exposure. In these cases the admission x-ray may be unremarkable only to show diffuse infiltrates many hours after the exposure. Long-term changes in the fine bronchioles and alveolar structure have been noted after continued exposures. These subtle changes may be noted on spirometry tests. For this reason, firefighters throughout the country are tested utilizing some form of pulmonary function test on their annual or biannual physicals.

Of course asphyxiants and irritants are not the only dangers noted with exposure to smoke and combustion gasses. Carcinogens are also prevalent in the smoke environment and are responsible for ending the careers of many professional firefighters. One study conducted at twenty-four fires, where firefighters wore air sampling devices indicated that within all twenty-four fires six known carcinogens were found. Among these were acrylonitrile, arsenic, benzene, benzophyrene, chromium, and vinyl chloride. These chemicals, in the levels found during a single fire, would probably leave no lasting effect on an exposed victim, but in the case of firefighters who may experience repeated and combined exposures, the danger of cancer is greatly multiplied.

Danger to Firefighters

Advancements in personal protective gear have allowed firefighters to gain access deeper into structures involved in fire. These advancements have permitted firefighters to make dramatic rescues and fire stops that in past days would have been impossible. As a result, the personal gear is becoming even more contaminated with toxins generated during fire fighting activities. Throughout the nation there is a move to periodically clean fire fighting gear to lessen the continued effects caused by wearing this contaminated gear. Unfortunately, many departments are not universally participating in this practice. Safe storage practices of contaminated gear have been addressed in standards throughout the fire service yet even today, during heightened awareness of contamination, many firefighters are storing their gear in clean clothing lockers, back seats of their autos, and even keeping items such as helmets and gloves in their food lockers. If we are to realize a change in the life expectancy of professional firefighters, these practices must be altered. Many departments are still requiring firefighters to hang their gear in the same place as the apparatus is kept where exhaust fumes penetrate the fabric day after day. This is another practice that should be remedied to lessen the long-term effects that firefighters have experienced throughout the history of the profession.

Changes in the use of self-contained breathing apparatus (SCBA) must also be addressed. Almost universally, fire services are requiring the use of breathing apparatus during fire fighting activities. Most firefighters believe that the danger

● CAUTION

To change the life expectancy of professional firefighters, the habits of storing gear in clean clothing lockers or back seats of vehicles, and of keeping gloves and helmets in food lockers, must be changed.

phase of gas production ends when the fire is no longer free burning. This belief causes many firefighters to remove their breathing protection during the overhaul phase. Studies have shown that for approximately one hour after extinguishment of a fire, the atmosphere within a structure can contain sufficient quantities of carbon monoxide and hydrogen cyanide gas to cause significant inhalation injury and even death.

Treatment

Treatment of victims of smoke inhalation should be aimed at increasing oxygenation to the cells. This can be accomplished by providing 100% oxygen via a non-rebreather mask on the breathing patient. If the patient is not adequately breathing or an oximeter reading indicates saturation less than 90%, intubation should be considered with positive pressure ventilation. If auscultation of the lungs reveals wheezing or other signs of airway constriction, an updraft or aerosolized solution of alupent or proventil should be considered. Other bronchodilators, such as epinephrine or theophylline that are cardiac stimulants, should be avoided as some products found in smoke are sensitizers to the myocardium.

Some literature suggests the use of a cyanide antidote kit because one of the toxic findings in smoke poisoning studies indicates cyanide as a significant toxicant. The cyanide antidote kit primarily relies on the conversion of hemoglobin to a non-oxygen-carrying compound, methemoglobin. By using the kit the rescuer is essentially causing an increased compromise to the respiratory system. Furthermore, some of the products found in smoke are nitrate- or nitrite-based toxins. These toxins convert hemoglobin into methemoglobin, which is the physiologic antidote to cyanide poisoning. For these reasons the cyanide antidote kit in its entirety should not be used to treat smoke inhalation patients. If it is determined that the fuel burning in an environment would generate cyanide, then the patient could be treated with only the sodium thiosulfate portion of the kit instead of methemoglobin stimulating portion of the kit.

Another consideration is that of hyperbaric oxygen (HBO). HBO has been proved to be a viable treatment for smoke inhalation. Because the primary toxins affecting the body are CO and cyanide, it stands to reason that HBO would work. The emergency responder dealing with a smoke inhalation patient should consider the possibility of transporting a patient to a hospital that can provide this treatment instead of transporting to a hospital that does not have the capabilities.

Decontamination is another issue that should be considered. Many books addressing this subject state that decontamination is not needed for the typical victim of a structure fire. Since the variety of chemicals found in these typical fires have already been discussed it is not reasonable to make this assumption. At the minimum gross decontamination should always be done on a victim of fire. This gross decontamination should involve the removal of all the contaminated clothing. The gross decontamination is performed for two reasons. First, it will lessen the exposure experienced by a patient by removing clothing that is holding contaminants against the skin. Many times, because of fire fighting efforts, this clothing is wet, increasing absorption of any chemical agents embedded in the clothing. Second, it lessens the secondary contamination suffered by the fire crews providing care, the EMS transporter, and the hospital staff. The probability of secondary contamination from a fire victim is minimal but does exist. It is better in this field to be safe than sorry.

SPECIALIZED RESPIRATORY EQUIPMENT

Overview

The injuries resulting from exposures to toxic materials can be very different from those encountered with normal medical illnesses, particularly in respiratory injuries where blood gas concentrations and blood conditions may be affected by the exposure. The equipment found useful in chemical exposures, in many cases, is the same equipment, or adaptations of equipment, used for medical emergencies. This section reviews some commonly used respiratory equipment as well as providing an overview of equipment that the reader may not be so familiar with. This section also addresses any specific considerations when using this equipment on a hazardous materials emergency.

Pulse Oximetry

Brief History   The first use of transillumination for medical assessment was done in 1935. In 1970 the current technology was developed that enabled the evaluation of color and opacity in a ratio to be assessed. The newfound technology was only used in research. During the 1980s the pulse oximeter was available for use in surgery, intensive care units, recovery rooms, and emergency departments. During this period, the units were still too fragile to carry on a prehospital care unit. In the early 1990s units were built to withstand the rigors and abuses of street use and became the only mechanical means for prehospital assessment of oxygenation.

Description   Pulse oximetry has been hailed as the fifth vital sign following blood pressure, temperature, pulse, and respirations. Although, it is a very useful tool, especially for patients involved in hazardous materials exposures, it cannot be used properly without knowledge of what the readings signify.

The pulse oximeter indicates, in a percentage rating, the amount of oxygen delivered to the peripheral tissues by the circulating blood. The oximetry reading may indicate changes in circulating oxygen much before there is any recognized change in blood pressure, pulse, or level of consciousness. In this respect, the medical provider may be forewarned of impending changes in the patient's condition prior to any changes in the traditional vital signs.

The oximeter unit works by determining a ratio between oxygen rich hemoglobin (oxyhemoglobin) and oxygen poor hemoglobin (reduced hemoglobin). This is done by evaluating a pulsating arterial bed in an appendage of the body. The most common evaluation sites used are the finger, toe, bridge of the nose, or ear lobe.

The unit utilizes a probe that contains two sensors and a light source. This two sensor unit is the most common type found for prehospital use. Other more sophisticated units have multiple sensors, but these are usually too sensitive for field use. Each sensor detects a slightly different color of light as it transilluminates the selected body part.

Oxyhemoglobin has a brighter and more transparent color than reduced hemoglobin which, in comparison, has a darker more opaque color. Once sensed, the oximeter calculates a ratio, based on brightness and transparency, and displays a read-out that represents the percentage of hemoglobin saturated with oxygen. The probe is designed to detect only a pulsating capillary bed. The pulsation is what identifies the oxygenated arterial side of the capillary bed and filters out the desaturated vascular side.

Oxygen saturation *does not* bear a direct relationship to partial pressure of oxygen as related in a blood gas study. Normal arterial blood has a partial pressure of oxygen between 80 and 100 mm/hg (PaO_2 80–100). Normal arterial oxygen saturation, measured by an oximetry unit, is between 97–99% (SaO_2 97–99%). Therefore, when an SaO_2 reading drops below 94% the provider should suspect respiratory compromise. A reading of less than 91% SaO_2, oxygen therapy with possible intubation should be considered. A SaO_2 of just less than 90% correlates to a PaO_2 of around 60mm/hg.

Although, the oximeter is recognized as a very useful tool, it has its shortcomings. It will only work on a patient with a well circulating (pulsating) capillary bed. If the patient is cold these area capillary beds may have a considerable decrease in circulation, making oximetry impossible. Warming the site will reestablish circulation and enable the oximetry to read the pulsations.

Specific Hazardous Materials Considerations   There are changes in hemoglobin that can affect the ratio measured by the oximeter unit. For example, carboxihemoglobin (HgCO), the combination of carbon monoxide and hemoglobin is seen by the oximeter unit as oxyhemoglobin (HgO_2). HgCO and HgO_2 have similar properties, both are very transparent and bright red in color. Thus, a person acutely exposed to increased amounts of carbon monoxide will fool the oximetry unit and may register an unusually high SaO_2 reading, when in reality the patient may be severely hypoxic.

Methemoglobin (MetHg), a chemically changed hemoglobin molecule, cannot carry oxygen. Methemoglobin is formed when a patient is exposed to organic nitrogen substances, like fertilizers. Methemoglobin is chocolate brown and obscures the light as it passes through the capillary beds. The oximetry reading reflects the opacity found in deoxygenated blood. The reading will be very low and may not reflect the true oxygenation present. Depending on the concentration of methemoglobin in the blood the oximeter may not register at all. Therefore, the oximeter reading should not be relied on in patients suspected of this type of poisoning.

■ **NOTE**

Particles less than 3 microns can gain entry into the lower airways, while particles at 1 micron or smaller can penetrate the alveolar space.

■ **NOTE**

Because HgCO and HgO_2 have similar properties, a person acutely exposed to increased amounts of carbon monoxide may register an unusually high SaO_2 reading on an oximeter unit and may actually be severely hypoxic.

■ **NOTE**

Oximetry readings, which reflect the opacity of deoxygenated blood, may be inaccurate in the presence of methemeglobin, which is chocolate brown and obscures light as it passes through the capillary beds.

The cyanide antidote utilizes nitrogen-based drugs to counteract the cyanide poisoning. The antidote converts a percentage of hemoglobin into methemoglobin. In these cases, the oximeter should not be relied upon for the judgment of how well the patient is oxygenated. Even methylene blue, a strong dye, utilized as an antidote for nitrite/nitrate poisoning, can obscure the vision of the oximeter probe and cause a transient decrease in the reading.

By recognizing the advantages, disadvantages, and the mechanism by which the oximeter unit functions, medical providers can truly utilize this piece of equipment to their advantage, when assessing a poisoned patient. One should remember that the oximetry unit is an assessment tool. Treatment should be based on an evaluation of all of the signs and symptoms presented by the patient. There should never be invasive treatments rendered based solely on what is read off the unit screen.

Bag Valve Devices and Field Ventilators Providing Positive Pressure

Airway adjunct equipment such as bag valve masks (BVM) with pressure exhalation devices, and field ventilators have found a useful place in hazmat medical emergencies. This section evaluates some of the tools available for the hazmat medical provider.

Description   To explain the pressure exhalation devices it is necessary to define certain acronyms that are addressed in this section.

• PEEP (positive end expiratory pressure). This technique is accomplished with the use of an exhaust device attached to the exhaust valve of a standard BVM (Fig. 5-13). It is also found on many of the disposable BVMs on the market today. The device is placed on the exhalation side of the device and usually has an indicator displaying settings from 0 to 15 centimeters of water. By turning the adjustment screw, the provider can set the exhalation device to the desired setting. It is

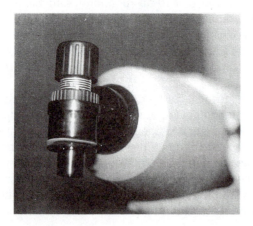

Figure 5-13 *A PEEP valve can be added to a BVM device as shown, or as a built-in part of a disposable device.*

recommended that a setting of between 4 and 6 centimeters of water be used for prehospital patients.

Although the device was manufactured for medical patients, it is useful for the treatment of patients exposed to any number of chemicals, from asphyxiants to irritants. Specifically, patients suffering from chemically induced pulmonary edema and hypoxic states caused by chemical poisonings.

At the end point of passive exhalation, instead of the intrathoracic pressure equalizing with the surrounding air, there remains a pressure of 1–15 centimeters of water (depending on the setting) higher than the surrounding barometric pressure. This positive pressure can only be accomplished when the device is used with a tight-fitting mask or an endotracheal tube.

• CPAP (continuous positive airway pressure). This technique involves the use of the same exhaust/exhalation valve as used on the BVM for providing PEEP. The difference is that CPAP is a constant pressure within the lungs that is greater than outside barometric pressure. The patient is provided with positive pressure on inhalation and during exhalation. The PEEP valve provides the positive pressure of 1–15 centimeters of water at the end of the exhalation cycle.

Positive airway pressure increases the oxygen tension in the lungs and accelerates the diffusion of oxygen through injured alveoli that may be edematous and filled with fluid as a result of a chemical exposure. Furthermore the positive airway pressure accomplished by using a BVM with a PEEP valve is believed to promote reabsorption of edematous fluid within the alveolar spaces, as well as maintaining open air passageways in the small bronchioles.

Several side effects have been noted with the use of positive airway pressures. First, there is a possibility of rupturing a weakened area in the lungs (bleb). The rupture would lead to a pneumothorax or subcutaneous emphysema, and ultimately a deterioration in the respiratory status.

The other problem that occurs as a result of increased airway pressure is difficult or even impossible to detect in the field. By providing higher respiratory pressure throughout the respiratory cycle, the interthoracic pressure also rises. The increase in interthoracic pressure works against blood returning to the heart. This condition will lead to circulatory embarrassment because venous return is impeded when venous pressure rises.

Field ventilators can be of great use for the nonbreathing patient. They can be equipped with these exhaust valves and be set to provide intermittent positive pressure ventilations, keeping rate and depth constant. Once set up and operating, the intubated patient can be ventilated, hands free. This allows the provider to assess and treat other aspects of the patient's exposures or injuries.

Hyperbaric Oxygen Chambers

Hyperbaric oxygen therapy (HBO) can be defined as a patient breathing oxygen while being exposed to a higher ambient pressure in a chamber. The theory and application dates back to the early 1900s when Dr. Haldane first studied the use of pressurized air and oxygen and their effects on the human body. Today, the science of hyperbaric medicine has become well established, treating certain difficult, expensive, or otherwise hopeless medical problems. Each year new applications for its use are identified. The mechanism of action provided by hyperbaric oxygen is twofold. There are those medical problems that benefit from an increase in pressure and those that benefit from increased oxygen in the bloodstream, tissues, and cells.

Unfortunately, there are restrictions on the use of hyperbaric medicine due to the cost of each treatment and the limited availability. Presently there are fourteen accepted uses of hyperbaric oxygen. The uses are determined and defined by the Hyperbaric Oxygen Society of the Undersea and Hyperbaric Medical Society. Of these fourteen uses there are only two related to hazmat exposures. It is recognized for the treatment of acute cyanide poisoning and carbon monoxide poisoning, and because of the close relationship to these two chemicals, smoke inhalation injuries. Studies have shown it is effective in hydrogen sulfide gas poisoning, organic nitrogen poisonings, and several other less common chemicals, but it is not normally used for these injuries. The hospitals are limited, by policy, on the use of the chambers because of insurance regulations. The insurance regulations follow the accepted indications previously listed by the hyperbaric society. That does not mean a chamber will never be used for any other hazmat injuries, but the process by which permission would be granted would take so much time that other treatment modalities could be well under way. We are personally familiar with the use of a hyperbaric chamber to treat a victim of hydrogen sulfide poisoning. Since the poisoning is so closely related to cyanide poisoning it was reasonable to expect the results to be favorable. In fact, in this case there were three victims of the same accident but only one received the hyperbaric treatment. One victim died, one had an extended stay in the hospital, and the patient treated with hyperbaric oxygen was released from the hospital within two days.

There are many places around the country with multiple patient chambers that are ideal for the treatment of patients suffering from a poisoning, complicated by the loss or decrease in level of consciousness. The patient can be placed in one of these chambers, with an attendant at her side, to render any aid needed during the treatment.

Most chambers are capable of about three atmospheres of hyperbaric pressure. The idea of providing oxygen to the respiratory system in a chamber is to increase the partial pressure of oxygen in hemoglobin and in solution within the blood. The increase in PaO_2 is done without an increase in bronchiole pressure as provided with a ventilator or bag valve device equipped with a PEEP or CPAP function. Basically more oxygen with less trauma. Unfortunately these chambers are not available in all areas of the nation.

The Fourteen Accepted Uses of Hyperbaric Oxygen Treatments

1. Acute air or gas embolism
2. Carbon monoxide poisoning

3. Crush injury, compartment syndrome
4. Acute Cyanide poisoning
5. Decompression sickness
6. Enhancement of healing in selected wounds
7. Exceptional blood loss anemia
8. Gas gangrene
9. Necrotizing soft tissue infection
10. Chronic refractory osteomyelitis
11. Radiation necrosis
12. Selected refractory anaerobic infections
13. Preparation for and preservation of skin grafts or flaps
14. Thermal burns

CARDIOVASCULAR ABNORMALITIES AND DYSRHYTHMIAS DUE TO TOXIC EXPOSURES

Several cardiovascular conditions can result from exposure to toxic material. Two types of cardiovascular dysfunctions are most commonly displayed as a result of poisoning: those affecting venous circulation or vascular tone and those related to inadequate heart pumping action. Serious cardiac dysrhythmias such as superventricular and ventricular tachycardias and bradycardias are the most commonly seen conduction malfunctions.

Description

Blood vessels are affected in many different ways. Thrombosis, excessive capillary leakage, internal bleeding, vasodilation, and vasoconstriction have all resulted from exposures to toxins. The most common vascular response noted is persistent hypotension and shock. The provider must make an effort to determine if shock is caused from vascular space expansion or volume depletion. In both instances there is a discrepancy between the capacity of the vascular space and the circulating blood volume. Determining the cause of the hypotension will guide the provider in corrective treatment.

Vasogenic Shock   Vasogenic shock is defined as an increase in vascular space. Vasogenic shock can be attributed to one of two mechanisms.

The first is related to a defect in the responsiveness of vascular smooth muscle to neuro or chemical stimuli. Acute nitrite poisoning results in the relaxation of the vascular smooth muscle causing profuse vasodilation and hypotension. The symptoms witnessed are very similar to those seen with the over-zealous administration of nitoglycerine.

The second can be caused from a decrease in activity of vasomotor centers of the brainstem. Usually due to a hypoxic cause the brain is unable to send neuro stimuli (signals) to the vascular muscles. This condition is usually the late result of an asphyxiant.

In the case of vasogenic shock, fluid challenges are of limited value. The true treatment would be the use of vasoconstrictors such as dopamine. Hyperoxygenation is also important to ensure a continued oxygen supply to the cells during the decreased circulation.

Hypovolemic Shock   Hypovolemic shock is defined as a critically small circulating blood volume. In chemical exposures, hypovolemic shock can be caused by abnormal permeability of blood vessels. This permeability causes leakage of blood or serum from the vessels into surrounding tissues, cavities, the lungs, and gastrointestinal system. Dehydration may be another cause of low volume shock.

Severe pulmonary edema can volume deplete the circulating blood volume causing a twofold problem. Initially, poor oxygenation will result from the decrease in oxygen crossing the alveolar capillary membrane. Next, the decrease in circulating blood volume caused from the leakage of fluids into the lungs reduces circulation and oxygen carrying capability. Severe pulmonary edema can result in the loss of two liters of fluid from the blood.

Increased pulse rate and orthostatic hypotension are early indicators of hypovolemic shock. Treatment of hypovolemic shock involves rapid volume replacement and oxygen therapy.

Heart Failure   Heart failure appears when pumping action does not keep pace with the circulatory requirements of the body. Heart failure, as a result of exposure to a toxin, is evidenced by several different syndromes.

Hypotensive shock (cardiogenic shock) is the result of a decrease in overall cardiac output. Definitive treatment usually involves increasing stroke volume by increasing contractility. The most common way of treating this condition is by digitalizing the patient, a treatment that should be done in the hospital setting under the direction of a physician. The prehospital treatment will be supportive. The use of dopamine is suggested because it will increase blood pressure and increase cardiac output by increasing preload.

Pulmonary edema results from an unbalanced cardiac output. The left heart is failing to keep pace with the right heart. The result is a backup of blood trying to enter the left side. This backup causes an increased pressure in the pulmonary circuit resulting in fluids being forced into the alveolar space.

The treatment is the same as if the condition were of nontoxic origin. Drugs used to decrease the preload and volume are useful. Lasix for diuresis, morphine to increase capillary pooling, and nitroglycerin to increase vascular space all work to decrease the pulmonary pressure.

Chemical agents may also impair cardiac output by producing many types of cardiac dysrhythmias. An EKG will identify the rhythms but there is no way ini-

Figure 5-14

Treatments for superventricular tachycardia stimulated from a toxic response may be difficult to break. The most common drugs carried in the field, such as Adenocard and Brevibloc, may not work to suppress this dangerous rhythm.

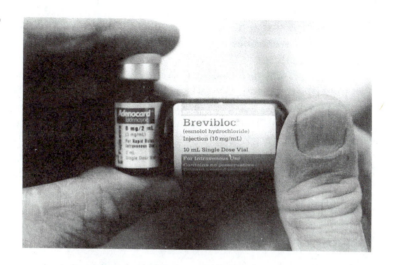

tially to determine if disorders are of toxic or nontoxic origin. Tachycardic rhythms such as superventricular tachycardia can be one of the most dangerous. If not treated, it may lead to a myocardial infarction and cardiac arrest. Chemicals like solvents or one of the halogenated hydrocarbons cause these tachycardias.

A tropical flower, Angel Trumpets, used by some drug abusers to make a hallucinogenic tea causes runaway heart rates of greater than 200 beats per minute. Ecstasy, a hallucinogenic amphetamine has also reportedly caused superventricular tachicardia. Both of these toxic responses are difficult to treat using the standard means to reduce heart rates.

Many drugs are available for treating these conditions (Fig. 5-14). Verapamil, Brevibloc, and adenosine are some of the most common used in the prehospital setting. Other treatment is supportive and geared to maintaining rate, rhythm, and oxygenation. In any case, tachicardias stimulated from chemical exposures are difficult to break and if broken, difficult to maintain.

HEAT-RELATED EXPOSURE AND INJURY

The emergency worker, whether representing the fire service or EMS, needs to recognize the importance of hydration while working in an abnormal environment. This section summarizes the effects of heat on workers exposed to high heat atmospheres while utilizing encapsulating suits or other protective gear.

The idea that workers can survive and produce more if kept properly hydrated in hot environments has been studied for many years. During WW II, Field Marshal Rommel realized he was dealing with extremely hot conditions with his German troops in the Sahara Desert. Rommel wanted to find what would allow his troops to function proficiently in this atmosphere. He devised an experiment that would indicate the best way to keep his troops hydrated. The experiment divided a segment of his troops into three teams—A, B, and C.

Team A was sent out to march as far as they could without consuming any water. They were able to reach about 10 miles before becoming exhausted and unable to continue.

Team B was told to march as far as they could but were given as much water as they desired. Team B was able to march approximately 16 miles.

Team C also marched as far as they could, but were weighed prior to starting. Once every hour the team was stopped, and their lost body weight was replaced with water. Team C was able to march 26 miles. This was a total of 260% further than team A.

Although this study was not conducted using exact scientific specifications, it brings to light the importance of hydration. It also gives some insight into forced hydration or drinking beyond the desire of thirst.

Physiology

The human body maintains a complex balance between heat loss and heat production. Heat is produced as a by-product of metabolism. Simply put, heat is generated from chemical reactions necessary for cellular work in the production of energy, sustaining life itself. Heat produced from cellular metabolism is lost or maintained by the process of radiation, conduction, evaporation, and convection. These processes are necessary for temperature homeostasis within the body.

Some important factors determine the rate of metabolism and therefore, the rate of heat production. First is the basal metabolic rate (BMR), which is the cellular production of energy and an increase of temperature caused by the cell's energy production. The BMR refers to the heat production of an individual while at rest but not asleep. The average BMR for an adult at rest is 70 calories per hour. Under stress, whether the stress is due to eating a large meal, emotions, or hard work, the metabolic rate can increase dramatically, adding further to heat production within the body.

External stressors are also a factor. High or low external temperatures also affect metabolic rates. Warmer ambient temperatures initially increase metabolic rates, but once the body core temperature increases, due to ineffective cooling, the metabolic rate decreases dramatically. This robs the body of energy, causing apathy.

Fluid is lost in a variety of ways on a second-by-second basis. The movement and loss of fluid is done to maintain homeostasis both for the osmotic balance of fluids within the body and for the maintenance of temperature. The gastrointestinal system loses about 100 milliliters per day, the kidneys 30–60 milliliters per hour, the respiratory system 15 milliliters per hour, but the major loss of fluid from the body occurs from the skin in the form of sweat. This loss can be as high as 3 to 4 liters an hour, even during rest in extreme heat.

During exercise, respiratory loss can rise ten times and sweating increases significantly. Two hours of strenuous exercise, such as running, firefighting, or hard work in an encapsulating suit causes the plasma volume to decrease 12–14% just within the first 10 minutes. For the next 110 minutes the loss of plasma volume is only 2–4%. The loss from sweating can be offset somewhat by water transference within the body. The redistribution of water helps maintain fluid volume within the blood stream.

cellular respiration
the use of food and oxygen to form energy (glucose + oxygen = energy [ATP] + heat + carbon dioxide + water)

glycogen
the material stored by the body that can be broken down into glucose when needed

pyloric valve
a valve found in the base of the stomach that regulates the quantity of food or fluid that reaches the bowel

For example, a by-product of **cellular respiration** (glucose + oxygen = energy [ATP] + heat + carbon dioxide + water) is water and can produce up to 1 liter. During exercise the body produces energy by metabolizing glucose. Once the blood glucose is depleted, the body draws energy from **glycogen** stores in the liver. The use of each gram of glycogen releases 3–4 grams of water. A complete depletion of glycogen stores releases up to 2 liters of water. Therefore, while exercising, the body can compensate by a redistribution of up to 3 liters of water into the blood volume. This compensatory mechanism is only somewhat helpful and works best in physically fit aerobically trained individuals.

When it is anticipated that a responder will be working in a hot and humid atmosphere, it is important to prehydrate those individuals. Because the absorption of water is a slow process, hydration must be started early. If not, the fight against dehydration becomes a losing battle.

The **pyloric valve** at the base of the stomach is a regulating mechanism allowing passage of nutrients and fluid into the intestines (this process is referred to as gastric dumping). Because the absorption of water into the bloodstream takes place in the small intestine, the pyloric valve becomes the deciding factor as to how rapidly consumed fluid gets into the bloodstream.

The pyloric valve at its completely open position only allows approximately 800 milliliters per hour to pass. Other factors slow the movement through the pyloric valve. Gastric emptying is significantly slowed when a fluid with a sugar concentration of greater than 3% is consumed. Most "power" drinks have a sugar (sucrose, glucose, fructose) concentration much higher than 3%. In fact, most have higher than a 5% concentration. Although, there is no nutritional benefit to plain water, it is absorbed 35–39% faster than power drinks. Since the glucose stores can be depleted along with important electrolytes, it may be necessary to consider a fluid replacement that can address both problems.

● **CAUTION**

A person working in a hot environment, like an encapsulating suit, can lose between 2,000 and 4,000 milliliters of fluid an hour, so prehydration and forced hydration are critical.

As stated earlier, if hydration is not started early then we will be fighting a losing battle. Remember, the greatest loss of plasma (12–14%) occurred in the first 10 minutes during two hours of exercise. Also, gastric emptying can only occur at a maximum of 800 milliliters per hour. That means that only a maximum of 800 milliliters can reach the small intestine where absorption takes place. Since a person working in a hot environment, like an encapsulating suit, can lose between 2,000 and 4,000 milliliters an hour, it is easy to see that prehydration and forced hydration is important. The importance is further enhanced when the ambient air temperature can reach 100°F, as it so often does in the South, Midwest, and West Coast.

Prior to placing an entry team member (one who enters a hazardous atmosphere) in an encapsulating suit in these conditions, it is suggested to preweigh and prehydrate them. This accomplishes two goals. First, it gives the entry team member a head start on hydration. Second, it gives the medical support personnel a guide to rehydrating the team member postentry.

A loss of 5% body weight between weigh-ins indicates a worker that may be close to a serious dehydration condition. For example, a 200-pound worker loosing 10 pounds (5%) postentry has lost greater than 3.5 liters of fluid. For a worker to totally recuperate from a fluid loss of this magnitude he would need approximately 4.5 hours just for his body to reabsorb that amount of fluid. Furthermore, this amount of fluid loss could indicate that this worker may be teetering on heat exhaustion or stroke.

The important point is that in extreme environments, prehydration is of utmost importance. During long incidents where workers are exposed to high temperatures for extended periods of time or are working hard in these extreme temperatures, it is important to monitor fluid loss and intake, forcing hydration if needed to maintain fluid balance and increase energy output of workers.

If emergency workers are only allowed to drink when thirsty, the battle may be lost and heat exhaustion or more tragic heat stroke may be the outcome. An educated observance by medical personnel on the scene may be the deciding factor as to whether the battle against dehydration is lost or won.

Acclimation

Acclimation may be an important aspect to consider when examining the likelihood of suffering a heat-related injury. The human body has the ability to make subtle changes to adapt to environmental conditions, even a hot environment. If emergency workers perform on a daily basis in a hot environment, their bodies adapt by shifting and storing a higher quantity of water for its expected use. Another part of adaptation is the ability to sweat. Sweating involves a muscular contraction to open and close the gland. A person who exercises these glands regularly, develops a stronger, more active muscle and as a result, sweats much easier than a person who is not often exposed to hot environments. Studies have indicated that persons acclimated to hot environments have a greater plasma volume, allowing them to lose more volume before becoming critical. These studies have also indicated that the acclimated person loses fewer electrolytes during the sweating process.

Every few years heat waves sweep across the northern states causing death to hundreds of persons living there. Yet, in the South temperatures range in the upper 90s to 100s with humidity factors around 90% without a great loss of life. Much of this is due to the acclimation of the persons living in warmer environments. Unfortunately the opposite is true when the cold waves sweep through the South. Understanding acclimation gives medical personnel insight into predicting effects suffered by emergency workers as well as citizens in their districts.

Determining Severity of Heat

There are several methods of determining the severity of heat an emergency worker will be exposed to. Three of the most common indexes are the wet bulb/globe

---

SAFETY

To avoid heat exhaustion or heat stroke, the fluid loss and intake of emergency workers must be monitored and intake forced, if necessary.

acclimation

the ability of the body to adapt to an environment

temperature index (WBGT), the temperature–humidity index (THI), and the humature (H) Fig. 5-15. All are useful to an officer who is concerned with the hostility of the environment his workers are exposed to.

The WBGT takes into account the air temperature, humidity, and radiant heat. The Marine Corps uses this index to determine if the weather is too hostile for training purposes. It is a bit more accurate in predicting the actual severity of the conditions an emergency worker is exposed to. The THI, on the other hand, takes into account the humidity, dry bulb temperature, and the dew point to calculate a realistic temperature, much like the wind chill factor. The humature utilizes the dry bulb temperature and dew point to calculate a realistic temperature. The humature is much easier to assess with only limited information. All of the heat indexes are useful in assessing how long a worker can be expected to function in the environment and how much recuperation time may be needed. During the summer months all EMS units should have a chart readily available for use in the field.

Effective Temperature   The temperature of saturated air given the same thermal sensation as any given combination of temperature and humidity, effective temperature (ET) is a direct ratio between ambient temperature and humidity.

Temperature–Humidity Index   Formerly called the discomfort index, the temperature–humidity index (THI) is calculated as follows:

$$\text{THI} = 0.4(T^* + T\text{w}) + 15$$

where T^* = ambient air in °F and $T\text{w}$ = wet bulb temperature.

Humature   Using dry bulb temperature and dew point, humature (H) is figured as follows:

$$H = T^* + (p - 21)$$

where T^* = ambient air in °F and p = the vapor pressure of ambient air in millibars.

Wet Bulb/Globe Temperature Index   Considering air temperature, humidity, and radiant heat, the wet bulb/globe temperature index (WBGT) is calculated as follows:

$$\text{WBGT} = 0.2T_g + 0.1T + 0.7T_w$$

where T_g = globe temperature, T = ambient air in °C, and T_w = wet bulb temperature.

Metabolic Thermoregulation

As stated earlier, the body uses several methods for maintaining the balance of temperature within the body. "The body temperature is dependent on a sensitive balance between heat production and heat loss" (Vance, 1985). Conduction, con-

RELATIVE HUMIDITY									
	10%	20%	30%	40%	50%	60%	70%	80%	90%
104	98	104	110	120	132				
102	97	101	108	117	125				
100	95	99	105	110	120	132			
98	93	97	101	106	110	125			
96	91	95	98	104	108	120	128		
94	89	93	95	100	105	111	122		
92	87	90	92	96	100	106	115	122	
90	85	88	90	92	96	100	106	114	122
88	80	84	85	87	90	92	96	106	109
86	80	84	85	87	90	92	96	100	109
84	78	81	83	85	86	89	91	95	99
82	77	79	80	81	84	86	89	91	95
80	75	77	78	79	81	83	85	86	89
78	72	75	77	78	79	80	81	83	85
76	70	72	75	76	77	77	77	78	79
74	68	70	73	74	75	75	75	76	77

TEMPERATURE IN F°

NOTE: Add 10 degrees when protective clothing is worn and add 10 degrees when in direct sunlight.

Humature	Danger Category	Injury Threat
Below 60	None	Little or no danger under normal circumstances
80–90	Caution	Fatigue possible if exposure is prolonged and there is physical activity
9–105	Extreme caution	Heat cramps and heat exhaustion possible if exposure is prolonged and there is physical activity
105–130	Danger	Heat cramps or exhaustion likely, Heat stroke possible is exposure is prolonged and there is physical activity
ABOVE 130	Extreme danger	Heat stroke imminent!!

Figure 5-15
Humature heat stress index.

vection, radiation, and evaporation are all methods the body uses to loose or, in some cases, gain heat.

Conduction   First, conduction is of limited use on the emergency scene. Conduction is defined as heat transferred to a substance or object in contact with the body. It accounts for approximately 3% of heat loss. Unless the emergency worker is submerged in cool water, conduction plays a limited role.

Radiation   Radiation is the loss of heat in the form of infrared heat rays. This form of heat loss can account for approximately 60% of normal heat loss. When the temperature of the body is greater than the surroundings, a greater quantity of heat is radiated from the body to cooler objects in the surroundings, such as floors, walls, and ceilings. Conversely, heat is gained by the body when the surroundings are at a higher temperature. The efficiency of radiation is demonstrated to firefighters when the heat from an actively burning fire is apparent from hundreds of feet away.

Convection   The circulation of cooler air around our bodies accounts for 10–15% of our heat loss. The more air movement provided around the body, the higher the heat exchange. This movement of air is termed convection.

Evaporation   Evaporation accounts for 20–30% of the body's heat loss that is generated when sweat is produced and vaporized absorbing heat. The rate of heat lost to water through evaporation is many times greater than the heat lost to air at the same given temperature.

The loss can be further described as a loss of 0.58 calories of heat for every gram of water that evaporates. This loss causes continued heat loss at a rate of 12–16 calories per hour. Under normal basal rates 150 milliliters of water will remove all the heat produced per hour.

When the body becomes overheated, large quantities of sweat are excreted to the surface of the body in an effort to cause evaporative cooling of the skin. This occurrence takes place when high core temperatures stimulate the preoptic area of the hypothalamus, which sends excitation signals via the autonomic nervous system to the sweat glands. Cooling of the skin directly effects the core temperature of the body. This cooling is done through conduction of heat from the blood supply found at the surface into the skin.

When the core temperature elevates, profuse dilation takes place in the capillary beds of the skin. During cool temperature conditions, the blood flow to these capillary beds may, in theory, be virtually zero. During high temperatures when core temperature rises, the capillary beds may profuse up to 30% of the cardiac output. It is easy to understand why the skin is termed the "radiator of the body."

Cooling vests were once a popular item for hazmat responders working in encapsulating suits. These vests were filled with cold water and worn over the thorax in an attempt to cool the core temperature. Many theories devel-

■ NOTE

Because of the
possibility that heat at
the core is retained
even if the surface is
cool, the use of cooling
vests may contribute to
heat emergencies.

oped from the use of such vests including one stating that these vests may in fact contribute to heat emergencies. The theory based its findings through the principle stated previously. Cooling of the core takes place on the fact that dilation of the surface vessels allows 30% of the cardiac output to rise to the surface where heat is given off. When a cool vest is worn, the surface vessels constrict holding in the heat at the core even if the surface is cool. It is because of theories such as these that cooling vests are no longer a popular item in the hazmat response field.

Heat generation in the body at rest, is a direct result of the basal metabolic rate. Unfortunately, when the core temperature rises, so does the metabolic rate resulting in more heat production. Under extreme conditions the body can excrete up to 4 liters of water per hour. This converts to 200 calories of heat loss or 28 times the normal basal metabolic rate of heat production. The increase in heat production can all too easily get out of hand. The limits of temperature the human body can withstand are directly dependent on the humidity. If the air is dry and there is sufficient air flow to allow evaporation, the body can withstand temperatures up to 150°F for an extended period of time (up to 2 hours). On the other hand if the air is 100% humidified, the maximum temperature withstood for a limited time is only 94°F. The maximum temperature safely withstood drops to 85–90°F if the person is involved in heavy work.

Effects of Heat within an Encapsulated Suit

Temperature, humidity, and solar load (full sun without the benefit of clouds, or of any shade) can greatly increase heat stress on all response personnel. The hostile environment presented by these factors will mostly affect those working in level A and B, however, can to a lesser or greater extent, affect anyone on the scene of an incident.

In order to appreciate the temperature that will be sustained within a suit, the adjusted temperatures as they relate to the solar loading must be examined. Although it is uncertain if this model will truly pertain to a level A encapsulated body, it will give some insight to what an entry team member might be experiencing. The following calculations have been used when determining solar loading and also in biological monitoring. It is felt that the same comparison can be made with the wearer of an encapsulating suit although it has not been proved to have a direct corrolation.

Within an encapsulated suit, a team member is affected, not only by the effects from solar loading, but also from the effects of heat generated inside of the suit. Examining the interior effects combined with the exterior effects, these calculations do not seem unreasonable.

The average temperature can be equated to 70% of the internal body temperature plus 30% of the surface temperature of the suit or:

$$\text{Average Temperature} = .7 \ (\text{internal temp.}) + .3 \ (\text{surface temp.})$$

Internal temperature is denoted as the core body temperature which is usually .6 to 1 degree higher than the oral temperature of 98.6. So let's work through a practical scenario. As the individual enters the suit in a relaxed state (normal body temperature without any environmental stress) and within a shaded area of approximately 85°F with a 80% humidity, the suit temperature would be roughly:

$$\text{Average Temperature} = .7(\text{internal temp.}) + .3\ (\text{surface temp.})$$
$$= .7(99.6) + .3(85)$$
$$= 95.2°F \text{ adjusted to humidity} = 97°F$$

Now let us assume that the core temperature remains the same for the first 5 minutes, however, once in the sun, the external suit temperature rises giving an internal temperature as follows:

$$\text{Average Temperature} = .7(\text{internal temp.}) + .3\ (\text{surface temp.})$$
$$= .7(99.6) + .3(90)$$
$$= 96.72°F \text{ adjusted to humidity} = 136°F$$

Another 5 minutes go by with the external suit temperature increasing to 100°F and internal humidity increasing from the ambient 80% to 90%:

$$= .7(99.6) + .3(100)$$
$$= 99.7°F \text{ adjusted to humidity} = 156°F$$

A few minutes later the core temperature starts to rise and the external suit temperature, due to intense sunlight, has risen to 110°F. Internal temperature could rise as follows:

$$= .7(99.7) + .3(110)$$
$$= 102.79°F \text{ adjusted to humidity} = 191°F$$

Another 5 minutes go by with a total of 20 minutes within the suit. Humidity has now reached 100% with internal temperatures around 102°F and solar load to 115°F:

$$= .7(102.79) + .3(115)$$
$$= 106.45°F \text{ adjusted to humidity} = 205°F$$

The foregoing discussion is hypothetical utilizing a heat stress formula incorporating humidity factors. The numbers show a dramatic increase in temperature once the suit is closed and work is performed associated with solar load. The normal core temperature for healthy humans is between 98° and 99°F. Although humans can withstand variations in this temperature that can span between 95°F up to 104°F without sustaining injury, once the core temperature reaches around 107°F, body temperature thermoregulation fails and heat stroke becomes imminent. This example highlights the importance of having a medical officer who is knowledgeable on the subject of heat stress.

Factors Contributing to Heat Emergencies

Heat Exhaustion   Heat exhaustion is one of the most commonly seen, but the least serious, heat emergencies. It is caused from the loss of fluid and sodium as a result of excessive sweating. The dehydration and sodium loss accounts for the symptoms generally seen by medical responders that include positive orthostatic vital signs with accompanying dizziness or syncope, headache, nausea and vomiting, diarrhea, and decreased urinary output, this combined with the history of working in a hot environment and possibly poor fluid intake. Treatment includes moving the patient to a cool environment and rehydrating with an IV of 0.9% sodium chloride or Lactated Ringers solution. If heat exhaustion is left untreated and allowed to progress, heat stroke may result.

Heat Cramps   Heat cramps result from a disproportunate loss of fluid and electrolytes from the body. In a hot environment, sweat, which is high in sodium, is suddenly lost. This loss results in the skeletal muscles suffering in a deficit of sodium causing intermittent cramping. The cramping usually occurs in the abdomen, legs, arms, and fingers. Treatment includes moving the patient into a cool environment and giving fluids. It is not usually necessary to replace sodium because the sodium has just been disproportionately lost and will be replaced once transported from other areas of the body.

Heat Stroke   Heat stroke is the complete failure of the thermoregulatory mechanism and can be observed by cerebral dysfunction. Symptoms include tachycardia followed by bradycardia, hypotension, rapid shallow respirations, seizures, decreased level of consciousness into coma. Heatstroke becomes imminent when core temperature approaches 42°C or 108°F. Mortality results from cerebral, cardiovascular, hepatic, or renal failure. The estimated mortality from heat stroke is from 50–80%. The wide variation is due to the chronic injures sustained from the initial insult that causes death at a later date.

As the core temperature rises 1°F, the metabolic rate rises approximately 6% further adding to the core temperature. Once the core temperature rises to 110°F, the basal metabolic rate doubles. Furthermore, when the hypothalamus becomes overheated, its heat-regulating ability becomes greatly depressed and sweating and capillary dilation diminish. This vicious cycle, unless abruptly interrupted, will eventually cause death.

The patient is often found unconscious. The onset of symptoms usually happens quickly and without warning. There is sudden collapse with a rapid loss of consciousness. Normally, the skin is hot and dry, although, there have been some reported cases of continued sweating. Central nervous system damage and cerebral

! SAFETY
The wide variation in the estimated mortality from heat stroke, 50–80%, is due to chronic injuries sustained from the initial insult that causes death at a later date.

edema, due to hyperthermia, is evidenced by seizures, confusion, delirium, and disorientation. If not rapidly treated, the patient loses bowel and bladder function and develops a deep coma.

Dehydration has a direct effect on the thermostatic setting of the hypothalamus. Those that become dehydrated actually have a higher core temperature. Therefore, it becomes much easier for those who are dehydrated to reach that point of no return due to the increased rate of metabolism and decreased rate of sweating. This further explains the need to maintain an adequate intake of water in hot surroundings. How would the temperature of your automobile be affected if the radiator was only half full?

When medical personnel deal with hazardous material team members who become exposed to hostile temperatures in an encapsulating suit, it becomes a high priority to develop a method of determining the hydration status of those involved. The best indicator of the hydration status of entry team members is by comparing weights prior to starting the evolution (Fig. 5-16) and continuing to monitor their weight throughout the exercise. The change in weight must be directly correlated to the loss of water. This loss can be replaced with forced hydration (hydrating until the entry weight is regained). Understanding the principles of the movement of water and the absorption within the body will aid medical personnel in determining the readiness of the team member to return to further suit work.

● **CAUTION**

Due to an increased rate of metabolism and a decreased rate of sweating, dehydrated individuals have a higher core temperature and are thus more likely to reach the point of no return without adequate intake of water in hot surroundings.

Figure 5-16
Calculating preentry weight and comparing it to postentry weight is one way to determine the hydration status of the entry team personnel.

The calculations below can be used when utilizing a direct weight replacement method of rehydration.

8.35 pounds	= 1 gallon
3,780 milliliters	= 1 gallon
1 pound	= 453 milliliters
3.78 liters	= 1 gallon
1 oz (fluid)	= 32 milliliters
15 oz fluid	= 1 pound

Temperatures within an encapsulating suit can reach well in excess of 100°F. Core temperatures of a suit worker can approach 104–105°F. Because heat stroke and the detrimental effects of heat stroke occur around 106°F, heat loss/gain, along with fluid retention is a paramount concern when dealing with hazardous materials incidents. This remains true regardless of the level of suit needed for entry.

COLD-RELATED INJURIES

When discussing injuries related to environmental factors, exposures to the cold should be discussed. This section deals with cold injuries related to environmental factors but concentrates on, to a much greater degree, injuries related to hazardous materials exposure, primarily acute frostbite injuries. During today's emergency service response, the use of rehab units and other resources to relieve crews from hostile environmental conditions, is more the standard than the exception. Therefore, cold injuries due to the weather have become less likely than ever before. But, instantaneous cold injuries from the release of cryogenic gases and other escaping compressed gases is a real possibility and is addressed in more depth.

● **CAUTION**
Instantaneous cold injuries from the release of cryogenic gases and other escaping compressed gases is a real possibility.

Types of Cold Injuries

Cold injuries are separated into two categories: those that result from freezing of tissue, namely frostbite, and those that involve nonfreezing injury to tissue, called chilblain and trenchfoot (also called immersion foot).

Chilblain   Chilblain is a mild form of cold injury and is usually associated with a long exposure to a cold atmosphere against the bare skin. The perfect scenario causing this injury is the exposure to blowing, wet, cold wind against the face or hands. The skin indicates damage by turning red, dry, and rough. Edema and skin sloughing are common later signs. The injury usually heals once the exposure to cold is stopped.

Trenchfoot or Immersion Foot   Trenchfoot or immersion foot has been a common wartime type injury. Both received their names from injuries commonly suffered during WWI, WWII, and the Korean War. Trenchfoot resulted from soldiers being pinned down in trenches with their feet submerged in cold water or mud (32–50°F) for periods of 10–12 hours. Immersion foot, at one time thought to be a different injury but later discovered to be virtually the same, was a result of ship-wrecked sailors spending days in lifeboats with their feet submerged in cold water. This type of exposure results in a lack of circulation to the extremity, causing a white or cyanotic appearance of the skin and later severe swelling. If the extremity was removed from the environmental condition and rewarmed, complete recovery within 10 days was the usual outcome. In wartime, unfortunately, the more common occurrence was to partially rewarm the injured extremity, then return to battle. This resulted in poor circulation, tissue necrosis, gangrene, and eventual extensive tissue loss and amputation.

Exposures to Cryogenics   Although emergency services has been related to combat situations, these types of environmental injuries would be rare to employees of the services. What is important to the emergency responder, when dealing with hazardous materials, is the understanding of acute freeze injuries as a result of the exposure to cryogenic materials. Cryogenic materials have the ability to penetrate most of the protective components used for encapsulating suits. The temperatures are so severe that even if the suit is not breached as a result of a frank exposure, tissue damage can still result because of conduction through the material and into the skin. These exposures can cause surface tissue injuries or transverse the skin and injure tissues well below the surface.

cryogenics
the field of science that deals with very low temperatures

What is a cryogenic? **Cryogenics** is the field of science that deals with very low temperatures. In the hazardous materials sense, cryogenic is described as a gas that exists in a liquid form through a process of refrigeration and compression. To define it even further, the temperatures of the cryogenic products must be addressed. Cryogenic temperatures range from absolute zero (in theory) as the lower limit, and −101°C (−150°F) as the upper limit. The gases usually stored or transported in a liquid cryogenic state are:

Oxygen

Hydrogen

Nitrogen

Neon

Argon

Carbon Dioxide

Helium

Frostbite Injuries   The injury that results from exposure to these cryogenics is usually an acute frostbite injury. Frostbite, which occurs as a result of tissue freezing, is a devastating wound that usually leaves scarred tissue and sometimes loss of extremities or appendages. Because the liquid portion of tissue is not water, but electrolytes mixed in water, the temperature that is required to cause a freezing injury is several degrees less than 0°C (32°F). The tissue must reach a temperature of −3° to −4°C for frostbite to be caused. The ambient temperature therefore, must be at least −6°C, as the tissue temperature is influenced not only by the external temperature, but also by the internal temperature. The most frequently injured areas, due to frostbite are the feet, hands, nose, and ears.

Once the tissue is exposed to the cold temperatures, ice formation results within the extracellular fluids. The formation of ice crystals increases the osmotic pressure in the extracellular area and draws fluid from the intracellular space. The final results may involve several structures. First, the cells suffer severe dehydration, and if the exposure is of long enough duration, **cell lysis** results. Next, the formation of ice crystals can disrupt the vascular structures and surrounding tissues. The final result, once the area is rewarmed, can include vascular rupture and cellular death.

Rewarming of the injured tissue causes red blood cell sludging that leads to thrombus formation. The already compromised vessels supplying blood to the injured tissue may become blocked and, as a result of the injury, begin to leak. These conditions lead to a decreased circulation and ultimately, **necrosis** of the tissue. Not unlike heart muscle injury, anything that compromises circulation to the area or decreases the concentration of oxygen in the blood will increase the area of injury.

Different tissues within the body respond to cold exposures differently. The tissues most sensitive to cold injuries include the nerves, blood vessels, and muscle. Moderately sensitive tissues include the skin and connective tissue, whereas very resistant tissues are the bones and tendons. Environmental conditions also factor in on the severity of the injury. An increase in relative humidity and wind conditions improves the chances for injury. The potential for injury is also increased when the tissue is exposed to volatile liquids such as gasoline. Wet clothing or contact with metal also increases the chance of cold injury.

Other physiologic and environmental factors should be considered when judging how severe the cold injury can be. Nutritional status is an important factor. If the cells are starved for nutrition, then they have little reserve to deal with a cold injury and, as a result, the tissue damage will be markedly worse. Studies have also demonstrated that race may also be a factor. Some studies have identified blacks as having as much as six times more susceptibility to cold injuries than do whites.

Drugs and medications can also have a detrimental effect on cold injuries. Any drug that causes a constriction of peripheral circulation or altered thermal adaptation may increase the chances, or contribute to the worsening of a cold injury. Vasoconstricting drugs such as caffeine or nicotine can cause hypoxia and rapid heat loss to an extremity. Vasodilators, like alcohol, contribute to a loss of core heat and cooling of the blood eventually leading to hypothermia. Sedatives such as alcohol, barbiturates, and narcotics decrease metabolic activity and ulti-

cell lysis
the destruction of cells

necrosis
death of tissue

mately body heat production. As the body core temperature decreases so does the ability to rationalize and problem solve, further contributing to a longer exposure to environmental conditions. Any elevation in activity increases core body temperature and lessens the chance of developing a cold injury.

Assessment   During the onset of frostbite the sensation of pain diminishes because the pain impulses are no longer sent from the injured nerve fibers. At temperatures less than 10°C the pain impulses virtually cease and are replaced with a feeling of tingling or numbness. The tissues appear blanched then become rock hard and develop a frosted appearance.

After rewarming occurs the victim will complain of throbbing pain for three days to several weeks, depending on the severity of the injury. Sometimes blisters occur on the injured tissue one day to a week later. The frostbitten tissue will eventually turn black. During the next 3–4 weeks the black tissue will gradually slough off and a defined border separating viable tissue and dead tissue will become apparent. As the healing process continues the dead tissue will continually separate and eventually a spontaneous amputation will occur.

Frostbite is rated in degrees, very similar to thermal burns. The evidence left by frostbite will determine the degree of injury suffered by the patient. For example, a frostbite injury affecting only the top layer of skin is deemed first degree. Second degree is a partial thickness injury. Third degree involves a full thickness injury, while a fourth degree demonstrates complete necrosis and tissue loss to the bone or underlying structures.

Treatment   The treatment of cold injury involves rapid rewarming. This can be accomplished by submerging the affected part in warm water of between 90° and 108°F for 20–40 minutes. This can be safely done in the field as long as there is no chance of refreezing and the water temperature can be maintained. If the extremity becomes refrozen after thawing, the injury can become much worse. The rewarming process is very painful, therefore analgesics should be given. During the rewarming process, a flushing of the injured tissue will be seen indicating that blood vessel dilation has occurred. This is not necessarily a sign that all of the tissue will survive. This assumption can not be judged until weeks after the injury, once the healing process is well under way.

Hyperbaric oxygen, as definitive therapy, has proven in some cases to decrease necrotic tissue formation and improve healing time. The use of hyperbaric oxygen on necrotic tissue healing is one of the recognized uses of hyperbaric oxygen. (See hyperbaric oxygen in the respiratory section.)

Scenario

You are the paramedic on Ambulance 5, an ALS unit that specializes in hazardous materials emergencies. You have been dispatched in conjunction with two other ALS units to assist a fire unit already on the scene of a small fire where several firefighters are reported down. Upon your arrival, the incident commander informs you that the fire involved a residential garage. The fire was knocked down in about 15 minutes with overhaul taking an additional 30 minutes. The initial complaints by the firefighters were brushed off as complaints due to the hot summer day. The first firefighters on the scene complained of burning of the eyes as they approached the house before they put on their breathing devices. Once the fire was knocked down and overhaul was under way, several firefighters removed their masks to complete the overhaul, two of them removed their jackets because of the 98° temperature and a reported humidity of 85%.

A quick visual assessment indicates that four firefighters are injured, all having shortness of breath, a productive cough, and complaining of burning in the chest. Two of the firefighters are cyanotic, have a decreased level of consciousness, and are displaying involuntary muscle jerks in both their legs and arms. All have reddened eyes and are tearing profusely.

Scenario Questions

1. Is decontamination of the patients immediately necessary? Why or why not?

2. If you have decided to decontaminate, what areas of the body would be targeted for extensive decontamination?

3. What routes of entry has the chemical probably taken?

4. Would you consider this a systemic poisoning?

5. Would it be more important to find out the chemical affecting the firefighters or to treat the symptoms displayed by the firefighters?

6. Could heat be a factor in the absorption of the chemical?

7. Could heat be a factor in the symptoms displayed?

8. When a full assessment is possible, what types of data should be gathered on these patients?

9. What specialized equipment may be needed for the treatment of these patients?

10. Do any special arrangements have to be made with the hospital prior to transport?

Chapter

6

Treatment Modalities

Objectives

Given a hazardous materials incident, the ALS provider should have the competencies necessary to provide advanced life support to persons injured by chemical exposures. Students completing this section are expected to recognize the pathophysiology of individual poisonings and to determine the cause for signs and symptoms displayed by the patient.

As a hazardous materials medical technician, you should be able to:

- Identify the routes of exposure, means of decontamination, pathophysiology involved in the exposure and resulting injury, signs and symptoms expected, basic and advanced life support for the injury, and the common places that these materials are found.

Carbon monoxide	Phenol and phenol-containing products
Cyanide and cyanide-related materials	Organophosphates and carbamate insecticides
Hydrogen sulfide	
Hydrofluoric acid and hydrogen fluoride gas	Antipersonnel chemical agents
Nitrites and nitrates	Radiation and radioactive materials.

This chapter addresses treatment modalities for specific poisonings. The treatments presented within this book are recognized acceptable treatments. We urge those of you following this book to obtain the approval of your local medical director before placing any of these procedures into service. A written set of protocols, located at the end of this chapter, can be used in their entirety or modified to meet your local needs.

In most cases of hazardous materials exposures, the health care provider will be treating signs and symptoms rather than a specific chemical exposure. This chapter not only outlines specific treatments for poisonings but also the pathophysiology, signs and symptoms, and the rational for each of the treatments.

It should also be stressed that treatment of contaminated patients be withheld until proper decontamination is done. By providing treatment to a contaminated patient the care giver may further risk causing contamination to himself or worsening the outcome of the patient by delaying the decon process. Whenever possible decontamination of the patient should be done on the scene. There should never be a time when a contaminated patient is transported to a hospital without properly protecting the attendants, the ambulance, and forewarning the hospital.

CARBON MONOXIDE POISONING

Carbon Monoxide

Properties: Colorless, odorless gas. Soluble in water, benzene, and alcohol. Specific gravity of 0.96716.

Explosive limits: 12–75%.

Uses: Organic synthesis, metallurgy.

History

Carbon monoxide poisoning and its symptoms were first described by John Scott Haldane in 1919. Dr. Haldane was an industrial medical doctor whose responsibilities included the care of workers building tunnels under waterways. Within the pressurized tunnels, crude gasoline equipment and torches were used. Haldane discovered that the sickness affecting the workers had to do with the gas, carbon monoxide. He also realized that giving the workers oxygen while under pressure (within the tunnels) caused the workers to respond faster than when the same treatment was done outside the tunnels. This thinking helped develop the hyperbaric treatment for carbon monoxide poisoning used today.

Lanesboro, Minnesota, June 26, 1992. Accidental death was the ruling of the county coroner in the carbon monoxide poisoning of a 37-year-old-man. The man died when he attempted to gas a rat that had moved into his attached garage. He apparently started his car and left it running in the garage. Eventually the gas seeped into the house causing a toxic atmosphere, killing the man.

Pathophysiology

Carbon monoxide (CO) is an odorless, colorless, and nonirritating gas. It is produced as a by-product of incomplete combustion and is used in manufacturing and laboratories (Fig. 6-1). Common exposures to this chemical involve smoking, combustion engine exhausts, faulty gas and kerosene stoves, space heaters, charcoal and sterno fires, and industrial facilities. The gas can only enter the body via the respiratory system where it crosses the alveolocapillary membrane and enters the bloodstream. Once in the bloodstream its effects are systemic.

Although CO is odorless and colorless, some victims of an acute exposure report an acidic taste in the mouth. It is important for those working in the fire or emergency medical services to have a high suspicion of the presence of the gas because of how it is generated or used and the signs and symptoms displayed by the victim. Year after year news stories about families being killed by malfunctioning heating systems within their homes are found in the paper and on TV. Smoke from structure fires generates an enormous amount of CO especially in the smoldering phase. Approximately 12,000 Americans die each year in fires; 80% of the deaths are attributed to the breathing of toxic gases including carbon monoxide.

Methylene chloride (dichloromethane) is a halogenated hydrocarbon found in products used in both industry and around the home. Most commonly methylene chloride is found in items such as paint strippers, insecticides, aerosol propellants, and Christmas tree bubble lights. Once it gains entry into the respiratory system it is rapidly absorbed and then slowly metabolized into carbon dioxide (70%) and carbon monoxide (30%). Because methylene chloride is fat soluble it concentrates in fat cells, specifically the liver, where it gradually releases, causing a long-term exposure to carbon monoxide. Exposures to paint strippers containing methylene chloride has led to the death of workers from carbon monoxide poisoning.

The symptoms of acute exposure may appear suddenly and without warning. The concentration of this gas within the body and the effects experienced are

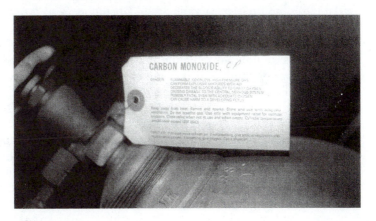

Figure 6-1 *Carbon monoxide is available in high pressure cylinders for industrial use. A leak from a cylinder of this type may go unnoticed until injury occurs.*

directly related to the physical activity and susceptibility of the victim and the percentage of the gas in the environment. There is presently no field testing devices capable of diagnosing carbon monoxide within the blood. Many times the signs, such as bright red skin, come late in the exposure. The victims rapidly become disoriented and lose muscular coordination. This is one of the contributing factors to the high loss of life, even in residential fires where the victims are well aware of their surroundings and escape routes.

carboxyhemoglobin (COH)

hemoglobin bound with carbon monoxide

Once CO is inhaled, it crosses the alveolocapillary membrane and enters the bloodstream. In the bloodstream it rapidly combines with hemoglobin, forming a strongly bonded compound called **carboxyhemoglobin** (COHg). Hemoglobin's affinity to CO is between 200 and 250 times as great as it is for oxygen. Therefore the bond formed with CO is also a much stronger bond than that formed with oxygen, making the problem even more difficult to treat. Carbon monoxide is also attracted to and binds with an enzyme found within the cell called *cytochrome oxidase.* This enzyme is necessary for cellular respiration and the production of energy. This physiology is not an early event in CO poisoning because unlike hemoglobin, cytochrome has a greater affinity for oxygen. The binding between carbon monoxide and cytochrome only takes place once there is a lack of oxygen supply to the cells as a result of CO-induced hypoxia. In fact oxygen's affinity for cytochrome oxidase is nine times greater than it is for carbon monoxide. Some toxicologists believe that this mechanism is responsible for the symptoms noted during carbon monoxide poisoning and not the widely believed physiology caused from the lack of oxygen transport to the cells.

Because of the distinctive manner in which carbon monoxide affects the body, it is classified as a chemical asphyxiant (prevents the usage of oxygen by the cells).

Carbon monoxide causes injury to the tissues by combining with hemoglobin (Hg) to form a non-oxygen-carrying compound, carboxihemoglobin. The first signs and symptoms displayed are caused from the organs with the highest oxygen consumption, specifically the central nervous system and the myocardium. The complaints are related to hypoxia suffered by these organ systems. The most commonly reported symptoms are headache, dizziness, and agitation, related to the effects of hypoxia on the CNS. Other symptoms such as chest pain can be related to myocardial hypoxia. Sudden exposures to very high concentrations may cause almost immediate syncope and cardiovascular collapse.

The sign of cherry red skin reported on carbon monoxide victims is directly related to the color of the victims' blood. Carboxihemoglobin is bright red and somewhat transparent. Once a victim is poisoned by CO, both her arterial and venous blood is bright red giving her skin the characteristic red appearance. This sign is not reliable and is sometimes late in coming. The absence or presence of this sign should not be relied upon to determine a field diagnosis.

This bright red and translucent appearance fools the pulse oximeter into reading carboxyhemoglobin as oxyhemoglobin. Therefore, the oximeter reading on a patient suffering from carbon monoxide poisoning should not be relied upon to estimate the oxygenation status of the patient. In fact, a patient severely poisoned with CO will read unusually high with an expected SaO_2 reading of 100%.

Carbon monoxide is also a normal end product of metabolism so expected levels of COHg are found in the blood of healthy individuals. These levels are usually approximately 0.85% for the healthy nonsmoking person. The level for moderate smokers is somewhat higher at 4%. Symptoms are not usually experienced by exposed individuals until the level is over 10%.

Several important factors influence the actual level of carbon monoxide within the blood of an exposed victim. Pregnant patients are an extreme concern to caregivers. Pregnant patients exposed to carbon monoxide have shown levels of carboxyhemoglobin twice as high in the fetus than that of the mother's level. The half-life of total carboxihemoglobin in fetal blood is around 15 hours and even with oxygen therapy carboxyhemoglobin takes five times longer to reduce.

Children also exhibit a sensitivity to the exposure of CO. They appear to be especially vulnerable because of their higher respiratory and metabolic rate and decreased volume of blood as compared to an adult.

Patients with chronic lung diseases, such as chronic obstructive pulmonary disease (COPD) or asthma, are also at a much higher risk. These patients already suffer from decreased oxygen-carrying capability and even a minor exposure can be devastating to this group.

Concentrations, exposure times, and the physical activity of those exposed all play a part in determining the severity of the effects. The annual death rate attributed to both accidental and intentional CO inhalation has reached over 3,500 in the United States.

● CAUTION

Pregnant patients, children, and patients with chronic lung disease are especially vulnerable to carbon monoxide poisoning.

Signs and Symptoms

The following concentrations of carboxihemoglobin elicit certain symptoms. These particular symptoms are generalizations and not found in all patients at the represented percentages. At best they are approximations of predictable events.

Carbon Monoxide Levels	Signs and Symptoms
0–10%	No symptoms
10–20%	Tightness across forehead and headache
20–30%	Headache and throbbing temples
30–40%	Severe headache with nausea, vomiting, and dim vision
40–60%	Coma and convulsions
Greater than 60%	Cardiovascular collapse and respiratory failure

Other symptoms caused during CO poisoning can mimic neurologic conditions such as multiple sclerosis, parkinsonism, bipolar disorders, schizophrenia, and hysteria. Cardiovascular effects are related to a rapid hypoxia experienced by the myocardial tissue. Those victims with previous coronary artery disease may suffer angina pain at carboxyhemoglobin levels less than 10%. In healthy younger patients, CO poisoning lowers the ventricular fibrillation threshold and if hypoxia continues, cardiac arrhythmias are the most frequent cause of death.

A carbon monoxide level in a structure fire rapidly reaches well over 10,000 ppm within only minutes. A firefighter not wearing or improperly using a self-contained breathing apparatus for 10 minutes can reach a carboxyhemoglobin level of 28%. A free burning fire generates less carbon monoxide than a smoldering fire. Most fire departments enforce airpack regulations during the active firefighting phase but do not enforce the use of airpacks during overhaul. In reality, the highest percentage of toxic gas is generated during the overhaul phase. Airpacks are warranted even more during this dangerous time.

Where Carbon Monoxide Is Commonly Found

Carbon monoxide can be generated from any number of processes. Below are listed some of the most common.

Internal combustion engines	Produce about 7% CO. Some literature suggests that in past years up to 50% of the annual suicides were committed using CO from internal combustion engines.
Faulty gas and kerosene stoves	Responsible for many accidental deaths each winter.
Charcoal and sterno fires	Use of these fires in unventilated areas has been responsible for a number of deaths in recent years.
Industry and laboratories	Some industries and laboratories produce CO, others use it in chemical reactions. It is available in high pressure cylinders for industrial use.
Fires	Not only produce CO but a number of other chemicals. CO is flammable and under the right conditions causes flashover or back drafts.
Methylene chloride	A widely used solvent in home and industry. Found in paint strippers, aerosols, fumigants, Christmas tree bubble lights, and fire extinguishers. When heated it produces carbon monoxide. Even patients who inhale the fumes of methylene chloride develop carboxyhemoglobinemia from the internal breakdown of the chemical.

Decontamination and Significant Danger to Rescuers

Decontaminating a victim of carbon monoxide poisoning is usually not necessary. The gas can only be inhaled, not absorbed through the skin. Usually the clothing of a victim will not hold enough of the gas to injure a medical provider, but it is always good practice to roughly decontaminate (remove the clothing) the victim involved in any type of exposure. Fire victims who have been removed from burning structures have many chemicals, including cyanide, formaldehyde, and hydrogen chloride, within their clothing. Therefore, it is very important to, at the minimum, remove the clothing of these victims.

Field Treatment

1. Remove the patient from the toxic atmosphere and administer 100% oxygen. If intubation or a bag valve device is used, attach a PEEP valve and use a setting of $4cm/H_20$ with 100% oxygen supplemented. Breathing room air COHg is halved in six hours (also referred to as half-life or T1/2). Breathing 100% oxygen reduces COHg T1/2 to 1.5 hours. Providing hyperbaric oxygen inside a chamber reduces the COHg1/2 to less than 1 hour.

2. Provide continuous cardiac monitoring and treatment of arrythmias with the consideration of the possibility that an acute myocardial infarction may have resulted from the poisoning.

3. Consider an osmotic diuretic to treat cerebral edema that resulted from the acute hypoxia. One ampule (25Gms) of D50 may be used but be cognizant of the rebound effect of dextrose.

4. If the CO poisoning is due to a suicide attempt, then Narcan should be used prophylacticly.

Definitive Treatment and Follow-Up Care

If a medical hyperbaric chamber is located within the area, all attempts should be made to transport the patient to that facility. Hyperbarics have been proved to provide the best reduction of carboxihemoglobin. At a pressure of 2.5 atmospheres and 100% oxygen the COHgT1/2 is less than one hour. Furthermore, the oxygen carrying capacity of the blood is significantly increased not only because of the reduction of carboxyhemoglobin but also because of the increased oxygen carried in solution.

CYANIDE POISONING

Hydrogen Cyanide and Cyanide Gases

Nomenclature: Also called prussic acid, formonitrile, NCN.

Properties: Soluble in water. It becomes a white liquid at temperatures below 26.5°C. Commercial material is 96–99.5% pure. Specific gravity (liquid) 0.688, vapor density 0.938.

Explosive limits: 6–41%.

Uses: Manufacture of acrylonitrile (vinyl cyanide), acrylates, adiponitrile, cyanide salts, dyes, chelates, rodenticides, insecticides. Many more listed in this section.

History

The use of cyanide as a poison dates back to the early Egyptians and Romans. It is well known because of its use for executions and homicides. Several more recent events involving cyanide involved the mass suicide of 900 people in Jonestown, Guyana, and the death of seven victims in Chicago from tainted Tylenol capsules. Today, smoke from residential fires contains toxic amounts of cyanide because of the synthetic carbon-containing compounds used in building materials and furnishings.

Karl Wilheim Scheele (1742–1786), a well-known Swedish chemist who was credited with the discovery of the elements chlorine, barium, molybdenum, tungsten, nitrogen, and manganese, also gained fame from isolating cyanide through a process using sulfuric acid and Prussian blue (blue iron ferrocyanide). This mixture he called prussic acid was in reality hydrocyanic acid. Scheele himself was killed by the hydrogen cyanide fumes generated when a beaker of the substance was dropped on the floor of his laboratory.

Pathophysiology

Cyanide is one of the most rapid-acting poisons. It gains access into the body most often through inhalation, but can also be ingested and absorbed through the skin and mucous membranes. It can cause death within minutes to hours, depending on the concentration, route of entry, exposure time, and the activity level of the victim. The speed at which cyanide gas works is evidenced by how rapidly (usually within a minute) a death row prisoner dies during a gas chamber execution.

Following an exposure, cyanide rapidly enters the cells binding with an enzyme, cytochrome oxidase. This enzyme, in its original form, is necessary for cellular respiration (the use of oxygen to convert glucose to energy—aerobic metabolism). By binding to cytochrome oxidase, cyanide causes a paralysis of the cells' ability to carry out aerobic metabolism. Without cytochrome oxidase the cell cannot utilize oxygen from the bloodstream. The process eventually causes the cells to attempt to function under anaerobic metabolism. The inadequate anaerobic state ultimately causes decreased cellular energy production, metabolic acidosis, cellular suffocation, and death. Interestingly, the half-life of cyanide in the body is only about an hour, but during a true exposure death takes place well before the body starts to detoxify or excrete the chemical.

Cyanide can cause death in an oral dose of less than 5 milligrams per kilogram. This dosage can be explained better by saying that a 150-pound patient will die ingesting a dose of about seven drops of concentrated cyanide. Inhalation of cyanide gas in a concentration of 0.3 milligrams per liter is fatal within seconds.

Many sources suggest the presence of a bitter almond smell (oil of bitter almonds, garlicky, and oniony also reported) may be noted either emitting from the patient's body or on his breath. The ability to detect a bitter almond smell is a sex-linked recessive trait and only 60–80% of the general population can smell the aroma. The remaining cannot detect the odor due to a sensory deficit. The deficit is greater in men than women by a ratio of 3 to 1.

Because a small amount of cyanide is present in various foods and also in cigarette smoke, the body naturally produces an enzyme that detoxifies it. The enzyme works through a chemical combination of sulfur and cyanide to form a harmless compound which is then excreted through the kidneys. Unfortunately, this enzyme is only present in small amounts and responds too slowly to assist in detoxifying cyanide during a true exposure.

Signs and Symptoms

The patient will present with a wide variety of signs and symptoms because cyanide poisoning affects virtually all the cells in the body. The most sensitive target organ is the CNS where the urgent need for oxygen is first sensed. Early effects can include headache, restlessness, dizziness, vertigo, agitation, and confusion. Later signs are seizures, coma, and death.

Serum Cyanide Levels	Signs and Symptoms
0–0.2	Normal (nonsmoker)
0.1–0.4	Normal (smoker)
0.5–1.0	Anxiety, confusion, unsteadiness, tachypnea
1.0–2.5	Headache, palpitations, dyspnea, depressed level of consciousness, atrial fibrillation, ectopic ventricular beats
2.5–3.0	Loss of muscular coordination, convulsions, reflex bradycardia, respiratory depression, coma
greater than 3.0	Apnea, cardiovascular collapse, asystole

Since it is impossible to determine the serum cyanide level in the field under an emergency situation, it might be more beneficial to recognize the early signs of cyanide poisoning so preparations can be made for the late signs especially as they relate to the respiratory and cardiovascular system.

Respiratory System Effects	
Early	Late
Tachypnea	Decreased respiratory rate
Hyperpnea	Respiratory depression
Dyspnea	Apnea and adult respiratory distress syndrome

Cardiovascular System Effects

Early	*Late*
Flushing	Hypotension
Hypertension	Acidosis
Reflex bradycardia	Tachycardia
Arterioventricular (AV) nodal or intraventricular rhythms	ST changes and cardiovascular collapse EKG-ST segment changes

Where Cyanide Is Commonly Found

Cyanide is commonly found in industry, either combined with other chemicals or in a relatively pure form. It is also used in commercial pest control. Probably the most common poisonings seen today by emergency responders involve victims of smoke inhalation. Many products in use today will emit toxic, cyanide containing gases when burned.

Approximately 150 plant species also contain a cyanide-producing agent called *cyanogenetic glycoside* (amygdalin). This chemical was produced and distributed on the market as Laetrile. Laetrile was marketed as a cancer treatment but was subsequently withdrawn after many deaths were related to its use.

Amygdalin by itself is a nontoxic chemical, harmless to the human body. Once amygdalin is mixed with an enzyme known as emulsin, the reaction yields hydrocyanic acid. Most seeds or pits of cyanide-producing plants contain both amygdalin and emulsin, therefore, just crushing these seeds and adding water releases the cyanide. As few as one pit or 15–60 crushed seeds can cause significant cyanide toxicity.

$$C_{20}H_{27}NO_{11} + 2H_2O = 2C_6H_{12}O_6 + C_6H_5CHO + HCN$$

amygdalin + water = glucose + benzaldehyde + cyanide

Following is a list of areas commonly noted as having or producing cyanide.

Industry	*Pest Control*
Metal cleaning	Insecticides
Metal polishing	Fumigants
Electroplating	Rodenticides
Metal heat treating	*Fires*
Synthesis of plastics and rubbers	Polyurethane Polyacrylonitriles
Soil sterilization	Nylon, wool, and silk
Fertilizers	*Clandestine PCP (phencyclidine) Labs*

Decontamination of Patients

Cyanide is a water soluble chemical. A victim exposed to cyanide in the liquid, solid, or gaseous form should be decontaminated by completely removing the victim's clothing, followed by a complete washing with water or a water base decon solution (mild detergent). Decontamination consisting of, at least, the removal of contaminated clothing should be done on all smoke inhalation victims. This step is necessary to reduce secondary contamination to emergency care providers, transporters, and the hospital staff.

Field Treatment

The human body has the ability to detoxify cyanide in small amounts, given enough time. The half-life of cyanide in the body is approximately 1 hour. The initial treatment then is to keep the patient alive with as much supportive treatment as needed.

Since respiratory arrest develops quickly, establish a good patent airway. Then as quickly as possible begin advanced treatment utilizing the cyanide antidote kit. The kit contains amyl nitrite, sodium nitrite, and sodium thiosulfate (Fig. 6-2).

The nitrites (amyl nitrite and sodium nitrite) convert hemoglobin into methemoglobin. Methemoglobin competes with cytochrome oxidase for the cyanide ion, actually attracting the cyanide away from the cytochrome oxidase. Methemoglobin is formed when the ferrous iron (Fe^{++}) within the hemoglobin is converted into ferric iron (Fe^{+++}). This change in valence of the iron atom within the hemoglobin attracts the cyanide ion, freeing the cytochrome oxidase to again participate in aerobic cellular metabolism. The last step is to infuse sodium thiosulfate, which acts as a cleanup agent by changing the remaining cyanide into a relatively harmless substance, thiocyanate.

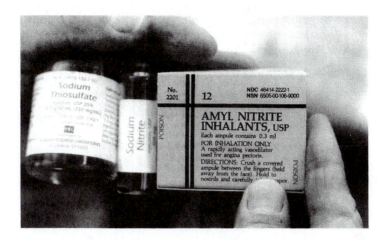

Figure 6-2 *Cyanide poisoning is treated by giving a series of drugs starting with amyl nitrite, sodium nitrite, and finishing with sodium thiosulfate.*

Although not approved by the FDA for cyanide treatment in this country, a drug called hydroxocobalamin (vitamin B12a) has been used successfully in Europe. The dose is four grams combined with eight grams of sodium thiosulfate. Hydroxocobalamin is available in the United States but in its FDA-approved packaging, it would require 400 vials to treat one patient.

The patient who survives an initial exposure must be closely monitored and admitted into the hospital intensive care unit. This precaution is necessary because of the late complications of acidosis, pulmonary edema, dysrhythmias, and neurologic deficits that may appear many hours after an exposure.

Although cyanide related poisonings in industry are rare, the large number of uses for cyanide related compounds make the potential great.

Dosages

Adult Dosages

1. Amyl nitrite perle inhaled for 15–30 seconds every minute while the IV is being established. If the patient is unconscious then the perle can be placed in the bag of a BVM and inhalation of the amyl nitrite can be used in this manner. If the patient is conscious, lay him flat and elevate the feet. Nitrites cause diffuse vasodilation and usually a significant drop in blood pressure. Amyl nitrite is only a first aid measure and converts only about 3% to 5% of the circulating hemoglobin into methemoglobin. An IV line should be established as soon as possible so that the sodium nitrite can be infused.

2. Sodium nitrite 300 milligrams per 10 cubic centimeters IVP slowly. Sodium nitrite also causes profuse vasodilation so the blood pressure of the patient should be watched closely. One adult dose of sodium nitrite converts approximately 20% of the circulating hemoglobin to methemoglobin. The percentage of methemoglobin should not exceed 40% or the patient's oxygen circulating capability will be severely compromised, a condition known as methemoglobinemia. The treatment at this point will be deleterious.

3. Sodium thiosulfate 50 milliliters of a 25% solution IVP over 10 minutes. Sodium thiosulfate acts to clean up by changing the remaining cyanide into a relatively harmless substance, thiocyanate.

Children's Dosages

1. The amyl nitrite dosage remains the same.
2. Sodium nitrite 0.33 milliliters/kilograms (10 milligrams of 3% solution per kilogram).
3. Sodium thiosulfate 1.65 milliliters/kilograms of the 25% solution.

Definitive and Follow-Up Care

Hyperbaric oxygen treatment is recognized as providing therapeutic treatment for cyanide poisoning. Any patient with an altered level of consciousness should only be considered for hyperbaric treatment if he can be placed in a multipatient chamber with an attendant at his side. The chambers are scattered throughout the nation and are becoming more popular. If you are familiar with a chamber in your area, consider transporting the patient to that facility first.

HYDROGEN SULFIDE GAS POISONING

Hydrogen Sulfide (H_2S)

Properties: Colorless, strong odor. Water and alcohol soluble. Specific gravity 1.189.

Explosive limits: 4.3–46%.

Uses: Purification of hydrochloric and sulfuric acids, analytical reagent.

History

Hydrogen sulfide poisoning is the leading cause of death related to toxic inhalation in the workplace. It is formed during the decomposition of organic material. It has the distinctive odor of "rotten eggs" that may only be present briefly under higher concentrations. One of the characteristics of hydrogen sulfide gas is its ability to cause paralysis of olfactory sensors (olfactory fatigue). The victim may only receive a brief whiff of the chemical before a total numbing of the smelling sense develops. That is why so many workers have fallen victim to the effects of this potent poison. Many times the fall of a worker in a confined space without prior warning, motivates an attempted rescue by a fellow worker, resulting in the death of both.

> Detroit, Michigan, June 1992. A worker was killed as he attempted to rescue an unconscious fellow worker. A 22-year-old plant employee fell unconscious due to fumes generated in the sewer he was cleaning. A 27-year-old co-worker attempting to rescue him was also overcome causing him to fall from a ladder. The 27-year-old worker later died.

Those who have spent many years working in the sewers or around low doses of the gas develop a chronic conjunctivitis that they have termed "gas eye" or "sewer eye." Some say that the workers use the severity of the conjunctivitis to determine if the affected worker should take a day off or at least work in an area without a concentration of the gas.

Pathophysiology

Hydrogen sulfide acts on the body as a respiratory irritant. In low doses it affects the upper respiratory system with bronchospasms, localized irritation, and edema and causes conjunctivitis and many other nonspecific complaints, such as dizziness, nausea, and headache.

In higher concentrations, the respiratory irritation goes deeper into the respiratory system causing chemically induced pulmonary edema. Susceptibility to these symptoms varies among individuals and may be increased due to multiple previous exposures, causing a sensitized-type reaction.

Both high and low levels cause cellular asphyxia in much the same way that cyanide does. Once hydrogen sulfide enters the bloodstream, it finds its way to the cells where it combines with the enzyme cytochrome oxidase. This action inhibits the enzyme from working to transport oxygen into the cell and participate in cellular metabolism. Thus, cellular respiration becomes very inefficient causing acidosis and eventually cellular hypoxia and death. Hydrogen sulfide is a stronger inhibitor of cytochrome oxidase than cyanide. For this reason hydrogen sulfide is considered to be somewhat more toxic than cyanide.

Hydrogen sulfide is slightly heavier than air with the vapor density of 1.19. It tends to lie in lower areas of a confined space or enclosed area. Of the deaths recorded, most occur at the scene. One study conducted indicated that 5% of all of the exposed victims died on the scene or arrived dead at the hospital. Victims who reach the hospital with vital signs usually survive unless other complications, such as hypoxic encephalopathy, ensue. Interestingly, blackened coins have been found in the pockets of workers poisoned with hydrogen sulfide.

! SAFETY
Because hydrogen sulfide gas numbs the sense of smell, death by poisoning from this gas in a confined space often claims both the initial victim and the rescuer.

Signs and Symptoms

The signs and symptoms presented by a patient who has had a significant exposure to hydrogen sulfide are almost exactly the same as a victim poisoned with cyanide. Both cyanide and hydrogen sulfide cause cellular hypoxia in all organ systems of the body, therefore, the symptoms can be very diverse. The organ systems that use the most oxygen are the first to demonstrate the symptoms. Organs such as the brain (CNS) and heart are affected early in the poisoning. Any deficits noted in these areas should stimulate the medical provider to immediately deliver care.

The early signs of hydrogen sulfide poisoning are a direct result of the hypoxic state suffered by the brain. The brain, not receiving oxygen, sends messages to the respiratory center to breath faster (tachypnea) and breath deeper (hyperpnea). The heart responds by increasing the blood pressure that results in reflex bradycardia. This is followed by AV nodal and intraventricular dysrhythmias due to the hypoxia.

The later signs are the result of prolonged cellular hypoxia, such as a slowing of the respiratory rate, hypotension, acidosis, and cardiovascular collapse. Hydrogen sulfide is one of the most rapid-acting poisons and can cause death within minutes. The emergency care provider must aggressively provide supportive care and ALS antidote as soon as possible if the patient is to benefit.

Where Hydrogen Sulfide Is Commonly Found

The most common sites for hydrogen sulfide are where any breakdown of organic material takes place, including septic tanks, sewers (both sanitary and nonsan-

Figure 6-3 *Hydrogen sulfide is one of the gases that gather in confined spaces. Testing of these confined spaces is mandatory before anyone enters the atmosphere.*

itary), wells, tunnels, and mines. It is also used in chemical laboratory experiments. It should be monitored for anytime a rescue of any type is done within a confined space (Fig. 6-3).

Decontamination or Significant Danger to Rescuers

Decontamination of the skin is not necessary due to poor cutaneous absorption. Removal of clothing should be sufficient to remove a chance of secondary contamination, although rescuers have been known to lose consciousness while giving mouth-to-mouth ventilations to poisoning victims.

! **SAFETY**
Rescuers have been known to lose consciousness while giving mouth-to-mouth ventilations to victims of hydrogen sulfide gas poisoning.

Field Treatment

The use of nitrites is recognized as an antidotal treatment for hydrogen sulfide poisoning. Nitrites attract the sulfide form, the cytochrome oxidase, and thus reactivate aerobic metabolism. This reaction forms sulfhemoglobin (very similar to methemoglobin), which is then quickly detoxified by the body.

Because the physiology is very similar to cyanide toxicity, the treatment is very much the same. Treatment consists of the following drugs and dosages.

Dosages

1. Amyl nitrite perle inhaled for 15–30 seconds every minute while the IV is being established. If the patient is unconscious then the perle can be placed in the bag of a BVM and inhalation of the amyl nitrite can be utilized in this manner. If the patient is conscious, lay her flat and elevate the feet. Nitrites cause diffuse

vasodilatation and usually a significant drop in blood pressure. Amyl nitrite is only a first aid measure and converts only about 3% to 5% of the circulating hemoglobin into methemoglobin. An intravenous line should be established as soon as possible so that the sodium nitrite can be infused.

2. Sodium nitrite 300 milligrams per 10 cubic centimeters IVP slowly. Sodium nitrite also causes profuse vasodilatation so the blood pressure of the patient should be watched closely. One adult dose of sodium nitrite converts approximately 20% of the circulating hemoglobin to methemoglobin. The percentage of methemoglobin should not exceed 40% or the patient's oxygen-circulating capability will be severely compromised, a condition know as *methemoglobinemia*. If methemoglobinemia develops at a level of greater than 40%, further treatment with nitrites will be deleterious.

3. Sodium thiosulfate is of no use with hydrogen sulfide poisonings.

Definitive Treatment or Follow-Up Care

Several sources suggest the use of hyperbaric oxygen to enhance the elimination of hydrogen sulfide. The treatment of hydrogen sulfide toxicity is not one of the identified uses of HBO and therefore, until significant proof is established recognizing its benefit, it will not become a part of standard care.

Hospital follow-up care should include monitoring the patient for aspiration pneumonia and late developing chemically induced pulmonary edema. Pulmonary edema may occur up to 48–72 hours after the initial exposure. Also, if nitrites were used as an antidote, a methemoglobin level should be drawn.

HYDROFLUORIC ACID BURNS AND POISONING

Hydrofluoric Acid

Properties: Colorless, clear, fuming liquid or gas (hydrogen fluoride). Water Soluble. Vapor density of 3.0. Specific gravity of 1.2.

Explosive limits: Will not burn. In contact with metal will form potentially explosive hydrogen gas.

Uses: Silicon and glass etching and frosting, semiconductor, plastic and dyes manufacturing, and rust removal agents.

History

Hydrofluoric acid (HF) is one of the strongest inorganic acids known. Because of its wide use and prevalence in industry, the probability of injuries are great. Hydrofluoric acid is also unique in the way it produces injury. Not only is it capable of producing severe corrosive effects on contact with tissue, but it also produces profound systemic effects that can be deadly.

Volusia County, Florida, September 1994. A tanker carrying 4,500 gallons of hydrofluorosilicic acid (H_2SiF_6) broke in half, spilling its contents onto the interstate highway. The spill covered several hundred yards of the popular roadway. As a result, hundreds of autos drove through the spill, which was actively fuming during the hot summer day. Approximately 100 people reported to local hospitals, complaining of injury from the chemical. Several were held overnight, the worst of whom was a police officer who had responded to the spill and initially directed traffic in close proximity to it. Several days after the incident all vegetation within a quarter-mile radius was brown or dead from the toxic fumes.

Pathophysiology

Hydrofluoric acid is an inorganic acid in the same class as hydrochloric and sulfuric acids, but unlike other inorganic acids, hydrofluoric acid maintains its strong ionic bond, allowing it to penetrate deep into the sublayers of skin and tissues. Most inorganic acids are strongly ionic and dissociate readily into ionic components. The injury typically produced by these acid burns is the result of the hydrogen ion breaking free (dissociating), causing a corrosive injury to the upper layers of skin. This type of injury produces immediate pain but tends to be superficial because of the coagulum formed by the injured tissue.

The burns caused from hydrofluoric acid may go unnoticed because of the lack of immediate pain. Hydrofluoric acid, conversely, maintains its strong ionic bond and slowly dissociates, allowing penetration into the subcutaneous tissue. Once deep in the tissue, the hydrogen ion separates from the fluoride and is left to cause the corrosive injury within these subcutaneous layers. The injury caused by hydrofluoric acid is twofold and therefore, does not stop there.

The fluoride ion seeks out and combines with all of the calcium and magnesium it can find. Bones and tissues provide the source of calcium and magnesium needed to produce calcium or magnesium fluoride. This process can result in severe decalcification and eventual destruction of the bone. Furthermore, the hydrofluoric acid that has not dissociated further penetrates tissues gaining access into the bloodstream. Once dissociated within the bloodstream, the hydrogen ion causes serum acidosis, and the fluoride binds with the calcium causing hypocalcemia.

Binding of the fluoride ion to the calcium found in the bloodstream causes an immediate threat to life. Skeletal muscle and cardiac muscle contraction is the result of serum hypocalcemia. The outcome of this dangerous condition is muscle tetany that can be followed by a sudden onset of asystole. An exposure of 2.5% of the skin surface to a 100% concentration, or a 10% skin exposure to a 70% concentration is enough to cause systemic poisoning and death.

The fluoride ion also bonds with calcium on the cell membrane and increases permeability to potassium, which leads to spontaneous depolarization along the nervous pathways, producing excruciating pain.

! SAFETY
An exposure of 2.5% of the skin surface to a 100% concentration of hydrofluoric acid or a 10% skin exposure to a 70% concentration is enough to cause systemic poisoning and death.

Weaker solutions cause an even more delayed onset. Lesions and pain may be delayed for up to 18 to 24 hours. One of the more common sites for this activity is the fingernail beds, which turn black with chronic low-dose exposure. The nail beds are one of the few areas lacking corneum stratum, the top layer of epidermis, which provides a limited protective barrier against chemical exposure. This lack of tissue allows hydrofluoric acid to penetrate easier and at lower concentrations.

Signs and Symptoms

Hydrofluoric acid can affect any surface of the body and can enter through all routes. The symptoms displayed by the patient depend on the area affected by the chemical.

Inhalation   If the vapor or gas is inhaled, burning and swelling of the oral mucosa and upper airways may be immediately noticed. The delayed symptom of pulmonary edema may start hours later and become very severe. Pulmonary edema has even been reported as late as 2 days after an inhalation injury. Inhalation provides an efficient route for chemical invasion into the circulatory system. Systemic reactions in this instance may have a rapid onset with more severe reaction.

Skin Exposures   If the chemical comes in contact with the skin, initially the patient may not have severe pain therefore may believe that the injury is only minor. In concentrations of 20% to 50% the onset of signs or symptoms can be delayed up to 24 hours. Strong concentrations of greater than 50% cause immediate, severe burning pain and evidence of acid burns much like that of hydrochloric acid. The skin may initially look blanched and feel firm to the touch. Up to 2 hours later, the skin will swell and turn gray in appearance. This initial injury will eventually turn into necrosed, blistered skin and develop ulcerations.

Eye Exposures   If hydrofluoric acid gains access to the eye, a sloughing of epitheleal tissue and cloudiness of the cornea may be noted. As with any acid exposure, the injury may be severe of not immediately irrigated. Like skin injuries, hydrofluoric acid causes a deeper and more severe injury to the eye than would be expected from other inorganic acids. Due to the limited surface area involved in an eye exposure and the natural flushing that takes place once an exposure happens, systemic reactions, if present, would be more subdued.

An inhalation, ingestion, or exposure of hydrofluoric acid on just 25 square inches of skin can cause severe hypocalcemia. If not recognized and accurately treated this condition can cause death. An EKG indicating a prolonged QT interval is diagnostic for hypocalcemia. In these cases IV infusion of calcium should be considered and guided by medical control.

Where Hydrofluoric Acid Is Commonly Found

Hydrofluoric acid is used in a number of industries usually involving manufacturing. It is used in the production of printed circuits, metal finishing, stainless steel, iron, and steel foundries, petroleum refining, and glass making (Fig. 6-4). It is also found in over-the-counter items such as cleaning compounds and rust removal agents. A form of hydrofluoric acid, hydrofluorosilicic acid, is used to fluoridate water in many communities.

Decontamination or Significant Danger to Rescuers

Rescuers must wear agent-specific protective suits and self-contained breathing apparatus. If ambulatory, the patient should be instructed to remove all clothing and move to a predetermined decontamination center.

Water is the initial decontamination solution of choice. If available, an epsom salt solution (magnesium sulfate) or calcium hydroxide (lime water) are the most effective decontamination solutions. Maalox or Mylanta can be applied topically. Tums also contain calcium and magnesium and can be crushed, added to saline, forming a slurry that can be painted over the wound. If the hands have become exposed, then the nails may be trimmed back to the nail beds to facilitate adequate decontamination of the hands.

Field Treatment

Hydrofluoric acid burns require immediate and specialized first aid, much different from other chemical burns. If untreated or improperly treated the patient can suffer permanent damage, disability, or even death. If on the other hand, the burns are properly diagnosed and proper treatment is rapidly started, the outcome is generally favorable. Treatment is geared toward tying up the fluoride ion, which essentially will limit and prevent further tissue damage.

Figure 6-4 *Hydrogen fluoride is used at many water treatment facilities for the fluoridation of water.*

Calcium gluconate is the drug of choice and is used in many different forms to provide the calcium at the point of injury. Extra attention should be taken when the respiratory system or large areas of the skin (greater than 25 square inches) are involved. A patient with hydrofluoric acid burns above the shoulders must be rapidly evaluated for injury to the respiratory system.

Dosage

Skin Burns

1. Once the affected areas are decontaminated, then the burned areas should be covered with calcium gluconate gel (Fig. 6-5).

   a. Mix 10 cubic centimeters of a 10% calcium gluconate solution into a 2-ounce tube of KY Jelly. This will provide a 2.5% gel.

   b. Continuously massage liberal amounts of gel into the burn site. Pain relief should be reported within 30–45 minutes.

   c. If calcium gluconate is not available Epsom salt (magnesium sulfate), magnesium containing antacids such as Maalox, Mylanta, or Tums can be used as a topical agent.

2. In the case of hydrofluoric acid burns, pain is an excellent indicator that the injury is continuing. If pain continues after calcium gluconate gel is applied then calcium gluconate infiltration is needed.

   a. Calcium gluconate in a 5% solution is injected subcutaneously in a volume of 0.5 milliliters every 1/4 inch into the burned area. This treatment should stop the pain.

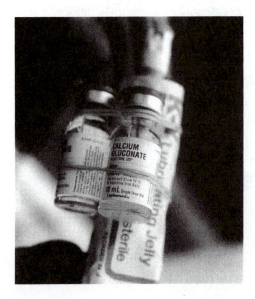

Figure 6-5

Hydrofluoric acid burns to the skin are treated by applying a slurry made from calcium gluconate and a water soluble jelly.

    b. Since pain is an excellent indicator that the free fluoride ion is still present to some degree. Injections should continue until the pain subsides.

Eye Burns

    1. Flushing of the exposed eyes should start immediately using large amounts of water or physiologic saline. While flushing is being performed, prepare calcium gluconate eye wash solution.

    b. Mix 50 cubic centimeters of a 10% solution of calcium gluconate into 500 cubic centimeters of normal saline IV solution.

    2. Connect bag, tubing, and Morgan lens together and flush tubing.

    3. Apply 1–2 drops of ponticaine, or other aqueous, topical ophthalmic anesthetic solution in the injured eye(s).

    4. Insert Morgan lens and start flushing at a moderate rate.

    5. Use care to protect other surrounding skin from runoff so secondary contamination does not take place.

Inhalation Injuries

    1. Start supportive respiratory measures, including positive pressure ventilations and 100% oxygen as soon as possible.

    2. Start nebulizer treatment when mixed and assembled.

    a. Draw 6 cubic centimeters of sterile water and 3 cubic centimeters of 10% calcium gluconate.

    b. Place solution in nebulizer and connect to oxygen. Provide effective fog.

    3. Assess and treat other burns.

Cardiac Symptoms of Hypocalcemia

    1. Provide continuous monitoring of the EKG, watching for prolongation of the QT interval.

    a. Muscle contractions or cardiac arrest should be treated with an IV bolus of 5 cubic centimeters 10% calcium chloride or 10 cubic centimeters of 10% calcium gluconate.

    b. IV calcium should be considered for any patient with exposures to hydrofluoric acid in a concentration of greater than 10%, over 5% or more of body surface area.

NITRITE AND NITRATE POISONING

History

Nitrites (NO_2) and nitrates (NO_3), organic and inorganic nitrogen compounds, are found in our environment in many different products and forms from colognes to paints and fertilizers. Although commonly found in both home and work environments, poisonings more commonly occur due to intentional misuse for recreational purposes. A pharmaceutical, amyl nitrite, a yellow liquid available in glass pearls wrapped in cloth much like ammonia capsules (found in the cyanide antidote kit) is one form of the chemical often misused. The user breaks the pearls and sniffs the capsule causing vasodilatation and a resulting "rush" or "high." On the street, these black-market capsules are called *poppers* or *snappers* because of the breaking noise they make. Nitrites are also available in liquid incense, room deodorizers, and men's cologne, under the trade names Rush (isobutyl nitrite) and Locker Room (butyl nitrite). Frequently, these items can be bought in head shops or shops specializing in drug paraphernalia. The production and sale of nitrites is primarily uncontrolled.

Most abusers fall into one of two categories. Juvenile drug abusers experiment with the nitrites to compound the effects of other drugs. The other group, typically males, most frequently use the drug to enhance sexual activity. These abusers report that the use of nitrites just before orgasm increases the intensity and duration of the orgasm.

The signs and symptoms of organic nitrogen poisoning may be subtle, but with a suspicion of the poisoning, knowledge of the pathophysiology, and good assessment skills, the poisoning can be rapidly identified and treated.

Pathophysiology

Exposure to nitrogen compounds can occur through absorption of the skin, mucous membranes, respiratory system, and gastrointestinal tract. The most efficient routes of access are through the gastrointestinal and respiratory systems. In industry, nitrogen compounds are found most commonly in the solid form, making inhalation of dust the primary route for poisoning in the workplace.

Once absorbed into the bloodstream these nitrogen compounds combine with hemoglobin and change the iron molecule, ferrous iron (Fe^{++}) into ferric iron (Fe^{+++}). The conversion of the iron molecule changes the hemoglobin into a non-oxygen-carrying compound called *methemoglobin*, a non-oxygen-carrying molecule. This change causes the poisoned patient to become hypoxic.

This condition of hypoxia caused from methemoglobin is termed *methemoglobinemia*. The color of the blood while in this condition changes from bright red to chocolate brown and is easily assessed during a blood draw. Even blood vigorously shaken in an oxygen rich atmosphere will remain chocolate brown in color.

Another diagnostic clue comes from the pulse oximeter. The oximetry unit works on the principal that oxygen-rich hemoglobin is bright red and transparent. Because of this principle, a patient suffering from increased methemoglobin, which is dark brown and opaque, will show an inaccurately low reading. Normally, a 1% methemoglobin level is found in healthy individuals. Significant signs and symptoms do not appear until levels at or above 10% are formed.

Nitrates and nitrites are also used medicinally because of their ability to relax smooth muscle thus causing vasodilating effects. Nitroglycerin, a nitrate, and amyl nitrite are two examples of the pharmaceutical use of these chemicals. Furthermore, the conversion of ferrous iron into ferric iron is used during treatment for cyanide poisoning. The cyanide antidote kit contains amyl nitrite and sodium nitrite that during a single dose can convert up to 25% of the hemoglobin into methemoglobin. The change of the electrical valence of the iron atom during this process attracts the cyanide ion out of the cell, allowing it to respire in a normal fashion.

Signs and Symptoms

The toxic effects of exposure to nitrogen compounds are exhibited in two different ways. First, the vasodilating effects are evidenced by the following signs and symptoms, which are usually short lived and often corrected with only simple treatments.

1. Throbbing headache and fullness of the head, are due to dilation of the menengeal vessels.
2. Flushing of the neck and face are signs of cutaneous capillary and vasodilatation.
3. Dizziness and syncope are due to cerebral ischemia related to profuse vasodilatation.
4. Tachycardia, sweating, and pallor are responses of the sympathetic nervous system to hypotension.

Second, evidence of methemoglobin presents with signs and symptoms of a different cause.

Percentage Relative to Normal Hemoglobin	Signs and Symptoms
10% to 15%	Mild cyanosis in extremities but usually no other symptoms
20% to 30%	Shortness of breath, changes in mental status, and changes in vital signs
Approximately 50%	Lethargy
More than 70%	Death due to hypoxia

Obvious cyanosis may be caused from low levels of methemoglobin. These low levels may not generate other cyanotic-type symptoms. On the contrary, cyanosis caused from hypoxia will cause symptoms such as confusion, anxiousness, and a severely dyspneic patient.

Where Organic Nitrogen Compounds Are Found

Organic nitrogen compounds are used in the manufacturing of dyes, paints, polishes, photographic chemicals, crayons, food preservatives, and fertilizers. For those who intentionally misuse the chemical for its pharmaceutical effects, nitrogen compounds are purchased in head shops or can be purchased in a number of over-the-counter room deodorizers.

Except for intentional misuse, fertilizers account of the highest percentage of nitrogen compound poisonings. Well water contaminated with fertilizers is often reported as a cause of nitrogen compound poisonings. Those most susceptible to poisonings of this nature are neonates. Fetal hemoglobin is more susceptible to oxidation by nitrates and nitrites. Furthermore, neonates also lack the natural ability to reduce the methemoglobin.

Decon or Significant Danger to Rescuer

If the poisoning takes place in a manufacturing facility, decontamination with an appropriate solution, usually a warm soapy solution, must take place to stop further skin absorption. If the contaminant is solid or a dust, the initial decon consideration should include removal of the dust prior to a wet decon. Wetting the patient who is covered with fertilizer dust may only increase absorption of the material. If burns are present, they should be lavaged with water or saline.

Fires involving fertilizers, paint, or dyes can all generate smoke concentrated with nitrogen-containing material. Firefighters have historically suffered from methemoglobinemia after fighting such fires. Even simple structure fires may generate smoke rich in nitrogen base chemicals therefore, victims of these fires should be suspected of having a portion of their blood hemoglobin converted into methemoglobin.

San Francisco, California, November 21, 1979. Firefighters responded to a structure fire in an area of downtown known for manufacturing and warehouses. The fire, apparently fueled by containers of illegally stored chemicals, grew to four alarms. A number of firefighters at the scene were overcome by smoke and exhaustion. One victim, a 48-year-old firefighter was transported to San Francisco General Hospital for evaluation. His signs and symptoms included acute respiratory distress, cyanosis, and lethargy. An exam showed no evidence of myocardial involvement. No improvement was noted after administering 40% oxygen. Blood collected from the patient revealed a dark brownish color. The local poison control center suggested that the symptoms were the result of methemoglobinemia and the patient received methylene blue as an antidote. Rapid improvement in the firefighter's status resulted and he eventually had a full recovery.

Field Treatment

Treatment of nitrogen compound poisoning initially involves the removal of the patient from the chemical then removing the chemical from the patient. This can be done using proper decontamination efforts prior to transporting the patient to the hospital.

1. If the patient is breathing give 100% oxygen uses a nonrebreather mask.
2. If the patient is unconscious, consider intubation and provide positive pressure ventilation with a PEEP setting of 4 centimeters/H2O.

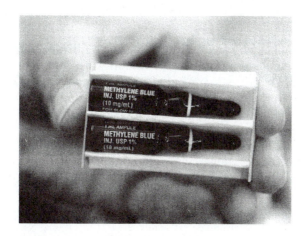

Figure 6-6 *Methylene blue is a dark blue drug that causes the conversion of methemoglobin back into hemoglobin allowing oxygen to be carried to the cells through normal means.*

3. Treat hypotension with positioning, fluids, and dopamine if necessary.

4. If the signs and symptoms support your suspicion of methemoglobinemia then give methylene blue 2 milligrams/kilograms over five minutes IVP.

Methylene Blue (Fig. 6-6) activates an enzyme methemoglobin reductase which then reduces $(Fe+++)$ ferric iron back into $(Fe++)$ ferrous iron and again enables the hemoglobin molecule to carry oxygen.

Definitive Treatment or Follow-Up Care

Some studies have indicated the use of hyperbaric oxygen. Hyperbaric oxygen increases the oxygen dissolved in the blood serum and lowers the levels of methemoglobinemia by decreasing the oxidation of hemoglobin. Follow-up or observation should be considered for any patient who has had a documented exposure to a nitrogen-based chemical known to generate methemoglobin. Delayed onset of methemoglobinemia has been reported up 16 hours after ingesting one of these agents.

PHENOL POISONING

Phenol

Properties: White, crystalline mass that turns pink or red if not pure. Soluble in alcohol, water, ether, petrolatum, and oils. Specific gravity of 1.07.

Combustible: Flash point at 172.4°F.

Uses: General disinfectant, pharmaceuticals, lab reagents, wood preservative. Others listed under Where Phenol Is Commonly Found.

History

Phenol/carbolic acid was first introduced as a disinfectant by Joseph Lister who gained fame and became known as the father of aseptic technique in 1865 with the use of phenol spray to disinfect the air in hospital wards. Dr. Lister theorized that many infections were spread from patient to patient by bacteria. The use of carbolic acid as a disinfectant greatly reduced the infectious rate between patients, confirming his theory.

This technique worked so well that later in the 1800s, 5% to 10% phenol was being used to bathe patients who entered the hospital to kill off any bacteria that would be brought into the hospital on their bodies. As a result of the popularity of this technique, phenol toxicity was first found in medical personnel. The toxicity resulted from medical personnel bathing patients, subjecting the hospital workers to chronic low-dose exposures.

Today its popularity in smaller percentages is widespread. For example, Campho-Phenique contains 4.5% phenol. Camphor and other similar substances interact with phenol, both to reduce its corrosive properties and to slow its absorption. Interestingly, phenol has also been used for facial treatments, coined *nonsurgical face-lifts* or *chemical peels*, where phenol is painted on the facial skin and a "controlled chemical burn" takes place, effectively removing the top layer of skin (corneum stratum) to reveal new, pink younger-looking skin.

An early antiseptic preparation containing phenol was Lysol. This Lysol has little resemblance to today's Lysol and was made by boiling heavy tar oils with vegetable oils in the presence of lye (sodium hydroxide). This preparation was very toxic and frequently used in the 1930s for suicide attempts.

Pathophysiology, Signs and Symptoms

Little is known about the physiologic mechanisms of systemic phenol poisoning except through the symptoms displayed by a poisoned patient. Phenol is not only absorbed through the skin but as a gas, dust, or fume can also be inhaled through the lungs. Symptoms develop in 5 to 30 minutes. Even low concentrations can cause devastating results. There are reports of children dying after the application of compresses soaked with 5% phenol solutions being applied over open wounds.

Systemic phenol poisoning exhibits a variety of symptoms. Initially, an excitation stage may be witnessed. Later, more intense symptoms appear, including profound central nervous depression, hypothermia, cardiac depression and loss of vascular tone, and respiratory depression. Convulsions are common and have appeared as late as 18 hours after exposure.

The initial excitation phase may or may not be seen. The excitation phase has been documented in animal studies but not evidenced in human experiences. It is believed that the danger stage of a phenol poisoning is not long lasting and generally passes after about 24 hours.

Phenol also causes an acid burn on direct contact to the skin. Like most acid burns, phenol denatures proteins found in the skin causing the formation of a coagulum and leaving a white to brown discoloration on the skin in the area of exposure.

Where Phenol Is Commonly Found

Phenol is most commonly found in over-the-counter disinfectants in very low concentrations. In high concentrations it has various uses in the hospital. The pharmacies usually dilute the concentration prior to sending it out for use by the physicians. Other areas of use are listed below.

Compound	*Use*
Amyl phenol	Germicide
Creosol	Antiseptic
Creosote	Wood preservative
Quaiacol	Antiseptic
Hexachlorophene	Antiseptic
Medicinal tar	Treatment of dermatologic conditions
Phenol	Outpatient podiatric surgery
Phenylphenol	Disinfectant
Tetrachlorophenol	Fungicide

Field Treatment and Decontamination Procedures

Treatment of phenol poisoning includes decontamination of the skin using copious amounts of water. This decontamination procedure is only somewhat helpful because phenol has limited solubility in water. Further decontamination must be done using alcohol, olive oil, mineral oil, or vegetable oil. If only small amounts of water are used, absorption will increase making the patient's condition worse. Care must be taken by the rescuer to ensure that secondary contamination does not take place. The use of correct chemical protective equipment is mandatory to prevent secondary contamination. Remember, an exposure of 5% phenol has reportedly caused death to children. Supportive measures such as assisting ventilations and controlling ventricular ectopy and seizures are done as the need arises. Continuous cardiac and vital sign monitoring in conjunction with respiratory supportive care with 100% oxygen are mandatory.

Definitive Treatment or Follow-Up Care

Insistence on transportation to the hospital for further observation is of vital importance. As stated earlier, systemic effects may last for up to 24 hours and seizures have been noted as long as 18 hours postexposure. Observation by medical staff should be ordered for up to 24 hours on any victim of a significant exposure.

PESTICIDES

Pesticides are a part of our everyday life. The use of pesticides have spread from their primary use on farms to extensive use in and around the home and office. Persons in the United States experience low-dose exposures to pesticides on almost a daily basis but most reported injuries from these toxic chemicals are to farmers, loggers, those applying pesticides, and migrant workers. Pesticides are usually divided into rodenticides, herbicides, and insecticides. Insecticides are further divided into chlorinated hydrocarbons, organophosphates, carbamates, and pyrethrins.

Rodenticides

anticoagulant
a chemical product that prevents clotting of the blood

Most rodenticides used today are the **anticoagulant** variety. These are used because the toxicity can be controlled for the size and susceptibility of the intended victim. Strychnine, thallium, and phosphorus are also sometimes used.

Anticoagulants   Scientists first became aware of the anticoagulation effects from these chemicals during the 1920s when cattle began to die of internal hemorrhage. It was found that cattle fed improperly cured hay from sweet clover developed a bleeding disorder (sweet clover disease) stimulated from the ingestion of a chemical bishydroxycoumarin found in the hay. In the 1940s coumarin was developed into a pharmacologic agent used to treat thromboembolic disease.

Warfarin was later developed as a more potent coumarin derivative and used not only as a medicine but also for rodent control. Because some rats developed a resistance to warfarin, even more potent anticoagulants were developed to work more efficiently. Chlorphacinone and superwarfarin produce a profound effect that lasts longer.

Route of entry: Most patients become poisoned by these agents through accidental ingestion. It is important to find out which type of chemical coagulant was ingested because chlorphacinone is about ten times as potent as standard warfarin and superwarfarin can be one hundred times as potent.

Treatment: As an emergency responder the field treatment is limited to the oral ingestion of activated charcoal. It is important to know that the maximum effects from the poisoning can be delayed as long as 36 to 72 hours. The hospital must assess the status of the patient by drawing serial prothrombin time (PT). Vitamin K is the antidote but dosage is guided by type of poisoning and the lab values.

Strychnine   Strychnine, a poison derived from an Asian tree *Strychnos nux vomica*, has the ability to block a neurotransmitter (glycine) of the inhibitory portion of the central nervous system. This results in hyperexcitability and usually seizures that can be brought on by only mild stimulation such as a touch or loud noise. The seizures usually occur while the patient is awake and aware of what is happening. Muscle spasms, hyperthermia, and metabolic acidosis can also result.

Route of entry: An oral route is the primary route of exposure to strychnine-based chemicals. Strychnine has found its way into water sources but more commonly children find that some of the agents resemble cracker and pretzel type mixes or breakfast cereal and ingest it purposely. The signs and symptoms usually occur 15 to 30 minutes after ingestion.

Treatment: Treatment involves the control of seizures. Gastric lavage or ipecac is not suggested because of the patient propensity to seizure activity. Paralysis once in the hospital is a consideration.

Thallium   Thallium is no longer commonly used as a rodenticide. Unfortunately, it has been used for suicides and murders because it is virtually tasteless and odorless. Because of this and its extreme toxicity, the sale and use of thallium is restricted. Thallium interferes with cellular respiration and also stimulates the depolarization of nerve pathways. After ingestion the peak signs and symptoms may appear as long as 24 hours later and are characterized by nausea, vomiting, abdominal pain, diarrhea, and gastric hemorrhaging. These symptoms are followed by coma and seizures. Poisonings of this nature also cause Mees' lines (white bands across the finger nails) days or weeks after the exposure.

Route of entry: Thallium poisonings occur as a result of accidental or intentional ingestion.

Treatment: Treatment from the emergency responder is supportive. In the hospital setting, a 1% sodium iodine gastric lavage has been proved as a useful treatment. This converts sodium thallium into thallium iodine which, because of it insolubility, is poorly absorbed.

Phosphorus   Many forms of phosphorus are available but the toxic variety is the white or yellow. The ingestion of phosphorus causes diverse systemic damage and includes the cardiovascular, hepatic, renal, and gastrointestinal systems. The signs and symptoms reflect the damage being caused from the poisoning, usually severe abdominal pains, gastroenteritis, and cardiovascular collapse. The patient may present with a garlic odor and the patient's stools and vomitus may glow in the dark.

Route of entry: Since phosphorus rat poison bears a striking resemblance to butter and is often spread on bread to entice the rodents, many times it is mistaken as just that, and eaten.

Treatment: Treatment by the emergency responder is supportive. Activated charcoal has proved to be useful and may be given by responders. Once in the hospital, a gastric lavage utilizing potassium permanganate, which oxidizes the phosphorus into nontoxic oxides, may be used.

Herbicides

Of the herbicide family, the paraquat, diquat and morfamquat are the most toxic and widely used. Each of these compounds has differing toxicities affecting different target organ systems.

Paraquat   Paraquat is extremely toxic and causes a failure of multiple systems. Although the systemic attack of paraquat involves the respiratory tract, kidneys, liver, heart, eyes, and skin, the pulmonary fibrosis is usually the cause of death. Paraquat is a strong corrosive and if ingested causes severe burns to the gastrointestinal and respiratory system. In commercial grades, as little as 10 milliters cause death. Once in the system, it is distributed in the liver, kidneys, brain, and concentrated in the lungs. Here it forms highly reactive superoxide radicals that kill lung cells and lead to pulmonary fibrosis. The formation of **pulmonary fibrosis** usually starts between 3 to 14 days after exposure. The fibrosis formation (superoxide radicals) is significantly multiplied if oxygen is given.

pulmonary fibrosis
unnatural growth of fibrous tissue in the lungs

Diquat   Although not as toxic as paraquat, diquat still attacks multiple systems to include the alimentary tract, kidneys, liver, testis, and eyes. The formation of superoxide radicals has not been reported but the onset of bilateral cataract is a chronic effect of the chemical.

Morfamquat   The least toxic of the dipyridyliums, morfamquat's systemic effect involves primarily kidney damage.

> *Route of entry:* Most deaths from dipyridyl herbicides are due to ingestion of an intentional or accidental nature. Fatalities have been reported from skin exposure to commercial grades of the paraquat. Inhalation toxicity is unlikely because the droplets are of such a size that they cannot reach the alveoli.
>
> *Treatment:* Field treatment includes supportive care and oral activated charcoal. If activated charcoal is not available, the ingestion of earth (clay) has been proved to be effective. Once in the hospital setting, gastric lavage is recommended if less than 2 hours after ingestion. Adsorption with activated charcoal, bentonite 7.5% in suspension, or fuller's earth in a 30% solution have all helped.

Insecticides

Insecticides are commonly divided into four types: chlorinated hydrocarbons, organophosphates, carbamates, and pyrethrins. All but the pyrethrin category are toxic to humans and pose a danger if carelessly used or intentionally misused. Pyrethrins are made from chrysanthemums and are only a danger to those who suffer from an allergy to the flower.

Chlorinated Hydrocarbons   Chlorinated hydrocarbons were the most commonly used insecticide in the United States from the 1930s into the 1960s. During the late 1960s the long-lasting effects of chlorinated hydrocarbons were first noticed. DDT is an extremely stable chemical, capable of maintaining its makeup even after harsh environmental conditions. For this reason, it was not a surprise that DDT found its way into water supplies, streams, and lakes. DDT was outlawed in the United States in 1972 after the chemical was linked to killing enormous numbers of fish and birds. DDT was responsible for almost causing the demise of the American bald eagles that were plagued with a defect in the egg shell, causing them to be brittle and breaking with just the weight of the mothers body. This genetic defect was said to be caused from the indiscriminate use of this pesticide. The effects were written about in the book *Silent Spring* (Carson, 1962), which brought to light the dangers of uncontrolled use of poisons such as DDT. Although DDT is outlawed in the United States, it is still used in other countries.

> *Route of entry:* Chlorinated hydrocarbons can enter through all routes. It has excellent absorption through the skin and is readily absorbed through the lungs and gastrointestinal system. Once in the body, the target organ is the CNS where stimulation is the primary manifestation. Muscle fasciculations, seizures, hyperthermia, and hypoxemia all may result from an exposure.

> *Treatment:* The treatment of an exposure to DDT or other chlorinated hydrocarbons is supportive. Decontamination is necessary because of continued absorption of the chemical through the dermal route.

ORGANOPHOSPHATE INSECTICIDE POISONING

History

Organophosphate injuries make up the majority of reported and treated insecticide poisonings in the United States. The Environmental Protection Agency (EPA) reported that over 80% of all hospitalizations in America, due to exposures of insecticides, were associated with organophosphates. Most of the poisonings occurred in California, across the Farm Belt, and in the South.

Organophosphates have gained such great popularity because of their effectiveness as insecticides. They are also considered relatively safe because of their unstable chemical structure. This instability causes the chemical to disintegrate in just a few days into relatively harmless agents. These chemicals also do not persist in body tissues or in the environment like other insecticides, such as DDT.

Organophosphates have mostly replaced the use of DDT as the insecticide of choice in the agriculture world. Probably the most toxic organophosphate insecticide is Tetraethylpyrophosphate (TEPP), which was the first synthesized during the mid 1800s. TEPP is still in use today but is not the organophosphate of choice.

Organophosphates first gained popularity during WWII when the German military developed and tested Parathion as a nerve gas. It was not until after the war that parathion gained recognition as an effective insecticide. Today many nerve agents are still in military arsenals across the globe. Sarin, Tabun, and Soman are the more notable military gases. All three are extremely potent organophosphate chemicals capable of inhibiting a lethal amount of acetylcholenesterase with a very small dosage. It is said that just a quarter of a drop of Soman on the skin can be lethal. In today's political climate it would not be unexpected to find radical hate groups using these chemicals for terrorism. In fact that was exactly the case in Tokyo during March of 1995.

Parathion is not available over the counter to the general public but is primarily used by the agricultural industry. It is one of the most dangerous of the commercially available organophosphates because of its ability to be readily absorbed through the skin, eyes, respiratory, and gastrointestinal systems. Malathion and Diazinon, on the other hand, are both considered to be much safer for home use because they have been synthesized to have poor absorption qualities and low oral toxicity.

> Tokyo, Japan, March 20, 1995. A terrorist attack left at least 8 dead, 17 critical, 37 serious, 984 treated with antidotes, 4,073 treated as outpatients with an estimated total of 5,510 Tokyo residents involved. These victims were exposed in a subway station to a chemical warfare agent named "Sarin." Sarin (methylphosphonofluoridic acid 1-methyl-ethyl ester, $C_4H_{10}FO_2P$ or $[(CH_3)_2CHO](CH_3)FPO]$ is a wartime nerve gas of organophosphate makeup that has a strong cholinesterase-inhibiting effect that is toxic by inhalation and absorption. The symptoms suffered by victims of this attack are similar to, but more severe than, parathion. Witnesses stated that a strong odor was present and described the odor as acidic. The Sarin was released in three areas on the trains and was set to discharge under the seat of government. The gas released before reaching its destination. Unfortunately, this type of terrorist attack is on the rise and attacks such as this are not confined to Europe or Asia but are a real threat here in the United States.

Large stockpiles of organophosphate-based nerve agents are stored throughout the United States. A great concern is that these containers, some more than 40 years old, will leak. At times these agents are also transported by train, trucks, and by plane. The potential for sabotage or accidental release is increased greatly during these transports.

Pathophysiology

Organophosphates can be either inhaled, ingested, or absorbed through the skin and eyes. Their effects on the body are systemic, affecting the nervous conduction

● **CAUTION**

Sarin, Tabun, and Soman (notable military gases) are extremely potent organophosphate gases capable of inhibiting a lethal amount of acetylcholinesterase with a very small dosage.

! **SAFETY**

The leakage of old containers of stockpiled organophosphate-based nerve agents and the potential for sabotage or accidental release during transport of these agents is cause for great concern.

pathways. In general, Organophosphates display a large range of differences in their toxicities, which are due to the differences in the chemical's ability to penetrate skin or absorb through the oral route.

Acetylcholine is the primary and most important chemical transmitter located at the synaptic junction of the parasympathetic nervous system. As an electrochemical impulse is transmitted through a nerve cell, it reaches a junction between two nerve cells. For the impulse to be transmitted to the next cell, acetylcholine is released. The acetylcholine then travels across the junction, stimulating the conduction at the distal end of the synapse, thereby continuing the electrical impulse. Once it has fulfilled its function an enzyme, acetylcholinesterase is released to break down the acetylcholine by hydrolysis into two inert chemicals. All of these chemical reactions take place millions of times a day in the body and only last a fraction of a second.

Organophosphates bind with acetylcholinesterase, inhibiting it from functioning. The results are an overstimulation and excitation of the nerve impulses because the synapse becomes flooded with acetylcholine. After the excitation phase, which lasts variable amounts of time, the nerve cell becomes paralyzed and stops functioning.

The nerves most prominently affected include the CNS, the parasympathetic nervous system, including a few sympathetic nerve endings like the sweat glands, and the somatic nerves. Therefore, the signs and symptoms displayed by a victim of organophosphate poisoning reflect the effect on these nerves.

Signs and Symptoms

The acronyms SLUD, SLUDGE, and DUMBELS are all used to describe the symptoms found in a victim of organophosphate poisoning.

Salivation	Salivation	Diarrhea
Lacrimation	Lacrimation	Urination
Urination	Urination	Miosis (pinpoint pupils)
Defecation	Defecation	Bronchospasm (wheezing)
	Gastrointestinal	Emesis
	Emesis	Lacrimation
		Salivation

Other symptoms noted because of the site affected are

CNS—Anxiety, restlessness, convulsions, absent reflexes, coma, respiratory and circulatory depression.

Somatic—Involve the striated muscles; fasciculations, cramps, weakness, paralysis, cardiac arrest.

Sympathetic—Tachycardia, hypertension

Parasympathetic—Sweating, constricted pupils, lacrimation, excessive salivation, wheezing, cramps, diarrhea, bradycardia, urinary incontinence, hypotension.

Where Organophosphates Are Usually Found

Organophosphates are found in large quantities anyplace in the United States where farming is an industry. They are also the pesticide of choice for home use but in quantities and concentrations much less than commercial grades (Fig. 6-7). These pesticides get to their destination by truck and rail, increasing the possibilities of an accident in almost any community.

Decon and Significant Danger to Rescuers

Typically, the patient presents with the strong scent of pesticides on his breath and all body fluids. Protection with rubber gloves and respiratory gear is suggested until the patient is stripped and decontaminated. Soap and water is usually enough to decontaminate, but if odor remains, a second washing of the skin with ethyl alcohol is recommended. All body fluids from the patient should be considered contaminated. The health care provider and transporter must exercise caution and use protective equipment throughout emergency care and transportation to the hospital.

Reason for Treatment

Atropine is the physiologic antidote for organophosphate poisoning (Fig. 6-8). It is administered in high doses while under close cardiac and vital sign monitoring. Great caution should be used in the administration of atropine to a patient who is

■ NOTE

Organophosphates are found in large quantities anyplace where farming is an industry.

● CAUTION

All body fluids from an organophosphate-contaminated patient should be considered contaminated, and extreme caution as well as protective equipment should be used throughout the emergency care and transportation of the victim.

Figure 6-7
Organophosphates and carbamates are incredibly popular especially in the farm belt and southern states.

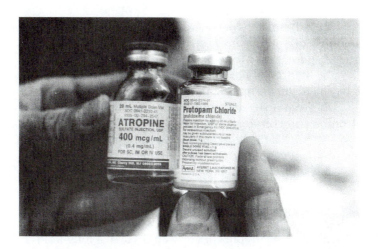

Figure 6-8 *High dose atropine is used on both organophosphate and carbamate poisonings. Protopam Chloride is only used in the case of organophosphate poisonings.*

hypoxic due to excess salivation and bronchial secretions or spasms. Giving high dose atropine to a hypoxic patient may cause ventricular fibrillation.

Atropine blocks the effects of acetylcholine while the body is naturally metabolizing the organophosphate. Atropinization must be maintained until all of the absorbed organophosphate has been metabolized and the body again produces sufficient quantities of acetylcholinesterase. Acetylcholinesterase is produced slowly (may take up to 30 days) in the body, causing the effects of the poisoning to linger well after any evidence of organophosphate has left the body.

One case of parathion overdose required the use of 19,590 milligrams of atropine over a 24-day period. In just one 24-hour period, 2950 milligrams were administered (Golsousidis and Kokkas, 1985).

Pralidoxime (Protopam, 2-PAM) is the next drug indicated for a victim of organophosphate poisoning. It is believed that an early dose of pralidoxime may decrease the need for such large doses of atropine. It is specifically useful in patients who present with muscle fasciculations and weakness. Protopam has three desirable effects:

1. Frees and reactivates acetylcholinesterase.

2. Detoxifies the organophosphate.

3. Has anticholinergic (atropinelike) effects.

Most authorities suggest that atropine and pralidoxime both be given to the organophosphate-exposed patient to combat both the present symptoms and delayed effects of the poisoning.

Treatment

1. Decontaminate patient to stop further absorption of organophosphate through the skin and respiratory system. Exercise care to protect aid givers from secondary contamination.

2. Administer high percentage oxygen with suctioning as needed. Large amounts of sputum may be present due to the acceleration of salivation. The excess sputum may become an aspiration hazard.

3. Start IV of normal saline with large cannula for rapid fluid replacement if hypotension becomes a problem.

4. Administer atropine 1–4 milligrams IV slowly every 5 minutes. Titrate to atropinization (drying of mouth, flushing, and dilated pupils). Atropine treatment may continue at the initial dose level for up to 24 hours, then if the patient is showing improvement the dose can be tapered off.

5. Administer protopam 1gm over 2 minutes IVP.

Definitive Treatment

Poisoning by organophosphates can involve a long-term treatment plan. As mentioned the acetylcholinesterase may be reduced in the body for up to 30 days. Therefore the treatment with atropine may theoretically continue for 30 days.

For a period of 24 hours, all body secretions should be handled as chemically contaminated waste. The victim's body should be washed free of the chemical on a periodic basis throughout this 24 hours.

CARBAMATE INSECTICIDE POISONING

Carbamates, a group of insecticides derived from carbamic acid, have effects on the body very similar to those of organophosphates. Like organophosphates, carbamates inhibit acetylcholinesterase causing a build-up of acetylcholine. They can enter the body through inhalation, ingestion, and dermal exposure. Unlike organophosphates that cause irreversible inhibition of the acetylcholinesterase, carbamates' combination with acetylcholinesterase is temporary and short lived. Carbamates also poorly penetrate the central nervous system, therefore, CNS depression and convulsions are rarely found in these poisonings.

Over-the-Counter Carbamate Insecticides

Some of the more common carbamates are sold under these trade names: Temic, Matacil, Vydate, Isolan, Furadan, Lannate, Zectran, Mesurol, Dimetilan, Baygon, Sevin.

Treatment

Initial treatment is supportive, such as maintaining airway, and decontamination. Atropine is again the drug of choice. The dosage is 0.4 to 2.0 milligrams IV, repeated every 15–30 minutes until atropinization appears. Usually, atropinization is only needed for 6 to 12 hours.

Pralidoxime (protopam, 2-PAM) is not indicated in the patient poisoned with carbamate. Given time, the carbamate-cholinesterase complex reverses, freeing the acetylcholinesterase.

EPA HAZARDOUS SUBSTANCE LABELING

Pesticides are regulated by the EPA. In order to give the public an idea on how hazardous a pesticide may be, the EPA requires that a label be attached to the container with information about the chemical. Four important items are required on this label:

1. The product name
2. The active ingredient
3. The EPA registration number
4. Signal words

The EPA registration number is very specific. It has a two- or three-section numbering system (00000-00000-00 or 00000-000). The first number set is the manufacturer's number. The second group of numbers is the specific product's number. If a third group of numbers is found this is a crop compatibility number, indicating the food product it can be used on. These numbers can be referenced during an exposure.

The signal words, Danger, Warning, and Caution are messages that identify the relative level of toxicity. Danger indicates the highest toxicity and caution the lowest.

Danger I	Highly Toxic
Warning II	Moderately Toxic
Caution III	Slightly Toxic
Caution IV	Relatively Toxic

ANTIPERSONNEL CHEMICAL AGENTS

Chemical antipersonnel weapons have gained popularity in both the general public and law enforcement because these agents are able to subdue persons without the use of extraordinary physical force. These chemical sprays offer a nonlethal form of protection that causes temporary extreme discomfort. These sprays are offered in three chemical formulas, each having similarities and peculiarities. They are CN (chloroacetephenone), CS (orthochlorobenzalmalononitrile), and the most popular, OC (oleoresin capsicum) commonly referred to as "pepper spray."

CN (chloracetephenone)

Synonyms: phenacyl chloride, alpha-chloroacetephenone, omega-chloroacetephenone, chloromethyl phenyl ketone, and phenyl chloromethyl ketone.

The effects from this chemical agent begin in 1 to 3 seconds and are characterized by extreme irritation to the eyes causing burning and tearing (lacrimal discomfort). Irritation to the skin is also common because the crystals stick to moist skin causing burning and itching at the point of contact. CN also causes upper respiratory irritation. These effects last between 10 and 30 minutes.

Both CN and CS are submicron (less than one micron) particles. They are extremely light and are carried to the target area in a carrier solution that evaporates quickly, dispersing the agent. Because of the light fine particle, both of these chemicals are prone to cross contamination between the victim and emergency response personnel or the victim and equipment. It is essential to realize how easily cross contamination takes place. The best way to deal with the hazard is to avoid the chance of it happening. Some police agencies still use these gases although much safer gases like OC are available and rapidly becoming the standard.

CS (orthochlorobenzalmalonitrile)

Synonyms: O-chlorobenzylidene malonitrile, OCBM, and military tear gas.

Signs and symptoms start in about 3 to 7 seconds and last 10–30 minutes. The effects reported are stinging of the skin especially in the moist areas, intense eye irritation with profuse tearing and blepharospasms. The burning also effects the nose and upper respiratory system. Some victims panic due to the feeling of shortness of breath and chest tightness. Victims describe effects as being ten times worse than CN. Some police agencies still use this irritant mostly for crowd dispersal.

OC (oleoresin capsicum)

OC has become the safest and most popular of the chemical agents. It is found in police aerosol sprays and over-the-counter agents. It is a non-water-soluble agent prepared from an extract of the cayenne pepper plant. The effects for OC start almost immediately when contact with the eyes occur. OC is not a submicron particle so access to the lower respiratory system is limited. The contact of OC causes immediate nerve ending stimulation but not an irritation. The effect from this chemical lasts between 10 to 30 minutes and usually leaves no lasting effect.

Signs and symptoms are usually inflammation of the eyes, nose, mouth, and upper respiratory system. Dilation of the blood vessels of the eye is common and burning causes lacrimation and blepharospasm. Because of increased tearing and upper respiratory mucous production, coughing and a choking feeling is common.

Extensive human studies conducted by the FBI have found no significant permanent reaction to the spray. Care should be exercised if any of the victims of the spray show signs of systemic sensitivity. Wheezing or decreased blood pressure should be treated through normal means along with insistence on transport to a medical facility.

Figure 6-9 *An easy method used to organize the hazmat drugs is to color code a drug box with each color representing a different type of poisoning. Then develop quick reference cards containing signs and symptoms of the poisoning with drug and dosage information on that poisoning. This allows the user to access information easily and stay within the approved protocols.*

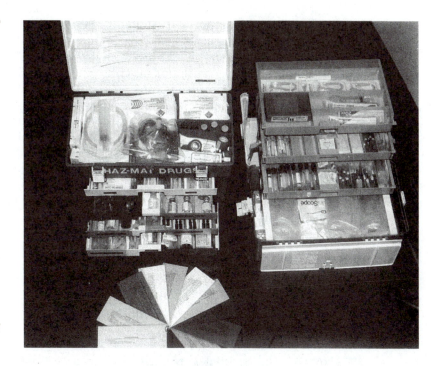

Treatment of Oleoresin Capsicum (OC) Spray

Since OC pepper spray does not cause notable tissue damage, treatment is aimed at relieving the pain associated with nerve stimulation. Irrigation of the eyes cools the nerves and relieves some of the pain but is inefficient in removing the agent. Furthermore, irrigation is difficult because the lids are initially in blepharospasms.

The use of an analgesic agent will remove the pain generated by the agent and also relax the blepharospasm. In most cases the analgesic agent will wear away at the same rate as the pepper spray. The most common analgesics used for this application are Alcaine, Ponticaine, or Opthalmacaine. These drugs do not provide any medicinal beneficial effect except for relief of pain. By relieving the pain, patients anxiety will lessen, blepharospasm will relax allowing the caregiver to perform a good visual exam and provide irrigation if it is needed. The biggest drawback to the use of these drugs in the field is their extreme heat sensitivity. Most manufacturers recommend that these medications be kept refrigerated or kept in an environment of less than 65°F.

RADIATION AND RADIOACTIVE EMERGENCIES

Radiologic emergencies can be addressed from two different aspects, those involving wartime activities and those that happen during peacetime accidents. This section primarily addresses peacetime accidents, although many of the principles involved can be adopted to wartime radiologic emergencies. If you wish more information on wartime emergencies contact the local Civil Defense Department.

When radiation emergencies exist, the term "radiation" refers to ionized radiation. Radiation is a broad term that describes energy transmission. Radio, sound waves, and visible light are also referred to as radiation, but normally do not constitute an emergency. Simply put, ionized radiation produces a charged particle that disrupts atoms and therefore is capable of causing harm to living cells and tissues. Ionized radioactivity can be produced by reactors but is also naturally occurring and can be found in and around the planet.

We are constantly being bombarded with minute amounts of radiation emitting from space. None of our senses can detect the presence of this radiation or for that matter radiation in general. Our ignorance and lack of understanding causes a fear of radiation.

Types of Radiation

There are three common types of ionizing radiation, known as alpha, beta, and gamma that are involved in radiation emergencies. There are other types of ionizing radiation, such as x-rays that will be discussed, but they are not normally involved in radiation emergencies.

Alpha Particles   Alpha particles are positively charged particles that are the weakest of the ionizing radiations. The particles are made up of two neutrons and two protons and are relatively large, moving more slowly than beta particles. Alpha particles can travel only about 4 inches from their source. These particles can be stopped by a thin piece of paper and therefore cannot penetrate even light clothing. If alpha particles come in direct contact with the skin they can penetrate only the top layer made up of dead cells so they pose little threat to the outside of the body. However, alpha particles pose a health hazard if they are inhaled, ingested, or enter the body through an open wound. Thus, they are an internal hazard.

Beta Particles   Beta particles can be positively or negatively charged particles, depending on the type of nucleus emitting particle. They are higher in speed, 7,000 times smaller (about the same mass as an electron), and have more penetrating power than alpha. Beta particles have much better penetrating capabilities and can readily penetrate skin and tissues. Most beta particles can travel about 30 feet with some traveling 100 feet in the air, and can penetrate 0.1 to 0.5 inches of skin. They can be blocked by a thin layer of metal or other dense material. Some particles can even be blocked by heavy clothing. Because the beta particles can penetrate and cause damage to the tissues, once striking or penetrating the skin they are termed internal and external hazards.

Gamma Rays   Gamma rays are not particles but rays of energy generated from a nucleus that is unstable due to excessive energy levels. The excessive energy is emitted as a photon or a pure pocket of energy. This electromagnetic energy has the most penetrating ability of the three radiations mentioned and is extremely dangerous. Gamma rays pass completely through the body causing damage through and through. Gamma rays are sometimes referred to as penetrating radiation. They are deemed to be internal and external hazards.

X-Rays   X-Rays are another type of electromagnetic radiation capable of penetrating the body tissue. They are produced on the site of use as the need arises. The technique used to produce x-rays involves bombarding a metallic target with electrons, which are then directed to the point needed. There is a danger of overexposure to x-rays but this danger is usually due to carelessness and not a radiation accident. If the x-ray equipment is involved in an accident, such as fire or explosion, there is little danger of radiation exposure.

Measuring Radioactivity

Roentgen: the measurement of radioactivity for gamma rays. It is the amount of ionization per cubic centimeter of air. It is a measurement of exposure and represents the amount of gamma radiation that produces two billion ion pairs in dry air. An ion pair is one positively charged atom and one negatively charged electron.

RAD (radiation absorbed dose): a unit slightly greater than one roentgen. This measurement is used to measure dosage. An acute exposure to 350 RADs kills 50% of the exposed population in 30 days (LD_{50-30})

REM (roentgen equivalent man): the amount of absorbed radiation. It is used as a measurement of biological effect. It represents the amount of absorbed radiation of any type that produces the same effect on the human body as one roentgen of gamma radiation.

Curie: a unit of measurement of radioactivity as compared to radium. It is the amount of radioactive substance that undergoes the same number of radioactive disintegrations in the same time as 1 gram of radium. More specifically it is the amount of a substance in which 37 billion atoms per second undergo radioactive disintegration.

Body Exposures and Expected Injuries

20 roentgens represents normal background radiation.

25 roentgens is the recommended maximum full body exposure.

50 roentgens indicates blood cell damage that is usually self-limiting.

100 roentgens causes blood cell damage and lasting illness.

200 roentgens causes 10% mortality and possible genetic damage.

350 roentgens causes 50% mortality.

650 roentgens causes 95% mortality.

Burns

200 RADs causes 1st degree burns.

500 RADs causes 2nd degree burns.

1,000 RADs causes 3rd degree burns.

Principles of Protection

Once a general understanding of radiation has been developed, the principles of radiation protection are easier to understand. The four basic principles to protect yourself, and any victims involved, is that of time, distance, shielding, and quantity.

Time   The time factor involving protection from radiation is simple: *Limit your time of exposure.* The longer the exposure to a source of radiation, the greater the damage to the body tissues. If an extended amount of time needs to be spent in an area affected by radiation, such as rescue or extrication, then a plan should be made to rotate personnel through the area, in effect limiting individual exposure.

Distance   Distance is the second protection principle. *Stay as far away as possible.* By doubling the distance away from the source, the exposure is decreased by a factor of four. For example, an exposure of 26 milliroentgen per hour at a distance of 3 feet would be reduced to 4 milliroentgen per hour at a distance of 6 feet. But, the opposite also holds true, and exposure of 8 milliroentgen per hour at a distance of 4 feet would be 32 milliroentgen per hour at 2 feet. Therefore, it is important to realize that the exposure to radiation is always greatly reduced by moving away form the source. (R = roentgen, a measurement of exposure to gamma or x-rays, mR/hr represents the amount of exposure to produce a certain amount of ions in one cubic centimeter of air in one hour. 1m/R = one milliroentgen or one thousandth of a roentgen.)

Shielding   Shielding is the third principle of protection. *Keep something between you and the source.* Shielding utilizes the fact that the denser the material, the more radiation is blocked by it. Lead is the most common shield used around x-ray equipment because of its high density. At the site of a radiation emergency this principle can be put to use using an automobile or a hill of dirt. As mentioned earlier, heavy clothing will stop all alpha particles and some beta particles. Lead shielding, although, not practical for rescue, is effective against gamma rays and x-rays.

It may be difficult to use this principle especially when performing a rescue. But, utilizing this and the other two protection principles, time and distance, will minimize the exposure.

Quantity   The last protection principle is that of quantity. Exposure is directly related to the amount of radiation being emitted from the source. Any means necessary to limit or reduce the amount of radiation emission will reduce the amount of exposure. It may be difficult to immediately reduce the amount of the spill (exposed radioactive material), but this principle also holds true for the victim, dirty with radioactive particles. REMOVE CONTAMINATED CLOTHING. The victim's clothing should be removed and bagged, then taken from the immediate area, therefore limiting the quantity of material. It should never be placed in the ambulance and transported with the patient to the hospital.

Location of Radiation and Common Sites for Accidents

Accidents involving radioactive material can be divided into six different categories:

1. Transportation of radioactive materials.
2. Medical uses of radioactive substances (Fig. 6-10).
3. Research facilities utilizing radioactive substances.
4. Industrial uses and nondestructive testing.
5. Nuclear reactor sites.
6. Isotope production facilities (radioactive fuel).

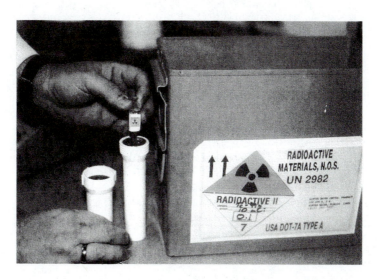

Figure 6-10   *Many of the radioactive substances transported through and around communities are for medical use. This type of radioactive source usually has a short half-life and is relatively harmless within a couple of days.*

Vindicator, a new company located in Mulberry, Florida, is the first facility to treat food for bacteria and microorganisms by exposing it to radiation. The radiation is supplied from a relatively strong source emitting gamma. So far the federal government has only allowed Vindicator to irradiate certain fruits. Recently however, there have been requests to start irradiating meat to kill E. Coli bacteria. If this treatment becomes popular, every community could have a facility of this type.

According to the Federal Emergency Management Agency (FEMA), fewer than 1,000 persons worldwide are known to have been involved in serious radiation accidents. Only 450 received medically serious doses, and there were only 21

Facts about Radioactive Emergencies and Injuries

Radiation Accident History, 1944 to 1991

Number of worldwide accidents	340
Persons involved	132,928
Significant exposures	3,037
U.S. deaths	32
U.S. accidents with death	14
Deaths worldwide	116
Russian deaths	33

DOE-REAC/TS Radiation Accident Registry

Radiation From Various Sources

Natural background radiation	U.S. average, 100 mrem/yr
Natural radioactivity in body tissue	50 mrem/yr
Air travel round trip (London–New York)	4 mrem
Chest x-ray internal dose	10 mrem per test
Radon in the home	200 mrem/yr (variable)
Man-made	100 mrem/yr

DOE-REAC/TS, Transport of Radioactive Materials, 1992

Chernobyl

135,000	evacuated
350	evaluated medically
1	died on the scene from burns
1	died on the scene from an explosion
203	hospitalized
29	died of thermal burns/ionized radiation
31	total died

JAMA, August 7, 1987, Volume 258

fatalities. In the United States there has never been an injury from radioactivity reported by an emergency medical responder or rescuer.

Types of Injuries

Victims of radiation accidents rarely immediately show signs or symptoms of radiation injury. The rescuer must be aware of the possibility of a radiation injury by evaluating the scene and looking for any evidence of a radiation exposure. Radiation injuries are found in one or a combination of either external irradiation, contamination, or incorporation.

External Irradiation   External irradiation is the injury that occurs when penetrating radiation (gamma or x-rays) passes through the body from an external source. The intensity of the injury depends on the amount of exposure to the radiation. Once the victim is externally exposed he does not become radioactive and can be handled by the rescuers without fear of receiving a radiation injury from the patient.

Contamination   When a patient becomes partially or wholly covered with radioactive material, it is referred to as a contamination injury. The contamination can be in the form of a solid, liquid, or gas. The contamination can either exist on the outside of the body or can enter the body through the skin, lungs, open wounds, or the digestive tract. Therefore, this injury can be an internal injury, external injury, or both. Rescuers handling the contaminated patient can, themselves, become contaminated and need to exercise the principles of time, distance, shielding, and quantity. If external contamination exists, decontamination of the patient must be done prior to transportation of the patient to the hospital.

Incorporation   Incorporation is the uptake or combining of the radioactive material by the tissues, cells, and target organs (the organs affected by the injury). Based on its chemical properties, the radioactive material seeks out the areas that easily incorporate the type of radiation material involved, such as radium to the bone or iodine to the thyroid. Medicinal radioactivity uses this principle to treat and diagnose these target areas. Incorporation cannot occur unless contamination has occurred and therefore, incorporation is always a combination of internal and external injury.

Irradiation   Irradiation (from gamma) of a part or even the whole body does not usually require immediate emergency treatment. This type of injury usually causes signs and symptoms days, weeks, or months after the injury takes place. Consequently a contamination injury (from alpha and beta) is handled as a medical emergency. Unless treated promptly the contamination will lead to an internal injury and eventual uptake into cells and tissues causing permanent injury sometime years after the incident.

Depending on the type of radiation, how much of the body was exposed, and the dosage determines how much radiation sickness the victim will suffer. The effects of radiation on the body are varied depending on the dosage. The skin may swell, blister, redden, flake, or itch. The breathing may be affected because of swelling or damage at the alveoli. Other widespread damage may be evidenced by permanent or temporary sterility, loss of menstruation, reduction of sperm count, damage to blood vessels, cancer, and genetic damage.

Recovery from large doses of radiation may require months or years. Recurrent or chronic problems such as chromosomal damage and reproductive difficulty may last a lifetime.

Rescue and Emergency Treatment

If information is received, either through the dispatcher or visual evidence on the scene, indicating a radiation accident, appropriate precautions should be taken. Usually the same general procedures as for other hazardous materials incidents holds true. Place the responding units a safe distance from the incident. Approach from up hill and up wind, placing emergency units out of any smoke, fire, dust, or gas emitted from the scene.

As with any hazardous materials incidents, the following agencies must be contacted and advised of the accident.

1. Local law enforcement
2. Local fire department
3. Local medical facilities
4. Any local individual or organization known to be trained in radioactive emergencies
5. Local civil defense office
6. CHEMTREC 1-800-424-9300

■ NOTE
Heavy clothing such as fire turnout gear, coveralls, coats, and jackets will provide some protection against penetration of alpha and some beta particles, but not against gamma rays.

Heavy clothing such as fire turnout gear, coveralls, coats, and jackets will all assist in stopping the penetration of alpha and some beta particles, however, this clothing will not stop the exposure to penetrating gamma rays. The best protection from gamma rays is dense metal such as lead, which has limited use in the field. Most radiation sources emit a combination of radiation that may include alpha and beta particles along with gamma rays. Using good judgment is the best protection.

If a rescue is needed, remember to use the principles of time, distance, shielding, and quantity. The best approach to a rescue is to divide the work to be done in the hazard zone between teams of workers, limiting the time spent in the hazard zone by each worker. The Federal Nuclear Regulatory Commission recommends that any individual at an emergency receive no more than 25 roentgens as a one-time, whole body dose.

The use of self-contained breathing apparatus (SCBA) is governed by the possibility of airborne radioactive particles found in gas, dust, or smoke. If it is determined that the rescuer needs a SCBA, then so does the victim. Arrangements should be made to provide one for him while he is being removed from the hazard zone.

The use of radiologic survey meters like the CDV-700 (civil defense meters) are recommended during initial entry into the hazard zone. These are not routinely kept on emergency medical services units. Stabilization of a victim should never be performed in the hazard zone. The patient should have a primary survey performed; any life-threatening procedures, such as opening the airway, control of hemorrhage, or placing the victim on a backboard, quickly performed; and the patient should be removed from the hazard zone. External decontamination on the scene is recommended and supportive care started.

If extrication of a victim is needed, there is no effect from radiation on machinery used to extricate. Disposable gloves can be worn to treat the victim and lessen contamination to the rescuer. If the patient is not breathing, artificial ventilations should be accomplished with the use of a disposable bag valve device. If the MAST suit or other reusable items are needed they should be used and decontaminated later.

As with other hazardous materials exposures, the medical responder must provide protection for herself and her unit. Once the stretcher is covered with a piece of visquen then a blanket is draped over the stretcher. The victim should be placed on the blanket and the blanket wrapped around him with only his head and one arm extending through it. These areas are left exposed to assist in assessing the victim's level of consciousness, vital signs, and respiratory status. Next, the hospital should be notified and advised of three important things.

1. The number of patients involved.

2. How many of these victims are known to be contaminated.

3. The medical condition of each victim.

Once the patient is transferred to the hospital staff, then the transporting unit should be placed out of service. A radiologic survey of the unit should be taken to determine the amount of contamination. If contamination exists, then decontamination must be done before the unit is placed back into service.

Appendix 6-1

HAZARDOUS MATERIALS MEDICAL STANDING ORDERS

Purpose

It is the intention of these standing orders to facilitate rapid medical intervention at the scene of a hazardous materials incident. These procedures are written in order to better define the responsibilities of the hazardous materials medical sector staff. These policies, although intended for the hazardous materials emergency, can be used on other scenes of poisonings when deemed necessary by emergency response staff.

Policy

In addition to the medical standing orders, the fire department shall recognize the following as emergency treatment for specific exposure conditions.

Description

The possibility of secondary contamination shall be recognized and measures taken to reduce the chance of such contamination. It is the responsibility of all individuals involved at the scene to take precaution to reduce secondary exposure. However, if an exposure has taken place, the following is a set of medical standing orders that have been authorized by the medical director to be used at the scene of a hazardous materials incident or during transport of an exposed victim.

General Treatment   Rapid assessment and initial medical practices are a necessity. High-dose oxygen concentration shall be delivered to the patient as soon as practical. (This may be started during decon.) The paramedic in charge shall contact the appropriate hospital as soon as practical and advise of the type of exposure and the number of patients involved.

It is imperative that the safety of civilian and emergency personnel be maintained while dealing with hazardous materials. Site safety includes barring entry into the hot zone without proper precautions, full protective clothing, and knowledge or permission of the incident commander. People who become victims while in the hot zone must be brought into the warm zone and decontamination effected before any medical treatment is performed. Rescuers must not become victims themselves by entering the hot zone, decontamination area, or warm zone without proper protection.

Never transport a contaminated patient!! Leave the contamination at the scene of the emergency. NEVER take it with you to the hospital!!

Special treatment modalities for exposure shall be initiated as soon as possible after decontamination.

If there will be extended operations on a hazardous materials incident, EMS personnel should notify the closest appropriate medical facility, advising the emergency department of the nature and extent of the operations. This alerts the hospital of the incidents that may require setting up a clean isolation treatment room and/or obtaining specific medications for the exposure treatment. The report should include specific names of chemicals involved, specific amounts, and the type of exposure expected (i.e., inhalation, skin absorption, ingestion, or injection). Determine if a toxicologist is available for consultation. Be sure to notify the hospital at the end of the incident so they can return equipment and personnel to normal use.

Drug Box Inventory

The following is a list of the standard hazmat drug box inventory. It shall be a second medication box carried and used in conjunction with the primary ALS box. This drug box shall be maintained specifically for hazardous materials exposures and poisonings.

Speciality Drugs	Methylene blue
Adenocard	Morgan irrigation lens
Alupent	Naloxone
Amyl nitrite perles	Oxygen
Atropine sulfate	Proventil
Breviblock	Ponticaine hydrochlorite
Calcium gluconate	Pralidoximine
Dextrose 5%	0.9% Sodium chloride
Dextrose 50	Sodium bicarbonate
Diazepam	Sodium nitrite
Dopamine	Sodium thiosulfate
Epinephrine	Thiamine

SPECIFIC TREATMENT PROTOCOLS

Carbon Monoxide Poisoning

(and all cases of altered mental status in the context of hazardous materials). Note: Unconsciousness may occur in exposure concentrations of 1.5% or greater and may cause tissue anoxia. Transportation to a facility with a hyperbaric chamber should be considered.

Description   Colorless, odorless, tasteless, nonirritating gas. Converts hemoglobin into carboxyhemoglobin, a non-oxygen-carrying compound causing chemical asphyxiation. Pulse oximetry will indicate an incorrect, unusually high oxygen saturation.

Treatment   Immediately administer 100% oxygen if conscious. If victim unconscious, consider intubation and PPV utilizing a PEEP setting of 4 centimeters of water.

Start IV of LR.

Administer Glucose 50% for cerebral edema, given in conjunction with, or followed immediately by 100 milligrams thiamine. Follow the 50% glucose with immediate hyperventilation and 100% oxygen.

If CO poisoning due to suicide attempt, give Narcan 2 milligrams IVP.

Aniline Dyes, Nitrites, Nitrates, Nitrobenzene, and Nitrogen Dioxide

Description   Commonly found in fertilizers, paints, inks, and dyes. Changes hemoglobin into a non-oxygen-carrying compound methemoglobin. Blood color changes from red to a chocolate brown color. Pulse oximeter will indicate an inaccurately low reading due to the opaqueness of the compound.

Treatment

1. Immediately administer 100% oxygen. If victim unconscious, consider intubation and PPV utilizing a PEEP valve set at 4 centimeters of water.

2. Start IV LR

3. If hypotensive, position patient, increase IV flow. If severe start dopamine.

4. Administer Methylene blue, 1–2 milligrams per kilogram IVP over 5 minutes. (Methylene blue may momentarily affect the pulse oximeter because of the opaqueness of the drug).

Cyanide and Hydrogen Sulfide

Description, Cyanide   One of the most rapid-acting poisons. Bitter almond smell to those without sensory deficit. Interferes with the uptake of oxygen into the cell and halts cellular respiration causing chemical asphyxiation. Pulse oximetry will accurately indicate an unusually high oxygen saturation due to the cells' inability to pick up oxygen from the bloodstream.

Description, Hydrogen Sulfide   Also known as sewer gas. Has a distinctive smell of rotten eggs but most dangerous when it can't be smelled. Formed naturally by the decomposition of organic substances. Heavier than air. Interferes with cellular respiration.

Treatment

1. Amyl nitrite perles broken and held on a gauze pad under the patient's nose. Allow the patient to inhale for 15–30 seconds of every minute. During the interval, the patient should breathe 100% oxygen. If the patient is not breathing, place the perles into a BVM and ventilate the patient.

2. If intubated, provide PPV utilizing a BVM and PEEP valve set at 4 centimeters of water.

3. As soon as possible start an IV of LR and immediately give:

4. Sodium nitrite 10 milliliters of a 3% solution IV over 2 minutes (300 milligrams). Monitor BP.

   Children—.33 milliliters per kilogram of a 3% solution over 10 minutes.

   Sodium thiosulfate—50 milliliters of a 25% solution over 10 minutes. Monitor BP.

   Children—1.65 milliliters per kilogram up to 50 milliliters over 10 minutes.

Sodium thiosulfate not given in hydrogen sulfide poisonings.

Organophosphate Insecticide Poisoning (OIP) and Carbamate Poisoning

Description   Pesticide can be inhaled, ingested, or absorbed. Once in the body it binds with the acetylcholinesterase initially causing excitation of the nervous conduction then paralysis. Commonly seen signs are salivation, lacrimation, urination, and defecation (SLUD). Can be lethal in less than 5 milligrams dose.

Treatments

1. Immediately give 100% oxygen to ensure tissue oxygenation.

2. Start IV NS or LR and give:

3. Atropine 2–4 milligrams IVP at 5-minute intervals until atropinization (mouth dries) occurs. There is not maximum dose.

   Use extreme caution in a hypoxic patient. Giving atropine to a hypoxic heart may stimulate ventricular fibrillation.

4. Pralidoxime (2-PAM, Protopam) IVP 1 gram over 2 minutes. Not used in carbamate poisonings.

Hydrofluoric Acid Burns and Poisoning

Description   The strongest inorganic acid known. Injury is twofold: causes corrosive burning of the skin and deep underlying tissue. Also, binds with calcium and magnesium of the nerve pathways, bone, and bloodstream. The results are spontaneous depolarization producing excruciating pain and cardiac dysrhythmias degenerating to cardiac arrest.

Treatment

Skin Burns

1. Immediately flush exposed area with large amounts of water.
2. Apply calcium gluconate gel to burned area.
   (Mix 10 cubic centimeters of a 10% calcium gluconate solution into a 2-ounce tube of water soluble jelly).
3. Massage into burned area.

If pain continues then

1. Calcium gluconate in a 5% solution is injected subcutaneously in a volume of 0.5 milliliters per square centimeter or every 1/4 inch into burned area.

Eye Injuries

1. Immediately flush eyes with any means possible.
2. Mix 50 cubic centimeters of a 10% solution into 500 cubic centimeters of NS IV solution.
3. Connect bag and tubing to a Morgan irrigation lens and infuse.

Inhalation Injury

1. Mix 6 cubic centimeters of sterile water into 3 cubic centimeters of 10% calcium gluconate.
2. Place solution in nebulizer and connect to oxygen to provide effective fog.

Phenol

Description   Also known as carbolic acid. Found in many household items and is commonly used as a disinfectant, germicide, antiseptic, and as a wood preservative. It causes injury much the same as other acids by coagulating proteins found in the skin. Systemic effects are seen throughout the central nervous system, evidenced by CNS depression including respiratory arrest.

Treatment

1. Decontaminate initially with large volumes of water then irrigate burned area with mineral oil, olive oil, or isopropyl alcohol.
2. Support respirations, control seizures, and ventricular ectopy with recognized means of treatment.

Chemical Burns to Eyes

Note: Watch water runoff so other parts of the body do not become contaminated (especially other parts of the face, ears, and back of neck). Eye burns are almost always associated with contamination of other parts of the face or body.

Treatment

1. Immediately start eye irrigation by whatever means possible.
2. Ensure all particulate matter or contact lenses are out of the eyes by digitally opening the lids and pouring irrigation fluid across the globe.
3. Prepare the Morgan lens by attaching an IV solution of NS or LR. Ensure that the tubing is full and a steady drip of solution is running from lens.
4. Apply 1–2 drops of ponticaine hydrochloride into the injured eye.
5. Insert the lens by lowering the bottom lid and inserting, then raising upper lid and placing the lens against the globe.
6. Adjust the flow so that a continuous solution is flowing from eye.
7. Continue irrigation until arrival at the hospital.

Bronchospasms Secondary to Toxic Inhalation

Wheezing due to exposure of the respiratory system to an irritant.

Treatment

1. Immediately give 100% humidified oxygen.
2. Issue an updraft of either Alupent or Proventil, 1 unit dose nebulized.
3. If wheezing continues administer terbutaline (Brethine) 0.25 milliliters injected SQ.
4. Repeat the dose after 30 minutes if needed.

Tachydysrhythmias

Superventricular tachycardia due to sensitization of a toxic exposure and CNS stimulants.

Treatment

1. Establish an IV lock and give;
   a. 0.5 milligrams per kilogram of Breviblock IVP or
   b. Adenocard 6 milligrams rapid IV push followed by 10 cubic centimeters of saline IVP.

Chloramine and Chlorine

Description   Chloramine is the mixture of over-the-counter bleach and ammonia. Forms an irritating gas that converts to hydrochloric acid in the lining of upper air passages. The mixture is toxic and flammable. The patient typically complains of a burning sensation to the upper respiratory system, coughing, and hoarseness.

Treatment   After the patient is removed from the atmosphere and appropriate decontamination completed give:

1. 100% oxygen via NRB mask.
2. Assemble a nebulizer and administer 5 cubic centimeters of sterile water.
3. If burning persists titrate half strength adult bicarb (3.75% or 4.2%) and administer 5 cubic centimeters through a nebulizer.

This is the only time a chemical will be neutralized in or on the body by field medical personnel.

OC (oleoresin capsicum) Pepper Spray and Other Lacrimators

Description   The patient will usually present with extreme burning of the eyes, nose, and congestion due to increased mucous production. Exam will find the patient suffering from increased tear production and blepharospasm.

Treatment   Since the agent does not cause significant tissue damage the treatment is aimed at relieving the pain caused by nerve stimulation.

1. Initially determine the history of the injury. If a determination can be establishing that the pain is caused from capsicum spray then the eyes should be immediately numbed.
2. Once it has been assured that the patient is not allergic to caine derivatives apply Alcaine, Ponticaine, or Opthalmacaine.
3. When the blepharospasm is relieved, a visual exam is performed to assess for trauma of the eye.
4. Assess for clear lung sounds and BP changes to ensure that a sensitivity has not occurred.

Scenario

You are a member of an ambulance responding to a chemical spill in a semiconductor manufacturer. Dispatch reports that there are 4 patients that have been removed from the structure by the hazmat entry team. Upon your arrival you notice that the zones have been set up and the last patient is just entering the cold zone from the decontamination corridor. All four patients are unconscious. The hazmat officer reports that a fork-lift inside the building collided with some shelving causing it to fall. The shelving stored several different chemicals. You are handed a stack of MSDS sheets that were provided by the manufacturing facility. Among the chemicals you notice a variety of acids (including 70% hydrofluoric, 10% sulfuric, and 20% acetic acid), 50% sodium hydroxide, formonitrile, and a solvent, methylene chloride.

Patient #1 was the driver of the fork-lift. She is a 35-year-old female who is unconscious but responds to painful stimuli. Further signs include flushed skin, BP of 180/100, pulse of 54, and the EKG indicates a first degree block with sinus bradycardia, pulse oximetry is 100%, and she is hyperventilating at 36 times per minute. There are no signs of trauma but a garlicky, almond smell has been reported by the rescuers.

Patient #2 is a 28-year-old man who has a 6-centimeter laceration on the forehead and several open wounds on the legs. Chemical burns are noted to both legs from the knees to the ankles. The entry team stated that he was removed from under the debris that fell

from the shelves. The wounds appear milky in color and the skin is slick to the touch with blistering noted at the outer edge. The skin's general appearance other than the burn is normal in color and temperature. The BP is 130/84, a pulse of 60, SaO_2 reading of 95%, and respirations of 32. An EKG indicates a prolonged QT interval.

Patient #3 is a 60-year-old man who was the supervisor in the shop where the accident happened. He was the first to respond and while he was pulling boxes off of patient #2 and attempting to drag him free of the debris he clutched his chest and dropped to the ground. His skin is pale, cool, and diaphoretic. His BP is 90/40 with a pulse rate of 40 and irregular. His breathing is shallow at 12 times a minute. An EKG indicates atrial fibrillation with frequent multifocal PVCs. Pulse oximetry indicates an oxygen saturation of 88%.

Patient #4 is an 18-year-old female. She was working in the front office when she heard the crash. She also responded to help. She attempted to move a 10-gallon container of liquid and upon lifting it from the leaning shelf the container failed spilling the fluid over her head, face, and down her chest soaking her clothing. She worked removing debris until she eventually lost consciousness. She presents with flushed skin, rapid breathing at 36 times a minute, pulse of 140, and a BP of 160/100. The pulse oximeter reads 100%. She still has a faint odor of solvent in her hair. An EKG indicates ST depression with occasional PVCs.

Scenario Questions

1. Evaluating the symptoms presented by patient #1 what do you expect to be the primary offending chemical?

2. What are some associated symptoms that could verify your diagnosis?

3. You have confirmed your diagnosis. Now

how would you treat the patient? What supportive care would you use (include any drugs and dosages)?

4. Evaluating the symptoms presented by patient #2, what do you expect the offending chemical to be?

5. Is an exposure of this amount of body surface area significant? Why?

6. You have confirmed your diagnosis, now how would you treat the patient? Include the supportive care, drugs, and dosages.

7. Is it possible that patient #2 is also suffering from the effects of multitrauma? If so explain.

8. Evaluating the signs and symptoms presented by patient #3 what do you expect the exposure to be?

9. Is it possible that patient #3 is also suffering from a medical condition? If so explain.

10. Could an exposure contribute to the exacerbation of a preexisting medical condition?

11. Evaluating the signs and symptoms of patient #4 what do you expect the offending chemical to be?

12. Does patient #4 need further decontamination?

13. You have confirmed your diagnosis. Now how would you treat the patient? Include any specialized equipment, drugs, and dosages if appropriate.

14. If this situation happened today, is your unit, agency, or district prepared to handle it?

15. What would you suggest is needed to bring your agency up to the standards needed for emergency medical hazardous materials response?

EMERGENCY PROVIDER PROTECTION

What are the issues for the next 10 years? How are we going to manage these issues and what will the laws predicate? In looking toward the future based on past experience, employee protection and exposure prevention are at the top of the list. Close runners up are the future training issues, policy and procedures, and the ever-increasing use of hazardous commodities.

When discussing employee protection we are often dealing with a tremendously large quantity of unknown information. In some areas we have a firm understanding of toxins and their effects, but other areas are only starting to come to the forefront.

What makes one person susceptible to one group of chemicals and the next person almost immune? Are there truly differences between male and female when it comes to exposure or are the differences based on human physiology? Does race, in terms of genetic makeup, make a difference? More importantly, how do we scientifically, morally, and ethically investigate these issues?

This section is truly the tip of the iceberg when it comes to overall personnel management. Many issues will intercede as we discuss what may be done in order to manage our hazardous materials incident. There is a fine line between what should be done in terms of ethical, moral values and what can be done according to law. Civil rights, firefighter bill of rights, Americans with Disabilities Act, equal employment opportunity issues, and Federal Register suggestions all affect our discussion of medical surveillance.

The protection of the emergency worker, a large component of resource management, is the common thread that links medical surveillance and the hospital hazardous materials issue. Although the issues that we explore in Chapter 7 seem different than those in Chapter 8, one must keep in mind the issues discussed. Management of these issues across the board is far reaching. Federal agencies are starting to trend toward disseminating information, rather than what we have seen in the past, which was enforcement. However, what the future holds in this area is unknown to all of us.

For hospitals to become self-sufficient if a hazardous materials event occurs requires training and active participation from all levels of management. The idea that an exposed patient may walk through the emergency room doors to an internal event that may lead to patient, visitor, and employee exposure is very real. The hospitals cannot rely totally on the emergency response agency(s). They, to a certain degree, have the ability to handle small events. Chapter 8 was written to enlighten the reader on the capabilities that medical facilities already possess. These facilities need to organize their resources and become involved with facility-specific training and continuing education in order to meet the above stated goals.

The hospital will ultimately receive the exposed patient. Although the concepts of fluid isolation are basically the same as chemical contamination, hospital employees historically have not seen (or been taught) the connection. This area of education for the hospital employee requires attention. Management, as well as employees, must be trained so that when the event occurs employee, facility, and community protection can be maintained.

An overall example of some of these issues, and one that is very controversial, can be highlighted with the problems claimed by the Gulf War veterans. In 1991 veterans of the Gulf War started to present with symptoms that to this day have only been identified as the Gulf War syndrome. Individuals from mechanics, soldiers, and military police, to the wives who remained on American soil, are all showing signs and symptoms of chemical overload. Mutagenic, teratogenic, secondary exposure, and allergic reactions are but a few of the conditions that are materializing within a small segment. (An estimated 45,000 soldiers of the 700,000 that were sent to the Middle East have developed symptomology of chemical exposure.)

In some cases, the exposure has only presented with nuisance disease processes; for others it has been individually debilitating; and for yet others their offspring have suffered, with reported cases of the offspring having hemangiomas (tangled blood vessels), spina bifida (an opening of the canal in the spinal column), hydrocephalus (water on the brain), impaired breathing, short-term memory loss, and even loss of limbs, to name a few.

Although the issues are more complex, given the fact of wartime activity and several foreign countries that are involved, most of the chemicals that these individuals were exposed to are chemicals that are made in the United States. Are emergency workers under the same toxic cloud as our veterans or, given enough time, will the same abnormalities seen in the veterans start to present in the emergency response population? Although the quantities and toxicities of the chemicals that emergency responders are subjected to are "diluted" as compared to chemicals used in military, incidents such as the sarin incident in Japan (see Chapter 6) and the Riverside, California, incident (see Chapter 8) identify the possibility of what can occur.

We don't know what the future holds within this discipline. We must train all who will become involved with chemically contaminated patients. The controls that can be applied and the direction of future goals as we see it today must become a part of normal operations. For the most part, our tool to track this is medical surveillance. Through the use of a stratified medical health database, quality training geared toward employee protection the future trends of worker's compensation can be molded today. It is up to emergency responders, emergency department workers, and the managers of each to see the potential problems we must deal with today in order to manage tomorrow's hazardous materials incidents and their effects.

Medical Surveillance

Objectives

*Before, during, and after the hazardous materials event, with the medical
director you should formulate the level of surveillance that should occur on as
many considerations presented. It is up to the local agency's management in
association with the resources personnel to meet on a regular basis in order to
evaluate local resources. You should be able to organize the components of a
medical surveillance program given the resources available within the local
response area.*

*When presented with a hazardous materials incident, you should be able to
identify the capabilities of the medical sector in terms of medical surveillance
that may be provided. In addition to the medical preplanning that should occur
and the long-term considerations of medical surveillance, you should compare
signs and symptoms with hazardous materials scene variables.*

As a hazardous materials medical technician, you should be able to:

■ List the pertinent information for analyzing the systems capabilities for:

Preincident medical surveillance.

On-scene medical surveillance.

Follow-up medical surveillance.

- Identify the foregoing components and how to manage such a program within the local system, using available resources.
- Describe the importance of the following and the surrounding considerations of each:

Preemployment physical	Cursory physical
Annual physical	Follow-up physical

- Discuss the federal regulations that identify a medical surveillance program.
- Describe the individual components in detail, of the preincident medical surveillance program.

Chest x-rays	Blood chemistry
Pulmonary function test	Complete blood count
Pulmonary diffusion test	Biological monitoring of known
Vision test	exposure
Auditory test	Urine analysis
ECG and stress ECG	Conventional physical exam.
Blood work	

- Discuss the need for interval medical histories.
- Describe the individual components in detail, of the cursory medical exam:

Blood pressure	ECG 12 lead
Pulse	Tympanic temperature
Respirations	Weight
Pulse oximetry	Forced hydration
Lung field auscultation	General sensorium
Other considerations	Blood drawing

- Describe the exit physical in terms of the previously listed components, discussing in detail each area of concern.
- Discuss the pros and cons of field blood draw analysis.
- Discuss the use of exclusion criteria.
- Identify the pros and cons of medical status form in terms of the law.
- Identify the components of the program review segment within the medical surveillance program.
- Identify the need for critical incident stress debriefing within the context of medical surveillance.
- Describe how you as a student may establish a medical surveillance program within your system utilizing existing resources while maintaining cost constraints.

A key element at a hazardous materials incident is the responding agency's ability to deal with entry and postentry physicals. When we hear or read about medical surveillance we usually think of the entry team only. However, the entry team, decontamination, and medical response team all may require some level of medical attention.

This concern for the safety of emergency responders is mirrored with all the federal regulations and consensus standards regarding this particular subject (OSHA 29 CFR 1910.120; EPA 40 CFR 311; Ryan White Act; NFPA 1581; and NFPA 1500). The reality is that all emergency responders and the possible victims of a hazardous materials incident need some level of medical attention. It is up to the medical officer and the medical staff to evaluate that need and respond appropriately.

The stresses at a hazardous materials incident are many and varied. The usual example of a stressor that comes to mind when thinking of medical support and their respective roles is a toxic chemical, although the medical needs and the stresses that emergency response personnel must deal with are more far reaching. The medical team members may find themselves performing cursory physicals, treating heat stress, low-level radiation, biological hazards, and the normal traumatic and medical events. This wide range of duties, which includes continuous monitoring and the associated medical reporting, describes medical surveillance. Medical surveillance is intended to maximize the health of the emergency worker while minimizing the health risks.

medical surveillance
the process by which the health of an emergency worker is observed, maximizing the long-term health benefits while minimizing the risks

Technically, **medical surveillance** is twofold. One, is the engineering controls that revolve around work safety. These are the physical controls that we can employ to reduce the possibility of an injury. These would include, but are not limited to, the personnel protective gear, decontamination techniques, medical monitoring of the entry and decon team, and physical exams, to name a few. The second part of this program is the administrative controls. Usually we think of these as standard operating guidelines, federal and state/ regulations, procedures, and consensus standards such as the NFPA. These controls or procedures must outline the type of approaches that are used prior to, during, and after the event. All are designed to reduce or eliminate potential injury.

There are four basic reasons why we need to monitor one's health status prior to, during, and after the hazardous materials incident:

1. To provide rapid and appropriate emergency care and treatment at the scene of the incident for response personnel and possible victims.
2. To associate the traumatic chemical event with possible future health effects.
3. To document the types of exposures that an employee has dealt with during his or her career.
4. To follow federal guidelines when dealing with hazardous substances.

OSHA advocates that a program that evaluates one's health be maintained by the employer. Although the exact content of the physical is under the direction of

Figure 7-1 *It is possible that mistakes will be made, for example, possibly a suit would fail while the responder is kneeling.*

the medical director, the process ensures that any work-related injuries can be documented and appropriate action taken in the future. For example, under 29 CFR 1910.134, OSHA states that "a medical evaluation and documentation associated with the appropriate training take place for those individuals that must during their employment use a respirator to carry out their assigned work." It is assuming that the environment one may be subjected to is too hazardous to be without respiratory protection. This pertains to those individuals who have a normal responsibility for entrance into a hazardous environment, confined space, or any other hazardous atmosphere in which a respirator must be worn. Training and suggested medical health databases should be maintained for all employees who, during the normal execution of their job, may be subjected to a hazardous environment (Fig. 7-1).

Furthermore, federal documents suggest that emergency department personnel are included within this requirement and the requirements of SARA III. The *Federal Register* does not state such a requirement; however, it suggests training, medical surveillance, and an understanding of such duties if involved with hazardous materials. Although the hospital staffs are not considered responders, the possibility of emergency department staff becoming involved with hazardous materials in terms of a chemically contaminated patient is quite high. California, for example, has suggested that the requirements of SARA III and all of OSHA apply to hospital staffs. These recommendations may become more stringent as time goes on due to the future developments of health care reform and the worker's compensation laws. However, as a general current trend the requirements are becoming less stringent, and the burden placed on the local medical authorities. In some areas of the country, the requirements are the state's responsibility. In total OSHA (and its representative agencies) has become an agency that supplies information on legislation and how to comply, rather than an enforcement agency.

There are several suggestions of management of hazardous materials medical surveillance from the *Federal Register.* The OSHA program outlines these suggestions:

■ **NOTE**

Training and suggested medical health databases should be maintained for all employees who, during normal execution of their job, may be subjected to a hazardous environment.

■ **NOTE**

Under the <u>Federal Register</u>, employers are required to maintain health-related records during employment and 30 years thereafter; all health records for 70 years in some states.

- Any firefighter who in the normal course of duty is not on a hazardous materials team must be a part of the medical surveillance if he or she is exposed to any hazardous material for 30 days or more within 1 year (the levels must be equal or above the OSHA standard for exposure) or there is a injury due to exposure.
- All members of the hazardous materials team must be part of medical surveillance.
- The frequency of these tests is not limited to the foregoing but rather is the responsibility of the examining physician.
- The examining physician has the right to go beyond the federal or state suggestions when in his/her opinion the worker's involvement is such that a health hazard exists.
- Within areas of concern air monitoring shall be employed and documented such that future exposure cause and effect may be established.
- The employer is responsible for the cost of such a program and is required to establish and maintain the program for all eligible employees. This may under certain conditions include any employee that during the work environment may be exposed to half of the OSHA exposure limit.
- The physical shall be done every 12 months unless the examining physician elects to evaluate the employee every 24 months, based on the exposure potential of the worker.
- Upon retirement or termination from the area of employment, a physical exam shall take place. In conjunction with the medical records, a job description of the employee should be maintained within the records to identify the working responsibilities of the said employee.

Documentation of the whole program plays a important part in medical surveillance. Under the *Federal Register*, employers are required to maintain health-related records during employment and 30 years thereafter. In some states, Florida being one of them, the state law requires employers to maintain all health records for 70 years after termination or retirement (this is usually found under the administrative codes or records schedules—Public Record Act).

NFPA has produced a standard that highlights firefighter safety and health (NFPA 1500). In this standard it is recommended that a health database (health database, medical surveillance, medical monitoring, and continued medical maintenance are synonymous for this discussion) be maintained for all firefighters for the duration of employment. The Ryan White Act, although primarily geared toward bloodborne pathogens, states that the medical records of the employee must remain confidential and within a centralized location. The same officer that is used to maintain the bloodborne pathogen regulation can be utilized for the chemical exposure program, i.e., the medical surveillance program. NFPA 472, NFPA 471, NFPA 473 also references the health and safety aspects of a hazardous materials incident.

Even if we ignored the federal and state laws that are in place, we are designing a program that protects all emergency responders. Remember, medical surveillance is not limited to just chemical exposures, it is also associated with heat and physiological stresses that are of concern, not to mention the biologicals.

PREINCIDENT PHYSICALS

When looking for prospective employees, one of the concerns that emergency services have is their health status. Most services have a retirement package and medical package that the future employee would like to have. This package is very expensive to someone. Whether it be the city, county, company, state, or federal government, someone is paying for the retirement and medical costs. As with any management that is responsible to reduce operating cost, it behooves the hiring agency to evaluate the health of any prospective employee.

Hiring Physical

The preemployment (prehire) physical is the most sensitive area of this program. Given the legal climate in which we live, employers must be careful not to discriminate when choosing the best possible employee. This area is highly volatile and sensitive when it comes to employment.

In 1990 the Americans with Disabilities Act (ADA) was passed. This act reminds employers of the responsibilities they have toward all employees. Its focus was to expand the laws surrounding individuals with disabilities. The scope of this law and the regulations that are represented within was built on the Rehabilitation Act of 1974 and the Vocational Rehabilitation Act of 1973. The requirements of this law and the issues that surround it are far reaching with intense social implications.

In general the ADA states:

- Discrimination against anyone who can perform the essential job function is prohibited.

- Unless it presents a hardship to the employer, the employer must make every attempt to give the employee adjustments to the job or job environment.

- Preemployment medical physicals are only allowed if the employer has given a conditional offer of employment contingent on the outcome of the exam. It is prohibited as an exclusion from employment.

Although the concept of having a qualifiable medical exam is in theory commendable for the evaluation of prospective employees in terms of a healthy individual and team, the ADA firmly prohibits any such action. Any medical exams must accompany a conditional job offer.

The key part to this medical investigation starts with the questions that are asked of the perspective employee. Each question should be designed to elicit a

● CAUTION

The ADA firmly prohibits the use of a medical exam for the evaluation of prospective employees; any medical exams must accompany a conditional job offer.

particular area of health care, maintenance, and family history while maintaining the rights of the individual. However, the questions must be asked in a manner that does not invade one's personal life, which may include his or her disabilities.

In most departments or agencies, by the time the prospective employee makes it to the prehire physical, all the tests that have evaluated job knowledge and performance have been passed. The concern now is redirected toward the health of the individual, once the conditional job offer has been made. In order for an agency to have a truly healthy employee, one may ask the questions of health and family health patterns. This will not prohibit hiring of this particular individual, but it will establish health trends. Once hired, internally this individual may not be eligible for the hazmat team or confined space team itself. These issues like the issues of physical abilities are well beyond the scope of this text. One must however, maintain the fundamental ethics of employment, along with the laws of society when selecting prospective team members.

Physical Exam

Medical History   The questioning should start by gaining a family history.

This history can provoke other questioning, however should be limited to the immediate family. Following are examples of the types of questions that may be asked:*

> Are your parents living?   Yes/No
> How old at time of death?
> When did your father die? Year___ Cause_____
> When did your mother die? Year___ Cause_____
> Do you have brothers and/or sisters? Yes/No
> How many brothers? _____ How old?_____
> How many sisters? _____ How old?_____
> What family diseases are there?._____
> Do/did you smoke? Yes/No How long____How much_____
> Does your brother or sister smoke? Yes/No
> Have you been immunized?_____
> Have you been immunized for hepatitis? _____
> What allergies do you have?_____
> What medications are you presently taking?_____
> What long-term medications have you taken?____

---

*These questions are provided only to demonstrate the type of information needed in order to establish a baseline of health information. They are not designed to meet the requirements of the ADA. Employers must have the medical history questionnaire designed by their respective human resource department in order to meet the ADA requirements.

What hobbies do you have?_____

What off-duty employment do you have?_____

What previous employment have you had?_____

What were your responsibilities?_____

Have you ever been injured on the job?_____

Injury off duty?_____Reason_____

Illness?_____Reason_____

Have you ever been acutely/chronically exposed to a chemical?___

If so, to what chemical?_____

Under what conditions?_____

What personnel protective gear was used?_____

This is just a sample of the questions that need definitive answers. An in-depth history that surrounds this individual is needed in order to establish baseline medical history. Off-duty employment is important from the aspect of what other exposures this individual may have or will encounter. Does this employment endanger the health of the employee?

What hobbies or activities occur away from the job? Hobbies can include the possibility of low-level exposure to chemicals. For example, an employee may have an off-duty job or hobby that requires the solvent methyl ethyl ketone. At the workplace the individual may use a ventilation system or respirator to protect himself from the chemical. Does this individual take the same level of precaution at home as he/she would at work? Some hobby glue contains cyanoacrylate. This glue bonds very quickly and releases a vapor. If this glue were used in the workplace, a respirator would be in order. However at home the same level of protection may not be used.

Our jobs sometimes change over a course of a lifetime. People move from one locale to another for a variety of reasons. For this reason a history of employment is needed so that possible historical exposures can be tracked. The chronological order of jobs is usually required on the application for employment. In association with work history and work practices from a management point of view, a chronological health history is also important from the medical surveillance angle.

Has this individual been exposed before? If so, to what chemical, and what was the exposure level? It is important to identify job responsibilities in all previous employment. Has the applicant been trained to the same expectations as present on-line personnel. If the individual was injured previously, was it due to a lack of concern or a lack of knowledge?

Medical Assessment   A health assessment should be well rounded in order to establish a good medical "view" of the employee. Although suggested for hazardous materials team members, this assessment is not required for all emergency responders. However, after reviewing the suggested standards, one may want to incorporate all or components of the following medical assessment, either annually, biannually, or as necessary. Some components may be performed to provide a health baseline at initial employment, then again only after retirement, termination, or if an exposure occurs.

The content of the physical is totally the responsibility of the medical control board. It is suggested that the individual controlling the program be a licensed physician or occupational hygienist with an understanding of the chemical exposure aspect to employee health.

A *chest x-ray* should be performed in order to establish any preexisting abnormalities. Exposure to a chemical can cause a minor injury and the employee may not remember an insignificant exposure. The damage of this "insignificant exposure" may have been substantial enough to result in a slight injury but may not have a high enough level of discomfort to present a concern to the person (or this individual did not recognize the cause and effect of the incident). This can be identified with a chest x-ray. In association with the x-ray, a pulmonary function test should be done. This test establishes the lung capacity of the individual and serves as the baseline of total pulmonary function. Each pulmonary test should encompass, but not be limited to, a forced expiratory volume (FEV), forced vital capacity (FVC), and the FEV to FVC ratio. This should be compared to normal values with respect to age, sex, weight, and height. In addition this establishes a well-rounded baseline of the individual. Total lung capacity, residual volume, forced expiratory flow, maximal expiratory flow, and functional residual capacity should all be calculated and recorded for annual comparison.

Some agencies have even gone as far as to require pulmonary diffusion tests in which a short-lived radioactive substance (xenon gas) is introduced into the lungs. The individual has a series of scans while the substance is being inhaled. Then a radioactive substance is injected into the vein and a second scan is done. The individual's lung capacity is monitored, and diffusion thorough the lung tissue is viewed. By scanning both the gas side of the lung (alveoli) and the vessel side (capillary bed), the lung can be evaluated for the scarred lesions. Poor diffusion areas can then be established.

Preexposure diffusion tests are usually not a part of the medical surveillance program. Most physicians in industrial medicine agree that this test should be limited to postexposure after the pre/post chest x-rays have been done, evaluated, and compared. The diffusion test is an additional tool for the physician to evaluate the extent of injuries after exposure to certain chemicals. The primary concern is the health of the respiratory system. This should be evident after reading respiratory injuries discussed in Chapter 5.

A *vision test* corrected and/or uncorrected should test for color perception, depth perception, and refraction. Some authors have even suggested that a test for the degree of night vision be conducted. As suggested by some physicians, a test for dyslexia may also be conducted.

The *auditory test* should reflect the hearing capacity of the individual. Levels that should be tested are 500; 1,000; 2,000; 3,000; 4,000; and 6,000 hertz. In association with the auditory test, adequate hearing protection is required by the employer in order to maintain the level of acuity that is required of the employee. The employee at all times must protect his or her hearing against high-frequency noise and noisy environments.

The *cardiovascular* system can be assessed by comparing the resting EKG 12 lead to a stress EKG. This graded assessment can give the cardiovascular baseline that is needed for employment. This comparative EKG is then used to collate the postexposure injury with the preexposure findings. Any suspicions about the electrical conduction system then can be further analyzed.

Blood chemistry should be done in order to establish a baseline for liver function or obstruction, kidney function, and complete blood count (fractional CBC). This should include platelet evaluation, differential, hemoglobin, and hematocrit. In some cases erythrocyte count may be indicated. Some documents allude to the possibility of freezing blood and storing it so that the blood before injury and the blood after a significant exposure can be analyzed. These research papers have also suggested that the blood at hiring be compared with the blood at the time of termination and/or retirement. By doing so, future trends in chemical exposure may be analyzed. However, we must mention here the ADA and its requirements. Freezing of blood, genetic testing, and drug testing are very controversial issues. Presently there is much public concern about the legitimacy of such medical screening endeavors. Most of the purely medical literature suggests that a sample of blood should be taken and frozen for analysis especially before and after heavy metal exposure. How this area will evolve is yet to be seen.

Complete workup including blood, urine, EKG, and neurological testing should be performed after all exposures of known problem chemicals such as pesticides, heavy metals, and aromatics (see list in post incident section for complete listing, page 332). In these cases a comparative analysis may provide information on exposure review.

Urine is used to test for multiple system function to include color and appearance, pH, specific gravity, and glucose levels. Protein, bile, sediment, and glucose tolerance are all additionally suggested.

The individual should receive the conventional age; sex; height; weight; diet; temperature; blood pressure; pulse; respirations; a head, nose, and throat evaluation; neurological responses (inclusive of reflexes); and evaluation of the musculoskeletal system. Also included should be an exam of the genitourinary system, abdomen, rectum, vagina, and testes. Integumentary, peripheral neurological, and vascular system evaluation are also suggested.

Annual Physical Exam

Annually the aforementioned physical exam should be performed on all personnel. As you may think, the testing in total can become quite expensive. The laws are not to create a financial burden to the employer but to provide a safety function. Depending on the facility and the exposure level, the complete exam may be repeated every 2 years rather than annually. At the time of this writing the only two emergency response groups that a medical surveillance program is applied to are the hazardous materials team members and the firefighter who has been exposed to a hazardous material (see OSHA list in the introductory part of this chapter).

However, what is suggested, and is of normal practice, is that a portion of the physical exam be performed every year. Every 2 to 3 years the critical physical exam be performed (including all medical components). For example, the traditional physical exam can be performed every year. Every odd year the pulmonary function test be given, and the 12-lead resting and stress test in the even years. By leveling out the monitoring, financial concerns and budgeting are also evened throughout. Depending on the injury, other tests may be needed. For example, every other year a chest x-ray is taken, and every 2 to 3 years a pulmonary function test.

Interval history (a questionnaire to identify new medical concerns, exposures, and/or illnesses) is taken during this process to establish any new problems the individual may have encountered between the original history questionnaire and the present. Interval history taking is useful to establish current norms. After an exposure and based on the information in the interval history interview, a diffusion test, specific blood testing, and periodic testing (testing every 3 months) may be indicated.

The entire monitoring program should have a single depository for all personnel. This area should be managed by a licensed physician or qualified occupational health practitioner. All records should be maintained for 30 to 40 years after termination or retirement. This time is dictated by federal or state laws or local ordinance (70 years in some states).

Under the Ryan White Act, one individual must be identified as the infection control officer (designated officer; DO). This appointed individual is responsible for the dissemination of medical information if and when the employee becomes exposed to a biological. This same officer should also be responsible for tracking chemical exposure and should have direct communication with the medical director and/or medical control board.

Another facet of medical surveillance is the wealth of information gained after a few years from the annual physical. If, during the annual physical, a health concern is found, the medical clinic can address the concern before the disease process takes hold. Another example is when a team member is exposed to a chemical at the scene of a hazardous materials event. In such cases, a routing process from the medical clinic to the receiving hospital should be in place in order to deliver the annual physical information to the attending physician at the time of exposure. This annual physical information is then compared to the tests that are currently being performed on the employee. From this information the attending physician can evaluate the level of damage that the individual has incurred. Future therapy and prognosis can then be formulated. This routing process needs to be available 24 hours a day, 365 days a year, at a moment's notice.

Interval exams such as the annual physical may be increased to every 6 months depending on the frequency of possible exposure and the current medical status of the patient. (Different than the interval histories, this area is termed *postincident physicals* in this text and is directed by the attending physician. They may be as often as every 3, 6, or 9 months or a variation thereof.) For those members that commonly respond to low-level exposure incidents, the annual physical associated with the entry physical should provide a well-rounded surveillance program. However for the crews associated with a variety of incidents within a shift, a higher frequency of surveillance may be provided.

TEAM MEMBER PHYSICALS

Annual physicals are necessary for all individuals involved with hazardous materials emergency response. According to the federal standards, the hazardous materials team is the only group that requires medical surveillance in terms of preincident planning. However, if the role of an employee includes special teams, such as hazardous materials, dive rescue, or confined space rescue, a component of or a complete program may be provided. These persons should have a medically designed program for the area of specialization.

Biological Monitoring

Beyond the annual physical, a few additional components may be necessary. In cases where an exposure has occurred, it may be necessary for the individual to go through extensive exposure monitoring. This monitoring process measures the biological fluids for the suspected chemical such that damage to a target organ can be identified early on. Chemicals, metabolites, and tissue sampling are but a few components to this type of long-term monitoring. In cases where the hazardous materials responder has the potential of becoming exposed, a different approach is used with components of biological monitoring and the annual physical in mind. For example, the hazardous materials technician should in addition to the routine physical have a heavy metal, cholinesterase, and aromatic hydrocarbon (included is PCBs) workup. This is done at the time of team selection. After the individual has spent 2 years on the team another analysis is performed (provided that the individual has responded to calls involving these materials or the suspicion of these materials is high). High-level medical surveillance should be done every 2 years until the emergency worker retires from the team. All medical information is maintained and analyzed for possible exposure effects.

The same procedures should also be performed for dive-rescue and confined space team members. The level of chemical exposure is far less than for the members of other hazmat teams, but the chance of contaminated water entry is quite high.

Any time that an exposure occurs the affected individuals should have a full medical workup (see Postincident Physicals later in this chapter.). This would include confined space and dive rescue operation within an contaminated environment.

! SAFETY

Interval exams, such as the annual physical, may be increased possibly as often as every 3, 6, or 9 months, or a variation thereof, depending on the frequency of possible exposure and the current medical status of the patient.

! SAFETY

In cases where an exposure has occurred, it may be necessary for the individual to go through extensive biological monitoring.

■ NOTE

Dive-rescue and confined space team members may also undergo biological monitoring.

■ NOTE

Exit physicals from the team are just as important as entrance physicals into the team.

Exit physicals from the team are just as important as entrance physicals into the team. From the time we enter the emergency services, we enter a work environment that has the potential for exposures to a variety of biologicals and chemicals. The magnitude of these exposures is what we are evaluating long term. In order to quantify these exposures, one must have an entrance physical before team activities begin, annually, and upon retirement from the team. From this information cause and effect can be analyzed.

Cursory Physical

There are no known tests to predict exposure. We cannot identify toxic-level effects beforehand. Because chemical substances can produce synergistic, potentiation, antagonistic, and additive effects, we want to ensure the highest level of safety. A part of this safety is, as a part of the total medical surveillance program, to perform a cursory medical examination at the scene of a hazardous materials incident (this same physical is performed at a incident that involves a confined space rescue).

We identify the product on hand, research its potential harm, look at its permeation and degradation qualities, and establish through research an effective means to mitigate the incident. A critical part of this process is to perform a medical examination in order to establish three important criteria:

1. To identify preliminary basic medical information. An informational source at the time of the incident, i.e., research of toxic values and associated health effects.

2. To outline an approach toward medical monitoring, which establishes testing against normal values. Is the surveyed individual capable of enduring the tasks of the incident? If not, do we have parameters that can eliminate the rescue worker from the incident, i.e., exclusion criteria.

3. To research medical surveillance criteria, which may include antidotal treatment.

▮ SAFETY

● Through an exam process, looking at such items as hydration, sensorium, and coordination, the physical character and preparedness of the emergency worker can be revealed.

Through an exam process we can test and judge the individual's physical preparedness. Looking at such items as hydration, sensorium, and coordination will give vital information on the physical character of the emergency worker. Remember we are at the scene of an incident that takes many personnel to mitigate. We do not need (and do not want) the additional medical problem of an injured or exposed rescue worker.

The medical physical at the scene of the incident is the responsibility of the medical command officer and is called the *cursory physical*. This exam procedure reduces the available response personnel, but can be done efficiently with one paramedic and two emergency medical technicians. This physical has two components, the *entrance physical* and the *exit physical*. This cursory exam should include but is not limited to:

- Blood pressure
- Pulse
- Respirations
- Pulse oximetry
- Lung field auscultation
- EKG 12 Lead
- Tympanic temperature
- Weight (before hydration)
- Forced hydration
- General sensorium
- Blood drawn (under specific guidelines)

The cursory medical examination is just as important as the annual physical, and in some cases possibly more so. The idea is to establish a baseline of information immediately preceding the incident. If an exposure occurs, normal values are available for comparison. Additionally, personnel preparedness can be evaluated in terms of exclusion criteria.

Heat stroke and exhaustion are common episodes at the scene of a hazardous materials (and confined space) incident. Medical command should be alerted to these potential problems and ready to treat them. In any case, normal biological functions can be derived from the information attained prior to scene entry.

Hydration is extremely important at any emergency incident, and even more so at the hazardous materials incident (Fig. 7-2). Acclimation to heat stresses usually takes anywhere from 3 to 5 weeks, during which time the individual has worked up to 90–120 minutes of heat stress exposure. This conditioning can

! SAFETY
Medical command should be alerted to the potential problems of heat stroke and exhaustion, and be ready to treat them.

Figure 7-2 *During briefing, medical surveillance can proceed, along with forced hydration.*

change the physical responses that must occur within one's body so that extreme heat conditions can be tolerated during the normal work day. But how many of us really are able to withstand that type of stress? More important, how many individuals work on their heat acclimatization potentials? In order to support our personnel, we must have some degree of heat acclimatization training. Associated with this is watching weight differences during workout periods and scene activity. A 4% decrease in body weight for the person that is in cardiovascular shape may not be threatening. However this loss of body fluid could be potentially hazardous to the mildly acclimated individual.

During any operation, entry, backup, and decontamination team members should have their sensorium evaluated constantly. By asking simple questions relating to person, place, and time, associated with counting serial numbers by threes or nines can give vital information of the level of alertness. Some literature has stated that any long-term (greater than 20 minutes) entrance into a hazardous environment should have a coordination/sensorium test done on the individual post- and preentry. Mental alertness can be established by observing the gait and the orientation of the emergency worker. Being alert and oriented to person, place, and time, associated with components of the field sobriety test (see General Sensorium in the next section) gives medical sector monitoring criteria.

A few articles and research data have suggested that an exclusion questionnaire be performed prior to the cursory medical, which would include questions geared toward specific medical history during that last 48 to 72 hours. It would pose such questions as:

> ! **SAFETY**
> Some literature has stated that any long-term (greater than 20 minutes) entrance into a hazardous environment should have a coordination/sensorium test done on the individual post- and preentry.

> ! **SAFETY**
> An exclusion questionnaire should include questions geared toward specific medical history during the last 48 to 72 hours.

- What type of medication have you taken in the last 72 hours?
- What quantity of alcohol has been ingested within the last 24 hours?
- In the last 3 to 4 weeks have you had any fever, nausea, vomiting, diarrhea? If so what was the diagnosis/treatment?
- If female, are you pregnant?
- Have we met any exclusion criteria for all baseline vitals.

The questions are, who receives this baseline medical, and to what degree do you take the exam? The entry, backup team decontamination and decontamination support, along with EMS should understand the process, and each member should give prior consent to the full cursory medical if and when the situation arises. The type of incident (i.e., level of protection, type of chemical) will dictate the level of entrance and exit physical. It will also dictate the number or diversity of personnel that will require the physicals. The entry, backup team, and decontamination team are the individuals who assume the greatest risk (in some cases the medical support team and transportation team receive the cursory physical). The actual entry into the a hazardous atmosphere has the highest potential for accidental exposure (it has also been found that decon team members stand a high risk due to the possibility of secondary exposure). These individuals should receive the highest levels of medical examination affordable to them.

Considerations of the Entrance Physical (Fig. 7-3)

1. *Blood Pressure.* Blood pressure is a measurement of working pressure (systolic) and the resting pressure (diastolic). This relationship is shown by the formula CO = SV × HR; B/P = CO × PR, where CO is the cardiac output. The CO is dependent on the heart rate (HR) times the stroke volume (SR). In other words, for every beat of the heart a finite volume of blood is ejected from the heart. Plugging this into our second formula, the cardiac output times the resistance (peripheral resistance, PR) gives us a blood pressure (B/P). The systolic is the working pressure of the heart, whereas the diastolic is the resting pressure of heart activity. This pressure allows us to recognize the individual's heart function and make a decision about entry (whether decon corridor or actual entry). We must concern ourselves with the systolic and the diastolic and how they compare with age/sex normal values and the normal value of the individual. Be alert for both hypotensive and hypertensive members. The blood pressure should be compared to normal values and or documented norms for that individual. An exclusion value would be a diastolic above 100 mm Hg. If we have a diastolic pressure of above 100 mm Hg, the heart during its normal resting phase is working as hard as it would during the working stage. Orthostatic blood pressure should be taken if an increase in pulse is noted. An increase of pulse rate or a decrease in blood pressure that is maintained for 2 minutes or longer should be an exclusion criterion requiring medical evaluation.

Be sure to obtain your blood pressure prior to the SCBA harness being placed on the individual. Some medium-build males (and some muscular females) with

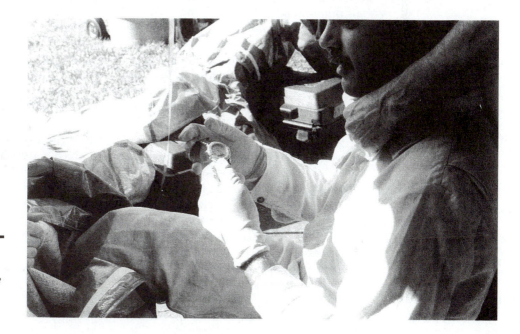

Figure 7-3 *All jewelry, watches, and rings should be removed prior to entry.*

a large thoracic cage have shown abnormally high BP. Although not supported by any medical research, it has been suggested that the weight of the SCBA while the wearer is sitting applies pressure to the subclavian artery, which in turn reduces the blood flow, causes a back pressure, and elevates the BP. Another suggestion is that the weight increases intrathoracic pressure, which increases BP. Whatever is happening, the key is to evaluate the entry team member prior to donning the SCBA and again after. This phenomena seems to only occur while the individual is in the sitting position. For this reason, watch your entry team members while they are in the hot zone, consistently checking for coordination and sensorium levels. Those individuals who had a problem during suit up, more than likely will have a problem if they get into the crouched position within the hot zone.

2. *Pulse.* At the same time you are taking the individuals pulse, be aware of the quality and rate of blood flow. As we become more proficient in our skills we also tend to become more complacent. The quality and rate of heart contracture can give you clues on possible hidden anxieties. An accelerated heart rate with a bounding pulse may just be a sign of high anxiety or (see orthostatics above) a fear that this individual may have been hiding. Compare pulse rates with normal values. Does the blood pressure relate to the pulse that we are getting? A pulse rate above 100 bpm at preentry is the exclusion point for the pulse criterion. Target heart rates should only be used to evaluate personnel under strenuous conditions. By taking 220 + the age of the individual times 70%, the target heart rate can be calculated ({220 + age} × .7). If target heart rates are reached during the entry operation, the individual should be excluded from the incident, and possibly for the entire day. Target heart rates should never be obtained during an hazardous materials or confined space emergency.

3. *Respirations.* Respiratory rate and auscultation should be done on all patients and our response team is no exception. Analyze the respiratory status of the emergency worker. Look for capillary refill, mucosa color, and general sensorium in terms of complete respiratory assessment. Remember that rate and depth of each breath is a component of the respiratory assessment. Any responder with a respiratory rate above 24 a minute prior to entry should be excluded from the incident.

4. *Pulse oximetry.* Pulse oximetry can confirm what you have already observed by evaluating the saturation of oxygen in the blood electronically. A comparison of the rate, volume of respirations, lung auscultation, capillary refill, and mucosa color all give you an understanding of the level of oxygenation your emergency worker is maintaining. SaO_2 below 93% (be very suspicious of 100% SaO_2, see Chapter 5) should be eliminated from the incident and medically evaluated.

5. *Lung field auscultation.* Auscultation is an important aspect of the cursory medical exam. If at all possible have the same individual who listened on the entrance physical listen on the exit physical. Continuity is important especially if an exposure has occurred. While auscultating the lungs, listen posteriorly and anteriorly in the bases and in the apex of the right and left lung field. Be alert to sounds such as rhonchi, wheezing, or rales, especially when assessing the team

! SAFETY
A pulse rate above 100 bpm at preentry is the exclusion point for the pulse criterion.

! SAFETY
Any responder with a respiratory rate above 24 a minute prior to entry should be excluded from the incident.

! SAFETY
Any responder with an SaO_2 below 93% should be eliminated from the incident and medically evaluated.

TARGET HEART RATES

(220 + AGE) × .7

AGE	Pulse Rates	AGE	Pulse Rates	AGE	Pulse Rates
18	167	33	177	48	188
19	167	34	178	49	188
20	168	35	178	50	189
21	169	36	179	51	190
22	169	37	180	52	190
23	170	38	181	53	191
24	171	39	181	54	192
25	172	40	182	55	192
26	172	41	183	56	193
27	173	42	183	57	194
28	174	43	184	58	195
29	174	44	185	59	195
30	175	45	185	60	196
31	176	46	186		
32	176	47	187		

member after exit or if an exposure occurred. Individuals who are affected with a cold or viral infection are not good candidates for entry, backup, or decontamination sectors.

6. *EKG 12 lead.* EKG 12 lead is becoming a normal assessment tool in the prehospital arena. Most advanced life support units now carry the equipment needed to take a 12-lead EKG. While 12-lead is now entering the EMS workplace so is 12-lead interpretation. Cardiology has been within the EMS environment for many years. Most seasoned response personnel have made it a "hobby" to read such strips. Now that it has become a job requirement, more EMS responders are learning 12-lead interruption. When assessing your entry team (sometimes including decontamination personnel) members, have the most experienced EMS worker read the 12-lead EKG and compare it with normal established EKG rhythms. An EKG lead I, II, III as a minimum is the normal capability for all advanced life support units. However, as 12-lead EKG becomes a greater part of the prehospital arena, this too is a need at the scene as part of the premitigation physical. Elevated U waves can be an indicator of dehydration or associated with other abnormalities that may be present without the benefit of symptoms. Any dysrhythmia that has been established as "normal" for that individual *must* be cleared by the medical director. If not cleared, all dysrhythmias are within the exclusion parameters.

SAFETY
Elevated U waves can be an indicator of dehydration.

SAFETY
All dysrhythmias are within the exclusion parameters.

7. *Tympanic temperature.* 98.6°F is the established norm for an oral temperature. The tympanic thermometer can give you an accurate body temperature. At a hazardous materials incident we are placing personnel in a suit that restricts the body's normal cooling potential. We are placing these people in an environment that creates an elevated body temperature without a mechanism for cooling.

The entrance temperature is to establish a baseline temperature for that individual that day at that particular time. When this temperature is compared with the exit physical temperature, you can get an idea of the severity of heat stress your emergency worker has endured. If there are *any* signs of heat stroke or the temperature indicates the possibility of heat stroke, aggressively cool your team member. Conversely, in atmospheres (and depending on the time of year) that present with low ambient temperatures, be prepared to warm the emergency responder.

Temperature at entry should be between 97.0° and 99.5°F (98.0°–100.5° core F). Any temperature above or below this range must be treated appropriately.

8. *Weight* (before forced hydration). The difference of body weight from the entrance physical as compared to the exit physical can give valuable information on the hydration status of emergency workers, including all entry team members, backup team, decontamination team, decon support, and at times EMS personnel and all support personnel. Weight can establish hydration levels. A 5% to 8% weight change indicates an extremely dangerous dehydration episode. A safe working percentage to establish dehydration/rehydration is 4% weight differential. Upon exit, weights must be taken and equaled once in rehabilitation. By dividing the exit weight by the entrance weight a quotient is given. Subtract that from 1, times 100, will give the percent weight loss. Be sure to evaluate your worker upon exit in the same dress that the worker was in at entry. A worker with a 4% difference in body weight should be observed. With a weight loss of greater than 4% aggressively treat your the team member for dehydration. At 6%, an IV of Lactated Ringers and transport to the closest medical facility must be considered and is highly recommended by many physicians. Upon exit, continue to evaluate their weight in relation to the volume of fluid over respective time. This area must be discussed with the medical director during the planning stages of the program.

9. *Forced hydration* (discussed in Chapter 5). Remember that acclimated individuals are able to withstand a higher degree of heat stress than those who are not. In either case it is important to "load" your team members with hydration fluids prior to the event. This can be done during suit up and briefing of the incident. In some departments it is a requirement of employment to maintain a hydrated state. A few departments have gone as far as mandating it during the summer months.

Caffeinated drinks, although preferred by some, are strongly discouraged. Becoming involved in a hazardous materials incident is highly stressful. Research by the International Association of Firefighters and those involved with CISD has indicated that caffeinated drinks add to the anxiety of the emergency worker. If a negative situation is occurring, the stress level is already high. By ingesting caffeinated products you may be performing a disservice to your peers. Caffeinated drinks are also diuretics, giving the emergency worker a variety of potential hydration problems.

10. *General sensorium.* General sensorium is an evaluation that begins at the entrance physical and does not stop until the individual is in rehabilitation. Here we are not only watching for the conventional alert and orientation to person, place, and time, but also the appropriateness of conversation, gait, and verbal response. Question the individual in relation to headaches, blurred vision, numbness, or ringing in the ears. Grip strength and touching the tip of the nose with arms extended, head back with eyes closed, should also be tested. A cognitive evaluation must also become part of the orientation process. Spelling a word backward and counting by 3's, 7's, or 9's (serial assessment) should also be administered. In any of these categories the individual can only miss one entire assessment. More than one is an exclusion variable.

While these individuals are working, observe their behavior and question whether it is normal, especially if an exposure has occurred.

11. *Blood drawn (under specific guidelines).* Drawing blood is only performed in specialized situations. The medical officer *must* weigh *all* the risks and benefits of such a procedure, remembering that veinipuncture is another route of exposure if decontamination is not adequate.

Drawing blood prior to entry is a debatable element of the cursory medical. Here we have to weigh the risks versus the benefits. For example, if we take blood to obtain a chemistry level, how many tubes of blood must we take? Does the chemical possess such a high risk that having an understanding of the entry personnel's liver, kidney, and blood chemistry prior to entry seem a reasonable undertaking? If we decide to obtain a blood sample only to have a current blood chemistry profile, does the benefit outweigh the risk of the puncture wound that we will create? The rationale is to have a blood chemistry profile prior to the entry in the event of the individual becoming exposed. Generally the annual physical can provide the baseline data that a doctor would need to decide on medical therapy. However, there are situations in which taking a blood sample prior to entry may be a consideration, for example, in the case of a completely unknown substance.

The other side of this coin is the quantity of blood that is required for the lab tests. We do not want to blood let our response team prior to entry. Some labs require 10 milliliter blood samples for each type of test. Most incidents do not require this level medical testing. The annual physical should be set up in such a fashion that these questions of baseline blood chemistry can be answered just by reviewing a medical chart. The following are general guidelines for the drawing of blood should the need arise:

1. Indications for blood to be drawn: In the event that the hazardous materials incident potentially poses an extremely hazardous environment, the medical safety officer may have blood drawn from the entrance and backup teams. This shall be limited to blood chemistry, however may be expanded after consultation with the medical director and/or staff toxicologist. Consideration criteria for the drawing of blood:

! SAFETY

In a general sensorium, missing more than one category is an exclusion variable.

● CAUTION

Blood should be drawn only under specific guidelines, and the medical officer must weigh all the risks and benefits of such a procedure.

! SAFETY

An annual physical can provide the baseline data that a doctor would need.

- Total unknown chemical involvement
- Any highly toxic chemicals that are identified through research of the chemical(s)
- CBC, heavy metal, or cholinesterase levels to be drawn after consultation with medical control, medical director, and/or toxicologist

2. Maintenance of drawn blood: Once blood has been drawn from an individual, it shall be labeled and placed in a cool environment. Each blood tube requires that the team member's name, time of blood draw, name of phlebotomist, and date be placed on each tube. The tubes must remain cool until they are turned over to the laboratory personnel. This can be accomplished by layering a cooler with ice and placing a terry cloth towel over the ice. The blood samples are then subjected to a ambient cold environment. At no time should the blood be frozen or have direct contact with the ice.

   a. Blood chemistry tubes must be spun down. Each laboratory has its own procedures for the spinning of blood. An example follows:

   (i). Let blood sit in test tube rack for 15 minutes after drawing within a cool environment.

   (ii). Spin in centrifuge for 7 minutes.

   b. Place the spun samples in a cooler as described above along with the other samples.

   c. Once the incident has been stabilized or injury has occurred, send blood work to the appropriate facility.

■ NOTE

While blood samples are stored in an ambient cold environment, at no time should the blood be frozen or come in direct contact with the ice.

12. *Other Considerations.* During the research for this chapter many articles and much reference material were reviewed. One discussion that has been presented as an addition to the on-scene medical surveillance is that of alcohol breathalyzer. Several hazardous materials teams have established protocol for the use of the alcohol breathalyzer at the scene during suit up. The rationale is to identify those individuals who are slightly under the influence or the night before may have had too much to drink. Although the intentions of this testing may be noble, the practicality and legality of this procedure may be questionable. For example some departments have suggested that a .03 on the breathalyzer is an exclusion of this individual, based on dehydrated state one may be in after and during the metabolism of the alcohol. Although the authors do not disagree with the premise, the fact of the matter is that depending on an individual's metabolism, medication intake, or diet, one may have a breathalyzer evaluation greater than .03 but those individuals are neither drunk, dehydrated, nor impaired. Then there are the ethical issues of what one does when he or she is off duty. Although beyond the scope of this text, these issues strongly affect this area of medical surveillance, and must be addressed by individual departments.

An additional test that could be provided at an incident is glucose testing. Most commercially used glucometers are accurate enough to identify team member's sugar levels. This testing may be especially important if a member is a diabetic. Under stressful conditions glucose levels can change, making this test more reasonable than the breathalyzer. As discussed under blood drawing, when using the glucometer, another route of entry has been produced. The risks versus benefits have to be weighed.

All of the preceding elements (with the exception of blood work and the considerations identified in 12) of the entrance physical should be performed at all scenes in which some level of protective gear is to be worn, including confined space rescue and dive rescue incidents within hazardous environments. This is dependent on the type, quantity, concentration, and state of matter the hazardous substance presents with. Medical control, the medical sector officer, and the incident commander must make the decision of the level of medical surveillance early in the incident. Additionally, standard operating guidelines can identify the normal response and the medical criteria that will be established given a set of predetermined parameters.

All personnel that may become involved with the chemical either directly or indirectly (secondary exposure) should have the cursory examination. The only difference is the blood work if such a procedure is considered valuable. Only entry and backup teams shall have blood drawn if the medical commander deems necessary.

A procedure must be in place for this decision-making process. For example, the medical officer, medical director, and chemist/toxicologist should be consulted. All three would have to be given all the facts of the incident and all must agree on the procedure.

Just like the annual physical, the cursory medical should be a part of the standard operating guidelines and should reflect the level of medical evaluation for each level of an incident. Basically we are looking at a full cursory medical for all entry team personnel, decontamination, and possibly EMS whenever:

- The health hazard is high. This includes all hazardous materials mitigating incidents in which protective clothing is worn.
- We establish a team concept for recon or mitigation.
- We cannot fully identify the product, or a combination of chemicals may produce an additional health effect.
- The medical command officer deems necessary.

Essentially we are building a mechanism for filing of paperwork identifying important information, the information being the medical quality of the rescue worker at the scene of a hazardous materials incident.

There will be times when entrance into a hazardous environment is necessary to perform an emergency rescue. When this condition occurs, the entrance physical may have to be waived. Then the annual physical and rehabilitation physical becomes important for future evaluation.

This whole program requires that the entrance and exit physicals be maintained within the personnel's medical record file (Fig. 7-4). Again, a routing mech-

Figure 7-4 *All entry physicals should be documented along with chemical and toxicological information. All information should be maintained for future exposure analysis.*

anism must be in place to file this paperwork in the appropriate area. This information should not be confused with the exposure record. The cursory medical exam only indicates the date, time, chemical involved along with the medical information. The exposure record states that an exposure occurred and future medical attention is needed. The exposure record may also accompany the worker's compensation application, depending on local and state laws and the rules and regulations of the department or agency. Location of these records varies from department to department. However, each agency should evaluate its respective needs for the maintenance of paperwork and the location of such.

This brings up the legality of the cursory medical status sheet. This "tickler" sheet is presented only to record the pertinent medical information at the scene. If, for example, all individuals who had a medical turned out to be fine, then there is no need to keep such a form. If a system would like to keep this information as a record of on-scene evaluations of the team members, then a copy of the report identifying each member must be made, while crossing out the information of the other respective members. Remember that in a court of law this may be considered a medical record, but in this case only one name and the appropriate information for that member is the only information that should be on the report. Another way of handling the information that would be needed is to place a hospital bracelet on each team member with the vitals and hydration information. The only reason that we are using the medical status sheet is to identify the large amount of medical information associated with the appropriate member for the duration of the incident, in case problems arise. If this member is transported either for hydration or due to exposure, a legitimate medical report will be generated.

At the scene of a hazardous materials incident we have the time and the manpower to perform all of the procedures that are necessary to carry out our job. At the scene of a fire we should (technically speaking) take the same precautions,

■ NOTE

In a court of law, a cursory medical status sheet may be considered a medical record. If a system keeps the record of on-scene evaluations of each team member, then only the individual's name and appropriate information should appear on each report.

Firefighters fight house fires everyday. At these fires (as with all fires) numerous hazardous chemicals are produced from combustion. The structure fire is one of the most dangerous chemical operations that fire departments perform. Entering an area of toxic materials with bunker gear and SCBA, the firefighter is subjected to a myriad of medical concerns. Firefighters enter this high-heat, chemical-laden environment with respiratory and skin protection, without having the opportunity of having a preentry physical being performed. In this setting it would be unrealistic to have the cursory medical. However, postentry physicals during rehabilitation should be a part of any fire ground operation. During this time B/P, pulse, respiration, and forced hydration should be performed as a minimum.

! **SAFETY**
● During a fire, the rehabilitation sector should conduct an exit cursory medical and increased hydration.

however, preentry physicals are not feasible, so annual physicals play a dynamic role in the preservation of fire fighting forces. The annual physical becomes the cursory medical in this particular scenario. During a fire the rehabilitation sector should conduct an exit cursory medical and increase hydration. These functions should become a part of the standard operating guidelines for all fire suppression services. An exit physical can provide the command staff with potentially readied forces with the practicality of medical evaluation during rehabilitation.

One must remember that the cursory physical information can be considered a medical record. It is a medical evaluation and is confidential. As an example, when exclusion criteria are applied to a team member, the commander of the operation does not have the right to know why another individual may be needed to complete the operation. This information falls under patient confidentiality laws.

Exit Physicals

As a part of the cursory physical we also have an exit physical. This physical is used to establish rehabilitation criteria or guide us through our treatment modalities. It is an additional tool to establish rehabilitation and compare with our baseline. Again, we are not sure how exposure can affect the human body. By acquiring extensive information at the time of the incident, we give ourselves and the medical community information that can be used later. In addition to this, by medically evaluating our personnel, normal stressors, such as heat stroke, can be responded to appropriately.

! **SAFETY**
● The exit physical can aid command staff in decisions about future personnel needs.

The exit physical is a part of the termination of the incident, but it can also guide the command staff toward future personnel needs. Generally when an incident arises and the entry team is asked to perform its duties, the exhaustion level that is reached in a short period of time is quite high. In some situations it would be advisable not to send the original entry team back in for an hour or until readiness can be ensured. However, most hazardous materials teams only have limited resources and manning. What is the team going to do between the initial entry and backup team response and the time it takes to properly rehabilitate these individuals? The exit physical can aid command staff in making such decisions.

The exit physical is composed of all the components of the entrance physical with the exception of drawing blood. This physical should be taken at 1 minute after decontamination and again at the 5-minute mark. At the 5-minute mark, all vitals should be at a minimum within 10% of the entrance vitals.

The exit physical should include:

- Blood Pressure
- Pulse
- Respirations
- Pulse oximetry
- Lung field auscultation
- EKG 12 Lead (taken at 5 minutes)
- Tympanic temperature
- Weight (before hydration) and during rehabilitation
- Forced hydration
- General sensorium

In any case that entry is the mode of mitigation, all entry team members and the associated decontamination team shall have the cursory medical performed on them. If the hazard is potentially high, all decontamination support members should also have this physical performed on them at exit. If medical command foresees the possibility of transport, then all patient contact personnel should also have a cursory medical performed.

We must be prepared at the scene of a hazardous materials incident. EMS personnel may, as a part of their SOG, dress-out the transporting unit. With this level of safety, EMS personnel should, after team needs are met, protect themselves and perform a cursory medical on themselves. Even if the EMS personnel are not required to treat or transport a victim or team member, the sheer fact of involvement may lead to medical problems such as dehydration. EMS, like the team members, must be aware of dehydration and fatigue at the scene of a hazardous materials incident.

Handling the scene of a hazardous materials incident takes planning, strategy, and tactics. In order to perform all the rudimentary functions without forgetting a component, data sheets should be generated. Just like the fireground tactic board, the medical status sheet gives EMS responders a systematic approach to the preentry and exit physicals.

Personnel should be familiar with all data sheets and their appropriate use. Additional copies should always remain in a file, somewhere on the hazardous materials response unit. A complete kit containing all the needed paperwork that would allow personnel to start the cursory medical should be placed in a conspicuous area of the response unit.

As with all scenes, documentation must be generated. This status sheet enables the response team to start the paperwork within the medical sector. If an

! SAFETY

EMS, like the team members, must be aware of dehydration and fatigue at the scene of a hazardous materials incident.

Hazardous Materials Medical Status Sheet

Incident Location: _____   Incident #: _____

Chemical(s): _____   DOT #(s): _____

TEAM MEMBER:

A: _____   B: _____   C: _____   D: _____

ENTRANCE	A	B	C	D		A	B	C	D
Blood Pressure					Oximetry				
Pulse Rate					EKG 3L 12L				
Respirations					Hydration cc				
Weight					Temperature				
EXIT (one minute)	A	B	C	D		A	B	C	D
Blood Pressure					Oximetry				
Pulse Rate					EKG 3L 12L				
Respirations					Hydration cc				
Weight					Temperature				
EXIT (five minute)	A	B	C	D		A	B	C	D
Blood Pressure					Oximetry				
Pulse Rate					EKG 3L 12L				
Respirations					Hydration cc				
Weight					Temperature				

Exclusion Factors Assessment: General Sensorium: A/O × 3; Serial 3s to 27; Finger to nose; Diastolic below 100 mmHg; Pulse less than 100; Resp. less than 24; Oximetry greater than 94%; EKG = NSR; Temp between 97 and 99.5°F

The above are norms. Assessments that do not meet this criteria must be evaluated. Exclusion factors must be done at 5 minutes exit.

Figure 7-5 *This cursory medical is done in the event that an individual becomes overcome due to heat stress, chemical exposure, or fatigue. It evaluates the status of the emergency response personnel. Exit weight must eventually equal the entrance weight.*

exposure should occur, preliminary paperwork has already been started, such that postexposure vitals can be compared with the preexposure physical.

Figure 7-5 is an example of such a form. Each form should be simple enough so that anyone possessing medical training can "plug" in the facts.

Figure 7-6 shows the routing of information at a hazardous materials scene.

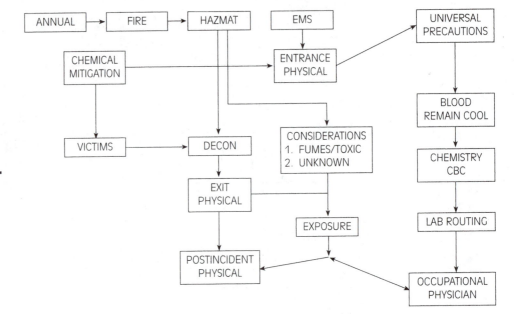

Figure 7-6 *The routing of information at a hazardous materials scene is very important. All entities must know the procedures involved.*

Postincident Physicals

If exposure occurs, a set of follow-up physicals is indicated. Again like most of any hazardous materials mitigation, preplanning must occur. Standard operating guidelines must address the likelihood of an accidental exposure. How to deal with it and who is responsible for postincident physicals must be addressed. Potential exposures are dependent on the products that are in your jurisdiction. While it would be nearly impossible to identify all substances, certain types and/or certain harmful chemicals can be identified as potentially hazardous. These substances along with any exposure should activate a mandatory postincident physical schedule.

The following is a list of extremely hazardous compounds that have the potential to cause severe health effects. Therefore, anytime these general chemical categories are encountered a postincident physical may be indicated, provided that the quantities and concentration of the material was significant.

- Aromatic hydrocarbons
- Halogenated aliphatic hydrocarbons
- Dioxin
- Organophosphate and carbamate insecticides
- Polychlorinated biphenyls
- Organochlorine insecticides
- Asbestos

- Cyanide salts and their related compounds
- Concentrated acids and alkalis
- Nitrogen-containing compounds
- Any compound that was in fume, dust, or airborne configuration

One must be realistic in the approach. For example, the laws that surround the removal of asbestos if one must replace old siding on a house are strict and the removal process is costly, lengthy, and quite detailed. However, the amount of airborne asbestos that one may be subjected to is nominal when an appropriate level of PPE is used, considering that the same monitors that are used in the asbestos removal industry would peg the needle off the scale if placed within a few blocks of any major interstate (asbestos is used within the brake lining, which is released during the application of brakes).

In general, the surveillance of these chemicals is such that if the quantities and concentrations are high, medical surveillance may be indicated. This level should be addressed during the planning stages of the program and evaluated as needed.

The postincident physical should take place immediately after the incident and every 3 months thereafter or under advisement of the attending physician.

! SAFETY

The postincident physical should take place immediately after the incident and every 3 months thereafter or under the advisement of the attending physician.

Performing salvage and overhaul during and after a fire is an act often performed in the fire service. During this stage of operations, firefighters are subjected to carbon monoxide along with other toxic gases (see Chapter 5). Although known to all firefighters, SCBA should be worn to protect the emergency worker. However, unfortunately, firefighters frequently do not don respiratory protection. As a medical surveillance support function, carbon monoxide monitoring should occur around the structure and with personnel.

Carbon monoxide monitors should be placed about the structure during salvage and overhaul operations in order to advise working crews of the toxic levels. During this time, these crews should have respiratory protection in order to perform task duties. Once the levels are reduced, then if needed, the operation can continue without the assistance of SCBA.

Personnel can be monitored by the use of colormetric tubes that are designed for measuring CO in expired air. The individual expires a breath of air into a plastic bag. The colormetric tube is then used to measure the amount of CO. Caution: specially calibrated colormetric tubes must be used for this evaluation. The air monitoring CO colormetric tubes are not calibrated for this function.

In either of these procedures, standard operating guidelines must identify the use and time when this type of air monitoring (see Chapter 11) will be performed. Who will perform the procedures? What are the options when personnel are above the OSHA exposure limit? What medical rehabilitation will be performed on those affected workers? These are a few of the issues that should be addressed within SOG.

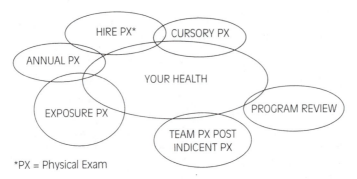

Figure 7-7 *Your health is impacted by the use of administrative and engineering controls. All must be in place.*

*PX = Physical Exam

This is continued until the next annual physical or until deemed unnecessary by the medical director.

If an unknown was encountered, after the incident the chemical(s) should be identified. Once this occurs, specific tests to identify target organs can be assessed appropriately. Comparison of sequential medical examination can identify the medical course of action. In any case, review of the incident should occur. This is done to identify all possible engineering controls and their use and how they worked under real time conditions (Fig. 7-7).

The amount of medical documentation required by law is increasing. At some point it may become the responsibility of the employer and the employee. In the meantime, one suggestion is to document within a standard diary, all of the calls that you as an emergency worker respond to. In this document, the place, time, and current conditions along with the chemical data could be logged, along with a diagram discussing the scene and command structure. A chronological order of annual, cursory, and postexposure physicals can be maintained within this book.

Medical Surveillance Comparison

Figure 7-8 is a comparison of all physicals an employee may have, each capturing a finite amount of information at different points within one career.

Chemical and Biological Exposure Log. Figure 7-9 is a sample format of a chemical exposure log. The log is set up to highlight your emergency response history. The first page identifies the medical surveillance that one would have during the course of employment. Items such as weight, vitals, and last physical are placed here. The log continues with an overall exposure record displaying the incident number, date and time, location, type of incident, and type of exposure. The next page identifies the exact type of exposure inclusive of bloodborne pathogens. The next page allows a brief exposure narrative that corresponds to the chronological record and exposure type record to be documented. If the incident was a chemical exposure, the next record set enables you to document all the components of the hazardous materials incident, a very detailed form for capturing all hazmat medical information. This form allows incidents notes and schematic of the scene. In total, this record allows the emergency responder to document all

! SAFETY
A chemical exposure log is set up to highlight an individual's emergency response and exposure history.

HIRE	ANNUAL	HAZMAT	CURSORY	POST
Medical Hx* Occupational Hx	Interval Hx	Interval Hx		Interval Hx
Physical Exam	Physical Exam	Physical Exam	Physical Exam	Physical Exam
Sex/Age norms	Sex/Age norms	Sex/Age norms	Sex/Age norms	Sex/Age norms
Temperature	Temperature	Temperature	Temperature	Temperature
Weight	Weight	Weight	Weight	Weight
BP	BP	BP	BP	BP
Pulse	Pulse	Pulse	Pulse	Pulse
Resp.	Resp.	Resp.	Resp. Pulse Oximerty	Resp.
Auscultate	Auscultate	Auscultate	Auscultate Hydrate	Auscultate
Vision	Vision	Vision		Vision
Audiometric	Audiometric	Audiometric		Audiometric
Cardiovascular	Cardiovascular	Cardiovascular		Cardiovascular
Neuro	Neuro	Neuro	Serial #	Neuro
Musculoskeletal	Musculoskeletal	Musculoskeletal		Musculoskeletal
Genitourinary	Genitourinary	Genitourinary		Genitourinary
Tests				
Blood	Blood	Blood		Blood
Chem	Chem	Chem		Chem
CBC	CBC	CBC		CBC

	Heavy Metals
	  Organophosphate
		Heavy Metals
		  Organophosphate
				Heavy Metals
				  Organophosphate

Liver/enzyme	Liver/enzyme	Liver/enzyme		Liver/enzyme
Urine	Urine	Urine		Urine
X-ray	X-ray(odd)	X-ray		X-ray
Pulmonary Function	Pulmonary Function	Pulmonary Function		Pulmonary Function Diffusion
EKG 12	EKG 12 Stress(even)	EKG 12 Stress(even)	EKG 12	EKG 12 Stress

*Hx = History

Figure 7-8 *The following is a comparison of all the physicals an employee may have, each capturing a finite amount of information at different points in one's career.*

types of emergency work and the details of the incident in an organized fashion. At the end of the document an area is left for the accumulative hours one may be exposed to a particular biological or chemical hazard (Fig. 7-10).

PROGRAM REVIEW

A periodic and systematic review of cases should be developed and used in conjunction with total system analysis. Comparison of personnel, equipment, and procedures is essential to determine trends. These trends may mark early signs of system failure, thereby establishing appropriate corrective measures.

Medical Surveillance Record

Name: _____ Fiscal Year: _____

Address: _____ DOB: _____

Phone: (___) - _____ SSN#: _____ Age: _____

Sex: _____

Basic Medical
(every three months):

	Date:	Date:	Date:	Date:
Weight (in kilograms)				
Blood Pressure (mmHg)				
Pulse				
Respirations				

Date of last physical: _____ Location of paperwork: _____

Information found within documentation: _____

SMAC: ____ CBC: ____ PT/PTT ____ EKG: ____ (Stress) ____ PFT: ____

place lead II
four second strip

Physician: _____ Phone: _____

In an Emergency: _____ Phone: _____

_____ Phone: _____

_____ Phone: _____

Personal Exposure Record

Biological and Chemical Exposure Log

This record is a valuable tool with regard to chemical and biological exposure. The information that is contained herein is confidential. The health and safety of an individual exposure is the primary concern of all firefighters and EMS providers. If found, please notify and/or contact the following:

This exposure record belongs to:

_____
(name)

_____
(address)

_____
(phone work/home)

Figure 7-9 *Personal exposure log.*

Chronological Exposure Record:

Incident #	Date/Time	Location	Type	Exposure
1				
2				
3				
4				
5				
6				
7				
8				
9				
10				
11				
12				
13				
14				
15				
16				
17				
18				
19				
20				
21				
22				
23				
24				
25				
26				

Exposure Type: Incident # _____

☐ Etiological Exposure: _____

Airborne: _____ Type: _____

Bloodborne: _____ Type: _____

Other: _____

☐ Radiological Exposure: _____

Time interval: _____

Alpha: _____ Beta: _____

Gamma: _____ X-Ray: _____

Ren: _____ MilliR: _____

☐ Chemical Exposure: _____

Solid: _____ Liquid: _____ Gas: _____

Powder: _____ Granular: _____

Mist: _____ Smoke: _____ Fiber: _____

Dust: _____ Liquid: _____ Cryo: _____

Chemical Name: _____

Commercial Name: _____

Type of exposure: _____

Absorption: _____ Injection: _____

Inhalation: _____ Ingestion: _____

Exposure Type: Incident # _____

☐ Etiological Exposure: _____

Airborne: _____ Type: _____

Bloodborne: _____ Type: _____

Other: _____

☐ Radiological Exposure: _____

Time interval: _____

Alpha: _____ Beta: _____

Gamma: _____ X-Ray: _____

Ren: _____ MilliR: _____

☐ Chemical Exposure: _____

Solid: _____ Liquid: _____ Gas: _____

Powder: _____ Granular: _____

Mist: _____ Smoke: _____ Fiber: _____

Dust: _____ Liquid: _____ Cryo: _____

Chemical Name: _____

Commercial Name: _____

Type of exposure: _____

Absorption: _____ Injection: _____

Inhalation: _____ Ingestion: _____

Figure 7-9   (Continued)

Chemical Exposure Record

Incident #: _____ Department alarm #: _____ Date: _____
Level of engagement: _____ Time of incident: _____
Sector assigned to: _____ Unit assigned to: _____
Work done: _____ Units on scene: _____
Weather: _____ Temperature: ___ F Humidity: ___ %
Wind speed/direction: _____ Temperature dewpoint: ___ F
Prehydration: _____ Fluid used: _____ Time: ___ Amount: ___
Posthydrate: _____ Fluid used: _____ Time: ___ Amount: ___
If entry team _____ Time in: _____ Time out: _____
  Level/type suit: _____ Suit #: _____
  SCBA used type: _____ SCBA #: _____
  Cursory medical completed: _____ (see below)
  Postincident medical completed: _____ (see below)
If decon team _____ Start time: _____ End time: _____
  Level/type suit: _____ Suit #: _____
  SCBA used type: _____ SCBA #: _____
  Cursory medical completed: _____ (see below)
  Postincident medical completed: _____ (see below)
Field medical preincident: _____
Blood pressure: _____ Pulse oximetry: _____ Blood drawn: _____
Pulse rate: _____ EKG: _____ Wt in: _____
Respirations: _____ Lung sounds: _____ Wt out: _____
Spill information
Product name (commercial): _____ CAS #: _____
Chemical name: _____ DOT #: _____
Solid: ___ Liquid: ___ Gas: ___ Cryogenic: ___
Vapors: ___ Dust: ___ Fumes: ___ Mist: ___
Exposure limits: _____ Reference material: _____
Monitored exposure limits: _____ Source: _____

Exposure Narrative Record

Incident # _____
(number corresponding to chronological exposure #)

_____
_____
_____
_____
_____
_____

Exposure Narrative Record

Incident # _____
(number corresponding to chronological exposure #)

_____
_____
_____
_____
_____
_____

Exposure Narrative Record

Incident # _____
(number corresponding to chronological exposure #)

_____
_____
_____
_____
_____
_____

Figure 7-9   *(Continued)*

Incident Reference Schematic

North ↑

Incident Notes

Incident # _____   Department Alarm #: _____

Date: _____

Time: _____

Figure 7-9   *(Continued)*

Chemical Exposure Summary

Date	Chem. Name	Quantity Available	Quantity Monitored	Exposure Time	Total Time

Total Time to Date

Incident Notes

Incident # _____

Department Alarm #: _____

Date: _____

Time: _____

Figure 7-9 *(Continued)*

SOUTHWEST FLORIDA PROFESSIONAL FIREFIGHTERS
LOCAL 1826 / I.A.F.F.
EXPOSURE REPORT

1. Firefighters Name ______ 2. Social Security # ______ 3. 1825 Dist. # ______
4. Incident Date ______ 5. Alarm Time ______ 6. Incident # ______

A MEDICAL / RESCUE EXPOSURE

7. Patients Name ______
8. Facility Patient Transported To ______
9. Transporting Agency ______
10. Name of Suspected or Known Disease: ______
11. Precautions Taken By You at Scene and After Exposure
☐ gloves ☐ face mask ☐ goggles ☐ protective gown
☐ washed exposed skin ☐ physician treatment (if this is checked complete *section F*)

How you were exposed	A. Topical Skin Contact	B. Needle Puncture	C. Direct contact to open wound	D. In Eyes	E. In nose or mouth
12. BLOOD					
13. URINE					
14. SPUTUM					
15. VOMITUS					
16. FECES					
17. OTHER					

B FIRE INCIDENT TYPE

18. ☐ Residental Fire 19. ☐ Industrial Fire 20. ☐ Vehicle Fire 21. ☐ Commercial Fire 22. ☐ Wildland Fire 23. ☐ Trash/Dumpster fire 24. ☐ Marine Fire 25. ☐ Explosion 26. ☐ Rescue 27. ☐ Spill
28. Other (describe in 1 or 2 words) ______ 29. Type of occupancy (1 or 2 words) ______

C LENGTH OF EXPOSURE BY FIRE STAGE / ACTIVITY

FIRE STAGE :	A <1 hr.	B 1-2 hr	C 2-3 hr	D 3 + hrs
30. Incipient				
31. Free Burning				
32. Smoldering				
33. Non-Fire Incident				

	A < 1 hr	B 1-2 hr	C 2-3 hr	D 3 + hrs
34. Extinguishment				
35. Empty / Ventilation				
36. Rescue				
36. Light Overhaul				
37. Heavy overhaul				
38. E.M.S.				
39. Investigation				

D SMOKE / CHEMICAL/ OTHER EXPOSURE

41. Smoke Conditions ☐ light ☐ Heavy ☐ None
42. Smoke Color ______

Chemicals Present	A Vapor Gas	B Dust	C Liquid Heavy Mist	D Light Mist	E Smoke	F Solid Powder
43.						
44.						
45.						
46.						

47. Type of Chemical Exposure to Above: ☐ Inhaled ☐ Ingested ☐ Skin Contact ☐ Eye Contact ☐ Other ______

E SYMPTONS

	AT INCIDENT	SYMPTON	AFTER INCIDENT	FOR HOW LONG
48.		Eyes Burn		
49.		Cough		
50.		Nose Bleed / Cough Blood		
51.		Nose/Lung Irritation		
52.		Nausea/Queasiness		
53.		Dizzy		
54.		Ears Ringing		
55.		Headache		
56.		Skin Rash/Burn		
57.		Unconscious		
58.		Other:		

F MEDICAL DIAGNOSIS

59. DID YOU RECEIVE MEDICAL EVALUATION OR TREATMENT AFTER EXPOSURE? ☐ YES ☐ NO
60. DIAGNOSIS ______
61. DOCTORS NAME ______ 62. FACILITY ______

G SPECIAL EQUIPMENT / DECONTAMINATION ☐ PASS ALARMS ☐ BUNKER PANTS ☐ HELMETS ☐ BUNKER COATS ☐ GLOVES ☐ HOODS ☐ SCBA'S ☐ BOOTS

63. WERE YOU SUPPLIED SPECIAL EQUIPMENT FOR THIS INCIDENT? ☐ YES ☐ NO DESCRIBE ______
64. WERE SPECIAL DECONTAMINATION PROCEDURES FOLLOWED AFTER THE EXPOSURE ☐ YES ☐ NO DESCRIBE ______

H CO-WORKERS AT TIME OF EXPOSURE

65. PLEASE LIST NAMES OF OTHER FIREFIGHTERS WORKING CLOSE TO YOU AT TIME OF EXPOSURE ______

I ADDITIONAL INFORMATION (YOUR FUNCTION AT INCIDENT, UNUSUAL CIRCUMSTANCES, ETC.)

66. ______

J 67. DATE REPORT FILLED OUT ______ 68. NAME OF PERSON FILLING OUT REPORT (PRINT) ______

69. SIGNATURE ______

Used by permission of Local 1826 IAFF.

Figure 7-10
Exposure report. Many unions are offering to their membership an exposure report that is used for exposure tracking.

We can address several items in order to start this review process:

1. Computerization of all personnel medical files, with continuous monitoring
2. Analysis of exposure versus incidents
3. Protective measures established from the computer monitoring

Computerization of all personnel medical files with continuous monitoring is one avenue we have to monitor our personnel. ADA, civil rights, patient confidentiality laws play into this consideration. However, in order for us to do this effectively consent from all personnel is required. Annual physical data along with cursory physical information can be placed into a relational database. From this medical trends can be analyzed. Everything from chemical contact, type and concentration of the chemical, temperature in the suit, and ambient air temperatures should also be placed in our database.

Analysis of exposure versus incidents is a capability of computerized recordkeeping. Once a database has been established, questions about the engineering or administrative controls can be asked. Under the guidance of industrial hygienists, toxicologists, and medical personnel, trends can be established and the problems surrounding the trend can be changed or modified.

Protective measures can be established once we know what controls are failing. Procedures can be analyzed for review and in some cases models of historical scenarios can test these new procedures. In every step of the way it will be important for the medical director to become involved with the surveillance program and the intricacies of such a program.

It is important that the attending physician (and medical director) is knowledgeable of hazardous materials operations, the training and the tactics employed. Exact frequency and contents of the exam is truly the responsibility of the physician, however, a current copy of the OSHA standards and NFPA documents will enable the physician to organize the appropriate surveillance program.

CRITICAL INCIDENT STRESS DEBRIEFING

During the early 1980s and into the present decade, stress management has been a focus of most all services thanks to the work of Jeffrey Mitchell and Grady Bray, who are noted for their research in determining the effects of stress on the emergency worker. In recent years, the emergency services have placed special emphasis on the management of stress and the related outcomes. In the early days, the term "burnout" was applied. Typically, this term was referring to those individuals who, during the course of their career, had been subjected to an event or several events that would change their life forever.

For some, the friendships and peer support helped them recover from the depression, while others sought out other methods to cope with the feelings and emotions, such as alcohol and drugs. Stress in life is something from which we as human beings are not going to escape, especially those involved in the emergency services.

The stress that any emergency worker can encounter is wide ranging. On one side of this double-edged sword are the anxieties that make this line of work so interesting, while on the other side is the reality of pain, emotions, and human suffering.

Each one of us differs emotionally. Some can handle stress very well and it is only with high levels of emotion that these individuals seem to crumble. Others have a more difficult time and seem to keep the emotions in, or talk about them freely with close friends.

We know that police work, EMS, fire fighting, emergency room nursing, and aeromedicine all can provide the stress that may interfere with the emotional status of the worker. Hazardous materials at any level is no different.

At the hazardous materials incident, chemical properties, environmental conditions, and sheer quantities severely impact the incident—all factors that we cannot control. As emergency workers we all have one thing in common, our physiological makeup is more or less the same. We all need and like to control the incident. We are taught from day one that control of the incident is a means to the end. We practice and encourage incident command systems which in turn organize and control the incident. However, at the hazmat scene, control is only temporary and can be lost at any moment. Chemical properties, environmental impact, and quantities of the materials that we are handling cannot be changed or sometimes even controlled.

We all have a higher or lower degree of obsessive-compulsive behavior or, in other words, we need control and perfection. Again, in training we are taught to be perfect all the time and beat ourselves emotionally when something occurs that was uncontrollable.

How do these simple traits affect our emotions at the hazardous materials incident? Why do they occur? We know by virtue of the dangers hazardous chemicals present that direction factors are not truly in control. We also know that the precision at which we must perform is at 100% perfection. We know something can go wrong. For these reasons and countless others, the hazardous materials team must have within its team structure a program that can ensure the emotional well-being of all members. In order to do so, a policy and procedure for hazardous materials (and confined space) medical surveillance must include the components of critical incident stress management. This type of need will not occur often which makes the necessity of such a program a vital component of the medical surveillance policy.

This policy should provide several levels of emotional support for the affected responders and their families. The first level is a group of peer personnel who have been educated in hazardous materials mitigation and the principles of critical incident stress debriefing (CISD). They should have experience with both the CISD format (defusing and debriefing) and be involved with a peer support group.

The second level of the program should be the CISD team. Here defusing and debriefing along with mental health professionals can establish criteria for when the defusing and/or debriefing will take place. The mental health professional must become involved with the hazardous materials response team, understand-

■ NOTE
Medical surveillance must include components of critical incident stress management.

■ NOTE
Several levels of emotional support should be provided for affected responders and their families.

ing all the components of defensive and offensive roles. Evolutions within encapsulating suits will give the mental health worker a firm understanding of what it is like to be a hazmat team member.

The overall structure of the CISD component to hazardous materials is the same as for all emergency scenes. First the defusing can take place at the scene or shortly thereafter. These defusings should be done by the individuals that have been trained in hazmat mitigation and CISD principles. They should have the highest regard for confidentiality with respect to the emotions and feelings of others. The defusing should consist of but not be limited to:

- An atmosphere of positive reinforcement
- Defusing process not longer than 60 minutes
- The session as informal and informative as possible
- No operational critiquing of the incident

If the incident warrants a complete debriefing, it should occur within 48 hours of the incident. This should transpire away from the incident site, station, hospital or work-related area. A neutral, nonreminding locale should be sought.

The support group should be made up of a broad spectrum of personnel, possibly organizing it by educational disciplines (i.e., EMS, police, fire, hospital), with

Figure 7-11 *All personnel should have some level of CISD after an incident. Decon, entry, backup, and the medical surveillance team may require debriefing after a stressful incident.*

the one common thread being the hazardous materials technician. Each group will need one or two mental health care workers and one to three hazardous materials peer supporters.

The debriefing process is within a positive reenforcing atmosphere, without the pressure from administrative authority (all rank and positions are equal). This session is formal and should consist of the following components:

1. Introductory phase
2. Fact-finding phase
3. Thought phase
4. Reaction phase
5. Signs and symptoms
6. Teaching phase
7. Reentry phase

Do not under any circumstance allow this to become an operational critique. This process should be followed whenever there is:

- Injury and/or death of a child or co-worker
- Exposure to emergency workers
- Injury, hospitalization, and/or death of an emergency worker
- Any of the above during night operations and/or large-scale incidents
- Incidents effecting family members
- Any incident which has excessive media
- Or as deemed necessary by personnel

The operational critique should be done 5 to 10 days after the debriefing.

CISD Textbooks for Further Reference

Critical Incident Stress Debriefing: An Operations Manual for the Prevention of Traumatic Stress among Emergency Services and Disaster Workers. Jeffrey Mitchell, Ph.D. and George Everly, Ph.D., F.A.P.M., Cheven Publishing Corporation, Ellicott City, MD.

Innovations in Disaster and Trauma Psychology: Applications in Emergency Services and Disaster Response. George Everly, Ph.D., F.A.P.M., Cheven Publishing Corporation, Ellicott City, MD.

Emergency Services Stress: Guidelines for Preserving the Health and Careers of Emergency Services Personnel. Jeffrey Mitchell, Ph.D. and Grady Bray Ph.D., Prentice-Hall, Englewood Cliffs, NJ.

HOW TO ESTABLISH A MEDICAL SURVEILLANCE PROGRAM

As one might suspect the expense of a medical surveillance program can become quite costly. Here, as with medical control, the medical director can play a significant role in providing some if not all the services required. The medical director can oversee the EMS functions of the department as well as the medical surveillance program.

The first goal to accomplish, which should be done by a committee, is to write down all needed components of the program, including resources that are already available to the agency. In unionized agencies, one member should be the union president and/or a representative. The various ranks and disciplines along with medical staff all should have direct input into the program. Only allow five to seven individuals in the committee, each having an area of responsibility and expertise. For example, the hazardous materials coordinator, medical director and EMS coordinator, middle management and high level management, along with a hazmat team member, would give a well-rounded committee.

The committee must identify goals and objectives of the program while identifying direct needs from a medical aspect that could be employed with the least amount of training and cost. Identify strengths and weaknesses of the system and look for avenues of correction. For example, the committee can contact other agencies that have medical surveillance programs, acquiring their SOGs. The committee then can identify those areas that would work within their system and incorporate within local guidelines.

Once the program is established, program review and system problem identification is a future goal consideration. At this point a new committee may be indicated or a subcommittee appointed to the goal of review. However structured, goals with realistic time frames are a must.

Utilize existing services within your system. Have a group of agencies working toward a common goal. Pooling resources and money can provide a strength to all agencies at the scene and during training activities. All personnel that would possibly respond to such an incident should be knowledgeable of the course of action that will take place prior to, during, and after the incident.

One area of consideration is that of a relational database. A dedicated computer must be purchased with the manpower to input the medical and incident information. At predetermined time frames, a correlated question to the database is asked, such as liver, kidney, blood abnormalities detected within the emergency response force? How many incidents of exposure to a particular chemical were encountered last quarter? Last year? For a program to exist under emergency situations, preplanning and a rehearsal must be a part of the training activities. For example, if your agency feels that blood chemistry is important under specific guidelines, then a cooler and centrifuge must be acquired. There must be preplanning for a medical director and/or toxicologist. Laboratories that can handle last minute blood work within a reasonable turnaround time are all considerations when planning for the medical aspects of a hazardous materials incident.

The exposure records and cursory medical evaluations must have a mechanism for the proper routing of this paperwork. All forms must be maintained and held in confidence. This, along with the computer analysis of the information, will present a challenge to any emergency response service.

Caution Issue

The issues that surround medical surveillance are far reaching. The ideas and controls presented in this chapter viewed the issues strictly from a long-term medical approach. However, these ideas and controls are not without their respective shortcomings. For example, the hiring physicals and the ADA (presented briefly in the beginning of the chapter) can present the management of any service with difficult problems when keeping medical monitoring in mind.

The whole premise of medical surveillance is that of health maintenance within a field that can have adverse environmental conditions. The issue of medical records and the maintenance of such records are issues that each individual service will have to find viable solutions to.

It was not feasible nor the intention of the authors to discuss all the issues that surround this complex and ever-expanding discipline. The information presented was gleaned from health surveillance programs within industry, suggestions by the United States Coast Guard, EPA, OSHA, NIOSH, and nationally recognized safety programs. The information was intended to place in order the progression of what is realistic and of normal practice within the medical/industrial hygiene industry.

The ideal reason for medical surveillance is not the political issues that surround us but rather the medical justifications for biological indicators of toxic agents. Based on our knowledge of analysis, the detection of physiologic change, and the findings of hypersusceptibility in individuals, we can make our jobs easier and safer for future generations of emergency responders.

Computerization of records, maintenance of the cursory physical and periodic interval monitoring are all issues that border on invasion of privacy, patient confidentiality, and the breach of civil rights. However, the major issues here are the concerns we have for emergency responders. Only through intensive research, monitoring and tracking of exposures, will we find the solutions to employee safety. Only through a program that identifies the issues and ethically incorporates these concerns toward administrative and engineering controls will we as a group have the ability to guide future health issues within the emergency services.

To this end we must realistically manage all of these issues with the health of the emergency worker in mind. These issues are hiring physicals, interval and annual physicals, interval histories, maintenance of medical records, confidentiality, and accessibility of medical records, abnormal findings and the effect on medical pensions, retirement, and worker's compensation, informing the worker of the abnormalities, and the certification for employment once an exposure has occurred to name a few. All have an overall deliberate impact on employee health. The importance is that the emergency worker is provided with the proper education, resource and management if the detection of a disease process occurs.

Summary

The topic of medical surveillance is far reaching. The documentation alone can be cumbersome. All hazmat team members understand the dangers that can be encountered at an incident. We as medical personnel must be in tuned to the management of the incident. Daily activity greatly affects the results seen on a hazmat incident. It has to do with the emotional and physical wellness of the individual.

All team members including of EMS personnel should be involved with a training program that increases their personal aerobic capacity and physical strength, at the same time watching what they eat in order to decrease body fat, serum cholesterol levels, and emotional stress.

Programs should include weight training, smoking cessation, drug and alcohol awareness, and the understanding of emotional impact. All have components within a well-structured medical surveillance program.

Time off from emergency response and the equipment needed to accomplish the above goals should be provided in order for the team members to attain the state of physical and emotional health that is required.

The one common thread that all surveillance programs should maintain and strive to perfect is the scrupulous examination of postexposure individuals. The main purpose is to capture the effects of an exposure. Examine all exposures and document any changes from the baseline information. This along with correlations between chemical exposures and the affect need to be investigated. We are not sure of the biometabolism—the behavior of these chemical exposures. When effects manifest themselves later a correlation can only occur if the event has been documented. There are many chemicals about which we know their behavior in acute high level exposure, but we know little about chronic low level exposure.

We must also protect our personnel emotionally by using principles practiced during critical incident stress debriefing.

Scenario

Your department has allocated a large portion of the hazardous materials response budget to the establishment of a hazardous materials medical surveillance program. The medical director and yourself have just returned from a class describing the fundamentals of such a program. The medical director is very concerned and has persuaded the chief of your department to allocate initial start-up and continued financial support. Both the chief and the medical director are concerned about the legal ramifications of such a program and the continued support structure.

You have been placed in charge of the development, implementation, and maintenance of such a program. In order to assist in your project you elicit the help of a mutual aid jurisdiction. They have just started to develop a similar program, but have been hampered by the lack of financial support from their department. It is your goal to develop, implement, and educate the response personnel in both jurisdictions, while implementing a cost effective program.

From your notes that were taken at the medical conference you have just returned from you start to develop your program.

Scenario Questions

1. How should prehire physicals be handled and what ramifications do the existing programs have?

2. Who could provide the annual physical and what are the components of such an exam?

3. What are the components of the cursory exam and how will this information be routed to the appropriate medical authority?

4. Considerations of the cursory exam:

   a. Elements of the exam are?

   b. Should blood drawing, blood alcohol, or blood sugar be a part of such an exam?

   c. What exclusion criteria should be established?

   d. If a team member becomes exposed what are the procedures?

5. How will you handle the review of such a program?

6. If there is a need for a critical incident stress debriefing who will handle this part of the operation and how will they do it?

7. What other considerations will you have that may have far-reaching ramifications?

chapter
8

Hazardous Materials Considerations for Hospitals

Objectives

At a hazardous materials incident where a patient has been injured due to an exposure and transportation to a medical facility is involved, or when a hazardous materials incident takes place within the hospital facility, the student must understand the complexities and special considerations of the effects a chemical may have in the health care setting. Furthermore, the student should identify the importance of developing a preplan, preparing the emergency department for chemical emergencies, and the importance of regional poison control centers to hospital staff when faced with contaminated patients needing emergency medical care.

As a hazardous materials medical technician, you should be able to:

■ Identify the goals that hospital receiving facilities must set for preplanning to be completed.

■ Understand key points to note for identifying patients who are injured by exposure to hazardous materials.

■ Give an overview of incident command when working within the hospital supervisory structure.

■ Recite step-by-step the considerations to be addressed when:

Receiving the call.

Setting up a decontamination area.

Directing the arrival of the patient.

Providing decontamination.

Selecting the proper level of personal protective equipment.

■ Identify the role of the regional poison control center when a hospital is faced with an in-house hazardous materials emergency or the arrival of a contaminated patient.

INTRODUCTION AND HISTORY OF HOSPITAL HAZARDOUS MATERIALS RESPONSE

In recent years hospital staffs have been challenged with increasing responsibilities, one of which involves dealing with chemical exposures. The industrial use of chemicals is on the increase and so are the related accidents. The ultimate landing field for the chemically injured patient is the hospital. Staff in the emergency department and those working in support areas must be ready when faced with the challenge of handling hazardous materials emergencies.

Not only are accidents happening outside of the hospital, but also within the hospital itself. Chemicals in the form of sterilization gases, radioactivity, special solvents, biohazards, and many more have found uses in the hospital setting. Therefore, the hospital staff may face the effects of chemical events happening outside of the hospital (external) as well as events taking place within the hospital (internal).

Although these two events result from completely different situations, the principles for handling these emergencies are somewhat similar. External events affect the hospital only when patients start arriving from the field. The emergency department may be faced with serious secondary contamination hazards from patients who walk in, drive in, or are carried in by both good samaritans and EMS transporting agencies. Internal events require rapid deployment of trained individuals to the scene to stabilize the event and perform any defensive measures necessary to limit the effect.

Hospitals store and use great varieties of hazardous materials. Hospitals are not so different from manufacturing or industry insofar as chemicals are just part of doing business. The major difference lies with the enormous possibility of life hazards found in the hospital setting. Hospitals house a concentrated number of

■ NOTE
Hospitals are not so different from manufacturing or industry in that the storage and use of a great variety of hazardous materials are just a part of doing business.

debilitated patients who may not have the ability to self evacuate or provide personal protection for themselves in case of a hazardous release. Therefore, this section was developed for use by hospital personnel and emergency responders.

Several important steps must be taken to ensure the safety of both staff and patients within the hospital if faced with a hazardous materials emergency. Preplanning is a necessary step to properly prepare for an emergency involving hazardous materials. Furthermore, writing and accepting a comprehensive policy and procedure to outline job duties in the case of an accident is not only important (and a component of 29 CFR 1910.120 under emergency response planning) but mandatory through the Joint Commission for the Accreditation of Hospitals (JCAH). Within this chapter many considerations are examined for inclusion into a written policy. The policy can be simple or complex enough to include step-by-step instructions for each participant. Keep in mind that no policy will work without occasional review and practice in the form of drills.

Riverside, California, February 19, 1994. On February 19, 1994, at 1946 hours, the Riverside Fire Department received a call for assistance to a cardiac patient in distress. What was thought to be a routine advanced life-support call turned out to become a horrifying hazardous materials event. Once on the scene paramedics found a 31-year-old Hispanic female in the terminal stages of ovarian cancer. The radio patch to the local emergency room described a cardiac patient in severe cardiogenic shock. The customary ALS modalities were enacted and the patient was transported to the hospital.

Once at the receiving facility, the patient degenerated into full cardiac arrest, and arrest procedures were started immediately. The appropriate ALS techniques were in place when arterial blood gases were drawn. A garlic/onion/ammonia odor emanated from the victim and a gasoline-type sheen was noted on her skin. While the procedures were underway the case took a bizarre twist, resembling a science fiction story.

The nurse drawing the blood soon collapsed with the doctor falling shortly thereafter. The two employees were evacuated and care started to focus on the staff rather than the patient. The condition of the doctor and nurse who collapsed progressed into severe respiratory distress, with episodes of failure. The physician in charge of the emergency room began care of the cardiac patient while other staff assisted in the emergency efforts of their co-workers. Another physician and respiratory therapist assisting in the arrest procedure soon became ill. At this point an evacuation of the emergency room was ordered and the fire department called. The 31-year-old patient was pronounced dead.

At 20:52 hours, fire department personnel were on scene and the hazardous materials team was responding. The hospital administrator, environmental resources, and the coroner were notified and a full evacuation ordered. All emergency room patients were transferred to local hospitals along with six hospital employees.

By 23:00 hours, the hazardous materials team made entry into the emergency room providing reconnaissance and air monitoring utilizing Level B protection. Combustible gas indicator, radiation monitoring, an assortment of colormetric tubes and oxygen monitors all provided a negative response to a hazard. The 31-year-old female was placed in a double-sealed bag and an airtight container for further evaluation by the coroner's office. The emergency room was decontaminated and turned over to the hospital by 03:00 hours February 20.

One could speculate on the problem that occurred in this Southern California community, however the facts are that six hospital employees became seriously ill from some chemical. Two of these employees were critical and may have sustained damage that will remain with them for the rest of their lives. It is truly unknown what exactly occurred that winter night in Riverside, California, however many speculate on the possibilities:

1. Because the patient was reported to have a garlic/onion/ammonia smell, organophosphates were once thought to be the culprit. If the physiology of organophosphates is reviewed, the signs and symptoms displayed by the hospital staff do not match those of organophosphate poisoning. This unknown chemical had to be fast acting, probably something that affects cellular metabolism.

2. Like many emergency rooms, the hospital places organic material down the sink. This material is later cleaned by the house-cleaning department. Many times an acid solution is used to decompose the material for easy passage down the drain. Given this scenario, the possibility exists of the development of significant quantities of hydrogen sulfide gas. Examining this as a plausible scenario the action of hydrogen sulfide could have produced the effects witnessed in the patient and hospital employees, however, why were only a few staff members affected. Furthermore, the odor of hydrogen sulfide is that of rotten eggs, not what was reported as an onion/garlic/ammonia smell as described by the nursing staff.

3. Recreational drugs associated with the cancer medication is a possibility, but is not probable. It is unknown if the patient was engaged in recreational drugs and no evidence has surfaced to indicate such. However, it is speculated that some illegal drugs in combination with the cancer medication could cause a chemical reaction to occur within the body and produce toxic gas. Looking at the possibility of the patient trying to relive her pain and suffering through such medication is purely speculation.

4. It has been known that cancer patents sometimes relieve their pain by using a folk remedy of dimethyl sulfoxide (DMSO) for antiinflammatory reasons. This chemical is rapidly absorbed through the skin into the bloodstream. Although not confirmed, the Lawrence Livermore National Laboratory conducted research on this possibility and found that DMSO in an oxygen-enriched environment forms dimethyl sulfone. Once cooled, dimethyl sulfone breaks down into dimethyl sulfate, an extremely dangerous and toxic substance.

According to the reference material, dimethyl sulfate has a faint onion smell and is absorbed rapidly by the skin and into circulation. It is estimated that less than 0.1 ppm can give sufficient absorption to cause serious, if not fatal, pulmonary failure.

This scenario, although unable to be reproduced in the lab, gives us the possibility of the chemical in question. The DMSO could account for the gasoline-type film noticed by the emergency room staff, upon arrival of the patient. The toxic qualities of dimethyl sulfate also accounts for the sudden almost fatal reaction of the health care providers. The onion/garlic/ammonia smell also could be caused by these two possible chemicals.

Many questions remain unanswered. The fact is that something did occur, several individuals were affected, and the possibility of it occurring again. . . .

PREPLANNING

Hospitals practice preplanning in the form of fire drills, medical emergency training, and disaster planning. The inclusion of the hazardous materials incident, whether it originates inside or outside of the hospital, should also be a part of pre-plan development.

Preparation for such a hazardous materials emergency must be a high priority for the hospital's management and staff. These plans should encompass a large spectrum of hazardous materials possibilities including as much self-reliance on handling the emergency as possible. Self-reliance is an important issue during hazardous materials emergencies occurring outside of the hospital when patients are involved. Patients arriving by ambulance and personal vehicles may not be decontaminated. Relying on the local hazardous material response team for decontamination may not be an option when the team is at the scene mitigating the event.

During the inception of hazardous materials teams, including hospitals within the incident plan, was generally overlooked. The plans that did exist addressed decontamination on the scene by hazardous materials team members followed by the transportation of a clean patient to the emergency department for treatment. The hospital staff then took over the care of the patient. This scenario has merit but often is far from reality.

When preparing a hazardous materials preplan, identifying the location of the nearest hazardous materials response team and its educational level may be the first step in identifying resources for the education of hospital personnel. During this educational process, a personal level of understanding between the hazardous materials response team and the hospital staff can grow. Such questions as the following should be discussed before the need arises:

Who responds to the hospital if an in-house hazardous materials emergency exists?

What level of training should the hospital staff have to initiate control of the incident?

What equipment is needed to be self-sufficient in case of a large hazardous materials disaster?

Understanding what to expect from field personnel can aid the hospital in determining what additional resources are needed in the event of an emergency.

Industrial sites and laboratories may have personnel that the hospital staff can rely on if a hazardous materials team has not been established in the area or if it is tied up on an incident and cannot respond. Outside industrial facilities many times have additional resources to offer besides trained personnel. Occasionally decontamination equipment and resource material are used in these facilities and can be transported to the hospital if an early request and prearrangements are made. Asbestos abatement crews, popular throughout the country today, can set up decontamination corridors and provide some respiratory equipment. All of these considerations can be addressed in a comprehensive preplan.

PROVIDING HOSPITAL TRAINING

In the early 1970s, hospitals expanded their services by providing training to the first paramedics. Hospitals in South Florida and California were the first to take emergency medical providers from ambulance services and fire departments and teach them advanced skills such as the insertion of intravenous lines, endotracheal intubation, advanced assessment skills, and terminology. All of these skills were taught to allow the emergency department staff and paramedics in the field to act as a team when dealing with critically injured or ill patients. Although expanded from the early days, the training of paramedics today still centers around the basic skills.

Today, the skills required to manage hazardous materials emergencies are practiced and refined by hazardous materials responders. The skills, including incident command, personal protection, and hazard control can be easily taught to the hospital staff by the supporting emergency services. The philosophies of incident command and hazard control are already in effect within the hospital but must be reorganized and evaluated for use during hazardous materials emergencies. It has become the duty of the emergency services to evaluate the needs within the hospitals and introduce to the staff the principles and practices of hazardous materials mitigation.

Planning for a hazardous materials emergency is the first step in identifying the educational needs of hospital staff. The educational level required to make sound, rational judgments may be beyond the level of current education found within most hospitals today. If a hospital becomes involved with contaminated patients and the educational needs have not been met, the incident could be disastrous. The time to learn about hazardous materials response and how that response is suppose to take place is not while the event is taking its toll. Instead, staff must know what to do and how to do it well before an emergency arises. For example, a hospital receiving a chemically injured patient who has been decontaminated on the scene and now requires medical care may seem like an inconsequential emergency. But, because field decontamination may not be thorough and because the patient may still weep evidence of the chemical, untrained medical staff could become affected by the chemical if not properly trained.

Training needs for emergency responders are defined by OSHA and the NFPA. They divide competency-based training into the categories of awareness, operations, technician, and specialist. Each build upon what was previously learned, adding further training and skill requirements. Although these standards do not address hospitals in particular, the competencies still have merit within the hospital setting when dealing with a hazardous materials emergency. For example, if a patient arrives at the emergency department requiring decontamination and the hospital staff wish to complete the decontamination, then the level of competency as described by OSHA and NFPA is that of operations with a concentration in decontamination. These training guidelines have merit within the hospital setting because hospital personnel faced with a contaminated patient may be met with many decisions affecting the health and safety of other staff members and patients alike. Learning safe and correct means of dealing with hazardous materials emergencies is a lengthy process that involves long training hours and competency testing that hospital training departments are not accustomed too.

For example, curricula for hospital employees would be all of the awareness-level training as it is outlined in OSHA 29 CFR 1910.120, NFPA 472, and 40 CFR 311 and most of the operational level as outlined in those documents, with perhaps a lighter concentration on those areas that would not be a factor in the hospital setting. Segments of the technician level should also be taught to the hospital staff concentrating on personal protective equipment and decontamination efforts. All supervisors should have a class that is geared to the management structure under a disaster situation, i.e., the incident command.

Training also identifies the need for additional resources required for an organized and systematic approach. An incident of significant magnitude (multiple patients arriving that are both decontaminated and grossly contaminated) will need additional resources that may not be identified. A resource list must be developed and kept current including support agencies so that the hospital can make notification in a timely manner. Emergency management offices including the Local Emergency Planning Committee (LEPC) and State Emergency Response Commission (SERC) can assist in this listing. This resource list and a communication network should be a part of the operating procedure with a standard format starting at the local level and continuing to the state level if the incident warrants.

INTERNAL HAZARDOUS MATERIALS INCIDENTS

Hazards

Hospitals are a hot spot of hazardous materials. Biohazards are found in almost every department, radiation is used throughout the hospital, mixed gases for anesthesia and sterilization are both toxic and flammable, and solvents used in the laboratories all pose a potential problem within the confines of a hospital. Dealing with these emergencies on a first responder level is not only important but necessary within a hospital.

Another word of caution to responders in the hospital setting. The magnetic resonance unit (MRI) poses an extreme danger to those who may wander into the magnetic field. The field is so strong, even in the off position, that it can pull a wrench from the hand and propel it many feet. In particular danger are those with any ferrous material within their body. Pacemakers, cranial plates, aneurysm clips, inner ear implants, metal joint replacements, or metal about the eye orbits are all possible targets. The magnetic coil is cooled by liquid nitrogen or helium that, if escaping, could cause severe cold injury and with their huge expansion ratios, can displace oxygen causing asphyxiation.

> **! SAFETY**
> The magnetic field of a magnetic resonance unit poses an extreme danger in that the field is so strong, even in the off position, that it can pull a wrench from the hand and propel it many feet.

> **! SAFETY**
> The magnetic coil of a magnetic resonance unit is cooled by liquid nitrogen or helium that, if escaping, could cause severe cold injury and, with its huge expansion ratio, it can displace oxygen and cause asphyxiation.

Hazardous Materials Response Brigades

One way to control hazardous materials incidents within hospitals is to establish a hazardous materials response brigade. Forming a response brigade must start with training. The team should receive the training necessary to guide them in recognizing hazardous materials emergencies and understanding the effects, locations, and inherent dangers of the chemicals found within its facility. Attending training with the local hazardous materials response team, including operations and technician level classes, where appropriate, will give the brigade an appreciation for the special problems encountered on incidents of a larger magnitude.

The duties of the responding team could be limited to isolating the incident and directing evacuation as needed. The team should mobilize by a coded message over the intercom system much like a cardiac arrest or fire is broadcast. (Code 90, code red, or code blue are all used by hospitals for these purposes.) A signal word or code (code green) can be given to mobilize the response team members from wherever they are in the hospital. All professions within the staff of a hospital can provide support. Maintenance personnel are useful in controlling ventilation and drainage within the contaminated area. Patient care staff can be used to move patients if an evacuation is necessary. Sanitation department can control spills by providing absorbent. The security department can control access into the area by guarding hallways and doorways that may lead into the contaminated area. The team can even be trained to respond and maintain control of the incident through the mitigation process. If at any time the brigade felt the incident was beyond its control, a contractor or hazardous response team could assist.

Zones

The team must be familiar with the terms *hot zone*, *warm zone*, and *cold zone*, so it can define these zones within the hospital until an advanced mitigation team can arrive to clean up or contain the incident.

Hot Zone   The hot zone is the area of the spill or release that contains the chemical and is the most hazardous. By closing doors, containing runoff, and controlling ventilation systems, the hot zone can be limited in size. No one should enter the defined hot zone without having complete chemical specific protection. Therefore, the in-house response brigade may only contain the release but not enter to clean up or fix the problem.

Warm Zone   The warm zone is the identified area that provides a buffer around the hot zone. If decontamination of victims is needed, it is within the warm zone that the decontamination is performed. All nonessential personnel are prevented from entering the warm zone. Only properly protected personnel can be within the warm zone. This usually consists of personnel protective equipment one level below that which is needed in the hot zone. For example, if hot zone entry requires Level A protection, then warm zone will require Level B.

Cold Zone   The cold zone is the area where equipment and the response brigade will be located. This area is referred to as a safe zone. Nonessential personnel should still be limited from entering the cold zone. This area is used for gathering equipment and discussing plans needed to mitigate the incident.

As a part of this team, laboratory specialists may be used to clean up small spills. These specialists are typically trained to handle only limited chemical spills. By enhancing their knowledge base and skill level, a resource within the medical institution can be further developed and used in a dual role. The spill quantities handled through the lab team would be small; anything beyond their capabilities would still warrant a defensive stance.

Each member of the team must be familiar with the policies of the hospital and know what part of the incident he or she is responsible for. A team leader should take command of the incident and provide the necessary guidance for the responding team members. The team leader also coordinates the arrival of the local hazardous materials response team or hazardous materials cleanup company.

Patient and team member safety is of utmost importance when dealing with an incident of this type. To understand the hazards facing the brigade, a resource or research team must first gain as much information about the product as possible. The material safety data sheet will provide specific information on the spilled chemical. The NIOSH pocket guide can also be referenced to provide toxicology information concerning the released chemical. The hospital library, poison control center, or emergency department computer bank may also supply additional information to present to the responding professional hazardous materials team.

Through team training and the enactment of policies, a response brigade can limit the effects of a minor hazardous materials release within the hospital. Without a team, the incident can easily become a toxic nightmare involving patients, staff, and visitors to the hospital. Liabilities may be multiplied if an incident of this type is not handled through proper training and preparedness.

EXTERNAL HAZARDOUS MATERIALS INCIDENTS

Identifying the Arrival of Contaminated Patients

After a hazardous materials accident there are basically two types of patients brought to the hospital: those who are injured but not contaminated and those who are injured and contaminated. If the patient was only exposed to a harmful chemical but is not presently contaminated, emergency workers can provide care without the fear of secondary contamination. This is usually the case with victims of gas exposures. On the other hand, contamination is usually the result when a solid (particulate matter) or a liquid becomes involved in the accident causing the patient's illness.

There is no great mystery to identifying contaminated victims when the scene obviously dictates the facts. These victims are usually a result of an industrial or transportation accident where hazardous materials are either in use or being moved from one site to another. An ideal situation involves EMS personnel calling, either by phone or radio, to advise emergency department personnel of the situation. With this notification, the hospital staff, if equipped, can properly prepare for the incoming patients. However, too often EMS providers transport contaminated patients without prior notice to the hospital. This situation may be the result of inadequate training by the EMS providers or because of confusing or inaccurate details gained on the scene. Many times EMS providers are called upon to transport patients that may not instantly show the evidence of contamination. Only after the providers get sick or further information is gained about the scene does the involvement of a toxic material become evident. Hospital staff should be knowledgeable about clues found on the scene that may alert them to the presence of hazardous materials. The staff should ask the EMS providers to describe the scene concerning:

- Type of incident
- Type of decontamination
- Estimated time of arrival
- Type of chemical
- Location within the community
- Wind direction and speed
- Number of patients
- Signs and symptoms
- Quantity of chemical
- Who is on scene

Hospitals also face the probability of contaminated patients either walking in or being carried in by other lay persons. In these cases, the damage and contamination to the hospital and staff may occur before any prior knowledge of the incident exists. A delicate balance between protection of hospital staff and provision of emergency medical care requires the judgment of a well-trained medical team to minimize the effects of situations of this type. Clues from two or more patients originating from the same workplace or geographic area complaining of similar symptoms, such as chest pain, nausea and vomiting, or sensation changes may indicate the possibility of exposure to a chemical and possible contamination. Other clues may involve breathing difficulties, coughing, eye and nose irritation, drooling, and headaches. Any of these symptoms may clue the astute emergency triage person to a hazardous materials emergency. If the patient has not been properly decontaminated, then the hospital staff is at risk for health effects from secondary contamination.

Walk-in contaminated patients are the biggest challenge to any emergency department. This type of incident accentuates the need for preplanning to identify rooms or areas into which contaminated patients can be taken to minimize the effects of secondary contamination.

Table 8-1 is a brief list of possible chemical injuries and causes that have previously been discussed within the text and are fairly common within any community. These are possible "red flag alerts" that should be recognized when these signs and symptoms appear. By no means is this a complete listing, however, it may cause enough suspicion by the health care giver to redirect questioning concerning the possible causes of the displayed symptoms.

Receiving the Call

When a call informing the emergency department of a chemical injury or contamination is received, important information must be gained by the call taker.

Table 8-1   *Selected chemical injuries, their symptoms, and possible causes.*

Sign/Symptom	Possible Cause	Chemical Possibility
Bronchial constriction	Vapor inhalation	Water soluble irritants
Coagulation necrosis	pH 3 or lower	Acids
Edema of the upper airway	H_2O soluble chemicals	Aldehydes
Hypotension	Vasodilation	Nitrates
Laryngeal edema/spasm	H_2O soluble chemicals	Acids/bases
Liquefaction necrosis	pH 9 or higher	Bases
SLUD	Cholinesterase inhibitors	Organophosphates/insecticides
Fasciculations	CNS stimulator	Solvents

First, if available, information must be gained about the type or name of the chemical involved. Next, the caller should be asked to describe in detail the signs and symptoms displayed by the patient. Other important information is the type of exposure, the amount of the exposure, and the amount of time the patient was exposed to the chemical. Finally, determine if the patient was decontaminated and to what degree the decontamination was done prior to transport. If the patient was not decontaminated but has not yet been transported, inform the transporting agency to provide some decontamination on the scene. (By simply removing the patient's clothing the majority of the contaminant can be removed. Remember clothing covers up to 80% of the body surface.)

The hospital should direct the transporting agency to the area of the facility where decontamination will take place. In some situations this will not be the usual emergency room, but rather an isolation room outside the emergency department or in some cases outside of the hospital itself (morgues are a consideration).

After the call is taken, the hazardous materials response plan must be enacted. Other departments, such as security and maintenance, should be mobilized. The nursing supervisor can coordinate the participation of other departments and act as the incident supervisor. Isolation of the incoming patients is the initial primary objective. Secondary exposure from a contaminated patient is a real possibility. Organizing an area for patient receiving (already identified in the preplan) should be the initial task for staff personnel.

The Patient's Arrival

When the victim of a hazardous materials incident arrives at the hospital by ambulance it should be directed to a previously determined location. A representative of the emergency department, either a nurse or physician, should meet the ambulance and determine if the patient is prepared to enter the department. If decontamination is needed, then hospital staff in proper PPE should provide a brief assessment of the patient's medical condition so that arrangements can be made to provide the needed care once decontamination is finished. If needed, airway, breathing, or circulatory support must be provided simultaneously with decontamination. During the initial assessment, gross decontamination (gross decon is the removal of all clothing and jewelry) can be performed if not previously done. Clothing, jewelry, or other articles (gold, silver, and leather are products that absorb chemicals readily) that may carry additional contaminates into the hospital must be removed before the patient is taken inside. All personal articles should be bagged and marked with the patient's name.

The ideal situation involves the arrival of a patient who was decontaminated on the scene. Although field decontamination can be very thorough, the emergency department *must ensure* that the patient is clean. If not sure, the hospital staff must provide a second decontamination process before taking the patient into a clean area of the hospital.

! SAFETY

**If decontamina-
tion is not done in the
field and a patient is
transported by unpro-
tected caregivers, then
the caregivers will also
need to be assessed
and possibly decontam-
inated.**

If decontamination was not done in the field and the patient was transported by unprotected caregivers, then the caregivers also need to be assessed and possibly decontaminated.

Providing a Decontamination Area

A decontamination area must be previously identified (Fig. 8-1). The area can be outside or inside but must provide several basic services. Privacy is important because all of the victim's clothing, if not already done, will be removed and placed into contamination bags. Warm running water is also necessary because the patient will be bathed from head to toe. Personal protective equipment must be provided to the decontamination staff in order to safely care for the patient. Finally, the decon area must be equipped with a means of containing all of the decontamination runoff. Runoff may become a hazardous material mandating its disposal as hazardous waste (depending on the chemical and the concentration. (In these situations the local Department of Environmental Protection should be consulted for proper disposal.) Areas outside of the hospital may work well for this purpose but are at the mercy of inclement weather that may limit the effectiveness of decontamination.

If the area chosen is inside the building, all heating, ventilation, and air conditioning must be shut off to limit the spread of any airborne materials. The inside area must also be near the outside doorway. Moving a contaminated patient through an occupied area can cause contamination to many other victims. The identification of this area is of utmost importance. Some hospitals have structured their emergency department to have the first room upon entrance assigned as the

Figure 8-1 *The hospital must identify a decontamination area where patients can be taken. The transporting unit should know this area or be advised of it while enroute.*

hazmat patient room, whereas others have dealt with the problems of isolation, heating and ventilation, and containment of the runoff by using the morgue as a receiving area. (Morgues have their own heating, ventilation, and runoff containment systems. They are usually isolated from the hospital and have a separate entrance. The drawback to using this location is that medical support is hard to mobilize to the morgue.) Yet other hospitals have built elaborate hazmat treatment rooms and support facilities.

All nonessential equipment must be removed from the decon and treatment room. Adding floor and wall covering provides a protective barrier and allows a more effective cleanup after the incident. Plastic sheeting, 6 millimeters thick, can be clipped or taped to walls and floors for this purpose. The entry way into the room should be covered with a layer of plastic along with all door knobs and light switches that would be used under normal conditions.

Security personnel should be in position to direct the arriving ambulance to the entrance closest to the decontamination area. They should also guard the area of the decon room to ensure that nonessential personnel remain in safe areas and do not wander into potentially contaminated regions.

The area chosen must be large enough to facilitate the decontamination of more than one victim. The room should also be marked as the HAZARD area with either banner tape or barricades so that unprotected personnel do not enter. All personnel working within this area must have the level of protection warranted for the chemical contamination being dealt with. A transition area (warm zone) should be marked and identified for use as a tool and equipment pass area. Only staff wearing personal protective clothing should be physically inside the warm zone.

Patient Decontamination

Decontamination has many purposes. First it limits the amount of chemical entering the patient's body. Removal of the contaminant from the patient's skin ultimately reduces the amount of chemical that gains access. Decontamination also lessens the chance of the contaminant affecting others near the patient. Secondary contamination to health caregivers and support personnel is a real hazard. By cleaning the patient and disposing of the contaminated articles, this hazard can be greatly reduced or eliminated.

● CAUTION

If the skin is too briskly washed, then the top protective layer, the corneum stratum, will be disrupted, allowing chemicals to enter at a faster and more efficient rate.

In most cases the decontamination solution of choice is water and a mild detergent (a pure type of soap is suggested to lessen the possibility of reactions occurring with some chemical contamination). The patient's skin should be gently washed, exercising care that the runoff is not contaminating other vulnerable parts of the body such as the eyes, nose, and mouth; underarms; open wounds; or groin area. These areas must be cleansed well because of their high absorbency. If the skin is too briskly washed, then the top protective layer, the stratum corneum, will be disrupted, allowing chemicals to enter at a faster and more efficient rate. Once the decontamination team can be reasonably sure that the patient is clean, then the patient can be moved to a clean room for definitive treatment.

Personal Protective Equipment (PPE) for Hospital Staff

The U.S. Environmental Protection Agency has identified four levels of protection for hazardous materials incidents. The levels of protection are A,B,C, and D. Hospital emergency workers, medical directors, and administrators must be familiar with these terms that are used by emergency responders. A full description of these levels is found in Chapter 2.

The emergency staff faced with triage, decontamination, or treatment of a contaminated patient must be familiar with the use and availability of PPE. If training in the specialized use of PPE is not conducted or is not available, then the hospital may have to rely on local hazardous materials teams to provide the decontamination at the hospital when a patient arrives.

Protective equipment is not foreign to hospital workers. It is used every day to protect workers from infections as a result of contamination or exposures to viruses and bacteria. Therefore, the thought of training hospital staff in the use of PPE for chemical contamination is not far fetched. In fact an investigation of areas within the hospital such as the engineering department, waste management, surgical department, and supply room will likely reveal that a wide variety of equipment is already available. For example, radiology departments in most hospitals have chemical-impervious gloves along with aprons. This equipment could be used with certain chemicals and provides the adequate level of protection. The emergency room staff may be faced with compiling the equipment so that a widespread search of the hospital does not take place once the patient arrives at the hospital. Training, equipment, and written policies are all necessary to adequately and safely handle the care and treatment of a contaminated patient.

THE ROLE OF POISON CONTROL CENTERS IN HAZARDOUS MATERIALS EMERGENCIES

When emergency responders think of poison control centers they usually imagine a patient suffering from an overdose of medicine or displaying symptoms caused by some household cleaner. These accidents are the specialty of poison control centers but not the only specialty. The regional poison control center is only part of an extended network of agencies and individuals with a base knowledge and expertise to provide valuable services to hospital emergency departments, EMS providers, manufacturers and transporters of hazardous materials, and hazardous materials emergency responders.

History

Poison control centers started in Chicago in 1953. The idea and usefulness of providing toxicologic information to the public grew, stimulating interest in many other hospitals throughout the United States. By the late 1960s more than 600 community poison control centers were on record. These centers were usually based within a local hospital emergency room and were staffed by personnel performing other services within the area until a call came in. All had varying degrees of expertise, some lacking much of the needed medical or toxicologic expertise as many of these centers were staffed by individuals working on a voluntary basis. These centers were haphazardly placed, some existing just down the street from others, serving virtually the same areas. Other large geographic areas had no access to any centers whatsoever.

Regionalization of the centers became a big issue and late in the 1970s the American Association of Poison Control Centers (AAPCC) established criteria and guidelines for regional poison control centers (RPCC). By the early 1980s, the goal of organizing regional centers was a reality. These new centers had to meet a minimum criteria to remain in the network. The American Academy of Clinical Toxicology and the American College of Emergency Physicians supported the AAPCC's efforts to set and improve the criteria of poison control centers. Today 35 AAPCC certified regional poison control centers exist but the need is estimated at approximately 70 strategically placed RPCCs to serve the United States.

Poison Control Centers Today

Poison control centers have always been thought of as providers of information when an accidental poisoning takes place. Mothers know that they can call poison control for information if their child accidentally ingests a substance not meant to be consumed. These are the traditional roles of the centers, but the role is expanding. Concerns for environmental safety and workplace exposures have caused poison control centers to expand their roles, providing services useful for those companies manufacturing and transporting hazardous chemicals.

During hazardous materials incidents, a regional poison control center can be contacted and given information about the chemical involved. The center can then reference the material and return information concerning health effects, decontamination, and recommendations for treatment. In some cases referrals can be made to other resources concerning containment, cleanup, and disposal.

The resource goes beyond receiving information in the field. The centers can then fax hard copy information to the local hospitals, news media, or hazardous materials team. In many cases, getting out true, factual information lessens the confusion, fear, anxiety, and anger experienced by the public. The information gained by the receiving hospitals can alert the staff to possible treatments for exposed victims or to potential medical problems.

In many areas of the country, poison control centers play an intricate part of the team efforts during mitigation of an incident. These specialized poison control centers have gained additional knowledge and resources out of necessity and circumstance. Those, located in industrial areas, major ports, or highway transmission areas, have become more involved in hazardous materials emergency response because of the increased need for information.

These specialized poison control centers must meet additional requirements including:

- Having full-time specialists trained and proficient in retrieval, analysis, and communication of medical, toxicologic, and chemical information.
- Being accessible for information assistance on a 24 hours a day, 7 days a week basis to public safety and medical personnel. They must be similarly available to the public during incidents.
- Having resources organized to permit rapid access to technical and clinical information dealing with hazardous materials.
- Being located in a medical facility with ready access to consultants from various medical, laboratory, industrial, and occupational health-related clinical, and scientific specialties.
- Having procedures and capabilities for record keeping and review.
- Having interest and capability to provide training to medical and other health specialists on matters related to toxicology and hazardous substances.

Although the number of poison control centers has decreased since the 1970s, their expertise and role has expanded greatly. No longer should these centers be viewed just for information on childhood poisonings but also for the whole realm of information they can provide to hazardous materials emergency responders. The local poison control center's phone number should be posted, next to CHEMTREC's number, in all units that respond to hazardous materials incidents.

Certified Poison Control Centers

These poison control centers, as of 1995, are certified by the American Association of Poison Control Centers.

Poison Center	*Emergency Phone*
Alabama	
Regional Poison Control Center	800-292-6678 or 205-933-4050
Arizona	
Arizona Poison and Drug Information Center	800-362-0101 or 602-626-6016
Samaritan Regional Poison Center	602-253-3334
California	
Fresno Regional Poison Control Center	800-346-5922 or 209-445-1222
San Diego Regional Poison Control Center	800-876-4766 or 619-543-6000
San Francisco Bay Area Regional Poison Control Center	800-523-2222 or 415-476-6600
Santa Clara Valley Regional Poison Center	800-662-9886 or 408-885-6000

University of California, Davis, Medical
   Center Regional Poison Control Center   800-342-9293 or 916-734-3692

Colorado
Rocky Mountain Poison and Drug Center   303-629-1123

District of Columbia
National Capital Poison Center   202-625-3333 or 202-362-8563

Florida
The Florida Poison Information and
   Toxicology Resource Center, Tampa
   General Hosp.   800-282-3171 or 813-253-4444

Georgia
Georgia Poison Center   800-282-5846 or 404-616-9000

Indiana
Indiana Poison Center   800-382-9097 or 317-929-2323

Maryland
Maryland Poison Center   800-492-2414 or 410-528-7701
National Capital Poison Center   202-625-3333 or 202-362-8563

Massachusetts
Massachusetts Poison Control System   800-682-9211 or 617-232-2120

Michigan
Poison Control Center   313-745-5711

Minnesota
Hennepin Regional Poison Center   612-347-3141
Minnesota Regional Poison Center   612-221-2113

Missouri
Cardinal Glennon Children's Hospital
   Regional Poison Center   800-366-8888 or 314-772-5200

Montana
Rocky Mountain Poison and Drug Center   303-629-1123

Nebraska
The Poison Center   800-955-9119 or 402-390-5555

New Jersey
New Jersey Poison Information and
   Education Center   800-962-1253

New Mexico

New Mexico Poison and Drug Information Center	800-432-6866 or 505-843-2551

New York

Hudson Valley Poison Center	800-336-6997 or 914-353-1000
Long Island Regional Poison Control Center	516-542-2323 or 542-2324(25)
New York City Poison Control Center	212-340-4494 or 212-P-O-I-S-O-N

Ohio

Central Ohio Poison Center	800-682-7625 or 614-228-1323
Cincinnati Drug and Poison Information Center and Regional Poison Control System	800-872-5111 or 513-558-5111

Oregon

Oregon Poison Center	800-452-7165 or 503-494-8968

Pennsylvania

Central Pennsylvania Poison Center	800-521-6110
Poison Control Center serving the greater Philadelphia metropolitan area	215-386-2100
Pittsburgh Poison Center	412-681-6669

Rhode Island

Rhode Island Poison Center	401-277-5727

Texas

North Texas Poison Center	800-441-0040 or 214-590-5000
Texas State Poison Center	409-765-1420 or 713-654-1701

Utah

Utah Poison Control Center	800-456-7707 or 801-581-2151

Virginia

Blue Ridge Poison Center	800-451-1428 or 804-924-5543
National Capital Poison Center	202-625-3333 or 202-362-8563

West Virginia

West Virginia Poison Center	800-642-3625 or 304-348-4211

Wyoming

The Poison Center	800-955-9119 or 402-390-5555

Summary

It is truly unfortunate that hospitals generally have been left out of the hazardous materials loop. The misconceptions that assume hospitals have the "knowledge" about such events are frightening. It is the opinion of these authors that each medical institution should contact its local hazardous materials team and train with them.

The first step for all hospital health-care providers is to enroll in an 8-hour awareness level class and consider training through the operations level. Much of what is taught within these classes does not directly correspond to hospital haz-mat however, understanding what is done in the field creates a greater appreciation for the local team and emergency responders.

Many of the principles of the awareness and operations level can be adapted to the hospital event. Hospital employees attending these classes must do so with open minds and a willingness to learn. Education is the key to a successful operation, whether in the field or within the hospital. As in medicine, a great deal of information must be learned as groundwork upon which to build a strong foundation. Once the foundation is solid with factual information, practical application can be learned and better understood.

The emergency department staff and support personnel must be trained in the techniques and considerations of decontamination, reference, treatment, incident command, and incident termination. These are courses traditionally taught to field personnel at the technician level. For the hospital to perform optimally once presented with a hazmat scenario, this information must be digested and applied.

In addition to the didactic information, practical skills must be practiced, tested, and reevaluated to maintain competency. What sounds easy in a book or class presentations may be very difficult in reality. Understanding the equipment, how to use it, and its limitations are extremely important. In the hazardous materials environment, it is even more important because you may not get a second chance. One major mistake with a toxin could lead to a permanent injury or even death.

Hospital personnel have a hard time admitting that they can not handle all incidents. For the most part, nurses and doctors are programmed to help those who are in need. This type of incident must have a slow, deliberate, predetermined, and thought-out goal prior to laying on any hands.

Scenario

You have been dispatched to the trauma center in your community at the request of the hazardous materials unit on the scene of an overturned tanker truck. Because you work in the hospital emergency room on your off days and are familiar with their routine and are also familiar with hazmat response, the team wants you to coordinate efforts with hospital staff in the preparation of receiving hazmat patients. The truck is located on an interstate highway mostly used by weekenders returning from a day at the beach. The information received from dispatch indicates that the tanker, carrying 4,500 gallons of high concentrate parathion to a local flying service where it was to be diluted and spread over crops, rolled over and ruptured, spilling the contents over the road. Many vehicles traveling home from the beach were held up in the spill for approximately 20 minutes before being directed by the local police to drive on through to reach their destination. The concerns of the hazmat team is that many people received a significant exposure and will probably report to the hospital once symptoms surface. Furthermore, the team was also treating two officers who became very ill after apparently standing in the spill while they directed traffic. It is Saturday, June 5th, the temperature is 85°F with a humidity of 78%.

When you arrive at the hospital, you notice a group of about fifteen individuals being held outside of the emergency room entrance by two nurses dressed in gowns and particulate masks. All of the 15 people are ambulatory but are displaying signs of organophosphate poisoning (excessive lacrimation and salivation). Most are dressed in swimming suits and tee shirts.

Scenario Questions

1. Is a decontamination system necessary for these patients?

2. If so where would the best place be to set one up? What equipment is needed?

3. Is the capture of runoff important for this emergency decontamination?

4. Will accessing the hospital's hazardous materials contingency plan give us all of the information needed to adequately decontaminate these people?

5. Knowing the possible routes of entry for the chemical may be important. What are some of the means available in the hospital to get this information?

6. Is a command structure needed to handle this hazmat event at a hospital?

7. What are some of the other concerns for the emergency room when contaminated patients arrive there?

8. Is the protective equipment used by the initial two nurses adequate? What should be added or deleted?

9. Utilizing the in-house resources, where can more advanced personal protective equipment come from?

10. Once these patients are through a washing process does this mean that they are free of contamination? Explain?

11. Is your hospital prepared to handle an incident of this magnitude?

12. Does your hospital have written SOPs and a plan to activate if contaminated patients arrive?

13. Does your hospital train with the hazardous materials response team so that an operation of this type is controlled and mitigated without total confusion or chaos?

section

5

SELECTED TOPICS

During hazardous materials response, there is always the unusual incident or an exception to the rule. If the public does not know what to do, they do know that, whatever the job, it will get done by emergency responders. But at times, even this inventive group can be stumped. Only their training and resourcefulness carry them through and allow them to complete the task. Hazardous materials events have a high degree of difficulty. Emergency responders are often bewildered by the changing events and unexpected outcomes. The knowledge presented in this section will assist the responder in controlling some of the unexpected incidents at hazmat events other than at a highway spill or in a manufacturing facility.

Selected Topics provides emergency responders with knowledge about bio-hazards, the dangers of clandestine drug laboratories, air monitoring, and confined space medical operations that, when combined with common sense, will allow them to complete their tasks safely.

These subjects may not be applicable to all areas of the country as they may be adequately covered under other training topics. They are however, areas that need special attention if they are a concern within the student's response district or geographical area. The subjects were placed within this section so that the student or instructor could choose those which will be of the most benefit within their particular geographic or service area.

chapter
9

Biohazard Awareness, Prevention, and Protection

Objectives

Given an incident involving biohazardous materials, the student should be able to identify unique hazards involved including the means of entry, virulence, infectiousness, invasiveness, and pathogenicity. The student should also be familiar with the regulations involving biohazardous waste, the Ryan White Act, and exposure reporting.

As a hazardous materials medical technician, you should be able to:

■ Identify the difference between bacteria and a virus.

■ Define biohazard.

■ Define infection control.

■ Identify the means of entry to include the:

Virulence	Infectiousness
Invasiveness	Pathogenicity

373

- Identify the key points to the Ryan White Act.
- Identify the steps to clean up of a biohazard site.
- Formulate a high level, intermediate level, and a low level disinfectant and define the use of each level.

During the winter of 1990 in northern Virginia, just outside of Washington, D.C., a virus started killing monkeys in a biological research center. Veterinarians examining the animals diagnosed the disease as simian hemorrhagic fever. Unable to cure the monkeys, the test animals continued to die. Further examination uncovered the true culprit causing the death of these monkeys—Ebola virus.

Ebola virus was first documented in 1976 in an area around the Ebola River in Zaire, Africa. There a group of village residents died an excruciating death from diffuse blood clots and massive hemorrhaging. The scientists investigating that site referred to the signs and symptoms as a "biological meltdown." The cause of death was a new virus named Ebola Zaire.

Just prior to the outbreak of disease at the Reston, Virginia, research center, a shipment of monkeys from the Philippines had been received. It is believed that these monkeys carried the virus into the research center where they infected the other animals. It was unknown where the animals contracted the virus.

The fear of human infection within the facility peaked when one of the workers started having flulike symptoms. Further tests proved that he was not suffering from Ebola virus. All of the workers at the facility were tested for the Ebola virus and several were found to be positive, although none ever experienced symptoms. It became apparent that the Ebola virus in Reston was a different strain than the Ebola Zaire and did not affect humans. This strain was later named Ebola Reston.

An Army biohazard response unit was called to decontaminate the research building. The team euthanized approximately 500 monkeys and later flooded the building with formaldehyde gas to kill any remaining virus. The building to this day remains closed.

DEFINITIONS

biohazards
communicable diseases that are a threat to the life and health of emergency response personnel unless precautions and protective measures are exercised

Biohazards are communicable diseases that are a threat to the life and health of emergency response personnel unless precautions and protective measures are exercised. In the 1990s the term *biohazard* became popular to identify those organisms that invade and infect a host's body. The risk of communicable diseases cannot be seen. However, understanding communicable diseases will help emergency response personnel know how they can better protect themselves and how infectious diseases are transmitted during patient care and biohazard cleanup.

In 1994, Dr. David Satcher, the director of the U.S. Centers for Disease Control (CDC), stated that because of the spreading genetic resistance, "our antimicrobial drugs have become less effective against many infectious agents, and experts in infectious diseases are concerned about the possibility of a 'post antibiotic era.' At the same time, our ability to detect, contain, and prevent emerging infectious diseases is in jeopardy." (Satcher, pers. Communic.) For many reasons biohazards are the hazardous material of the future. Each year new biological agents or new strains of old biological microorganisms are affecting the population. For example, in 1976 a common soil bacterium, later named *Legionella pneumophila*, found its way into an air-conditioning system in a Philadelphia hotel. The result was an airborne lung disease that killed several American Legion members attending a conference there. AIDS started in the United States in the late seventies and is now responsible for the deaths of more than two million people worldwide. Drug-resistant tuberculosis (TB) is on the increase and is a real threat to emergency medical responders. This particular bacteria is spread through the air and is especially dangerous in highly populated areas. During the nineteenth century TB killed one forth of the European population. The Ebola virus that attacks humans has not yet found its way into the United States but it appears to be only a matter of time. It was recently found, not in a remote jungle village, but in the booming metropolis of Kikwit, Zaire, with a population of 600,000. These are just a few examples of what an emergency medical provider is faced with today. The future does not look better.

Our environment is filled with tiny microscopic living things called microorganisms. The living organisms that affect the providers of emergency care are divided into two categories: viruses and bacteria. *Bacteria* are single-celled microorganisms that are plantlike but lack chlorophyll. *Viruses* are smaller than most bacteria and live on or within other cells. A viral disease causes destruction of the host cells as a result of the parasitic action of the virus.

These microorganisms develop a relationship in humans that can be beneficial or detrimental to our health. The host–parasite relationship, as it is sometimes called, works in a beneficial way in the digestive system, where the bacteria, *E. coli*, work to break down food for digestion within the intestines. Sometimes the same microorganisms that are normally beneficial cause disease.

Infection is the result of either bacteria or virus gaining access into the body and multiplying. This invasion overwhelms or bypasses the defensive barriers always present in a healthy host. Infections are caused by primary or opportunistic pathogens.

Primary pathogens can, unaided by other opportunities, invade a healthy body, bypass defense mechanisms, and establish an infection. They are aggressive and can enter through many different routes. Measles and gonorrhea are examples of primary pathogens.

Opportunistic pathogens are usually unable to penetrate the defense mechanisms found in a healthy individual. Unlike the primary pathogens, opportunistic pathogens cause infection or disease by taking advantage of broken defense mechanisms. This breakdown can be as simple as an open wound or as complex as a compromised immune system. The breakdown of defense mechanisms occurs during injury, preestablished illness, drug and alcohol abuse, old age, and so forth. A staph infection is an example of an opportunistic infection. Staph is commonly found in our everyday environment and does not normally affect us adversely. However, staph often infects those with a compromised defense mechanism. Vulnerable populations such as those in prisons, nursing homes, homeless shelters, and day care centers, provide the source of diseases for everyone else. These areas are the same ones that have daily visits from emergency medical responders.

The Need for Infection Control

The need for strict infection control has increased as the risk of contracting a communicable disease has increased. Before 1980 there was not much concern about emergency response personnel contracting a communicable disease. The concern for personal protection was stimulated from a significant rise in hepatitis B virus (HBV) and human immunodeficiency virus (HIV) causing an increased risk to emergency response personnel. The number of HBV cases has increased annually along with the number of HIV cases. Although, HIV provokes a more fearful response, HBV poses a greater risk to emergency response personnel. According to the CDC, there is also a rise in several other diseases in which a risk to exposure is significant. Exposure to diseases like chicken pox, herpes viruses, rubella, diarrhea, hepatitis A, hepatitis C, influenza, measles, meningitis, mumps, and tuberculosis have all been documented. Diseases can be transmitted much like toxic hazardous materials as described in Chapter 5. With the changing needs of the health-care environment due to communicable diseases, the area of biohazardous cleanup may shift toward the hazardous material teams. There are many documented cases of a hazardous materials team being called to clean up after particularly bloody scenes. The trend may move into patient care where the establishment of hot, warm, and cold zones may be used to ensure strict infection control against contagious diseases.

LEGAL ISSUES OF BIOHAZARDS

Regulation of Biohazards

The requirement for infection control by OSHA was first voluntary in 1983 with a set of guidelines designed to reduce the risk of occupational exposure. In 1987, OSHA published notice for the proposed rules on Bloodborne Pathogens Standards. With numerous hearings and testimonies from expert witnesses, physicians, and other health-care workers, the publishing of the final rule on Bloodborne Pathogens Standard in the *Federal Register* was published on December 6, 1991. The requirement for infection control is mandated by OSHA for private entities. Government agencies, whether local or state, are not required to adopt the OSHA plan but must have a plan in effect with guidelines as stringent as OSHA's. Emergency response personnel who work for paid and career agencies of local or state governments must be covered under a plan, as the risk factors are the same for every emergency response person who comes in contact with communicable diseases.

OSHA and NFPA 1581 procedures are focused on "viruses" and "bacteria" as communicable diseases and how they are transmitted, even though OSHA and NFPA 1581 have set standards as to what equipment and procedures should be used to reduce contamination. Occupational exposure to bloodborne pathogens (29 CFR Part 1910.1030) establishes standards of protection listing the primary methods of protection as training, engineering and workplace controls, immunization against HBV, and the use of universal precautions.

universal precautions

measures taken when approaching the treatment of all patients as if they are infected with a contagious disease

Universal precautions is the treatment of all patients as if they were infected with a contagious disease. In some cases, this may be the use of gloves alone, other times a particulate mask, gloves, eye protection, and a gown may be indicated. In every case, the use of precautions helps emergency caregivers protect themselves in case it is learned later that the patient is contagious.

Probably the most common method of universal protection of emergency response personnel is wearing gloves and washing hands, which is the minimal level of protection. To reduce the risk of exposure to emergency responders, it is necessary to implement a different approach to patient care. In hazardous material incidents, different levels of entry and zones are established in which the level of protection and the number of personnel who make contact with the hazard are limited. Each hazardous material incident requires different levels of protection from level D to level A. For instance, in the hot zone of a hazardous material incident, the number of personnel and level of protection is established according to the nature of the incident. The same should be established regarding patient care or biohazard cleanup, realizing that some incidents will require more personnel in order to do a proper job. Therefore some persons should be used as support only, handling the equipment necessary, from the warm zone into the hot zone without making direct contact with the emergency responder. Any personnel not required to do primary patient care or biohazard clean up would remain in the cold zone.

NFPA 1999

The NFPA has developed a standard for protective clothing and has established guidelines for the manufacturing and use of medical protective equipment. Briefly outlined, the standard is as follows:

The standard NFPA 1999, *Protective Clothing for Emergency Medical Operations*, is mainly concerned with penetration by viruses. NFPA 1999 was established and published in August 1993 as a new performance standard for protective clothing for EMS and fire service personnel. The standard covers watertight integrity, physical dimensions, tensile strength, elongation and modulus, puncture resistance, small parts dexterity test, and bacteriophage penetration resistance test. From these standards medical protective equipment are designed for use in biological hazardous environments.

The Ryan White Act

The Ryan White Comprehensive AIDS Resources Emergency Act of 1990 sets requirements for notification of emergency medical responders by the hospital. The act allows emergency workers to find out if they have been exposed to a patient who is presently diagnosed with a specific communicable disease. The diseases outlined in the act are hepatitis B, non A/non B hepatitis, pulmonary tuberculosis, bacterial meningococcal meningitis, rubella (German measles), and HIV.

The employer of the emergency response personnel is required to name a designated officer to facilitate communication between the emergency response personnel and the receiving hospital. The notification takes place either by routine or by request of the designated officer.

Routine notification is done by the receiving facility to the designated officer any time the emergency response crew transports a patient who has an airborne communicable disease. This notification must take place within 48 hours of the diagnosis and includes any patients who expire enroute or shortly after arrival at the hospital.

Request for notification can be made by any members who attended to the patient on the scene or any of the transporting personnel where exposure to an infectious disease may have occurred. The request is made through the designated officer to the receiving facility.

Once a request is made, the designated officer reviews the case and notifies the receiving facility. The receiving facility then reviews the case and notifies the designated officer of one of three results.

1. An exposure did take place.
2. An exposure did not take place.
3. There is insufficient information to determine if an exposure took place.

Unfortunately, the Ryan White Act does not authorize the mandatory testing for communicable diseases/infections of patients suspected of causing a significant exposure. Any additional testing (testing other than medically necessary for diagnosing the present complaint) of patients may be done only after an agreement is reached between the patient and the doctor and then done at the requesting agency's (or person's) expense.

MEANS OF ENTRY

Reducing the Transmission of Communicable Disease

Communicable diseases are transmitted by direct and indirect routes. Both forms of transmission involve either viruses or bacteria, which must be in a sufficient quantity. Blood-to-blood is the most direct method of contracting either viruses or bacteria. Other means of transmission are extremely efficient, such as blood to mucous membrane. For example, the exposure of a patient's blood or other body fluids into the mouth or eyes is a danger that emergency responders must be cognizant of and protect themselves from.

Viruses are transmitted in a host, such as blood or body fluids (referred to as *bloodborne* or *airborne*) and cannot multiply outside of a living cell. Once the virus enters the body, it begins to reproduce, whereas bacteria are transmitted outside of a host on objects such as equipment and can multiply outside of the body, not needing a host to multiply or reproduce. Some pathogens can enter and affect the body only through open wounds, others can gain access if inhaled, and others only if swallowed or exposed to other mucous membranes. Emergency medical care providers should understand the different routes of entry so that protective measures can be taken to prevent the transmission of disease.

Virulence

Virulence refers to the amount of a particular microorganism that causes an effect in the host (the toxicity of the pathogen). The smaller the dose (fewer microorganisms) needed to infect a patient, the more virulent the organism is. Virulence is based on three effects: infectiousness, invasiveness, and pathogenicity. *Infectiousness* is the ability, once a pathogen gains access into the body, to initiate and maintain an infection. *Invasiveness* is the ability of the pathogen to progress further into the body once an infection is established. *Pathogenicity* is the ability of the pathogen to injure the body once the infection is established. Because of the extended use of antibiotic medicines and the ability of microorganisms to change their resistance to them, the virulence of many of the more common infections has become much greater.

Prehospital emergency care providers are at a higher risk of becoming infected with pathogens simply because of the uncontrolled environment in which they work. Emergency medical providers often work on scenes that have poor lighting, are grossly contaminated with blood and other body fluids, and may be extremely unsanitary. Many times these providers meet patients who are uncooperative, unruly, or violent. Bloodborne and airborne pathogens pose the highest risk in these environments.

When concerns about communicable diseases are discussed, usually human immunodeficiency virus (HIV) or hepatitis B virus (HBV) are the topics. Unfortunately airborne diseases are also on the rise today. Tuberculosis bacterium (TB), a disease that was on the decline for years, is now becoming a danger of epidemic proportions. In 1990, 25,500 new cases of TB were reported in the United States and since 1985 there has been a marked increase in the rate of new cases of TB, reversing a previously noted 30-year downward trend (Popular Science, 1996). Prehospital care providers must learn to protect themselves against the invasion of these diseases.

Concerns about tuberculosis have escalated to the point that regulations governing the use of filter masks have been enacted to protect medical caregivers. These regulations, OSHA 29 CFR 1910.134 and 42 CFR part 84 subpart K both set guidelines for the design and use of filtered air masks with the ability to filter particles small enough to prevent the inhalation of tuberculosis bacterium. In 1994 the CDC set guidelines addressing respirators for occupational exposure to TB and specified standard performance criteria for respirators upon exposure to TB. The minimal acceptable respiratory protection meeting the criteria is the HEPA (high efficiency particulate air) filters (Fig. 9-1). These criteria include:

- The ability to filter particles 1 micrometer in size in the unloaded state with a filter efficiency of >95% (i.e., filter leakage of <5%) given flow rates of up to 50 liters per minute.

- The ability to be qualitatively or quantitatively fit tested in a reliable way to obtain a face seal leakage of <10%.

- The ability to fit the different facial sizes and characteristics of health-care workers, which can usually be met by making the respirators available in at least three sizes.

- The ability to be checked for a face piece fit, in accordance with OSHA standards and good industrial hygiene practice, by health-care workers each time they put on their respirator (Fig. 9-2).

Respiratory protection (HEPA or respirators certified under 42 CFR part 84 subpart K) for employees exposed to TB is required under the following circumstances:

- When workers enter rooms housing individuals with suspected or confirmed infectious TB.

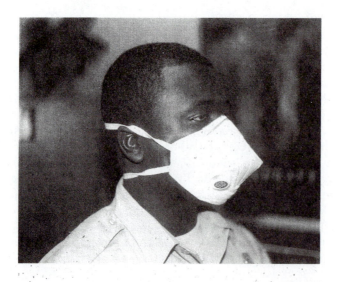

Figure 9-1 *A properly fit HEPA mask will protect the wearer from tuberculosis bacterium or other particles as small as 0.2 microns.*

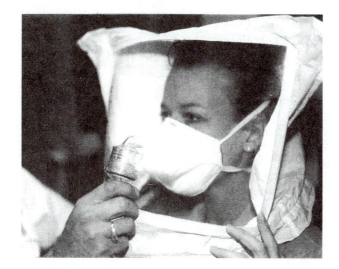

Figure 9-2 *To test for a proper fit, the wearer is placed in a hood and an irritant is infused. The wearer is then asked to speak and move his or her head in several directions.*

- When workers are present during the performance of high-hazard procedures on individuals who have suspected or confirmed infectious TB.
- When emergency medical response personnel or others transport, in a closed vehicle, an individual with suspected or confirmed infectious TB.

Documentation of proper fit testing must be completed on all wearers (Fig. 9-3). The fit test involves the use of an irritant smoke or saccharin (other odorants are also used). Two media for testing are needed in the event that the person being tested is not able to smell one of the testing agents.

● **CAUTION**

Never use any strong cleaning agents (including full-strength household bleach or chlorine bleach products) with the intent of killing bacteria or viruses on skin thought to be contaminated, because such chemicals damage the first layer of epidermis and worsen future contamination.

EXPOSURE TO COMMUNICABLE DISEASE

If exposed to communicable disease while cleaning up biohazard waste, remove all contaminated clothing and personal protective equipment and clean the exposed area. The area should be scrubbed with soapy water. If this type of cleaning is immediately impractical then use waterless antimicrobial cleaning agents or an alcohol product. Afterwards, and as soon practical, wash the area with soapy water. If the area exposed does not allow for scrubbing, such as a mucous membrane, then remove as much of the material as possible and irrigate with water repeatedly to remove the infectious material. This includes vigorous irrigation of the eyes if they are exposed to blood or body fluids.

Never use chlorine bleach products or any other strong cleaning agents on skin that may have been contaminated, believing that it is the best way to kill bacteria or viruses. These cleaning agents work well on porous and nonporous surfaces to kill bacteria and viruses, but the effects are different on the skin. Household bleach (sodium hypochlorite, $NaClO$) damages the first layer of epidermis and worsens future contamination. Damage to the epidermis, as a result of

HEPA FIT TEST REPORT

A.   EMPLOYEE: _____   DATE: _____

   EMPLOYEE ID: _____

   EMPLOYEE JOB TITLE/DESCRIPTION: _____

   _____

B.   EMPLOYER: Fire Department _____

   LOCATION/ADDRESS: Station # _____

C.   RESPIRATOR SELECTION: Half-mask, HEPA filtration _____

   MANUFACTURER: Health & Safety, Inc. _____

   NIOSH APPROVAL NUMBER: _____

   MODEL: HEPA _____

D.   FIT CHECKS:

   NEGATIVE PRESSURE   ☐ PASS   ☐ FAIL

   POSITIVE PRESSURE   ☐ PASS   ☐ FAIL

E.   FIT TESTING:

   ☐ QUANTITATIVE   ☐ SACCHARIN SOLUTION   ☐ IRRITANT SMOKE   ☐ ISOAMYLACETATE

   QUALITATIVE   QUALITATIVE   QUALITATIVE

   FIT   ☐ PASS   ☐ PASS   ☐ PASS

   FACTOR: _____   ☐ FAIL   ☐ FAIL   ☐ FAIL

   COMMENTS: _____

   _____

   _____

F.   EMPLOYEE ACKNOWLEDGEMENT OF TEST RESULTS:

   EMPLOYEE SIGNATURE: _____   DATE: _____

   TEST CONDUCTED BY: _____   DATE: _____

Figure 9-3 *HEPA Fit Test Report.*

significant exposure
direct blood-to-blood contamination through exposure by a contaminated needle stick or skin puncture by other sharp contaminated instruments

bleach or other strong cleaning agents, is evidenced by the liquefaction of fats in the skin layer (saponification), similar to the effects of an alkali burn. The damaged skin has a slick or soapy feel and allows bacterial contaminants to enter much more easily. There is no remarkable documentation of an exposure to intact skin causing a communicable disease, but it is reasonable to suspect that a bacteria or virus could gain easy access if the skin is injured in this way.

Significant Exposure   Normally defined as exposure by a contaminated needle stick or sharp puncture by other instruments in which blood-to-blood contamination takes place, a **significant exposure** may also be an exposure where blood or body fluids come in contact with an open wound, mucous membranes, or eyes. Each emergency health-care agency must have a written protocol to follow if an exposure takes place. OSHA's 29 CFR 1910.1030 Occupational Exposure to Bloodborne Pathogens, identifies rules that must be followed on a federal level. Even some state government agencies have produced documents regulating exposure protocol. If a significant exposure has occurred during cleanup or patient contact, the exposure must be documented and follow-up care identified.

On-the-Scene Biohazard Cleanup

Who is better equipped and trained to clean up scenes of biohazardous materials than the hazardous materials team? In many areas of the country the hazardous materials team responds to scenes where large amounts of blood and body fluids are spilled, to provide decontamination and ensure the area is safe from contamination. The risk to emergency responders or even hazardous materials teams involved in cleanup can be even greater than doing patient care, whether cleaning up a clinic that had a spill or a traumatic scene in which blood, body fluids, and/or tissue are present.

The highest level of protection must be adhered to. The scene should be handled similar to a hazardous materials scene where unknown toxins are present. The cleanup teams should use the following guidelines in making an assessment and handling a scene:

- Assess the potential risk present.
- Secure the scene or area from unprotected personnel.
- Identify the area or areas that are the highest risk.
- Wear the proper level of protection for the task at hand, such as double gloving, masks, gowns for areas when splashing or spattering of blood or body fluids is likely during cleanup.
- Wear gowns that are viral penetration resistant.
- Use a solution of household bleach 1:10 for porous surfaces or large visible areas of blood, body fluids and/or tissue and 1:100 for smooth surfaces for decontaminating spills of blood or body fluids. The solution must be mixed no more than 24 hours prior to use.
- Dispose of blood, body fluids, and tissue properly.
- Wash hands before and after any contact with equipment possibly contaminated with blood, body fluids, or tissue. Wash hands even after removal of gloves.
- Cover all open wounds of the skin, such as chaffing, abrasions, lacerations, and cuts before cleanup begins.
- Clean all visible blood, body fluids, or tissue from reusable equipment and place inside a container until decontamination using antimicrobial solutions can be accomplished.

With each incident, emergency response personnel risk possible exposure to communicable diseases. The risk during an incident is dangerous enough without the added risks of exposure during cleanup after the emergency care. The effects from biohazard exposure may be acute or chronic and can be carried by the emergency responder on his body to cause secondary contamination. Great care must be exercised to reduce initial and secondary contamination which can only be accomplished by following a regimented program of decontamination.

Biohazard Decontamination

Biohazard decontamination is the use of physical or chemical means to remove, inactivate, or destroy bloodborne pathogens on a surface or item to the point that it is no longer capable of transmitting infectious particles and the surface or item is rendered safe for handling, use, or disposal. Universal precautions must be observed throughout the decon process to ensure protection against contact with potentially infectious blood, body fluids, and tissue.

Some disinfecting solutions can cause deterioration of protective clothing. For example, isopropanol (isopropyl alcohol) causes elongation of latex gloves to the point of tearing and enhancing penetration of liquids through the gloves pores, which become larger. A heavier latex glove should be used when decontaminating scenes or equipment. A pair of heavy latex gloves, such as those used in commercial dish washing, are recommended over the top of a pair of exam latex gloves. Not all gloves are impervious to bacteria and/or viruses, so good hand washing should be done after any decontamination procedure.

Decontaminate all contaminated surfaces with an appropriate disinfectant. Disinfectant solutions such as bleach solutions should be mixed daily. The ratio should be 1:100 at a minimal and 1:10 as the strongest solution. Inspect and decontaminate all bins, pails, cans, and equipment that may have been contaminated. Make sure all broken glass and sharps are picked up using "mechanical means" such as a brush and dust pan, tongs, forceps. They should not be picked up directly with the hands.

Sharps and broken glass must be placed in containers that are closeable, puncture-resistant, leakproof, and red with a biohazard warning label. If a secondary container is placed inside the primary container, then it must meet all the same requirements as the primary container.

Place all potentially infectious material in a container that prevents leakage. The only exception to this is when placing clothing or other material that may be used for evidence into plastic bags. The plastic bag will deteriorate, evidenced by a process of putrefaction. Blood, body fluids, and/or tissue should first be soaked up with paper towels or a similar material.

Several commercial products on the market are used for cleaning up body fluids and/or blood. The product causes the body fluid or blood to gel into a solid material that can easily be picked up. To decontaminate large areas of blood and body fluids, a pump household sprayer can be used. Fill the sprayer with a 10%

solution of household bleach and hold the spray nozzle 6–8 inches away from surface and spray until thoroughly wet. Allow 10 minutes minimal contact time for the chlorine bleach to work.

Decontamination of biohazardous substances is done just like decontamination of a chemical agent, usually in one of two ways. First the substance is removed by washing it off of the surface. The second way is to neutralize the substance. In the case of biohazardous agents neutralization consists of killing the microbes that may be contagious.

Levels of Decontamination   Four different levels of decontamination are indicated for killing the microbial agents: sterilization and high-level, intermediate-level, and low-level disinfection. Only the last three apply to biohazard clean up in the field.

- *Sterilization.* Sterilization destroys all forms of microbial life including high numbers of bacterial spores. Sterilization can be accomplished by using steam under pressure (autoclave), gas (ethylene oxide), dry heat, or immersion in an EPA approved chemical sterilant for a prolonged period of time (usually 6–10 hours). It is not feasible for the field but is most often used in a controlled setting on equipment for surgery.
- *High-Level Disinfection.* High-level disinfection destroys all forms of microbial life except high numbers of bacterial spores. Used for reusable equipment that comes in contact with microbial organisms, this method uses hot water pasteurization (176°F–212°F) for 30 minutes, or exposure to an EPA-registered sterilant with an exposure time of 10–45 minutes or according to the manufacturer's label.
- *Intermediate-Level Disinfection.* Intermediate level disinfection destroys mycobacterial tuberculous, vegetative bacteria, most viruses, and most fungi, but does not kill bacterial spores. This method uses an EPA-registered hospital disinfectant or chemical germicide that contains a label indicating tuberculocidal. Hard surfaces may be cleaned with germicides or solutions of chlorine bleach (1:10 solution, approximately 1/2 cup of bleach per gallon of water). Intermediate disinfection is used for surfaces that come in contact with, or are visibly contaminated with, blood or body fluids. The surfaces must be precleaned of any visible material before using a disinfectant.
- *Low-Level Disinfection.* Low level disinfection destroys most bacteria, some viruses, some fungi, but not mycobacterial tuberculosis or bacterial spores. The method uses an EPA-registered hospital disinfectant containing a label not indicating tuberculocidal. Chlorine bleach (1:100 solution, 1/4 cup per gallon of water) can also be used. This excellent cleaner is used for routine cleaning where there is no visible contamination.

Interior or exterior surfaces, such as floors, woodwork, sidewalks, or roadways that have become contaminated or possibly contaminated with bacteria or viruses should be cleaned with an environmentally safe disinfectant or agent. In order to disinfect any area effectively the equipment first must be thoroughly cleaned. Equipment indicated for this purpose that should be carried by hazardous material teams for biohazard cleanup includes but is not limited to:

Disposable latex gloves, such as surgical or examination, single use only.

Heavy duty gloves, such as heavy latex used in commercial dish washing.

Leather fire gloves worn over disposable latex gloves for handling sharp glass or rough surfaces. (May have to be discarded after use as the decontamination of leather is difficult if not impossible.)

Derma Plus, skin antimicrobial protectant, may be used by personnel according to manufacturer's instructions. (May not be able to decon.)

Full disposable gowns, such as Tyvek, Tyvek Polyethelene, Tyvek Saranex.

Particulate mask rated for <5microns. (TB nuclei is 1–5 microns.)

Waterless hand cleaners.

Disinfectant solutions (commercial solutions and bleach).

Sharps containers of assorted sizes for disposable needles.

Fluid spill kits for cleaning blood and body fluids.

Sponges, disposable towels, and absorbent materials.

Sponge mops.

Brushes, brooms.

Buckets for disinfectant, 1–5 gallon with lids.

1–5 gallon lawn and garden spray containers with adjustable nozzle.

Forceps, tongs, flat shovels.

Regulated Waste

Regulated wastes are the materials that are potentially infectious, such as liquid or semiliquid blood, body fluids, and/or tissue. Regulated wastes include any contaminated sharp object, blood-soaked clothing, bandage/dressings, and equipment. They may contain contaminated materials that would release or could possibly release a potentially infectious material if compressed, such as a towel saturated with blood. Items that have dried blood or body fluids caked on them also have a potential to release contaminated materials if compressed. All regulated waste must be disposed of by a licensed disposal contractor.

TRAINING DEMONSTRATION

A simple way to demonstrate the principle of transmission is to use a black light and ultraviolet spray to show how viruses or bacteria are transmitted from direct and indirect transmissions. The demonstration shows the need for personal protective equipment, as well as the need to establish zones similar to a hazardous materials incident. Both reduce the possibility of transmission of communicable diseases.

During a training event, the ultraviolet powder can be sprayed or brushed on equipment, cards, or even a mannequin. Unknowingly, the emergency response personnel are exposed to the powder when they are allowed to touch the items during the training session. If equipment or cards are used, have the student pass them around so each student has a chance for exposure to the ultraviolet powder.

If a mannequin is used, set up a scenario such as doing a primary survey, that allows each student to take part with hands on. Once the scenario has reached the point where each student has been exposed to the ultraviolet spray, turn off the normal lighting and replace it with a black light. The black light will show how direct and indirect contamination such as viruses and bacteria can be transferred from item to item. The scenario demonstrates the need for proper hand washing and frequent glove changing, the need to establish different zones, and to limit the number of personnel making patient contract or cleaning up a biohazardous scene.

Equipment Needed for the Demonstration

Black ultraviolet light—fluorescent 15 watt light

UV spray or powder

Summary

Increased biohazard awareness has changed the way emergency responders provide care. The days of emergency caregivers responding from one medical emergency to another after only washing their hands is over. Many diseases await the unprotected worker so that procreation of the microorganism can continue. As microbiologist Barry Bloom of the Albert Einstein College of Medicine put it, "The life ambition of a bacterium is to become bacteria (plural). Their only goal is survival in a very hostile world of immune responses, environmental threats, and competition" (Popular Science, 1996). Emergency medical providers must protect themselves and exercise great care when working in conditions favorable to infection. Biological agents are very much like toxic chemicals; many are invisible and use the responder's weakness to gain access into the body where the results, over time, may be devastating.

Scenario

It is 3 A.M. and you have just received a call to the local homeless shelter where there has been a recent outbreak of tuberculosis. The call this evening is for another patient complaining of shortness of breath. Just last week you transported a patient who complained of fever, night sweats, and shortness of breath. Later you were advised that the patient was infected with TB and the infection was active. Being concerned for your own health you reported the exposure and were tested with a PPD (purified protein derivative). So far the tests have been negative and the thought has passed until now. In the homeless shelter, several hundred residents sleep in close proximity to one another. Many are in poor health due to a lack of medical care.

Your experience dealing with this group tells you that many communicable diseases have been found here from AIDS to hepatitis. Unfortunately, the symptoms of TB seem to activate at night stimulating many emergency responses to the facility.

When you reach the patient, he is profusely sweating and coughing. He appears in poor general health. Gathering a history reveals that the patient has been living outside for the past 6 months and is a heavy drinker and smoker. A physical exam reveals warm, wet skin with poor turgor. Vital signs are: BP 140/88, pulse 108, respirations at 24, and a SaO_2 of 88%. Lung sounds indicate rales in both bases with rhonchi in the upper airways, slightly diminished in the right lower lobe.

Scenario Questions

1. What precautions should be taken with this patient?

2. At what time during this call should these precautions be taken?

3. Is there a concern with transporting this patient?

4. After transporting this patient should the equipment be cleaned? How?

5. What solution should be used to clean items such as the stretcher? BP cuff? Walls of the unit?

6. How does tuberculosis enter the body?

7. Before you were able to protect yourself, the patient coughed in your direction causing sputum to strike your face and clothing. How would you clean your skin? Clothing?

8. Is this exposure significant?

9. Do you have the right to be informed if the patient later tests positive for a communicable disease?

10. Do you have the right to be tested for the disease if the patient tests positive?

11. Does your department have an infection control policy? If so, are you familiar with the documentation needed after a significant exposure?

12. Can you return to the hospital tomorrow to find out what was wrong with the patient? If not, how do you find out?

chapter
10

Clandestine Drug Laboratories

Objectives

Given a hazardous materials incident involving a clandestine drug laboratory, the medical responder should be familiar with the processes involved in making illicit drugs. The responder should also be familiar with the chemicals involved and the types of injuries that may be incurred by the operator of the lab, police, fire, or other official responder who ventures in without knowledge of the dangers involved with such laboratories.

As a hazardous materials medical technician, you should be able to:

■ Understand the scope of the problem involving the manufacturing of illicit drugs throughout the United States.

■ State the most common types of booby traps found in conjunction with these laboratories.

■ Identify possible health and safety hazards found at clandestine drug laboratories to include but not limited to:

Chemical hazards: acids and bases   Metal poisons

Flammables   Poisons

Respiratory irritants

■ Identify the most common types of labs found and the dangers presented by each.

Amphetamines and methamphetamines

Phenocyclidine (PCP)

HISTORY

clandestine drug laboratory
an illegal laboratory where drugs are manufactured

Clandestine drug laboratories, illegal laboratories where drugs are manufactured, have dramatically increased over the years. The Drug Enforcement Agency (DEA) reported that in 1981 only 184 laboratories were seized in the United States. That number grew continuously until in 1989 approximately 807 laboratory seizures took place. The years following 1989 found a steady decrease in laboratory seizures, probably based on many factors. Some believe that stringent laws on the purchasing of chemicals, associated glassware, and labware was a factor. Others think that the lab operators just got smarter as to the ways of law enforcement agents. Probably the most viable reason was the increased popularity of crack cocaine. Crack cocaine was easy to find and relatively inexpensive. The process used to produce it is not as dangerous as other drugs and it can be made in any kitchen with just over-the-counter ammonia and a microwave or baking soda and a pot on the stove top.

■ NOTE
Because methamphetamines provide long-lasting effects more cheaply than cocaine, large cities are experiencing a much greater incidence of methamphetamine-related emergencies.

Cocaine prices have steadily increased and it only provides a high for a short period of time. Methamphetamines provide effects that last ten times longer at the same price. Because of this, large cities are experiencing a much greater incidence of methamphetamine-related emergencies. In fact the DEA reported a dramatic increase in laboratories in 1995 and predicted that the case load would triple by the summer of 1996. Deaths due to methamphetamines have also increased dramatically in the last few years. In Los Angeles, California, from 1992 to 1994 the death rate due to these drugs increased 400% and in Phoenix, Arizona, it increased 500%. It was further found that when these labs were discovered by law enforcement agents 75% of the time children were in the same house or structure as the laboratories.

These laboratories have a potential for causing havoc on emergency response units. Not only are they a unique type of toxic chemical generator but they may also be filled with other dangers rigged by the criminals conducting the operation. This chapter discusses some of the unique dangers associated with illegal drug laboratories and provides emergency responders with some insight as to how the drugs are manufactured.

One of the most dramatic stories associated with illegal laboratories is the recent manufacturing of a drug called fentanyl. A self-taught chemist named George Marquart designed the laboratory and made large amounts of the drug that was sold on the street as a synthetic heroin (bagged and sold as

Tango and Cash and China White). By the time DEA agents found Marquart it was estimated that the drug was responsible for 300 deaths.

The manufacturing of a drug such as fentanyl is rare. The most commonly found clandestine laboratories manufacture three drugs: methamphetamine (82%), amphetamine (10 %), and phencyclidine or PCP (2.5%). These three drugs made up 95% of drug lab seizures by the DEA in 1990. However, other types of laboratories are increasing, in part, because drug lab operators are attempting to circumvent the controlled substance laws by creating a "designer drug" that would not meet the criteria for prosecution.

CLANDESTINE LAB OPERATORS

! SAFETY
Individuals operating illegal drug labs have no concern about the well-being of emergency responders, and many protect their labs with dangerous booby traps or by rigging the labs for destruction on discovery to prevent prosecution.

It is important to keep in mind the mentality of those operating these illegal laboratories. These individuals are committing felonies by operating such a lab and could care less about the well-being of emergency responders. Many protect their laboratories by placing booby traps around their operations. The unsuspecting responder may fall victim to such items as pungi boards filled with protruding nails, explosive devices, or poisonous gas, not to mention firearms used by the operators. Sometimes laboratories are rigged to be destroyed to prevent prosecution in case law enforcement personnel discover their existence.

It is simple to imagine how emergency providers could end up on the site of a clan lab and virtually walk into disaster. It is estimated that 30% of the drug labs are discovered as a result of explosion or fire. Because of the unsafe conditions under which these labs are operated many more may be discovered because of medical emergencies or even a simple nuisance call such as suspicious odors.

It would also be dangerous to assume that the labs are operated by individuals who know about chemistry or are knowledgeable about the chemical processes that they are creating. Operators of drug labs range from those with little or no chemistry background who are just following a cookbook recipe to chemists who possess doctoral degrees.

The chemicals and equipment can usually be bought or acquired easily. One book currently available called The Construction and Operation of Clandestine Drug Laboratories (Nimble, 1986) identifies alternative ways to construct the apparatus necessary to set up an illegal laboratory. It also points out other means of obtaining chemicals used in the process of manufacturing the drugs. Another legal book that assists the would-be drug manufacturer is The Anarchist Cookbook (Powell, 1971).

SCOPE OF THE PROBLEM

Illegal drug labs can be found almost anywhere. They have been operated in private residences, motel and hotel rooms, apartments, house trailers, houseboats, campgrounds, and commercial establishments. Clandestine labs set up in motor homes and trailer trucks are mobile and hard to trace. One clan lab operator even went through the trouble of burying an entire school bus about 10 feet below ground. Entrance into the bus was made through a small tunnel and hatch in the roof. The bus was converted into a meth lab. The discovery of this lab was not only a chemical hazard but also a confined space nightmare. Clandestine labs are virtually found nationwide and, with the exception of just a few states, most of the country has felt the effects of these laboratories.

In 1988 four states accounted for 78% of all the drug lab seizures in the United States. California's labs with a total of 355 represented 44%, Texas had 19% with a total of 151 labs, and Oregon and Washington combined to represent 15% or 120 labs. Many other states had drug lab seizures in the double digits.

The type of drug lab that is in operation, whether it is manufacturing stimulants, depressants, hallucinogens, or narcotics, determines the type of chemical or toxic hazard faced by the emergency responder. The drug lab type also dictates the type of contaminated patient the responder may have to treat in the case of a laboratory accident.

CLUES TO LOOK FOR

There are some common clues to look for when approaching any structure. Some will take practice and a suspicious nature to recognize. The list may not contain all of the signs but it will identify some of the most common ones found at the lab site. Emergency responders must have an in-depth knowledge of their response area and must be keenly aware of the possibility of a lab's presence.

The important signs to watch for in structures include:

Windows always covered or painted

Windows and doors secured with bars

Unusual pipes or duct work coming from windows or walls

Evidence of chemical containers or glassware

Unusual odors or tastes including bittersweet, ammonia, acetone, cat urine smell, or a metallic taste in the mouth

Persons going outside of the structure to smoke

Continuously running fans in inappropriate places

Portable generators for outdoor sites

Stressed vegetation

BOOBY TRAPS

There are many types of ingenious booby traps used by the drug manufacturers. Some of the more common listed in the available material include:

Foil Bombs   Foil bombs use a 2-liter soft drink bottle. The bottle contains 2 inches of muriatic acid with a twisted piece of aluminum foil attached to the top. The cap is placed back on the bottle to create an airtight seal. When the bottle is accidentally knocked over by an unsuspecting police officer or emergency medical responder, the aluminum reacts with the acid, building pressure until an explosion occurs. The explosions are of pressure origin but the expanding gas is filled with hydrogen and if near an ignition source can create a fire explosion. These bombs have been known to explode with such force that portions of the roof and windows of a house have been blown off.

Other Bombs   Other bombs used by more sophisticated and knowledgeable lab operators are triggered by radio waves. Again, when unsuspecting law enforcement or emergency medical personnel enter the area and trigger their portable radios, an explosion results. For this reason the use of portable radios within a lab area should be prohibited. Other triggering devices of explosives have been discovered wired to refrigerator doors and light switches.

Chemical Reactions   The most common chemical reaction used by lab operators is the tripping of a small amount of acid, such as HCl or nitric acid into a plate of cyanide salt, instantaneously producing deadly cyanide gas. Both of these chemicals are common chemicals used in the production of drugs.

Trip Wires   Trip wires are rigged to any number of items. The most deadly is an antipersonnel device that has been found in clandestine lab operations that involves tying a monofilament line to the trigger of a shotgun. These devices can be placed either inside the laboratory or surrounding structure or even on the outside property to keep unwanted persons away from the structure.

Other Antipersonnel Devices   Other apparatuses have been reported, such as pungi type devices made of nails driven through boards or fishing hooks hanging from tree limbs at the level of the eyes. Just about anything that could cause harm to or slow down law enforcement agents is employed by these drug manufacturers.

Emergency medical responders or fire personnel should never stop (unplug) the reaction process of a working lab without knowing what reactions are taking place. The premature stopping of some reactions may cause spontaneous combustion resulting in a fire or explosion. A chemist should be called in by the DEA to analyze the process and determine what can be shut down and when. If the situation prohibits immediate DEA involvement, then the area should be evacuated and the process stopped.

HEALTH AND SAFETY

Types of Chemical Hazards

precursors

the initially purchased chemicals used to make illegal drugs

reagents

the chemicals that are used to change the precursors into illegal drugs

solvents

the chemicals used to dissolve the mixture of precursors and reagents

When a working laboratory is discovered usually three types of chemicals are involved. These are referred to as the **precursors**, the **reagents**, and the **solvents**. These chemicals in any of the stages of the process can cause acute or chronic harm to those working in the area. They almost certainly have already caused physiologic harm to the illegal lab operators at the scene because of their chronic exposure. Illegal drug lab operators historically have a history of sloppy operations where less than safe practices are conducted.

These chemicals are made up of a variety of respiratory irritants, acids, alkalis, flammable materials, and water soluble metal poisons. Health problems such as skin or tissue burns, respiratory injury and failure, cancer and circulatory system malfunctions, and even death, are associated with chronic and acute exposures to these chemicals.

As some of these chemicals become harder to purchase, the clandestine lab chemist will find other substitutes to take their place. This type of creativity can make drug labs even more dangerous in the future. Chemicals used in drug labs are too many to list but following are some of the more commonly found chemicals:

> **SAFETY**
> **Health problems such as skin or tissue burns, respiratory injury and failure, cancer, and circulatory system malfunctions, and even death, are associated with chronic and acute exposures to the chemicals used in illegal drug labs.**

Bases

Sodium hydroxide (NaOH)—Destroys tissue on contact. Vapors cause severe respiratory injury and devastating eye injuries. Reacts with acids and water creating enough heat to ignite flammable or combustible materials.

Methylamine (CH_3NH_2)—An irritant to eyes, skin, and the respiratory system. Flammable and reactive with mercury.

Piperidine [c-$(CH_2)_5NH$]—Causes skin burns, highly toxic with dermal contact and ingestion. Flammable, reacts vigorously with oxidizing material. Fumes released contain toxic nitrogen oxides. Water soluble.

Acids

Hydriodic acid (HI)—Causes irritation to the skin, eyes, and mucosa. Releases toxic fumes of iodides when heated.

Hydrochloric acid (HCl)—Causes irritation to the eyes, skin, and respiratory tract. Water soluble affecting the upper respiratory system. In higher concentrations causes pulmonary edema. Contact with common metals forms explosive hydrogen gas.

Nitric acid (NHO_3)—Causes irritation of the eyes, skin, and mucous membranes. Prolonged exposure leads to pulmonary edema, pneumonitis, bronchitis, and dental erosion.

Sulfuric acid (H_2SO_4)—Causes damage to all tissues. Severe eye injury may result if exposed. Reacts violently with water, generating heat. Attacks metals, releasing explosive hydrogen gas.

Hydrogen chloride gas (HCl)—Hydrochloric acid in the form of gas. Will form hydrochloric acid when in contact with moisture within the respiratory system, mucosa, or sweat.

Flammables

Diethyl ether ($C_2H_5OC_2H_5$)—Inhalation causes CNS depression, dizziness, euphoria, and unconsciousness. Very flammable with a flash point of −49°F. Unstable peroxide crystals can form when exposed to air. Can explode when mixed with chlorine gas. Non-water-soluble.

Petroleum ether—Inhalation causes CNS depression, respiratory irritation, and giddiness. Highly explosive when exposed to heat. Produces CO_2 and CO.

Ethanol (C_2H_5OH)—Vapors irritating to mucosa and respiratory tract. Causes drowsiness, ataxia, and stupor. Flammable with a flash point of 55°F. Adding alcohols to high concentrations of hydrogen peroxides forms a powerful explosive that can detonate by shock.

Isopropyl alcohol [$(CH_3)_2CHOH$]—Vapors cause irritation to respiratory tract and eyes. Causes drowsiness, ataxia, and stupor. Flammable, fire produces irritating or poisonous gas. Reacts with oxidizing agents and hydrogen.

Acetone [$(CH_3)_2CO$]—Inhalation causes CNS depression, sedation, and coma. Prolonged exposure to skin causes irritation and dermatitis. Explosive when heated or exposed to oxidizing agent.

Coleman fuel—Combustible fuel. Inhalation of fumes causes irritation and CNS stimulation, muscle fasciculations and seizures.

Respiratory Irritants

Acetic anhydride [$(CH_3CO)_2O$]—Acute irritant to the eyes, skin, and respiratory tract. Contact with liquid or vapors causes tissue necrosis. Flammable, and toxic vapors. Soluble in water forming acetic acid.

Hydriodic acid—Causes irritation to skin, eyes, and mucosa. Releases toxic fumes of iodides when heated.

Methylamine—Irritant to the eyes, skin, and respiratory system. Flammable. Reactive with mercury and can cause explosive reaction.

Hydrochloric acid—Causes irritation to the eyes, skin, and respiratory tract. Water soluble affecting the upper respiratory system. In higher concentrations causes pulmonary edema. Contact with common metals forms explosive hydrogen gas.

Metal Poisons

Mercuric chloride (HgCl$_2$)—Corrosive and nephrotoxic. Inhalation causes irritation to mucosa, respiratory tract, abdominal pain, vomiting, diarrhea, drowsiness, ataxia, and coma. Can explode with friction or heat. Fire produces mercury fumes. Reactive with phosphorus, arsenic, silver salts, sulfides, acetylene, ammonia, alkali metals causing explosion.

Lead acetate [Pb(C$_2$H$_3$O$_2$)$_2$]—Renal toxic and neurotoxic with CNS symptoms. Nausea, vomiting, irritability, restlessness, and anxiety.

Poisons

Cyanide (sodium or potassium)[NaCN or KCN]—Extremely toxic. Absorbed through skin or respiratory tract. Not combustible. Interaction with acids releases highly flammable and toxic hydrogen cyanide gas. (See cyanide section in Chapter 6.)

Turning off the Process

Although it is not recommended to turn off the chemical process involved in making drugs, emergency responders may be met with having to make that decision based on a risk versus benefit analysis. It would be very easy to state that the process should never be turned off by emergency responder's but realistically the term "never" often does not hold true. Most resources recommend waiting for a chemist from the DEA to arrive and decide how to turn off the laboratory. Other options include calling chemists from local colleges or universities to determine the best option for stopping the process. All of these alternatives should be exercised before making the decision to intervene on the process.

■ NOTE
It is commonly recommended that a DEA chemist decide about how to turn off an illegal lab, but chemists from local colleges or universities can also be called upon before making the decision to intervene on the process.

If the determination to stop a process has been made by the on-scene commander after evaluating all other options, several safety steps should be taken to deactivate the chemical process in progress. The DEA suggests a method that should safely accomplish the shut down.

- Examine and determine if heating or cooling is taking place.
- Some reactions involve the heating of a chemical and then condensing utilizing tap water. In these cases remove the heat and allow the glassware to cool before turning off the water.
- If vacuum or gravity filtration is occurring allow this process to finish.
- If compressed gas is being used in a reaction, it should be first shut off at the cylinder top, then the regulator should be shut down.
- If vacuum is used within the system, the system should be slowly brought back to atmospheric pressure then the vacuum pump turned off.

- If there is an exothermic reaction (producing heat) taking place, it should be left until the process is completed then the reaction cooled to room temperature.

If a fire department is called on to assist the local police or DEA officials, it is important to keep in mind that a clandestine lab is three things:

A law enforcement action

A hazardous materials emergency

A crime scene

To successfully mitigate a lab incident all three items must be considered. Preserving evidence is of utmost importance and must be a consideration of the responding hazmat team.

Unfortunately, the possibility of causing harm to other emergency responders such as firefighters or emergency medical personnel is all too real. In many cases, the dangers may be unavoidable due to the clandestine nature of the labs, but in the cases where emergency responders suspect or even know that a lab is involved, it is pure survival to take the utmost care when approaching the scene.

Types of Labs and Processes

Amphetamines and Methamphetamines

To manufacture amphetamines or methamphetamines a substance called phenyl-2-propanone (P-2-P) must be made. It can be manufactured by three different methods. Each method uses different chemical agents in the precursor and reagent stages. Some of the processes use solvents.

■ NOTE

Producing amphetamines and methamphetamines by the benzaldehyde/ nitroethane method is a popular process because none of the chemicals are monitored or controlled. A lab operator can purchase them without drawing attention to his or her activities.

1. *Phenylacetic acid/acetic anhydride method.*   The chemical process involves the use of two precursor agents, phenylacetic acid and acetic anhydride, and a reagent material, anhydrous sodium acetate or pyridene. When the process is finished it nets about 40% of the amount of the precursor agents. The finished P-2-P is light amber and has a fruity smell and an oily texture.

2. *Benzaldehyde/nitroethane method.*   This process involves the precursor agents of benzaldehyde, which has the odor of almonds, and nitroethane. The reagents are butylamine, ammonium acetate, acetic acid, iron filings, and hydrochloric acid. The solvent of ethanol is also used in a washing and drying process. This method yields both P-2-P oil and orange P-2-P crystals that have a cinnamonlike smell (a pleasant odor). This process is popular because none of the chemicals are monitored or controlled so a lab operator can purchase them without drawing attention to his or her activities.

3. *Phenylacetic acid/lead acetate method.*   This method involves the heating of two precursor chemicals, phenylacetic acid and lead diacetate. During the reaction both acetic acid (vinegar) and P-2-P are produced. The reaction pro-

duces P-2-P that is dark orange. When this P-2-P is used to produce methamphetamine, the results are "dirty methamphetamine" or "peanut butter crank," as it is called on the street. The total yield of P-2-P from this process is about 25% of the starting weight of the phenylacetic acid.

4. *Hydriodic acid/red phosphorous reduction of ephedrine method.* The precursor chemical is ephedrine hydrochloride powder or tablets. The reagents are red phosphorous and hydriodic acid. The solvents are ethyl ether, freon 11, 12, or 13, hexane, and white gas. If this process is allowed to go dry then the red phosphorous can be converted into white phosphorous and self-ignite. This method is the most dangerous but the yield of P-2-P is the highest at 78% of the starting weight of the ephedrine.

Methamphetamine can then be produced by a chemical reaction that utilizes P-2-P, methylamine, alcohol, aluminum foil, and mercuric chloride. The reaction is exothermic (producing heat) but usually does not create enough heat to produce combustion. The yield of methamphetamine is about 80% of the starting weight of P-2-P.

Amphetamine can also be made via the benzaldehyde/nitroethane method. The precursors are benzaldehyde and nitroethane. The reagents are butylamine, ammonium acetate, and lithium aluminum hydride. The solvents used in the process are diethyl ether and tetrahydrofuran. The process yields amphetamine HCl. Using sulfuric acid as a titrating agent yields amphetamine sulfate.

Slang names for pharmaceutical-type amphetamines include Black Beauties, Blackbirds, Black Millies, Bumblebees, Chalk, Christmas Trees, Copilots, Dexies, Eye Openers, Hearts, Jelly Beans, Lightning, Nuggets, Oranges, Pep Pills, Roses, Thrusters, Truck Drivers, Turnabouts, Ups, Wake-Ups, and Go Faster. The most commonly used slang is Beans, Bennies, Speed, and Uppers.

Nonpharmaceutical or clan lab produced pills are known as Cartwheels, Cross Roads, Double Crosses, Mini Bennies, Speed, and Ups.

Phencyclidine

The most popular way to produce phencyclidine (PCP) is by the "bucket" method. The first stage (bucket) contains piperidine mixed with cyanide (either sodium or potassium). The cyanide is first dissolved into water then the piperidine is added to the mixture. If the step is done backward, there will be a release of cyanide gas because of the strong basic characteristic of the piperidine. The second stage (bucket) contains cyclohexanone mixed with sodium bisulfite. The two buckets are then mixed together.

Once this mixture stands, piperidinocyclohexanecarbonitrile (PCC) is formed as a solid at the top. It is removed and washed in water. Then the powder is dissolved into Coleman fuel or petroleum ether (solvent). To this solution phenyl magnesium bromide (PMB or grignard reagent) is added. This reaction yields PCP base, which is sold on the streets as a solid or a liquid in small single-dose bottles. If a pill form is desired, hydrogen chloride gas or hydrochloric acid is added to form PCP hydrochloride. (Base + HCL = PCP hydrochloride.)

The passing of the piperidine reporting act has reduced the availability of piperidine to illegal buyers. Piperidine is commonly used in industry as a curing agent for rubber, in epoxy resins, and as an ingredient in oils and fuels.

Summary

In any case involving the discovery of a clandestine lab, the goal is to escape injury. Removing the hazards, cleanup, and mitigation of the incident should be left to the professionals who are trained and practiced in these situations. The purpose of this section is to make emergency responders aware of the unique hazards of these laboratories. It also provides important information to hazmat teams who may be called upon by their local law enforcement agencies to analyze and remove any acute hazards. The ultimate responsibility is placed on the shoulders of the DEA who will provide the expertise for evidence handling and cleanup.

Scenario

You are manning a fire department-based ALS nontransport rescue unit. You have just arrived on the scene of a shooting involving a 22-year-old male. Police, who were on the scene when you arrived advise you that the patient is in the living room. When you reach the patient, a strong odor of solvent is noted on his clothing. The gunshot wound is to the lower leg and is not life threatening. While you are splinting and dressing the wound and making the patient ready for transport one of the police officers asks you to assess something she found in the back bedroom. When the ambulance personnel arrive you transfer the patient care to them and meet with the police officer.

The police officer directs you to the bathroom off of the master bedroom. The bathroom contains several large glass bottles, heating elements, condenser tubes, and collection containers. A strong odor of solvent and an irritating acid type taste is noted when you breathe. Having trained with the hazmat team you recognize that this is probably a drug laboratory. Furthermore, you advise the police that the laboratory appears to be functioning. Knowing the inherent dangers of a clandestine laboratory you advise the police to evacuate the building and call for the hazmat response team.

Scenario Questions

1. In addition to the obvious chemical hazards, what other kinds of hazards may be present?

2. Is it important to question the injured patient about the process?

3. Because you entered the cooking room and were able to detect the process by smelling the air, is this considered an exposure?

4. Is it true that small labs of this type usually do not possess a hazard?

5. If the patient advises that methamphetamine is the drug being prepared, what are some of the chemical hazards?

6. Is there a danger of a fire at a laboratory such as this?

7. If you are asked by the police to turn off the chemical process, what is your reply? Is this a decision you are prepared to make?

8. Should the bomb squad respond to the scene? Why?

9. If this emergency response happened to you today, whose responsibility is it to transport the chemicals away from the scene and who will dispose of them?

10. If this took place in your district, once the chemicals are confiscated and transported, whose responsibility is it to clean the house to ensure safety?

chapter

11

Air Monitoring

Objectives

When presented with an incident, the student should be able to define, employ consideration factors, and demonstrate initial monitoring procedures as they pertain to scene stabilization.

As a hazardous materials medical technician, you should be able to:

- Describe the four hazardous environments that air monitoring is directed toward.
- Discuss how meteorological conditions affect the monitoring process.
- Describe the difference in general air monitoring techniques and the considerations thereof.
- Describe mathematically the association between parts per million and percent.
- Discuss the physical and chemical properties that affect air monitoring.
- Discuss oxygen deficiency monitors.

    Limits of oxygen deficiency and enriched.

    Oxygen gradients.

    Procedures when utilizing the oxygen deficiency monitor.

    How the oxygen deficiency monitor works.

    Limitations of oxygen deficiency monitors.

403

- Discuss colormetric tubes.

    Mechanics of operation.

    Procedures when utilizing colormetric tubes.

    Three measurement methods.

    Limitations of colormetric tubes.

- Discuss photoionization detectors.

    Mechanics of operation.

    Relative response patterns.

    Limitations of the photoionization detector.

- Discuss flame ionization detectors.

    Mechanics of operation.

    Limitations of the flame ionization detector.

- Discuss combustible gas indicators.

    Mechanics of operation.

    Response curves or reference charts.

    Limitations of the combustible gas indicator.

- Discuss radiation detectors.

    Mechanics of operation.

    Limitations of the radiation detector.

- Discuss the level for personal protective equipment for use with air monitoring.

Air monitoring enables the rescue worker to gather information about the environment. It gives the medical sector information on the atmosphere in which the entry and backup team will enter. It establishes the known environmental conditions that they could be subjected to if a breach of the encapsulating suit should occur. This data collection assists the emergency worker in decision making when it comes to treating entry team members and/or victims of the incident. This tangible information about the chemicals that he or she is dealing with safeguards the rescue worker and the public.

However, air monitoring is not without its own limitations and pitfalls. One must practice, practice, and study to become proficient in the skill. This chapter was not designed to give the reader all the information about air monitoring techniques; it was written so that response personnel can understand the use, limitations, and interpretation of the data gathered. Unfortunately, most individuals feel that they can pick up one of the currently used instruments and use it. Some fall back on reading the manual at the time of the incident. Others rely on another worker to interpret the data for them. In all of these cases, the safety of the worker is severely compromised.

Basically these individuals are learning the capabilities of the instrument when anxiety is high. How much of the read information will they remember? This would be like asking a surgeon to perform an operation that he has never performed, read about it prior to surgery, and be expected to execute the procedure flawlessly. An impossible task.

Emergency workers must understand the tools we use prior to the event. Many responders do not fully understand the equipment and its operation. The possibility of misinterpretation or misuse can lead to health and safety concerns. On the other hand, 29 CFR 1910.120 (h) states specific requirements for air monitoring procedures. In short, the federal document states that whenever an employee is subjected to the possibility of a toxic or hazardous environment, then air monitoring shall exist.

GENERAL PRINCIPLES

Within the hazardous materials environment (including confined space) we have four basic considerations when dealing with air monitoring devices:

1. Oxygen concentrations or oxygen-deficient atmospheres can have a devastating effect on unprepared responders. This condition can be as simple as the displacement of oxygen by other gases or a depletion of the oxygen concentration. Examples of this would be hydrogen sulfide, which can displace oxygen and could cause further harm, or immediate health hazard of asphyxiation as seen with nitrogen gas displacement (or any other inert gas). Ventilated areas that have been pressurized due to inappropriate ventilation techniques or unventilated areas in which the concentration of oxygen falls below 19.5% are yet other examples.

2. Toxic atmospheres in which the chemical is a vapor, mist, fog, dust, fume and/or aerosol are within this category. A toxic atmosphere is that environment that is immediately dangerous to life and/or health. This environment may not be obvious, yet the location of the scene leads us to believe that the atmosphere may have potential harm. An example of this is any confined space or at the scene of a fire, where toxic gases may also be released.

3. Explosive conditions are obviously important. Several air monitoring instruments on the market are not rated for entry into these types of atmospheres. One must have an instrument that has the Factory Mutual Research Corp. (FM), Underwriters Laboratory (UL), or a recognized European certification on the instrument in order for the instrument to be truly intrinsically safe. Remember, if we have a toxic atmosphere, the explosive environments are above these levels. Without a clear understanding of these instruments and their interpretation of the data, the result could be devastating. Oxygen-enriched atmospheres are one example. If the instrument is not intrinsically safe and if it utilizes a heating element to sense, the instrument in itself can become the ignition source.

4. The presence of radioactive isotopes within any community is on the increase. Industry, medical institutions, and research laboratories are just a few examples of occupancies that have significant quantities of radioactive materials. Alpha and beta particles are respiratory hazards. Gamma waves can cause damage to tissues and organs. Fortunately, radioactive materials are usually not affected by meteorological conditions. Here time, distance, and shielding are the protective guidelines.

METEOROLOGICAL CONDITIONS

Weather conditions can affect our air monitoring procedures. Wind speed, and direction, temperature, humidity, temperature dew point, and barometric pressures all can affect the conditions we need to monitor. These factors can greatly affect the chemical and physical properties of the chemical(s) we are evaluating.

Wind

Wind speed and direction can affect your scene in two distinct ways. First and most obvious is the spread of contaminates downwind. If the direction of the wind changes, support and command would need to be moved, which loses time and uses manpower. Command staff should consider the possibility of wind direction change and set up the initial command and support functions in an area that will not require moving.

Second, increases in wind speed affect the dispersal of the contaminate, which may require evacuation conditions or a defend in place. During the fall and winter months, the increase in wind speed can create an additional medical consideration, hypothermia. In this case all personnel should be protected from the wind for any long periods of time. All encapsulated personnel will require the wind blocked, and possibly heated areas for rehabilitation will be needed.

The freezing of the air monitoring instrument can affect the readings, functions, and thus the data read out at the incident site. As the speed of the wind increases, the temperature will drop. We have all felt this on a winter morning when it was still; then the wind picks up and we find that we did not dress warmly enough. Instruments are no different; operations may be disturbed by the temperature drop. All instruments are rated and function at optimal temperatures; malfunction can occur due to temperature drops or severe temperature increases. Although the ambient temperature is within the limits of operation, the wind chill factor may bring the temperature well below the functioning levels.

■ NOTE

All instruments are rated and function at optimal temperatures; malfunction can occur due to temperature drops or severe temperature increases.

Temperature

Temperature can affect chemicals in two ways. First, the chemical's properties can change. An increase in temperature increases the vapor pressure (see Chapter 3). An increase in vapor pressure can give us a higher probability of reaching our dangerous lower explosive limit (LEL). This increase in temperature, which in turn increases our vapor pressure, can also increase the possibility of permeation of the chemical through our instrument so as to saturate the mechanism and give false readings. If the vapor pressure of the substance is normally significant, for instance, above 760 mmHg, the resultant vapor pressure increases our respiratory hazard and possibly data collection.

Decreased temperature can also greatly affect the team members on scene, especially during lengthy operations. The greater the temperature drop or increase, the greater the effect on our emergency personnel and the instrument's data.

Humidity

The greater the humidity associated with temperature increases, the greater the frequency of heat stroke. Decontamination can also become increasingly difficult as the humidity increases because dusts or particulate matter are sensitive to humidity changes. This may change the decontamination process to a certain degree and affects our monitoring.

Air moisture can also mix with vapors of a corrosive chemical. These vapors can cause the injuries discussed in Chapter five. In addition to the medical hazards that accompany humidity increases, the corrosivity of the gas can affect the electrodes within our monitor, giving a false reading or none at all.

Air monitoring will also become increasingly difficult. Increased humidity can cause moisture within the monitoring device. The filters and burning chambers can become saturated, distorting the instrument's ability to accurately read the environment.

Temperature Dew Point

frost point
the temperature at which atmospheric water freezes

Temperature dew point is that temperature at which the moisture in the air becomes visible as fog. Actually it is the water saturation within the air at a certain atmospheric pressure. As the temperature drops below freezing then the term **frost point** is applied.

Dew point temperature and the ambient temperature have a gradient. The larger the numbers (the difference between dew point and the temperature), the less saturated the air is with water. However, if the difference is small, the saturation is high. Expect fog in situations that create a decreasing temperature toward the dew point. Fog starts to appear within 4–5 degrees of dew point.

● CAUTION
Expect fog in situations that create a decreasing temperature toward the dew point.

The same problems that we had with temperature and humidity will become evident, except the problems of water saturation within the instrument occur at a faster rate. At the same time visibility at the scene will be lost. During nighttime operations as well as foggy conditions, stress levels increase because of this lack of visibility.

Barometric Pressure

As the atmospheric pressure decreases, the evaporation of the chemical increases. Likewise, as pressures increase, the potential for evaporation decreases. Realistically, the fluctuation of barometric pressures are so slight that on-scene personnel will not realize the change. However, if a weather front is about to move through your area or has already passed, the consideration of atmospheric pressures should be a part of the research sector's responsibility. The questions they need to find answers to are: How will this slight increase or decrease affect the chemical properties? What type of resultant hazards are going to be perpetuated from this change? Will my instruments give accurate readings under present and future conditions?

Interference Gases

interference gases
gases similar to those being tested that the instrument may misinterpret

The air monitor cannot distinguish between chemicals. It cannot identify the chemical in question. You, as the hazmat medical technician must be able to understand the information and based on that information, make a reasonable guess. However, at times gases that are similar to the gases that are being tested may be present. These substances are called **interference gases**. For example, suppose you are using an instrument that is designed to respond to hydrogen sulfide. Under calibration and field testing, the unit responded appropriately. In the environment that you are evaluating mercaptan is present (mercaptan is a sulfur derivative, thiosulfate). Your instrument may register as having hydrogen sulfide in the environment. Depending on the concentration of the interference gas and the gas being tested for by the instrument, the instrument will respond to very high levels or very low. The interference that registers a high reading is called the *positive interference*, while the low readings are a result of *negative interference*. The latter could be devastating to a hazmat team.

The type of monitoring needed (or wanted) can also impact our surveillance decisions. There are basically two types of monitoring techniques. The first is *direct* or *real time monitoring* and the second is *sample collection* or *sample analysis*. With the first, we are directly analyzing the environment at a specific point in time. Direct reading instruments (DRI) give rapid, immediate information about the atmosphere at the time of monitoring. This type of monitoring theory can be a gain of information or a hindrance. DRIs have several inherent limitations, which will be the focus of our discussion.

The second method, sample analysis, requires a laboratory on scene or an available contingency lab. The downside here is the availability of such a facility. The laboratory analysis in most cases can be used after the fact to define the chemical in question. However, the cost for this type of assistance to our operation can sometimes become prohibitive.

In either case, the primary objectives are the same. We want to qualify the toxicity, deficiency, explosivity, or ionization potentials of the chemical(s) in question. Air monitoring can indicate the need for medical monitoring, follow-up, and protection. It can identify the zones of responsibility, personnel needs, and the working environment.

This chapter's intent is not to give you a complete understanding of monitoring techniques, only a general overview of what is expected and realistic when we are concerned with air monitoring.

Laboratory sampling and the instrumentation that would be used either on site or within a facility is beyond the scope of this text. Rather DRIs are the focus of our discussions along with a general knowledge base. Even within this category of air monitoring, a limited discussion is presented. Advanced techniques or instrumentation are mentioned for clarification only. Please follow your current and local standard operating procedures, guidelines, rules, regulations, and manufacturer's recommendations.

DIRECT READING INSTRUMENTS

There are many different manufacturers of DRIs. How they are used varies from manufacturer to manufacturer. Limitations can also become a factor between different DRIs. However, no DRIs identify the chemical; they only detect possible hazards that may or may not be present and this only happens if the operator understands what the instrument is displaying.

DRIs can only measure and/or detect a limited and very specific array of chemicals. In addition to the limitation of detected chemicals, the percent or parts per million is also a limitation of use. Most DRIs presently on the market can only read as low as one part in a million. False readings are a norm for these types of monitoring instruments. One must understand the physical workings of the instrument and the interpretation of the data. This interpretation must be done by a knowledgeable response person. Numbers are useless unless we know how to apply them and what they mean. In order to understand each instrument one must comprehend the language of meters and the data information they will give.

concentration

the relative percentage of the monitored gas in the atmosphere

Concentration is common term used with air monitoring, however, what does it mean? The concentration, or relative percentage, of a chemical is its potency. This ability of a chemical to do harm is what we want to measure. It may be in parts per million (ppm), parts per billion (ppb), or percent (%). Particle size has an effect on the concentration of a chemical, for example, the difference between a vapor and dust particles. As we saw in Chapter 3, molecules can have a variety of shapes. These shapes have inherent "active" sites that can affect the concentration, especially within an organism (see section on metabolism of chemicals in Chapter 4). Likewise the state of matter that the chemical is in also affects the concentration. For example, if we have a substance that is a solid at room temperature, the airborne concentration is very low, however, it may still be toxic (some organic compounds have VPs of 0.1 at average room temperature). In contrast, if the chemical has a vapor pressure of say 1000 mmHg, the gas is extremely volatile.

What does a part in a million mean (Fig. 11-1)? What does that piece of information actually give us? Parts per million (and parts per billion) is a measurement. It shows a relationship between the parts of a substance and the air, liquid, or solid

PARTS PER MILLION (PPM)

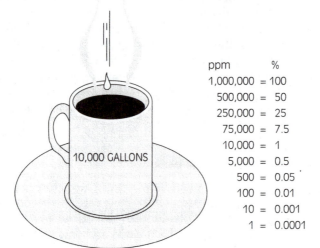

ppm		%
1,000,000	=	100
500,000	=	50
250,000	=	25
75,000	=	7.5
10,000	=	1
5,000	=	0.5
500	=	0.05
100	=	0.01
10	=	0.001
1	=	0.0001

10,000 GALLONS

Figure 11-1 *An example of a part in a million is a drop of cream in a 10,000 gallon cup of coffee.*

that the substance is in. An example of a part in a million (ppm) would be an ounce of cream in 10,000 gallons of coffee, or one percent is equal to 10,000 ppm. An example of a part in a billion (ppb) is a drop of cream in 22,000 gallons of coffee, or one percent is equal to 10,000,000 ppb. As you can see, the quantities are very small. At this point it is important to remember that most all DRIs can only measure as low as one part in a million.

Percent is a mathematical ratio based on 100 total parts, or one in a hundred. For example, 50 parts in 100 is 50/100 = .50, denoted as 50%. Parts per million has a direct relationship with percent. One hundred percent is a million. Fifty percent is 500,000 ppm. Likewise 25% is 250,000, so 1% is 10,000 ppm. For example, naphthalene (our solid that has a vapor pressure) has an LEL of .9%. That means at 9,000 ppm we have reached our lower explosive limit. (A little side note here, the IDLH for naphthalene is 500 ppm or .05%, a toxic range well under our LEL!)

Physical and Chemical Properties That Affect Air Monitoring

In Chapter 3 we explored chemical behavior at great length, partly to give you the groundwork for toxicology and air monitoring. Here we see the same information discussed but with a direct application to the problems inherent with air monitoring.

Boiling point is the temperature at which the atmospheric pressure and the vapor pressure of the liquid are equal. At this point the liquid changes from the liquid state into the gaseous state. Hazardous materials that are in the liquid state possessing a low boiling point usually have a high vapor pressure. In addition to being a fire hazard, this type of material is a significant respiratory hazard. The high boiling point liquids have low vapor pressures. These liquids need an active energy source to convert them from the liquid state to the vapor state, such that the end gaseous product would have to be forced, during a fire for example. Those products that have high boiling points are relatively safer than those with comparatively lower boiling points.

From an air monitoring standpoint, the liquids at room temperature that have a boiling point greater than the ambient temperature have considerably lower vapor emission. Those liquids that have a boiling point below the ambient temperature have a significantly higher vapor emission. Suppose our product is a liquefied gas such as propane or butane (LPG are gases that are pressurized into liquids). If released from the storage container, the molecules escape the liquid state and gas is formed, giving us a product to monitor. If the temperature is below the boiling point, we can't expect our metering device to give us any information.

During operations in which the temperature drops, static charges may accumulate within the air, giving false readings from instruments dependent on electrical differentials. This can occur in two ways: (1) the electrical charge increases or decreases the resistance over the electronic sensor, and (2) if the instrument utilizes IC chips for memory, the circuit may lose memory. These are remote cases, because these types of instruments are usually intrinsically safe, however these situations have occurred with static buildup as the only solution to the problem that has occurred.

The *state of matter* can be solid, liquid, or gas. Form may also be the color of the liquid or gas, or even the size of the fragments, such as in the powder form, crystalline, or as in dust. Solids must emit a gas before the solid will burn (acknowledging the fact that an ignition source must be present). The liquid must also emit a gas before ignition takes place. However, if the gas that is emitted from the chemical is above or below its flammable range, the gas will not become involved in combustion. Rather, a toxic atmosphere may exist.

Temperature and humidity can greatly affect our chemical, especially if we are involved with an aerosol. By strict definition, an aerosol is a airborne suspension of solid and/or liquid particles. Dusts, fibers, and condensation are all examples of aerosols. Moisture can adhere to these particles creating problems with the monitoring of such a particle. Differences in parts per million (thus %) can be seen in such atmospheres with high humidity.

Warm and hot air have the potential to hold significant quantities of moisture. The air monitoring devices that we employ are not calibrated within these environments; normally they are evaluated at room temperature and average humidity. All instruments that are in high humidity or very dry climates should be considered for recalibration.

With instruments that utilize an electrochemical sensor, condensation and temperature increases (as well as low temperature) can clog the semipermeable membranes of the sensor's surface, giving lower readings than what is actually present. In foggy atmospheres (see temperature dew point), sensing can also become skewed.

Very high or very low temperature can affect the monitor's readout, especially with digital displays. In the intense sunlight the liquid crystal display (LCD) will turn black, rendering the display (thus your information) inoperable. The user of real time instruments cannot see the atmospheric level of the chemical. In cold environments the reaction time of the LCD is very slow, so our monitoring strategies must compensate for this delay.

Melting point is the temperature at which a material changes from a solid to a liquid. As stated previously, a solid is easier to manage than a liquid. If a material has a low melting point, it could be expected to become a liquid. This material could possibly even become a gas if the boiling point is also low. Liquid materials can pose problems in containment and/or confinement, along with decontamination.

The melting point of some aerosols can affect the monitoring system's sensor either from a saturation standpoint or a physical melting point. In either case, the filament that senses current is affected, in turn affecting the instrument's readout.

Specific gravity is a comparison of weight between volume of water and the material being tested. The material is in a liquid or solid form when considering the specific gravity. Water (being 1.0) is compared with the material. If the tested material is less than 1.0, then the material will float. If the material is greater than 1.0, then the material will sink. When dealing with liquid materials, we must know if the product will float or sink.

Specific gravity is also sometimes called specific density. These numbers change with temperature. Most all chemicals are referenced to 20° C. *Do not* confuse the specific gravities with the vapor densities of gases.

Vapor density, like specific gravity, is a comparison, but this relationship compares air to the gaseous material. Air is given a density of 1.0. If the relationship indicates a number greater than 1.0, then the vapor will drop or settle. If the comparison yields a number less than 1.0, then the vapor will rise, creating a vapor cloud. This vapor may or may not dissipate. Vapor density is one of the properties considered when dealing with plume dispersion models.

Vapor pressure is the pressure of a product within a container as it boils off or is forced to "evaporate." Each atom within the material bounces about until it reaches enough velocity to escape the liquid. Once this molecule has escaped the liquid form, it is surrounded by air molecules, giving it the gaseous state. It is the movement of these atoms in the gaseous state that is measured as vapor pressure. All liquids have a vapor pressure.

Based on the vapor pressure we can calculate the potential concentration in parts per million within a container or confined space, provided that the chemical is at normal ambient temperature. By multiplying the vapor pressure by 1300 we reach an estimated concentration in parts per million. Some of the literature refers to this as space concentration. This can only be calculated for a commodity within a confined area.

Pentachlorophenol (PCP), commercially used as a fungicide, clandestinely as a drug, has a vapor pressure of .0001 mmHg. If we multiply the VP by 1300 we have .13 ppm (1300 × .0001 = .13 ppm). The IDLH is given as 150 milligrams per cubic meter. How can we compare the parts per million with the milligrams per cubic meter?

Most testing laboratories do not give the concentration in ppm. Instead they look at the weight within a sample. By using the following formula we can calculate the ppm.

$$\text{Parts per million} = \text{mg/m}^3 \times 24.45/\text{molecular weight}$$
$$= 150 \text{ mg/m}^3 \times 24.45/266.4$$
$$= 13.7$$

The calculated IDLH for PCP is 13.7 ppm. Under normal conditions, the vapor pressure does not give us IDLH conditions, however, a toxic environment may still exist.

Solubility is the ability of a material to dissolve within another material. Certain materials are miscible in any proportion, some are not. Being miscible refers to the ability to dissolve into a uniform mixture. The dependent factors here are the polarity of the material and the concentration of those materials.

Whenever we talk about a solution (or solubility) we are discussing a homogeneous mixture. All the parts of the end mixture are composed of the same material.

Flammable range is the range of concentration of a gaseous chemical that produces combustion. Within this range, the gas mixes with air and burns. Under this range, the gas-to-air ratio is too lean to burn. Above the range, the gas-to-air ratio is too rich, so again combustion will not occur.

The LEL is that minimum concentration that will ignite if a source of ignition is available. The upper explosive limit (UEL) is the maximum concentration that will ignite. Between these numbers is an ignitable atmosphere.

Different chemicals have a variety of different ranges. Chemicals that have very narrow ranges have a lower chance of ignition than do the compounds, which have a wide range of flammability. As a rule, the wider the flammability range, the greater the chance for combustion. Remember, just under the flammability ranges we may have the toxicity range. Air displacement is always a concern.

Our discussion of the organization of air monitoring is based on the four basic environmental considerations of monitoring:

1. Oxygen deficiency
2. Toxic atmospheres
3. Explosive atmospheres
4. Radioactive ionization

Oxygen Deficiency Monitors

Toxic and oxygen-deficient atmospheres have been known for several centuries. Until recently, miners have employed crude yet effective methods to recognize the hazardous atmosphere. They kept small birds or rodents in cages to warn of the deadly environment. When the animal had respiratory difficulty, the site was evacuated.

Although early workers in hazardous environments used animals, they did not understand that they were monitoring for toxic environments as well as oxygen-deficient atmospheres. One limitation to their method (and a problem with the chemical testing discussed in Chapter 4) is that the worker is moving more air than the responding signal animal. While the worker is laboring under biological demands, the animal is silently sedentary, which explains why this method at times did not work for the mining crews.

Since the first animals were used in mines, air monitoring has become more sophisticated. Today's instruments are a bit more reliable, yet they have limitations.

Some of the instruments on the market are combination devices. These instruments monitor two or more gases, but like the single oxygen monitor, they do not identify the atmosphere. All the monitors do is alert the user to the oxygen (or other gas) deficiency. This deficiency could be due to poor ventilation or because the oxygen concentration is below normal, poor maintenance of the instrument, instruments that are not calibrated properly, or displacement of another gas can all lead to activation of the monitor.

There is oxygen all around us. The normal concentration of ambient air is 20.7%–20.9%. If we look at this mathematically we can see that the difference in oxygen concentration is 0.2%. This translates to approximately 2,000 ppm. But because oxygen represents one-fifth of the air, a decreased reading as stated previously actually represents 10,000 ppm. This difference occurs over hundreds of miles so physiologically we do not "feel" this difference. Air pressure has a lot to do with the concentration of oxygen in air. The column of air that represents our understanding of atmospheric pressure differs as the altitude changes. For example, the oxygen concentration at sea level is approximately 20.9 %. At around 5,000 feet above sea level we have a concentration of 17%–18% oxygen. If we were to use an oxygen deficiency monitor calibrated in Denver (approximately 5,000 feet above sea level) in Orlando, Florida, the oxygen concentration would read above the 21% mark. Highly unlikely in ambient air, without an oxygen source.

OSHA has set the lower limit of oxygen to be 19.5%. Some monitors have an audible alarm when the oxygen concentration drops below the 19.5% mark, others do not. With the monitors that do not have an alarm, it is up to the emergency response personnel to observe the needle fluctuation. If the needle is not watched, the atmosphere will be interpreted as being unaffected. Other units have a needle "hold," such that when the monitor recognizes a fluctuation the needle remains at the level of sensing until it is manually readjusted.

Your monitor should have an audible alarm that continues to sound even if the atmosphere returns to normal or the needle should stay until it can be readjusted. This will alert the emergency responder that for a moment the ambient atmosphere was oxygen deficient (or displaced). NEVER turn off the audible alarm

! SAFETY While many instruments can monitor two or more gases, single gas monitors don't identify the atmosphere; they can only alert the user to a gas deficiency.

! SAFETY OSHA has set the lower limit of oxygen in ambient air at 19.5%.

! SAFETY Never turn off the audible alarm function of a monitor, and be sure to reset after the alarm has sounded.

function and reset after the alarm has been sounded. Find out why the alarm or needle fluctuation occurred.

Most of the instruments on the market read oxygen concentrations from 0% to 25%. Only a few read to 100%. Above the 21% mark is considered an oxygen-enriched atmosphere (OSHA 23.5%; 29 CFR 1910.146). This environment is potentially explosive.

The instrument is basically a sensor that "looks" for the oxygen molecule. A pump or hand squeeze is used to introduce the atmosphere into the instrument. The readout may be digital or analog. If the instrument has a remote feed sampling line, this enables the operator to evaluate the atmosphere at a selected height or depth. However, caution should be taken when utilizing remote lines of any length. The instrument will read the atmosphere as it enters the instrument. It will take time for the gas that was just introduced into the remote sampling line to make its way through the length of the tube and into the instrument. The longer the remote feed, the longer the time. If, for example, we have the oxygen deficiency instrument hanging from our waist without a remote feed, the instrument is only reading the atmosphere waist high. If we were to use a 4-foot extension to check head level and above, the responder would have to walk slowly, giving the instrument time to evaluate the environment at the time and place of inspection. A reaction time of a few seconds (sometimes as long as a minute) occurs. As with all monitoring systems, the responder must understand the equipment enough not to give false or inaccurate readings. In our previous example, if the responder only checked above his or her head and was walking slowly enough, the atmosphere being checked is only those points that were selected above head level. Anytime you are monitoring an environment, always calibrate the instrument in a "clean" atmosphere prior to entry. In other words, test the background for the levels and compare those levels with the environment in question. Be sure to move the instrument slowly enough to get accurate readings.

Test the atmosphere in a circular or repetitive fashion above, below, and at head level. Move slowly enough so that full atmospheric monitoring can be reasonably evaluated and don't stick the probe into any liquids. If the probe is placed into a liquid, it may travel up the tube and into the instrument, damaging the monitor.

Oxygen sensors are electrochemical devices. They have two electrodes, one to sense and the other to count (Fig. 11-2). The sensing electrode maintains a constant electrical potential, while the counting electrode measures the electromagnetic potential. This potential is directly proportional to the partial pressure of oxygen (or whatever chemical the sensor is set up to interpret). The sample is introduced through a semipermeable membrane to the solution within the sensing electrodes. The electrodes in association with the oxygen present set up an electrical gradient, which is then measured by the instrument, and through an amplifier, is displayed as a percentage of oxygen.

This movement of the oxygen molecule through the membrane and into the solution causes a reaction. Remember from Chapter 3 under reactions, that electrons are gained or lost. Here in the solution bath a small electrical charge is pro-

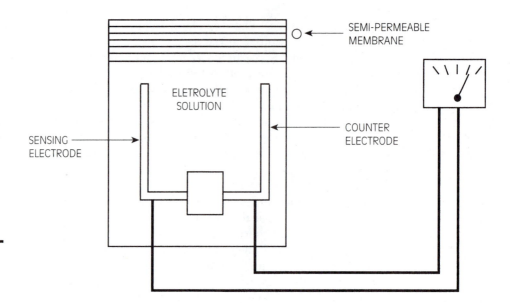

Figure 11-2 *A schematic of the oxygen sensor.*

duced. This "current" that is produced by the reaction is directly proportional to the oxygen in the air, which is why we can have one instrument that can evaluate the environment for several different gases (such as hydrogen sulfide and carbon dioxide and carbon monoxide). In these instruments there are several electrochemical sensors, one for each chemical that the instrument can detect. In each case the electrochemical sensor creates an electron current that is proportional to the contaminant within the atmosphere being tested.

Ozone, chlorine, hydrogen sulfide, fluorine, carbon dioxide, and bromine all can neutralize the alkaline electrode sensor bath. When this occurs, a false positive can be observed. When the alkaline bath has been neutralized the sensitivity of the instrument has also been reduced. Whenever a molecule of similar size (and reactivity) to the sensing molecule is introduced, a false reading occurs. For example, mercaptan, ethyl alcohol, ethylene, propane, sulfur dioxide, and nitrogen dioxide are all gases that interfere with the carbon monoxide and hydrogen sulfide electrochemical sensor.

There are several limitations of oxygen-deficient instruments:

- They are sensitive to atmospheric changes, including temperature, humidity (and dew point), and atmospheric pressure.

- They are not to be used in an oxygen-enriched atmosphere because the sensor will become saturated.

- High concentrations of carbon dioxide, hydrogen sulfide, or halogens can cause sensor failure, producing false readings.

- Temperature can affect the sensor. Operating ranges of 33°F to 120°F are common. *Do not* let the sensor freeze or come in contact with "polluting" gases.

! SAFETY
Ozone, chlorine, hydrogen sulfide, fluorine, carbon dioxide, and bromine all can neutralize an alkaline electrode sensor bath.

Three basic occurrences will be interpreted by the instrument as an oxygen drop: (1) poor ventilation within a confined space, (2) the displacement of the oxygen from another gas and, (3) sensor failure due to contact with an "acid" gas (halogens are the usual cause although exhaled air can cause "acid" gas. Never test the sensor with exhaled air).

Although we may be unsure at the time of monitoring what the true cause may be, a decrease in oxygen to the 19.5% level is hazardous. Below 19.5% an SCBA or SAR must be used (see Chapter 12). Essentially a drop of 1.5% oxygen (21% to 19.5%) concentration will translate into 75,000 ppm oxygen drop.

These types of sensors have a life span of approximately 6 to 12 months. Once the sensor is removed from the shipping wrap, the sensor is constantly monitoring the air. By the virtue of atmospheric pressure (760 torr) oxygen is constantly bombarding the sensing bath. In addition, high temperatures and humidity can greatly affect the sensor bath, eventually leading to sensor failure.

Toxic Atmospheres

Colormetric Tubes   Colormetric tubes are useful for identifying toxic atmospheres. They have a variety of names: indicator tubes, detector tubes, and associated manufacturers brand names. All can be used to gain rapid information, measuring the air contaminates (Fig. 11-3).

Unlike other DRIs, a measurement of a known contaminate can be made. The limitation is that one must know what chemical is present (so in a sense it is not giving us a known from a complete unknown). Detector tubes will not determine the chemical from a variety of chemicals. These tubes can detect a chemical or group of chemicals that react in a similar fashion.

The tube itself is glass, contains reagents, and is sealed hermetically. These reagents will react in the presence of a specific chemical or chemical group. The noted reaction is viewed through the glass tube as a "stain." On the side of the tube there may be graduations, upon which the "stain" can be read and concentration interpreted. If graduations are not placed on the tube then a measurement tool is provided and a sliding scale to denote parts per million and/or the percentage of various compounds.

If graduations are present they may be in millimeters or parts per million. In the case of the millimeter scale, a chart is used to reference the conversion from millimeters to parts per million. Some tubes only indicate the chemical reaction to a chemical group without giving any concentrations variables.

Figure 11-3 *An example of a direct reading colormetric tube.*

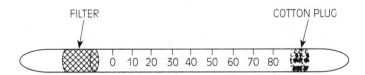

FILTER   COTTON PLUG

0 10 20 30 40 50 60 70 80

This type of testing is a concentration of the chemical in question at the time and location of the sample air. It will not tell you the whole hazardous environment or the boundaries thereof.

To retrieve an air sample using a colormetric tube, an aspirating or vacuum pump is required. The selected tube(s) are broken at each end and placed in the pump according to the manufacturer's specifications (place the correct end into the pump; most tubes are unidirectional). Each chemical and manufacturer has a reference chart as to how much air should be drawn through the tube. After placing the tube in the pump and referencing the number of "strokes" that are appropriate to gain a sample, place the instrument in the environment to be tested and the pump or vacuum will draw the air through the tube. Start the air draw and wait until the air has been displaced in the tube with the sample air. The reagents within the tube will change color or stain. This stain as compared to the measurement gives us the concentration. If the tube has a measurement graduation on the side of the tube the percentage or parts per million can be read from the tube. If the tube requires a measurement ruler with a corresponding concentration chart, then an extra piece of equipment must accompany the detector tube setup. If the tube only detects the presence of a certain chemical group, then the stain is only showing you that this chemical group is at the location of testing and at that specific point in time.

In all testing utilizing this type of detection, you must remain in the same location until all the air that was displaced by the pump occupies the appropriate space. This instrument will only accurately respond to the chemical in question at that particular location. Each location that requires testing also requires another tube.

In general, detector tubes indicate measurement in three ways:

1. Color comparison
2. Concentration conversion
3. Direct reading.

With color comparison, the tube is used after the sample has been procured and compared with a color on a reference chart. Here there will be a specific number of pump strokes for each chemical tube. The color intensity is compared to a chart, or group of tubes that have been produced by the manufacturer. The closer the color is to the comparison chart, the higher the degree of accuracy. At best this method can be off as much as 100 ppm.

Concentration conversion uses a scale in millimeters along the side of the tube. A conversion chart is required to convert the millimeter length of the stain to the concentration within the atmosphere. Without the chart, a conversion to parts per million cannot take place.

Direct reading is the simplest method and does not require an additional piece of equipment such as the comparison chart. The side of the tube indicates the parts per million in graduations. Much like the other two, the sample has a specific number of pump strokes for each chemical or chemical group. For brevity,

! SAFETY

When using a colorimetric tube, remain in the same location until all the air that was displaced by the pump occupies the appropriate space.

some of the graduations may require a multiplication factor in order to attain the parts per million.

Colormetric tubes are not without limitations. The accuracy of the tube is greatly affected when there are environmental changes. Temperature, humidity, barometric pressure, light, and shelf life all affect the chemical reaction that takes place in the tube.

Most tubes have a temperature range at which they can be used. Ideally the temperature of the calibrated tube is 68°F with 50% humidity. In a range of approximately 25° above and below this temperature the accuracy is not affected. However, some tube manufacturers have conversion tables for humidity and atmospheric pressures. The reagents within the tube may be more sensitive than one is lead to believe.

Light and shelf life are considerations if purchase of this method of detection is being contemplated. Again the manufacturer has specified shelf lives, usually 2 to 3 years being the average. During this time all stock should be rotated and shielded from all light sources. If for any reason the reagent in the tube seems to be stained prior to use, this tube should be discarded and another tube (possibly different stock) should be used. Never place the tubes in a refrigerated area for short-term storage. This will degrade the reagent giving possible false readings. However, long-term storage could be maintained by refrigeration.

If after the sampling process, the tube indicates an uneven stain, this tube should be discarded and another tube used for resampling (possibly from different stock). Sometimes the contaminate is in the air, however, the parts per million (percentage) is not high enough to give a reading (actually there is not enough chemical in the environment in order to react with the reagent within the tube). This does not mean that the air is "clean." This is simply telling you that there is not a significant level of contaminate to react with the reagent within the location at the time of test.

The reagent within each tube can react with a chemical or a specific chemical group, especially when dealing with complete unknowns. The reaction between the contaminate and the reagent produces the stain. The stain needs some "development" time. This can range from one minute to half an hour, depending on the chemical or chemical group.

The limitations of colormetric tubes include:

- Absence of stain does not mean that there is not a contaminate in the air being monitored. It may not be within the range of measurement.

- Detector tubes have an error factor of 50% or higher, due to meteorological factors, storage factors, and/or shelf life.

- Development for some chemicals may be long, thus extending the time of scene operations.

- Environmental factors greatly affect the accuracy of the detection. In cold weather, the reagents react more slowly; in hot weather the reaction is increased. In both cases the reads are false.

! SAFETY
Never place tubes in a refrigerated area for short-term storage because the reagent will degrade and possibly give false readings; however, long-term storage could be maintained by refrigeration.

- If the number of pump strokes are not accurate for the conditions of the environment and the chemical in question, the readings may be false.

- Other contaminates within the test environment can give especially high readings or low readings in certain conditions.

- Response personnel's visual perception of the reactants within the tube may vary.

Some manufacturers produce a chemical tube array, called *dosimeters*. These tubes are placed on the worker and a constant pumping of the atmosphere is brought into the tube cluster. This method is used as a personal monitoring device usually in low level contaminate atmospheres, when the chemicals of exposure are known. It can also be used with the manufacturers pump for general chemical group identification. Here several tubes of the chemical that are thought to be within the test environment are placed in a tube array. Each have the same pump strokes. The array can test for a specific chemical or chemical groupings.

Photoionization Detector (PID)   Both the photoionization (PID) and flame ionization detectors (FID) can be used to detect toxic atmospheres, however, like all monitoring devices, these units are not without some limitations.

The photoionization detector uses the ionization potential of a material. Humidity, other gases, dust, and particulate matter all can interfere with the operation of the unit, thus giving an inaccurate reading of the atmosphere.

This instrument is extremely useful in the detection of gases at low concentrations, usually where the air monitoring for leaks, low level contamination, and toxic information is helpful for operations. However, like the combustible gas indicator (CGI) a calibration gas is used in this monitor. The same basic problems that we see with the CGI occur with the PID. The range of detection (0.1–2,000 ppm) allows for a more accurate analysis of the environment, because of the principle that this monitor works on.

We all know that chemical reactions can either give up an electron or gain one. Atoms that lose/gain electrons are called ions, which are charged particles. The tendency of any material to gain/lose electrons varies from substance to substance. In the oxygen deficiency monitor a solution was used to formulate ions. This occurs as a chemical reaction between the oxygen molecule (or other gas if we have a sensor for that "other" gas) and the solution bath. We also know that because these particles are charged, they will move toward an electrode of the opposite charge, thus creating a current. This current can then be measured and displayed (usually in volts) on our monitor display.

In the PID the same basic principle occurs, however, the solution bath is now replaced with an air chamber. At either side of the air chamber are two electrodes. The particles collect on these electrodes and form a current. This current is then measured and displayed (Fig. 1-4). The degree of current has a direct relationship to the number of ions present.

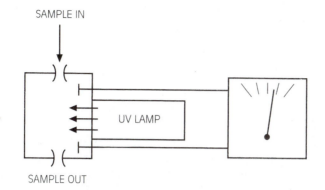

SAMPLE IN

UV LAMP

SAMPLE OUT

Figure 11-4 *A schematic of a photoionization detector.*

Normally in nature, energy must be conveyed to the atom to make it lose or gain an electron. Within the PID an ultraviolet light is used to create the energy required to make it an ion. This whole process is called *photoionization.*

Now that we have generated an ion we have to measure these ions in a meaningful way. All substances can absorb energy, some to a larger degree than others, nonetheless absorption takes place. We can measure this absorption by "watching" the level of energy required to make the substance an ion. This amount of energy that is used to formulate an ion is called the *ionization potential* (IP). This potential is the "fingerprint" of that substance and is measured in electron volts (eV). The higher the amount of energy required to produce an ion, the higher the ionization potential and of course the electron volts. Conversely the lower the ionization, the lower the required energy.

Typically, organic compounds have IPs less than approximately 12.0. Gases such as nitrogen, oxygen, and carbon dioxide have IPs greater than 12.0. For example, if your instrument has an ultraviolet light that produces maximum energy at 11.5 eV, the atmospheric gases cited above will not ionize, so they will not be detected. This is not to say that they are not there; just that they were not ionized, thus not measured. One must understand and be able to discriminate the operation in order to use the appropriate instrument.

Most PIDs on the market use isobutylene as a calibration gas. The lamp variety is 11.8, 11.7, 10.6, 10.2. 10.0, and 9.5 eV. The highest on the market is 11.8. The numbers are specific and are a result of the material that is used within the lamp. Back to chemistry!

Ultraviolet light is created within a lamp by producing an electrical charge and bombarding an excitable substance with this energy (in some instruments radio waves are used). When the electricity is generated and the atom of the excitable compound is showered with energy, the outer shells of the atom rise, then fall. This movement of the electrons generates ultraviolet light. Once focused the UV is then used to ionize the chemical in the air chamber. In order to get the variety of electron volts, a variety of chemicals are used:

9.5 eV	Magnesium fluoride
10.0 eV	Calcium fluoride
10.2 eV	Hydrogen
10.6 eV	Nitrogen
11.7 eV	Argon
11.8 eV	Lithium fluoride

The instrument must be calibrated in the factory and periodically during its lifespan (minimum of once a year, every 6 months is practical). The factory readings are produced by introducing the calibration gas to the instrument (typically isobutylene or benzene). The electron volts of the known gas and the lamp are adjusted to become "calibrated." When the calibration gas passes through the UV beam, ions are formed and current produced. During this whole process a potentiometer is used to adjust the output of the instrument in relation to the lamp with the particular gas. Once this sensitivity has been determined, the instrument is said to be calibrated.

We know that a reading of 8.4 eV is read as 8.4 ppm of the calibration gas. What is it if it is some other gas? We know that there is a direct relationship between the meter response and parts per million. In order to translate the information we must have a factory calculated response factor table or relative response (RR) patterns. These numbers when multiplied by the meter response seen on the instrument will give the concentration of the product in parts per million. Some instruments display the meter response in parts per million but this is not truly accurate.

So, for example, the relative response patterns for a 10.2 eV lamp are:

	IP	*RR*
Benzene	9.25 eV	1.0
Phenol	8.69 eV	0.78
Acetone	9.69 eV	0.63
Hexane	10.18 eV	0.22
Ammonia	10.15 eV	0.03

If we had a meter response of 4 meter units (even if the instrument states parts per million, remember it is parts per million to the calibration gas only) and think we have an atmosphere of ammonia, the ppm would be ($4 \times 0.03 = 0.12$ ppm) .12 parts per million. Here because of our limits of sensitivity this concentration may be questionable depending on the manufacturer's specifications. Because of the instrument's limitations, the PID must be used in conjunction with other monitoring systems. When the PID is utilized an oxygen monitor, CGI and FID should be used together.

Several instruments on the market contain a small microprocessor, within which are the relative response patterns. Here the instrument adjusts for the monitored gas in relation to the calibration gas. The use of charts and reference material is reduced. However, accuracy may also be reduced.

Limitations of the photoionization detector include:

- Humidity, dewpoint, and condensation all affect the UV lamp within PIDs for two basic reasons First, the IP of water vapor is 12.6 eV. Energy is not absorbed but rather scattered into defused light, lowering the energy used to ionize the gas. Second, when a lithium fluoride lamp is used, the lamp itself absorbs the water vapor, decreasing the amount of emitted light.

- Corrosive gases will etch the lamp, thus lowering the light potential.

- Any high concentration of gas will affect the PID. PIDs are designed for extremely low concentrations. Remember our concentration limitation 0.1 to 2,000 ppm.

- In high temperature atmospheres, condensation about the lamp is common.

- Dust can obscure the lamp, very much like condensation.

- Lamp not cleaned after each use can cause light to obscure.

Flame Ionization Detector In the flame ionization detector, the air is ionized by a hydrogen flame. Because it uses a flame to detect (ionize) molecules, oxygen must be present. If the atmosphere is oxygen enriched, this detection monitor should not be used.

Again, the flame ionization detector (FID) is a low-concentration detection device. The typical range is 0.2 to 1,000 ppm, which have a IP of 15 eV or less (for organic substances). Inorganic compounds are not "sensed" with FID. In other words, compounds such as hydrogen sulfide and carbon monoxide do not respond.

The principle of operation is similar to the PID, however, instead of light being the ionization energy, a hydrogen flame is used (Fig. 11-5). Like the PID and CGI, a calibration gas, commonly methane gas, is used at the factory.

Again the units of instrument measurement are that of the calibration gas. They are sometimes inappropriately enumerated as parts per million. This is true but only for the calibration gas. Instead, the units can be reported as units or calibration gas equivalents. Typically they are just referred to as units.

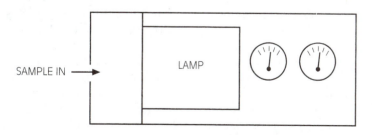

Figure 11-5 *A schematic of a flame ionization detector.*

SAMPLE IN →

LAMP

The meter readings do not give you concentrations, rather they state a possible presence. The relative response patterns are given in efficiency ratings, which are a collected group of data that suggests the degree of response of the instrument to the particular chemical.

The limitations of the flame ionization detector are:

- They cannot be used for inorganic gases.
- Their sensitivity to aromatics is very low.
- Some functional groups can reduce the detector's sensitivity.
- They require a highly skilled individual to operate and read them.

Explosive Atmospheres

Combustible gas indicators, or CGIs as they are sometimes called, utilize a Wheatstone bridge circuit as their sensor. Within the circuit there are two filaments; one filament is called the sensor filament and the other is the compensating filament. When the unit is in the on position, the filaments heat up. The gas being tested is pumped over the filaments. As the gas passes the sensor, the filament heats up. The compensating filament is a "control" for the system. It will "adjust" to conditions such as temperature and humidity (to a point). The difference in resistance (actually the decrease flow of electrons within the circuit due to the burning and heating of the sensor filament) between the sensor and the compensating filament is translated by the circuit to a meter readout (analog or digital). Obviously, here we are testing the atmosphere for the flammability component.

All CGIs have to be calibrated to a gas. The gases primarily used are methane, propane, and natural gas, but some are calibrated to pentane and hexane. Whichever gas is used, one is used as a consistent calibration gas. Essentially the heating of the sensor as it relates to the calibration gas gives us the reading. Some gases burn hot while other burn cooler. Each gas burns in relation to a specified resistance. It is this known resistance to the known calibration gas that makes it possible for us to say that the instrument has been calibrated.

CGIs are used to test the LEL within an atmosphere. They have different scales of measurement: parts per million, percent of gas, and the percent of LEL are some of the various types. The percent of LEL is the most common found on the market. Because of the calibration gas used and the scale of measurement, response factors must be used in order to interpret the meter reading.

The interpretation of the atmosphere that is the same gas that was used as a calibration gas is the easiest and the most accurate. If methane was utilized as the calibration gas, when in an atmosphere of methane at the LEL, the meter will read 100% (assuming that your meter reads percent of gas). The LEL of methane is 5.3%, while the UEL is 14%. At the LEL through the UEL the meter will read 100%. As we approach the UEL and above, the needle may fluctuate or peg to zero then back to 100%, and may then go back to zero. As you can see it is a percentage of the LEL that is read on the meter, not a direct reading of the concentration. You can also see that if the responder was not observing the needle, fluctuation may not have been noticed unless of course the instrument has an alarm or needle monitor.

If, for example, we are measuring a possible atmosphere that we think is methane, and our calibration gas is also methane, our meter will read 2.5% (actually it will read 2.65% or 50% of 5.3%). This is telling us that the instrument measured a 50% concentration of methane present due to the heating or combustion of the gas on the sensor filament. The greater the difference of the sensing filament to the compensating filament, the greater the difference or change in resistance. Change of resistance is directly proportional to the concentration of the gas. Because the LEL is 5.3%, then 50% of the LEL is approximately 2.5 (2.65%).

In atmospheres that are different, the LEL of the calibration gas is taken and multiplied by a response factor. Actually the instrument does not know if it is the calibration gas or another type of gas. All it can do is display the temperature differences (resistance) between the filaments within the Wheatstone bridge (Fig. 11-6). The resultant meter reading (or temperature difference) is the concentration of the gas as it relates to the calibration gas. From here we need a chart that describes the response curves or conversion from the calibration gas to the gas in question, thus giving the appropriate meter reading for the gas in question. Obviously one must know the gas being tested in order to fully monitor the environment appropriately.

■ NOTE

Some gases burn hotter than the calibration gas, and some cooler.

As we have seen, some gases burn hotter than the calibration gas and some cooler. At the factory, the manufacturer will introduce a variety of gases and measure the temperature (resistance) of each gas, from which the conversion factors or response curves are drawn. Actually it is the instrument's response to the gas as compared with the calibration gas in a graphical or numerical chart format. These two response curves are compared, the unknown to the known. So if we are measuring a gas that we know burns hotter than the calibration gas, then the meter reading will be greater than what is observed with the calibration gas. If the gas that we are monitoring burns cooler than the calibration gas, then the meter reading is less than the concentration within the environment tested. This works

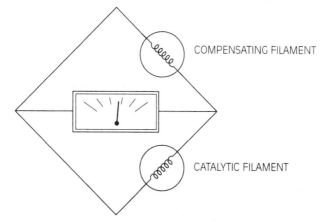

COMPENSATING FILAMENT

CATALYTIC FILAMENT

Figure 11-6   *A schematic of the Wheatstone bridge circuit in a CGI.*

out logically: A hotter burning gas takes less of the material to give high resistance, whereas a cooler burning gas requires more of a substance.

Without the reference charts or response curves, the CGI will not give you the information in a meaningful format. However, several monitors on the market have microprocessors that have been preprogrammed with this type of information, such that what is read off the display is the percent or parts per million after conversion.

In some cases an interference gas may be present. Here we may see fluctuations of the display because of the difference in filament heating. Usually a consistent drop is an indication of an interfering gas. Chlorine is a good example of an interfering gas. Here you will see (depending on the concentration of Cl_2) a sudden increase then a sudden decrease and/or a sudden increase and then stabilization.

As far as working in these atmospheres, the EPA recommends that at less than 10% of the LEL, the work may continue, but with caution. Above 10%, monitoring must continue and work is done on an as-needed basis. In excess of 25% of the LEL, the area must be evacuated.

Even though when we are well under the LEL, we are still in the middle of a chemical's toxic limits. If we have a reading on the CGI of 10% and above we are likely to be within the IDLH or ceiling levels of the gas in question.

The limitations of combustible gas indicators include:

- Oxygen enriched atmosphere will heat up the filament faster giving a false high reading.
- In oxygen-deficient atmospheres (less than 10%) they will not function properly.
- The filaments are sensitive to contaminates such as heavy metals. Organic lead vapors, sulfur compounds, and silicone, to mention a few, may corrode the filaments.
- High voltage will cause needle fluctuations or zero readings.
- Response curve factors must be used if the unit is applied to gases other than the calibration gas. In other words, you must know what gas you are dealing with in order to utilize the CGI to its optimum.
- High humidity will mask the sensor, giving a slower response or a false reading.
- Interference gases may skew the reading.

Radioactive Atmospheres

Radiation Defined   Radiation is electromagnetic energy that has the ability to destroy human tissue. Radiation is classified into two types: ionizing and nonionizing radiation.

As we discussed in Chapter 3, ionization is a rearrangement of the components of the atom. With nuclear ionization the same basic principle holds true, however, we are now also discussing the subatomic world. In general, the atomic unit is electrically "balanced." When the atom loses an electron, for example, the resultant unit is positively charged. This positively charged unit is in constant search for an electron in order for it to become "balanced" again.

With radioactive materials this same activity basically occurs, with one exception. The nuclear material has a "natural" occurrence of excess energy. The only way that this material can become "balanced" (and stable) is to release the excess energy. The process by which the energy is released is called *decay* (sometimes called *nuclear decay*, or disintegration). This energy is called *electromagnetic particles*, each type having its own name and characteristics.

Alpha particles are constructed of two neutrons and two protons. The structure is positively charged and slow moving. They also do not travel very far because they are positively charged, in search of an electron. They easily find an electron to combine with, thus forming a "balanced" or stable structure (this structure is the helium atom).

With alpha particles the shielding required could be your clothes, a newspaper, tyvek suit, or intact skin. The distance of travel is very short and at most is a few inches (4 inches) in undisturbed air.

Beta particles are basically the same as alpha in terms of being a "particle" (you will see why this concept is important when we discuss gamma). This particle is an electron that is smaller than the alpha particle, with average speeds of travel around the speed of light. These particles are dangerous and are considered to be both internal and external radiation hazards to human tissue (plants and animals included). The limitations to time of exposure, greater distance (average distance of travel from the source is 30–60 feet, with penetration, depending on the material .1 to .5 inch) and shielding apply to this particle. Several of the monitoring devices respond well to the beta particle.

Gamma radiation is a little different than what we have discussed so far. Here we have not a particle per se, but pure energy. It can be thought of as a wave of energy, very much like radio waves. Gamma radiation has a short wavelength that travels at the speed of light. The distance of travel is dependent on the parent material, however, due to its normal velocity and nonparticle structure, penetration is very deep into all materials. This is a very destructive type of radiation.

Radiation Detectors   In order to monitor these different types of radiation we must have a specially designed metering system. The most common radiation meters are the Geiger Counter and the dosimeter.

To be completely accurate the Geiger counter is called the Geiger-Mueller counter or detector (GM). The tube that senses the radiation is actually two chambers: one filled with an inert gas (helium or argon) and the other filled with a halogenated hydrocarbon. The radiation is allowed to pass through these gases by a window in the sensing wand. The radiation produces ionization of the gases and a measurement is taken. Depending on the model used, alpha, beta, and gamma radiation can be detected. Some detectors measure radiation from x-ray sources.

■ NOTE
Double the distance, the exposure rate decreases 1/4 of original exposure.

The dosimeter is a pencil type (or credit card form) meter that accumulates radiation. The rescuer needs this meter to monitor repeated exposure. X-ray technicians wear this type of monitoring device to determine the levels of exposure that they may have over a period of time.

Our environment always has a certain level of radiation present. Background radiation must be assessed prior to entry of the suspected environment (as with all hazardous environments). As stated in Chapter 5 the normal background level of radiation is approximate 100 mrem per year. Within body tissue, 50 mrem per year occurs and one airline flight from London to New York results in 4 mrem. So as you can see a constant level of radiation surrounds us.

All the topics of science have a multitude of terminology, and radiation ionization is no different. It is an attempt very much like we saw in toxicology to quantify dose. However, unlike our discussion in toxicology, we do know relatively more of the radiation exposure consequences than we know about pure chemical exposure.

Several units of measurement are used when dealing with radiation. The rad (radiation absorbed dose) is the most widely used measurement qualifier when relating radiation to body tissue. The strict definition of a rad is .00001 Joules of energy absorbed by 1 gram of material.

Our main concern with radiation is the effect it may have on living tissue. With the rad we have a direct absorption factor. As we have already discussed, radiation energy comes in different packages (and we did not mention x-rays, positrons, and neutrons). In order to quantify each radiation package, we have to have a factor that will qualify each dose. Quality factors are then used to determine dose rates (equivalence). The rem (radiation equivalent for man) and the roentgen (R, an exposure unit based on the time frame of usually 1 hour) are examples of this dose equivalence.

The roentgen is considered to be the oldest unit of measurement for radiation. This unit is most used in monitoring gamma radiation. However radiation is an active quantity. One unit that "looks" at this quantity is the curie (Ci). Several different units are used by the scientific community to describe radiation units, each having an equivalence to each other (for example, 1 sievert equals 100 rem, a sievert or Sv is a coulomb per kilogram, which is equal to 3876 R). However, we can describe all of what we have talked about in one easy sentence:

$$1 \text{ rem} = 1 \text{ rad} = 1 \text{ R under emergency circumstances.}$$

A maximum level of exposure under emergency conditions is the age of the individual $\times$ 1 rem or 100 rems. The EPA has an action level of 1mR/hr with a maximum annual dose of 5 rem. When the level of radiation has reached levels greater than 1mR/hr a potentially hazardous environment exists.

The limitations of radiation monitors include:

- The type of radiation (alpha, beta, gamma) cannot be determined.
- Sensitivity varies with model and manufacturer.
- Many environmental factors may give false readings.

PERSONAL PROTECTIVE EQUIPMENT

Any time an individual enters a potentially hazardous environment, personal protective equipment (PPE) must be worn. If the individual is involved with the air monitoring task, the minimal level of protection should be Level B. Level B is that level of protection that provides a minimal yet conservative level of respiratory and skin protection. At scenes where air monitoring is taking place, one must also think about the compatibility of these products with the equipment that we are using to protect us. For example Level B should light a bulb over our head with a picture of an individual in a protective garment with the air-pack over the garment. According to the standard with level B this would be fully appropriate. However, if the chemical we are dealing with can impregnate our breathing apparatus, what good is the level of protection if we do not also protect our equipment? This also holds true for air monitoring devices. In some environments the equipment itself will fail, giving us erroneous information or leaving the responder unprotected.

Nonetheless, all scenes need to be evaluated in order to establish the appropriate level of PPE. In some cases the responder may be charged with the determination of the hazardous environment. The perimeter of the event may be established by those individuals, in level C, as an example. Development of air monitoring strategies must be done before the incident and trained for in the appropriate fashion.

From the medical aspect of air monitoring, all individuals involved with patient care should have some level of monitoring to ensure that the decontamination practices that were employed truly cleaned the patient. Many times the decontamination of the patient is not done by necessity. In these cases the possibility of secondary contamination to the unaware and unprotected rescue worker is quite high. All victims who have been exposed to a hazardous material should be monitored in order to establish the need for possible further decontamination and or rescue worker protection.

The back of an ambulance is, for all practical purposes, a confined space. Proper air monitoring should be done and the appropriate personal protection should be worn. Continual assessment of the environment is essential to good patient care. Along with air monitoring, protective equipment wraps should be employed to ensure safe decontamination of equipment.

Whenever air monitoring takes place, whether it is within the exclusion zone (hot zone), contamination reduction corridor (decontamination, warm zone), treatment sector (support, cold zone) or ambulance, personal protective equipment must be worn (Fig. 11-7).

Whenever air monitoring takes place in any environment, a systems check of the equipment is essential, including a background monitoring for average (sometimes referred to background testing) chemical levels. For example, when monitoring for carbon dioxide, a background level should be done to compare with the CO_2 within the environment to be tested.

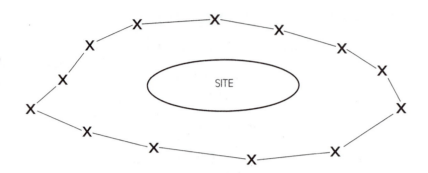

Figure 11-7 *Monitor around the site to find the acceptable levels. This area may end up to be a circle, box, or any geometric shape.*

calibration

a controlled exercise in which the full range of an instrument is analyzed

testing

verification of the basic operation of an instrument

All equipment must be tested with air monitoring devices that are calibrated periodically. This not only protects the responder, ensuring the equipment is ready for use, but also protects the agency against litigation and liability. Most manufacturers recommend that a calibration occur prior to entry, ensuring that the instrument is reading correctly during current conditions. Let us define the difference between *calibration* and *tested*.

Calibration is a controlled exercise in which the full range of the instrument is analyzed. The results show known chemicals with known responses. These responses are compared to the manufacturer's recommended responses under the chemical concentrations that are subjected to the device. This may include appropriate response to the testing gas(es) with appropriate reporting information, alarm point(s), and response times. The calibration tests must be done by qualified, competent individuals, trained by the manufacturer and to an extent "certified" as the department's technician. The results of the tests must be logged and filed for future reference. A system by which equipment is rotated, calibrated, and repaired must all fall into this area of system operations.

Testing is done at the scene, at the beginning of a tour of duty, or anytime that a device is required to be expeditiously evaluated. It is a method by which you as the user are verifying that it is operating within normal limits. It is not a calibration of the instrument. Any operator, not necessarily a certified technician, can do a test.

All monitoring tasks inclusive of the previous discussion should be well outlined within standard operating guidelines. Each individual within the team and support personnel should know who the calibration/repair personnel are, how calibration takes place, and when.

Summary

Air monitoring has come a long way since miners first took a bird or small animal into their work environment to today when we monitor hazardous materials sites, confined spaces, and the normal workplace with microprocessor technology. All monitoring is an attempt to make the work area safer. The strategies of air monitoring are many and the concerns are real. The personnel who are responsible for the air monitoring at any incident must have a well-developed knowledge of the instruments they use and their limitations. Selection, proper functioning, and testing of the equipment is essential if accurate and precise information is to be gathered.

It is our job as emergency responders to educate ourselves as to how to use air monitoring devices. Some of the instruments that are now currently being used and are considered "safe" are not suitable for the operations in which they are being employed. All instruments are machines and these machines do not think. So it is up to the user to understand, interpret, and modify his/her sampling strategies in order to gain the substantive information required at such incidents.

Scenario

You are on a scene of a hazardous materials incident. You have been placed in charge of the air monitoring sector. So far the information that has been given to you has been very sketchy. The wind is from the southwest at 2 mph, with a temperature of 88° and a humidity factor of 50%. Given the following questions, how will you appropriately monitor for contaminates?

Scenario Questions

1. How will the given weather conditions limit your air monitoring process, or will they?

2. If this incident occurred in the winter with temperatures of 40° and wind speed of 5 mph, how would monitoring change?

3. If this incident occurred in the summer with temperatures of 95° and humidity of 80%, without any wind, could this affect the process and how?

4. Given the above scenario with chlorine as an interference gas, what would be your plan of action?

5. Describe the limitations of the following:
   - Oxygen deficient monitors
   - Colormetric tubes
   - Photoionization detectors
   - Flame ionization detectors
   - Combustible gas indicators
   - Radiation detectors

6. What level of personal protection should one be in when providing air monitoring?

7. Discuss the considerations of air monitoring in terms of patient contact.

Chapter

12

Confined Space Medical Operations

Objectives

Given a confined space incident, the student should be able to identify the appropriate level of entrance along with rescue and safety procedures.

As a hazardous materials medical technician, you should be able to:

- Identify the laws surrounding permitted and nonpermitted entrance into the confined space.
- Define confined space and OSHA's four criteria for permitted entrance.
- Discuss scene management as it pertains to confined space entry.
- Discuss the atmospheric hazards that may be present, identifying each concern and associated hazards.
- Discuss the physical hazards that may be present, identifying each concern and associated hazards.
- Describe the similarities between the hazardous materials accident and the confined space response.

When researching material for this chapter, we found nothing on the medical approach to a confined space operation. Luckily, (or unfortunately) the problems that surround the hazardous materials incident are also found in confined space operations. There is a high degree of danger, such that a knowledgeable approach is a necessity. They are costly, labor-intensive operations, and rescue operations can occur within a hostile environment.

This chapter is not a complete reference on confined space operations. It will enlighten the reader on the general principles, medical aspects, and the philosophy behind such operations. It is an attempt to identify a systematic medical approach to the problem. As time passes and new objectives are identified, additional information will be known and a more refined approach will manifest itself, providing a greater degree of safety.

Just like hazardous materials response, the confined space response is on the rise. In general more than 10,000 accidents occur annually, most of which are not reported. In those 10,000 accidents more than fifty individuals will lose their lives. It is estimated that of the 500–600 minor accidents, 2% result in some sort of injury. Those accidents that occur without loss of life have tagged to them over 5,000 lost days of work (a conservative estimate).

The management of a confined space rescue is a new and complex issue. Although the laws that have been enacted (29 CFR 1910.146) have established training, very few incidents actually occur within any one jurisdiction (compared to all emergency responses). For this discussion industry, business utilities, and manufacturing plants are all synonymous, such that they are all businesses that have a confined space on the premises.

OSHA believes that this standard and its safety theories will prevent the tragedy of worker-related accidents. As our society becomes more involved in industry, each community will be faced with the complexities of such a rescue. Most of industry follows the standard, but in many areas only as a paper response. Most large companies follow the standard implicitly, but for many smaller companies complacency is the normal course of action.

In many areas the industrial community depends on the fire department's ability to handle confined space rescue. For many that require a permitted confined space rescue, the industrial fire brigade may manage the incident. However, this translates into money. Although the fire service may not be fully prepared to handle such an operation, the manufacturing company's position is to call the fire department for assistance or, in some instances, for complete management. According to the standard, the industrial community must contact the fire department and ensure their training and response abilities. Unfortunately in many areas of the country there is a situation of not knowing what the right hand is doing while the left hand is busy engaged in something else. Both business and the fire service must work together to ensure that all safety mechanisms are in place.

As with all advanced rescue operations, preplanning and extensive training foster safe operations. The standard is very specific in that all emergency workers involved with this type of rescue must have a minimum level of training, which is identified in the OSHA standard as follows:

1. Annually, the confined space rescue team must simulate a rescue operation within one of the preplanned areas or in an area that would be representative of the preplanned spaces (Fig. 12-1).

2. Each member must have a minimum of CPR and standard first aid. Within the fire service first responder and emergency medical technicians are commonplace. However, as you will see, more advanced training may become beneficial on such a scene. In these cases advanced life support may become an essential component of the operation.

3. Entry into a confined space can be defined as once the rescuer breaks the plane into the space, entry has occurred. The rescuer shall have the appropriate level of personal protection and must provide air monitoring (see Chapter 11).

For all practical purposes, in many areas of the country the fire department has become the contractor for emergency services of business. Industries' premise is that they can call the local emergency agency for the necessary assistance, thus depending on the fire department. For the most part this is an assumption and does not necessarily mean preplanning has occurred. Fire departments must have a plan of action. These standard operating guidelines must identify all known confined spaces (sewer pipes, business, manufacturing plants, etc.) and make con-

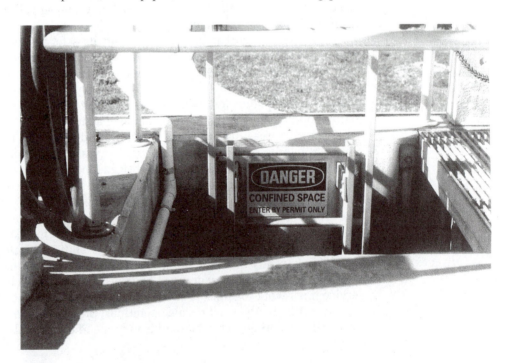

Figure 12-1
Preplanning will help assist in identifying those known confined spaces.

Figure 12-2 *Not all confined spaces are properly identified. Always beware of the space you are entering.*

tingency plans for possible unknowns (such as well pipes) (Fig. 12-2). However, once the fire service becomes involved with the emergency, a myriad of regulations apply. The fire department must follow the rules, regulations, and/or SOG that ensure a safe operation and are in compliance with the OSHA regulations (including any on-premise rules and regulations). For example in addition to 1910.146 (confined space) OSHA's 1926.956 may also apply (Construction Standards), or if in a sewer manifold 1910.268 (standard for working in manholes), or at the shipyard 1915, sub-part B. The following is an sample of the standards that may apply to the confined space rescue:

29 CFR 1910.252 Welding Operations

29 CFR 1910.261 Pulp Paper and Paperboard Mills

29 CFR 1910.262 Textile Operations

29 CFR 1910.263 Bakery Equipment

29 CFR 1910.268 Telecommunications

29 CFR 1910.272 Grain Handling Facilities

29 CFR 1910.940 Open Surface Tanks

29 CFR 1915 Maritime Standards

29 CFR 1926.650-651 Excavations, Trenching, and Shoring

29 CFR 1926.800 Underground Construction

29 CFR 1926.956 Underground Lines

How can we manage this problem? To do so, we have included topics that are strictly (so to speak) confined space operations in the tactical sense so that the reader becomes aware of the magnitude of this operation, the high degree of tactical decisions, and the associated problems. General outlines for basic and advanced life support have been deleted with the exception of specific logistical operations. Always follow local medical protocol, accepted medical customs, and above all preplan your medical approach when it comes to patient care. There will be times when certain medical practices will work, and other times when innovative techniques will have to be employed. In these situations experience, standard operation procedures, and good old-fashioned common sense (this is seasoned with direct working knowledge, research, and preplanning) will have to guide you toward a successful outcome.

You as the emergency responder must decide how to manage such an incident. In order to do so, the establishment of working SOGs, associated with solid preplanning and training will have to be done prior to system implementation. A "think tank" approach to the problem solving and decision making prior to the event is a good start in order to establish SOG and training protocols. When it comes to confined space operations (as with all hazardous materials incidents, EMS and emergency work in general) one must always take the time to establish safe working practices.

CONFINED SPACE DEFINED

As with any new obstacle, the problem must first be defined. The basic definition of a confined space (throughout industry and OSHA; the verbiage changes somewhat) is any container that has the potential of occupancy. Just about all the emergency scenes that are responded to within any community can fit into this vague explanation. However, in order to be precise, without the chance of ambiguity, OSHA has established criteria for the confined space. This space must meet all of these criteria for it to be called a "confined space." Once the area has been identified as a hazard area (remember that one of the basic concepts of hazardous materials is to identify) other laws, such as permitted entry may apply. The criteria for a confined space are as follows:

1. The size and the configuration of such a space must be such that it is large enough for a person to bodily enter to do assigned work.

2. The egress and entrance is limited. In other words, the confined space has a restricted access point.

3. The space itself was not designed for continuous human occupancy.

Once we have identified an area as a confined space, the question of permits arises. There has been much discussion on the feasibility of emergency crews acquiring such a permit for entry. Is it needed? Who requires it? These questions depend on the type of potential hazard present. Preplanning with the facility is the key to identification of permits needed. The responsibility of the permit, and the person(s) role within the question of responsibility, all must be identified before the fact.

According to the OSHA standard there are four basic criteria that establish the need for a permit:

1. The space has an atmosphere that is hazardous or has the potential to become hazardous.

2. The internal configuration of the confined space is such that a person may be trapped within the space, or may lead to asphyxiation, by inward converging walls, floors that slope inward and downward to a tapering smaller cross-section.

3. The material within the confined space can or has the potential possibility of victim engulfment.

4. Any other recognizable hazard that may cause serious safety and/or health effects is present. (Technically this could apply to our ambulance that we are going to drape for patient transport.)

An actual or potential hazard(s) that would cause death or injury does not exist in nonpermitted spaces. It is up to the emergency response official in association with the facility representative to identify the hazards, if any, and classify the space as a permitted or nonpermitted confined space. If, for example, initially the space is deemed as a permitted space (hazards present), but during initial setup all the hazards were controlled, the space (such as a straight sewer with positive pressure ventilation) can be reclassified as a nonpermitted space. There is a tactical advantage to reclassifying a space during any operation. As one may suspect, a permitted confined space operation has a list of very specific procedures. Once reclassification occurs, the hazards have been controlled, key elements, such as number of personnel, specialized equipment, and coordination activities, are no longer needed. One word of CAUTION, do not downgrade the operation from a permitted operation to a nonpermitted one for the sake of decreasing the incident. At the permitted level of confined space, a high degree of personnel and equipment is required. This would be analogous to a Level A entry within a hazardous materials incident.

The only way a reclassification can occur is if all hazards have been removed for the entire duration of the incident. As with all emergency operations, the level of entry and the downgrade or reclassification all must be documented, with rationale.

As we read these definitions we can see that the same recognition, identification, and implementation of response is seen in the hazardous materials incident (Fig. 12-3). The methodology of analyze, plan, implement and evaluate is present both in the confined space operation and the hazardous materials incident. In the hazardous materials incident we have elements of personnel safety, scene considerations, scene sectorization, and victim removal. The same exact set of problems are at the confined space rescue. One must be able to identify and evaluate

SAFETY
Do not downgrade an operation from a permitted operation to a nonpermitted one for the sake of decreasing the incident.

Figure 12-3 *As with hazardous materials incidents, confined spaces can impact on a service's resources.*

lockout/tag-out
a method used to eliminate electrical and mechanical hazards

> ! **SAFETY**
> Lockout/tag-out gives a rescuer the isolation component required to make the space safer.

the projected needs of the incident, set up scene safety, and provide a component of rescue operations For example, the cursory medical needs for entry and exit are roughly the same in the confined space operation as in the hazardous materials incident.

Within the realm of confined space rescue, strictly speaking, isolation of the incident is an important role of the first response personnel. **Lockout and tag-out** gives the rescuer the isolation component that is required to make the space safer. The spaces we are referencing here are generally found in industry. Many industrial processes require people to enter spaces that fit our definition of a confined space. The working conditions may necessitate welding, cleaning, and/or repair of equipment. To perform these tasks one must enter the space and perform the task. Electricity, mechanical movement, or stored items may be within this space. In order to provide a safe working environment, each employee has a tag that identifies the individual entering the space. This tag is then locked to the on/off switch of the machinery, in the disconnected position. This on/off switch must be capable of completely isolating energy from the equipment that is being repaired, cleaned, or maintained. The power to this particular piece of equipment and/or the surrounding equipment remains in the off or disconnected position. The tag identifies the worker(s) within the confined space. Doesn't this sound familiar as far as personnel accountability? NFPA 1500 section 3.5 states that at all emergency scenes, a system by which fire department personnel must be accounted for must be enacted. Most departments have taken this a step further and have placed within the incident management system a personnel accountability system. In confined spaces we are adding the component of additional personal safety by locking out (turning off all the power or mechanical movement) the machinery and tagging the machinery for those who enter.

At the hazardous materials incident we assign tasks and functions to the personnel. Whether these tasks are a part of standard operating guidelines or they are resource tasks, the same premise occurs, personnel accountability. At the hazmat incident we have grown accustomed to identifying the scene by using zones or sectors. In the confined space operation scene, identification and management is also utilized. The confined space operation, like the hazmat incident, can use the concept of hot, warm, and cold zones (Fig. 12-4). These zones have different definitions but the same general organization.

For example at the confined space operation, the hot zone is the confined area. This area has either a toxic, explosive, oxygen-enriched or oxygen-deficient atmosphere, (or has the potential of becoming a toxic, explosive, oxygen-enriched, oxygen-deficient atmosphere), mechanical hazards, or space constraints. It is an area where the number of rescue workers is limited. The warm zone is that area where decontamination will be performed, if needed, and in some cases the safety officer will be positioned there. The cold zone is where communications with the rescue workers, support, medical support, and incident command are located, to name a few. The safety and sectorization is similar to the hazardous materials incident

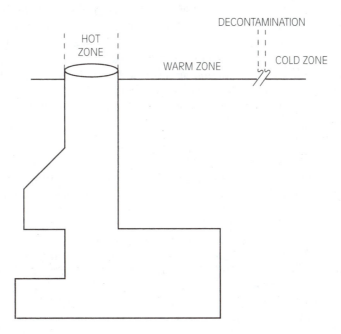

Figure 12-4

Establishing zones at a confined space is just as important as with the hazardous materials incident.

There are four general categories of hazards within the confined space, each having its own set of responsibilities and considerations. Each must be well known to all involved with the incident. These areas of concern are:

1. Scene sectorization
2. The confined space itself
3. Rescue personnel
4. The patient/victim

SCENE SECTORIZATION

Hot Zone

The hot zone is the confined space itself. It can possess toxic gases, an oxygen-enriched atmosphere, an oxygen-deficient atmosphere, temperature gradients, poor lighting, electrical hazards, mechanical hazards, and/or engulfment hazards, or it has the potential of possessing any of those hazards.

Within the confined space lighting becomes an issue as does temperature. Lighting should be of such quality that it provides a safe illuminated area. This illumination may increase the ambient temperature so ventilation becomes yet another consideration Ventilation of the space becomes important. Lighting can produce heat, which in turn may increase heat stress on the rescue worker. Lighting brings up the problem of explosive or potentially explosive situations (Oxygen-enriched atmospheres would fall into this category. Remember what happened during the Apollo program.) However, be careful when performing ventilation within the confined space. Pressurization of the space may place a once-diluted chemical into a concentrated pressure pocket, thus creating a hazard rather than removing one.

Basically the hot zone is that area defined as the confined space. The opening into this area is the demarcation of the exclusion zone. In some settings this area may have to be increased a few feet in all directions, for example an earthen shaft. If equipment and personnel are around this opening, the mouth of this shaft may collapse.

Warm Zone

The warm zone is the area around the hot zone, a "buffer" zone around the immediate hazard. Limited access of emergency responders, inclusive of the support

On January 27, 1967, NASA was performing a flight simulation of the Apollo 1 command module. During the simulation, fire swept through the command module, killing Col. Virgil I. Grissom, Lt. Col. Edward H. White, II, and Navy Lt. Comdr. Roger B. Chaffee. The fire was caused by an electrical arc within the command module. At the time the command module was enriched with oxygen.

teams and the visual safety manager may be just outside of this area. If hoisting operations are required, this area should have a hoisting technician, in addition to the safety officer (entry supervisor). This obviously depends on the type of incident.

If the situation warrants a decontamination corridor, the contamination reduction zone should be placed at the outermost edge of the warm zone, far from collapsible openings that surround the confined space.

Decontamination should also have a medical/rehabilitation sector toward the outermost area. Here medical surveillance, treatment, and rehabilitation can be performed.

Cold Zone

The cold zone should be an appreciable distance from the hot zone. Here the support teams to the entire operation, entry preparation, medical support and transport, incident command, and reference all are placed.

The incident command structure should provide input to the decision-making process at all levels. In such an operation, tasks, cognitive knowledge, and physiological factors all enter into the maintenance of a safe incident.

Roles and Responsibility

The entry team will have the highest degree of task-oriented knowledge. Each task will have to be practiced in training many times and rehearsed in the mind of the rescuer before implementation. This requires a disciplined level of cognitive information about the rescue, hazards involved, and the management of the victim.

On the psychological level the rescuers have a great deal of stress placed on them. This is very similar to the hazardous materials team member in a full encapsulation ensemble with claustrophobia being a physiological battle to overcome.

Communication between the visual safety officer (or the entry supervisor, ES) and the entry team should be on a hard-line communication link (there are manufacturers of rescue ropes that have incorporated a communication line through the center of the rope, thus reducing the second cable during operations yet maintaining high level communication). If at all possible the safety manager (ES) and the rescuer are always in visual contact. When this is not possible, an additional rescuer between the safety manager (ES) and the primary rescuer may be an alternative consideration.

Just as in the hazardous materials incident, a decontamination sector must be established prior to team entry. Personnel within this corridor must be briefed on the possible hazards of the material(s) in question and the appropriate solutions. At the end of this corridor, medical personnel with minimal protection should stage (this is dependent on the chemical that is involved).

All personnel who will enter the hot and warm zones must receive a cursory medical examination. Concern for hydration and heat stresses should be a big

Figure 12-5   *A confined space may have elements of hazardous materials, emergency medical care, rope rescue, and dive rescue.*

part of this medical preparation. The medical stress and exposures within a confined space are the same as those at the hazardous materials incident. Forced hydration with long rehabilitation periods are dependent on the heat stress incurred. At a minimum, all entry personnel must have plenty of hydration fluids, which can be monitored through weight readings on the entry team during the cursory medical and at periodic intervals.

A continuous medical evaluation should also take place—blood pressure, pulse, respirations, weight, and forced hydration at a minimum. Depending on the chemical, this exam should be expanded to the cursory medical that was discussed for the hazmat event.

Upon exit of all rescue workers, a full exit exam as discussed in Chapter 7 should be done. Heat stress and claustrophobia are the two major stressors at a confined space operation. Each must be addressed prior, during, and after the incident.

The confined space incident is an unusual and dangerous operation to manage (Fig. 12-5). The lines of communication between task players and management need to be open. Because of the wide variety of confined space operations, a procedure of roles and responsibilities should be thought out prior to the event. Pre-planning of all known "spaces" and potential ones should have standard operating procedures drawn up.

Incident Management System

The incident command system (or the incident management system as it is starting to be referred to) is a vehicle by which the management of all incidents are organized. At the confined space incident it is imperative to have such a structure in place before the rescue attempt. This management system identifies one person as the scene commander. He or she should be of strong character, knowledgeable of the tasks to be accomplished, and direct in his or her conduct. This commander is necessary at the onset of the incident all the way through completion. He or she is basically the facilitator of the incident who ensures accountability and safety. As with all rescue operations, the primary goals are to remove the victim(s), stabilize the incident, and conserve life and property.

The functions of this commander are to identify all tactical objectives in the order that they can reasonably be attained, to develop an organized structure so that tasks may be accomplished in a timely manner within that "safe" environment, to maintain communication between task function and management, and to act as liaison between the rescue workers and the management staff of the business. During this whole process, evaluating the tasks accomplished and identifying goal objectives is an ongoing process. Anticipating future needs and addressing these issues constantly becomes a part of this process.

In order to have a direct link to the task objectives and safety of the scene, communications to the visual safety officer or entry supervisor is a must. The visual safety officer is the person within the warm zone who observes the rescue operation. He or she has the responsibility of providing a safe operation in the confined space. This person needs to be identified in SOGs as the individual who will be the eyes and ears of the incident commander and who has the authority of the incident commander regarding safe operations. In other words, this individual has the overall responsibility of shutting down the operation if necessary.

The prime responsibility of the visual safety officer is to observe the rescue operation through visual and audible means. He or she will have a direct hard-line communication, radio communication, and visual observance between the rescuer in the hot zone and the entry supervisor in the warm zone. In addition to the observance of the operation, other sectors will notify this individual as to the completion of sector establishment. Each sector has a sector officer to manage and complete the goals of that particular area. After the sector officer has notified the entry supervisor (visual sector officer) of his or her establishment, the identification of sector tasks are within the sector itself. The overall goal of each sector is the responsibility of the incident commander. Once the sectors have reported establishment, the responsibility of that sector is transferred to the incident commander.

Initially the visual sector officer is in control of the confined space. The officer's responsibilities include briefing of the entry team, medical surveillance on entry, backup, and possibly decontamination of team members. As soon as entry is made, the total responsibility shifts to the incident commander. The visual safety officer's primary concern then becomes the immediate safety of the entry rescue team.

Support individuals such as hoist controller, supplied air manager, and/or remote air monitoring technician, are sometimes required in close proximity to the confined space entrance. This individual works closely with the visual safety officer, however, has a chain of command directly to the incident manager.

The decontamination team is located on the outermost perimeter of the warm zone. Here, just as in the hazmat incident, the goal is to rid the victim of any contamination. Again the possibility of secondary contamination is high (depending on the chemical). Gross decontamination (removal of all clothing) can provide between 60% and 80% reduction in contaminates, thus a lower possibility of secondary exposure. Medical monitoring of the patient should start as soon after decontamination as possible. Just as at the hazmat scene, oxygen therapy can start within the hot zone and continue through all sectors. Remember that if a contaminated atmosphere was encountered, the oxygen bottle and oxygen device will have to be decontaminated as it passes through each sector. Or each sector will have its own oxygen delivery system to give to the patient while the patient is in that particular sector. *Do not* provide oxygen in atmospheres that will create an oxygen-enriched environment, or in a flammable atmosphere. Oxygen and petroleum products *do not mix!*

The warm zone itself, other than the decontamination corridor, is a buffer zone to keep personnel and equipment at a distance. It is a secured area, discouraging any and all personnel and equipment placement.

The cold zone is where the support functions of the incident stage: Medical treatment, rehabilitation, reference, incident command, and transportation sectors are located here.

THE CONFINED SPACE

The confined space can pose a magnitude of hazards. It may have a toxic, explosive, oxygen-deficient, or oxygen-enriched atmosphere. Mechanical and electrical hazards may also be present. Temperature, ventilation, and lighting are all associated problems that pose an additional risk to this type of rescue operation.

Air monitoring is *always* a necessity when establishing the entrance criteria. Before entry is made into the "space," air monitoring must be done. This air monitoring must be tested at a minimum of four feet into the space. This procedure should take into consideration all voids, holes, and hidden areas (Fig. 12-6).

Chapter 11 presented a brief discussion of air monitoring. As seen in that discussion, instruments have a variety of limitations. One must always be aware of the limitations or what the numbers are telling you within the hazardous atmosphere. Continuous monitoring *must* start before entry and continue after all personnel have left the confined space.

! SAFETY
Air monitoring is always a necessity when establishing entrance criteria.

! SAFETY
Continuous air monitoring must start before entry and continue after all personnel have left the confined space.

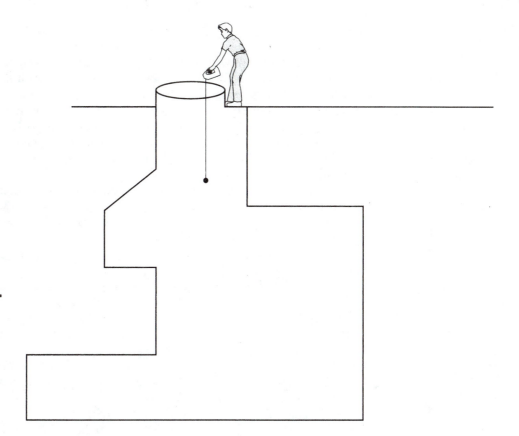

Figure 12-6 *When monitoring, take into consideration all possible voids within the space. Monitor the environment at all levels.*

There are two basic hazards hidden within the confined space that a rescuer will encounter. They may be encountered singly or in combination.

1. Atmospheric hazards
   - Oxygen-deficient atmosphere
   - Oxygen-enriched atmosphere
   - Flammable atmosphere
   - Toxic atmosphere

2. Physical Hazards
   - Electrical hazards
   - Mechanical hazards
   - Structural hazards
   - Physical space constraints

Oxygen Atmospheres

Normally we find approximately 20.7%–20.9% (20.95% is considered the average) oxygen in ambient air. OSHA has established that an oxygen-deficient atmosphere is below 19.5% and an enriched atmosphere is oxygen concentrations above 23.5%.

As we read in Chapter 3, oxygen is the first element in the Group 16 (Group VIA). Its atomic weight is 15.99994 or 16 and its density is 1.429 g/l. It is found as a diatomic molecule (O-O, or O_2), which will support combustion. If the concentration of oxygen is greater than the typical 21%, this enriched atmosphere will intensify combustion.

Oxygen is not only used for typical combustion but also as a source of energy on the cellular level. Within the body a maintenance level of oxygen must always exist. We usually think of this maintenance level as a percent of oxygen, but it also has to do with the partial pressures.

In the atmosphere we have between 20.7% and 20.9% oxygen, or roughly 21%. At sea level (14.7 psi, 760 torr, 760 mm Hg or 1 atmosphere) we see a pressure that is partial in relation to the total air volume (air having 21% O_2, 78% N_2, and 1% other gases). If we add all the pressures (Dalton's law) we would have pressure of oxygen at about 160 mm Hg ($PaCO_2$ = 40 mm Hg). As we breathe and air is brought into our lungs, the upper portion of the respiratory tree humidifies the air and in turn the oxygen. This reduces the partial pressure of oxygen to about 110 mm Hg within the alveoli ($PaCO_2$ = 40 mm Hg) and 110 mm Hg within the blood. If the partial pressure of alveoli oxygen should drop to 60 mm Hg the patient is termed as severely hypoxic.

Oxygen-deficient and oxygen-enriched atmospheres can both cause severe health problems. When the atmosphere is enriched (OSHA's definition of enrichment is greater than 23.5%), the partial pressure of oxygen within the blood increases, causing pulmonary irritation, reduced vital lung capacity, headache, dizziness, and chest pains. It is unknown if the enriched atmosphere causes a higher oxyhemoglobin saturation or if the phenomena is from the percentage of oxyhemoglobin in association with the dissolved state of oxygen within the blood plasma.

We do know that when one breathes enriched atmospheres, a percentage of the oxygen attaches to the hemoglobin and a percentage is in solution within the plasma, resulting in oxygen not bound with the hemoglobin. In long-term, high-pressure exposures (8–10 hours), intra-alveolar hemorrhage, edema, and hyperplasia of the alveolar walls have been noted. One source noted that this condition resulted in alveolar collapse over shorter time frames.

There is a direct relationship between the severity of the damage and the length of the exposure when given high concentrations (greater than 22%). If the partial pressure of oxygen is high and the length of exposure is over a long period of time (at high pressure) ophthalmologic examinations should become part of the postexposure and annual physical. Eustachian tube occlusion that resulted in pressure on the sinus causing headaches has also been documented.

Certain drugs, such as Chloroquine (Aralen), used for malaria treatments and rheumatoid arthritis and phenothiazine derivatives, used as antipsychotics, seem to increase the possibility of oxygen toxicity regardless of pressure.

On the opposite end, an oxygen-deficient atmosphere can also cause severe health effects. At 13%, any work that is performed by the body becomes extremely difficult. Mucosa becomes cyanotic, there is poor capillary refill, and the level of consciousness is minimal. At 10% cell death starts to occur with complete unconsciousness at 6%. Death is imminent if oxygen or resuscitation is not provided. Within 10 minutes, cell death occurs with irreversible damage.

The cells of the body require oxygen for metabolism. If oxygen is not available, some of the body cells continue to function but at reduced levels. This minimal percentage of oxygen that can sustain life is between 15% and 19.5%.

The causes of oxygen deficiency within a confined space are many. The rusting of the internal metal structures can cause a deficient atmosphere. Combustion, welding, cutting, and cleaning solvents all can cause the atmosphere to become deficient. It has also been noted that bacteria growing within confined areas can also cause this type of atmosphere (Fig. 12-7).

In the oxygen-enriched atmosphere, the health hazard is not our only concern. High concentrations of oxygen provide an excellent environment for combustion. The explosion hazard and increased explosion potentials when associated with other volatile chemicals only increase the risk.

In Chapter 11 we discussed oxygen monitors and the electrochemistry that enabled them to work. Typically, the sensors are a lead-(sometimes zinc) sensing

Figure 12-7 *Confined spaces need not be below grade. A stand-alone structure may present the same problems as the confined space or hazardous materials incident.*

electrode, with the counterelectrode made of platinum (a galvanic cell). These electrodes are placed in an alkaline bath (usually water and potassium hydroxide). The electrodes then measure the partial pressure of oxygen in air.

Temperature can affect the solution and electrodes. Normal operation is between 32° and 120° F. Below 32° the instrument will read slowly or not at all because of the solution freezing. Atmospheres with high water content (high humidity or approaching dew point) can cause error with this instrument due to the diluting effect that the water vapor has on the sensing device. So, for example, within a confined space that has a humidity factor close to 100% and a temperature above the ambient temperature, your air monitoring instrument will read low, possibly as low as 19%. An oxygen-deficient atmosphere! Although the reason may or may not be totally identified as a water vapor dilutant effect, consider the space as a hazard and dress appropriately (respiratory protection via hard-line or SCBA).

Any atmosphere that has an oxygen content lower than 20.7% should be considered oxygen deficient or at a IDLH level. All confined spaces should be considered oxygen deficient unless proven otherwise. We studied in Chapter 11, all the interference gases this type of monitor could read. These gases may give false readings on the "true" atmosphere. Therefore, if the deficiency cannot be controlled, high-level respiratory protection should be used.

Flammable Atmospheres

Is it explosive, flammable, or combustible? This question always comes up when teaching hazardous materials or confined space rescue. A possible end result—death of the worker—is the same. However, we must understand the difference in order to make sound decisions.

Basically we have a chemical reaction. In general the reaction is as follows:

$$\text{Oxygen (Oxidizer)} + \text{Fuel (Reduction agent)} = \overset{\triangle}{\text{Heat}} \text{ products}$$

We normally see this as the fire triangle, and to explain the burning of some products, the fire tetrahedron. How ever you think of the process, always remember that a chemical reaction is taking place.

Combustion is the term that we apply to the reaction when we see light and heat emitted from the reaction. Usually the reaction is with an oxidizer and not just oxygen. This is not to be confused with combustible liquids, which have a level of reaction related to temperature (flash point). For combustible liquids the flash point must be above 140° F and below 200° F. The other term that is sometimes confused is *flammability*. To be flammable is to have the ability to burn (or combust) easily. With flammable liquids it again has to do with the flash point. In this case, the liquids with flash points below 141° F are considered flammable.

The next time you look at a fire, notice that the flame itself is just above the fuel. This is easily seen with flammable or combustible liquids (Remember, the only difference between these two types of liquids are the flash point temperatures.

If we were to preheat a combustible liquid we would reach the flash point sooner.) Above the surface of the liquid is the flame. It is the vapors that are under reaction. The vapors above the flame also have to do with the vapor pressure of the material. If we have a material that gives off a considerable amount of vapor, then the potential for fire is high. On the other hand, if vapor pressure is low, the gaseous state is not easily reached (boiling point low, vapor pressure is high). Thus preheating must take place. This whole concept has to do with the material giving off vapors. Once an ignition source is found these vapors will sustain, flash, or ignite. In a confined space the vapors are contained, pressure will be high (i.e., vapor production and vapor absorption into the liquid material may reach equilibrium), creating a higher vapor possibly and a rich atmosphere.

Thus flash point, fire point, and ignition temperature are all related. The flash point of a material is that temperature which produces vapors sufficient enough to have a "flash" of fire with a source of ignition. Below this temperature the material does not produce enough vapor.

Above flash point is the fire point. Here we have a condition in which burning continues because the material is producing sustaining vapors, if the temperature creates the vapors (highlighting flammable liquids). If enough heat energy is present to ignite the liquid we call this the *ignition temperature*.

How does all this relate to the real world? Consider for example the confined space of a 50,000 gallon tanker. In order for it to be classified as "empty," (remember a container is never truly "empty") a residual amount of commodity remains within the container. A residual amount is termed as 1%. One percent of 50,000 gallons would be 500 gallons of commodity. Depending on the chemical, a significant degree of vapor will be produced, creating the possibility of an explosive atmosphere.

● **CAUTION**

A residual amount is termed as 1%.

We must not forget that if acidic and alkali solutions are allowed to come into contact with each other, a reaction takes place. This reaction may in turn produce a flammable gas, (steel container with sulfuric acid will produce hydrogen gas!). During the cleaning process of these containers solvents are used. The reaction with the residue and/or the tank lining itself may produce a gas (which can be oxygen displacing, toxic, and/or flammable).

One other point to consider. Some tanks are allowed to sit for long periods of time. During this time organic matter within the container may accumulate (an example of this is a tank with the opening left open, such as sewer lines, pipes, or utility vaults). From this organic matter, eventually decay will occur (another chemical reaction!), producing methane and hydrogen sulfide gas, to name a few.

The explosive hazards (flammable limits) are those areas of concentration that provide a range of ignitable gas if an ignition source is present. At the lower end of the scale we see the lowest concentration. We call this point the *lower explosive limit* value (LEL). Moving up the scale we see another point that provides the highest concentration of the gas that could ignite, called the *upper explosive limit* value (UEL). Below or beyond this range the gas is too lean (not enough of the gas) or too rich (too much of the gas) respectively.

Some materials have a very broad range while others have a narrow range. The chemical that has a wide range is extremely dangerous (alcohols aldehydes and ethers). If high vapor pressures are associated with this wide range, consider the gas a volatile hazard.

With the idea of flammable limits we must also integrate temperature into our possible scenario. In Chapter 3 and again in Chapter 11 we spoke about flammable limits and touched on the relationship that temperature and pressure play (Fig. 12-8).

We can think of this concept easily when discussing confined spaces. Within this type of environment the degree that temperature plays is important to consider. As temperature increases, the gas within the container expands and vapor pressure increases. At the same time flammable ranges may also expand. During times when the temperature drops, the reverse would be true.

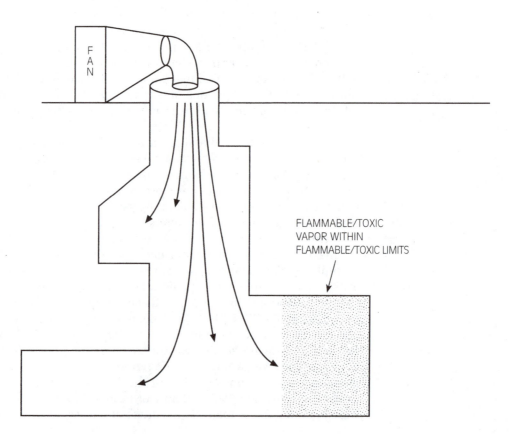

Figure 12-8 *By pressurizing a confined space you may in fact move the vapors down the space and create a toxic or flammable environment.*

We must not think of the flammable ranges as absolute values. These are reasonably stable figures, but always be prepared to encounter the flammable range below (or above) the published numbers.

Not only are we concerned with vapor pressure, temperature, and LEL values but vapor density plays a role. As we remember vapor density is a weight ratio of a chemical gas to that of the equivalent volume of air. The vapor density of any gaseous chemical can be calculated by dividing the molecular weight of the gas by 29. Within confined spaces water vapor may be a consideration with our calculation.

$$\text{Vapor Density} = MW + (18)/29$$

where 18 is the molecular weight of water and 29 is the molecular weight of air.

Dusts, suspended particles, aerosols, mists, and fumes all should be considered when entering a confined space. Each can present a magnitude of potential health and safety problems. Size, shape, and degree of suspension in relation to the chemical makeup of the gas all have an impact on the potential flammability. For example, the size and shape of a particle translates to the level of surface area the chemical has exposed. The smaller the particle, the higher the degree for a fast reaction, i.e., the more surface area.

In a flammable atmosphere (besides the personnel protection one must have) any ignition sources must be controlled. Electrical equipment, static electricity, electrical arcing, open flame, sparks, and hot surfaces can all provide the thermal energy that could start the reaction, possibly an explosion.

Under these environments intrinsically safe equipment is a must. The Wheatstone bridge within a combustible gas indicator must be isolated from the gas that is being tested. In order to do so the instrument has an enclosure about the instrument that can contain and withstand an explosion within the monitor if one should occur. This enclosure ensures that the reaction does not spread into the environment. The standards for this type of equipment can be found in the National Electrical Code, as defined by the National Fire Protection Association.

Within confined spaces the explosion potential is present if the space contains a flammable liquid or has had a volatile chemical stored within. We could discuss blast potentials and the terminology of explosion, however, the point of our discussion would be lost. The point here is the pressure that is created within a container and surrounding areas if such a reaction would take place. Let us consider an example of our problem and the resultant dynamic pressures that could result within a confined space given a few knowns. A practical application of this is that clandestine labs, confined spaces, transportation accidents, and fixed storage facilities can present an explosion hazard. Blast injuries can occur whenever there is a sudden release of energy related to an explosive incident. The rapid expansion of material going into a gas phase that move outward creating a pressure wave is the technical definition of an explosion.

Suppose we have a container that has a volume of 4,500 m^3. This container is filled with air and an explosive limit of butane. We can represent the rate of pressure increase as:

$$\text{Maximum Rate of Pressure} = K_g/V^{1/3}$$

$$= \frac{92 \text{ bar} - \text{m/sec}}{(4500)^{.333333}}$$

$$= 16.509 \text{ bar/sec or } 239.389 \text{ psi/sec}$$

This is a hypothetical situation and purely an experimentally determined number, however, at this pressure increase there is potential for container failure and building failure, not to mention the biological harm that would be incurred if emergency workers were within the container, container entrance, or surrounding area.

Of course some considerations would play into our fabricated scenario, such as, the mixture of butane and air may not become a total mixture, thus reducing our cloud of ignitable gas. The container or building may not be able to withstand the maximum overpressure of the explosion, or the mixture is not in the total space of 4,500 m³. However, this demonstrates the potential pressure buildup at an emergency scene.

When an explosive detonates, a wave of pressure moves from the center out in all directions. (Deflagration has a speed of blast wave of <3,300 feet per second and detonation has a blast wave speed of >3,300 feet per second, depending on the reference source, deflagration 1,100 ft/sec and detonation 6,600 ft/sec. In either case one is real fast and the other a little slower. To the end user the difference will not be known.) This initial shock wave is dependent on the type of explosive, confinement of that explosive, and the oxygen present (oxidizers). This pressure wave continues out in every direction until the energy has been released and equalized. The pressure wave can reach well above 1000 psi. Think of it as several hundred tons striking objects at the speed of 15,000 mph. Or a locomotive with freight moving 15,000 mph! Atomization of material, including all biological material, will occur close to the detonation. Total disintegration is not uncommon.

primary wave
a pressure wave as it moves through the air

secondary wave
a wave created when the air around a blast rushes in to fill the vacuum created during the original explosion

The pressure wave as it moves through the air is called the **primary wave**. The **secondary wave** is created when the surrounding air around the blast rushes in to fill the vacuum created during the original explosion. Because air is rushing into an area in which a vacuum has been created, the secondary pressure wave is longer in duration.

The effects of this sudden release of energy can be devastating. If the patient was far enough away from the explosion so that disintegration did not occur, one of the following four general categories of injuries will be seen, each of which has its own target organs:

- The first category is from the initial blast wave. The injuries involve the hollow organs and the interfaces between these organs. The gas-filled organs are compressed and develop a gas–pressure exchange causing injury, commonly called an *implosion injury.*

- The second type of injury is referred to as a *spalling.* This injury occurs when the pressure goes through a high density organ then strikes a low density organ, resulting in tissue violently spalding or flaking off of the low density tissue.

- The third injury involves organs that are attached or are in close proximity to each other. The primary pressure wave produces a shearing effect to organs such as the gastrointestinal tract, ear drum, and surrounding bones, lungs, and central nervous system. The secondary pressure wave causes material to fly around. This free-flying material causes traumatic injuries to the patients in the area. Blunt trauma, lacerations, abrasions, puncture wounds, penetrating injuries, and incisions are all seen. The tertiary effect that explosions have is due to deceleration injuries. Once the explosion occurs, the victim may be thrown in a direction of the blast wave travel. The target organs are abdominal viscera, lungs, nervous system, and skin.

- The fourth injury is due to the toxic gases, dusts, and fire created by the explosion. Similar to radiation emergencies, distance and shielding are the protective measures. The farther away you are and the more shielding you have from the blast wave, the safer you are.

Obviously the larger the explosion, the larger an area of concern. Victims should be treated for multiple trauma and deceleration type injuries. Management of this type patient, if contaminated, will be very difficult. An explosion that occurs within a confined space is more damaging than if the explosion occurred outside of the space. All patients, even if they are not showing signs and symptoms, should be evaluated and treated. Hospital observation may be warranted and continued for up to 12 hours.

● **CAUTION**

To have a margin of safety, entry is prohibited in all confined space environments that have an LEL greater than 10%.

In order to have a margin of safety, entry is prohibited in all confined space environments that have an LEL greater than 10%. However, under the flammable range lies the toxic values. Personnel protection must always become a part of the rescue attempt.

Toxic Atmospheres

As with all hazardous materials (including confined space operations), one must always become familiar with the chemical, physical, and toxic properties. Reactions that would not necessarily occur in a controlled laboratory setting can and do occur in the real world. The quantity of material that can occur within any one incident can become potentially hazardous. Chemical identification (research) as seen in the hazmat incident is also required at the confined space operation. However, we are not only concerned with chemicals, but with gases that are released during normal operations that can produce a toxic (and sometimes flammable) atmosphere (Fig. 12-9). Welding, painting, cleaning, and abrasive blasting can all produce a toxic atmosphere (see Table 12-1).

As we have discussed, chemicals have a variety of health reactions. Some can be acute and others chronic. Nonetheless, respiratory, skin, and ocular protection is always a high-level consideration, just as in the hazmat incident.

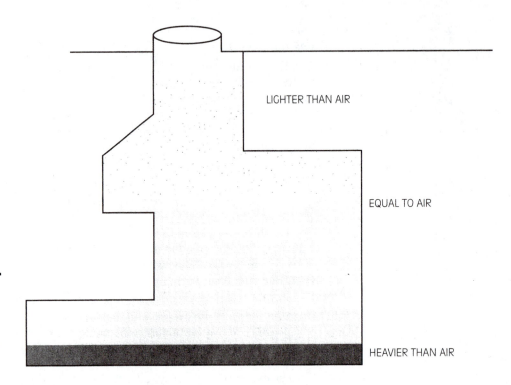

LIGHTER THAN AIR

EQUAL TO AIR

HEAVIER THAN AIR

Figure 12-9 *Beware of different chemicals that could "layer" within the confined space.*

There are four general categories for toxic gases. They can be classified as a strict asphyxiate, or as a combination of effects. Generally we can classify these chemicals into anesthetic, asphyxiant (chemical and simple), displacement, and/or irritants. However gases are categorized, all personnel within the hot zone must wear SCBA.

Electrical Hazards

Electrical hazards can occur at any emergency scene. When we talk about hazardous materials and confined space not only are we concerned with the potential electrical injury but also the possibility of an ignition source. For example, the confined space may not have electricity within it however, the motor within the space may ultimately be driven by electricity, manifesting a mechanical hazard. In all confined space operations, one must investigate the possibility of electrically driven motors, lighting, and/or any energized equipment.

Once the equipment has been identified, a lockout/tag-out procedure must take place. Lockout and tag-out is an approach for making the space safe during entry operations. Lock out means the equipment is locked in the off position. Tag-out identifies the personnel within the space.

Table 12-1 *Common chemicals encountered in confined spaces.*

Chemical Name	Formula	Density	PEL (ppm)	TWA (ppm)	STEL (ppm)	IDLH (ppm)	LEL (%)	UEL (%)	IP (eV)	Medical Notes
Ammonia	NH_3	0.5967	50	25	35	500	15	28	10.18	Pulmonary Edema
1,3 Butadiene	C_4H_6	0.650	1000	1000	1250	20,000	2	12	9.07	Irritant
Butane	C_4H_{10}	2.046	—	800	—	—	1.9	8.5		Asphy.
2-Butanone (MEK)	C_4H_8O	0.805	200	200	300	3000	1.4	11.4	9.54	Pulmon.
Carbon Dioxide	CO_2	1.557	5000	10,000	30,000	50,000	not flammable		13.77	Asphy.
Carbon Monoxide	CO	0.814	50	35	200	1500	12	74	13.98	Asphy.
Carbon Tetrachloride	CCl_4	1.589	5	2	20	300			11.47	Carcin, poisonous
Chlorine	Cl_2	2.486	1	0.5	1	30	not flammable		11.48	Irritant
Chlorobromomethane	$CHClBr$	1.93	—	200	250	5000	not flammable		10.77	CNS
Cyclohexane	C_6H_{12}	0.7206	300	300	375	10,000	1.3	8.0	9.88	Irritant
n-Hexane	C_6H_{14}	0.660	500	50	1000	5000	1.1	7.5	10.18	Irritant
Hydrogen	H_2	0.069					4.1	75	15.43	Asphy.
Hydrogen Chloride	HCl	1.268	5	5	5	100	not flammable		12.74	Irritant
Hydrogen Sulfide	H_2S	1.19	10	10	15	300	4	44	10.46	Resp.
Methane	CH_4	0.554					5	15		S. Asphy.
Nitrogen Dioxide	NO_2	1.58	5	1	1	50			9.75	Pulmon.
Nitric Oxide	NO	1.04	25	25	35	100			9.27	Resp.
Oxygen	O_2	1.429								
Ozone	O_3	2.144	.1	.1	.3	10			12.52	Resp.
n-Pentane	C_5H_{12}	0.6163	1000	120	750	15,000	1.4	7.8	10.34	Resp./Skin
Phenol	C_6H_6O	1.071	5	5	10	250	1.7	8.6	8.50	Liver/Kidney
Phosgene	$COCl_2$	1.432	.1	.1		2	not flammable		11.55	Pulmonary Edema
Propane	C_3H_8	1.56				20,000	2.1	9.5	11.07	CNS
Sulfur Dioxide	SO_2	1.45	5	2	5	100			12.30	Irritant
Toluene	C_7H_8	0.866	200	100	150	2000	1.2	7.1	8.82	CNS

Sources: *The Merck Index, 11th Edition, Niosh Pocket Guide to Chemical Hazards, Air Monitoring for Toxic Exposures, Threshold Limit Values for Chemical Substances and Physical Agents*; ACGIH; *CHRIS Manual*; USCG.

Note: When conflicting information was found in the sources, a conservative number was used. Densities are in the common state of matter, i.e., if normally a solid then the density is relative to a solid, if normally as a gas, then the density is relative to the gas. Medical notes are the most common and acute injuries found. These injuries in some cases are chronic as well.

Mechanical Hazards

Mechanical hazards are those that could cause trauma to the occupant. Lockout and tag-out procedures must take place. The mechanical equipment may or may not be driven by an electrical source. Pneumonic and hydraulic devices, common in industrial processes, need to be isolated and controlled so that mechanical injury will not occur. When pneumatic and/or hydraulic driven devices are present, all lines to the device should be bled and locked out to ensure the safety of all occupants within the confined space.

Structural Hazards

The construction of the confined space and the materials stored within the space can lead to traumatic injury. For example, in a grain silo, the grain can engulf the victim in the same manner as quicksand. On several occasions this very type of accident has occurred, resulting in death. In most of these cases, it was estimated that it took less than 20 seconds for the complete burial of the victim.

Although most confined spaces such as manholes, utility vaults, and the like have construction standards that are to be followed, sometimes these standards are bypassed, or the construction occurred prior to the confined space standard, or the space is not at the time considered a confined space. Nonetheless one must be on the lookout for potential hazards that may cause injury.

PERSONNEL CONSIDERATIONS

Entry Personnel

One of the first size-up considerations is that of atmospheric testing. We must know what the container has in it before any research or testing occurs. What is in it? How much? Has a chemical reaction taken place? Do we have a victim? What is the potential of this person still being alive? What type of physical hazards do we have? What type of mechanical or electrical hazards are there? Is it an explosive, toxic, or oxygen-deficient/enriched atmosphere?

Atmospheric testing is one of the first size-up considerations, however, as you can see several decisions must be made in conjunction with each other. Once the questions have been answered, then entry can be made.

If a potential victim is present, the same criteria for emergency entry as stated in the Chapter 2 holds true here also. Only under extreme situations should a blind entry be made (this would take into consideration lockout and tag-out with air monitoring and full Level B protection).

Safety Entry Supervisor

The primary responsibility of the safety officer (or entry supervisor) is that of visual observance under the heading of safety. He or she has the authority to stop the operation at any point. It is up to the ES to decide whether entry will take place based on the information that is given him or her from the initial size-up. The decision to enter must be based on several integral pieces of information. First, research of the confined space should lead to a specific hazard or a possibility of various hazards. Based on this information the ES should direct air monitoring and establishing a pattern for the monitoring. This pattern should be documented as the monitoring continues. Initially the air is tested several feet from the entrance to the space then slowly toward the space's opening.

The next phase of monitoring should be performed just inside the confined space's opening and in a circular pattern within a few feet of the opening. The personnel performing the air monitoring along with the ES should have Level B protection if an unknown or toxic possibility is present. Once the entrance has been investigated, the entrance remote air monitoring should continue toward the victim. Considerations for heavier-than-air and lighter-than-air gases as well as equal weight in air gases should be tested for. After the space has been completely examined from an air monitoring standpoint, appropriate control measures can be taken. Ventilation is one method for the control of the atmosphere.

● **CAUTION**

When ventilating a confined space, be aware of vapor pressurization within the space.

When ventilating a confined space, be aware of vapor pressurization within the space. For example if you had a heavier-than-air vapor present, however, below the toxic and flammable limits, ventilation may in fact push the vapor farther into the space. The concentration of the vapor may increase, reaching its toxic and flammable limits from this pressurization (see Fig. 12.8, page 450).

Ventilation may remove the vapor from the confined space while this movement may place it close to personnel operating in close proximity to the opening. Air monitoring in and around the space must start upon arrival and continue through the termination phase. Ventilation and air monitoring should be the last to conclude.

At times the response of the rescue personnel must be rapid in order to save a victim's life. In situations such as this it is the responsibility of the ES to put as many safety measures in place as possible. Here the entry team must, as a minimum, perform the rescue in Level B. The ES shall monitor conversation between the entry personnel, watching for inappropriate behavior, conversation, and signs of fatigue.

! **SAFETY**

● **A cursory physical must take place before entering a confined space.**

Just as we saw in the hazardous materials entry, a cursory physical must take place (see Chapter 7). Claustrophobia and hydration are the two most prevalent stresses experienced at the confined space operation.

Training, confidence building, and knowledge of the equipment and space can prevent the possibility of the claustrophobia from getting out of hand. For this reason we must train in an area that is representative of the spaces we are expected to enter.

If at any point a rescue worker starts to display the signs of claustrophobia, immediately this individual must be sent away from the incident to rehabilitation. He or she is then replaced to continue the operation. Even the most seasoned rescue worker can become claustrophobic given the right set of conditions.

The phobic reaction is an unconscious displacement of worry and distress. The distress is in (or sometimes out of) proportion to the outside stimulus, in this case a space that may be small and confining. Some of us can control this anxiety. Others at certain times may seem fine and at other times may not be able to perform. It is currently thought that caffeine can increase anxiety levels so it is suggested that caffeine products not be distributed at any emergency incident. The team should not look down upon claustrophobic individuals but support them to return when they feel it is appropriate.

In all situations once termination has been completed a defusing should start with a 24-hour debriefing of the incident as a must. The emotional impact that this type of rescue can have on an individual can lead to retirement from the team. These individuals have trained extensive hours prior to the event. Do not lose someone by not taking care of their emotional needs. Time and money have been invested in the rescue workers, and it is our job to take good care of them.

! SAFETY
Emergency workers must recognize the importance of hydration while working in the abnormal environment of confined spaces.

The emergency worker also must recognize the importance of hydration while working in this abnormal environment. The human body maintains a complex balance between heat loss and heat production. Heat is produced as a by-product of metabolism. Simply put, heat is generated from chemical reactions necessary for cellular work in the production of energy. Heat produced from cellular metabolism is lost or maintained by the process of radiation, conduction, evaporation, and convection. These processes are necessary for temperature homeostasis within the body. In the confined space environment, the heat within the protective clothing (same is seen in the hazardous materials incident) reduces the ability of the body to cool itself. Radiation, convection, evaporation, and convection are all working against the rescue worker.

External stressors are also a factor. High or low external temperatures also affect metabolic rates. Warmer ambient temperatures initially increase metabolic rates, but once the body core temperature increases, due to ineffective cooling, the metabolic rate decreases dramatically. This robs the body of energy causing apathy (and anxiety). The confined space may have higher than ambient air temperature and humidity factors. Along with the level of protection one must don and the activity level of the worker, we see a severe increase in body temperature.

Fluid is lost in a variety of ways on a second-by-second basis. The movement and loss of fluid is done to maintain homeostasis both for the osmotic balance of fluids within the body and for the maintenance of temperature. The important point is that in extreme environments, prehydration is of utmost importance. During long incidents where workers are exposed to high temperatures for extended periods of time or are working hard in these extreme temperatures, it is important to monitor fluid loss and intake, forcing hydration if needed to maintain fluid balance and increase energy output of workers.

If emergency workers are only allowed to drink when thirsty, the battle may be lost and heat exhaustion or, more tragic, heat stroke may be the outcome. An educated observance by medical personnel on the scene may be the deciding factor whether the battle against dehydration is lost or won.

General fatigue is yet another problem the entry supervisor must be on the lookout for. Emergency workers, because of their physiological makeup have the

SAFETY Rehabilitation of emergency workers at predesignated time intervals (30–40 minutes maximum suggested depending on level of PPE) is a must.

SAFETY Limit time frames of rescuers within the space.

tendency to overexert themselves. Rehabilitation of the worker at predesignated time intervals is a must (30–40 minutes maximum is suggested, but this is dependent on the level of PPE). Here the cursory medical (weight in particular) can be used to evaluate the worker (Fig. 12-10). Watching vitals before and after the entry can establish the medical criterion.

Confined Space Medical Status

Incident Location: ______________ Incident #: ______________

Chemical(s): ______________ DOT #(s): ______________

TEAM MEMBER:

A: ________ B: ________ C: ________ D: ________

ENTRANCE	A	B	C	D		A	B	C	D
Blood Pressure					Oximetry				
Pulse Rate					EKG 3L 12L				
Respirations					Hydration cc				
Weight					Temperature				
EXIT (one minute)	A	B	C	D		A	B	C	D
Blood Pressure					Oximetry				
Pulse Rate					EKG 3L 12L				
Respirations					Hydration cc				
Weight					Temperature				
EXIT (five minute)	A	B	C	D		A	B	C	D
Blood Pressure					Oximetry				
Pulse Rate					EKG 3L 12L				
Respirations					Hydration cc				
Weight					Temperature				

Exclusion Factors Assessment: General Sensorium: A/O × 3; Serial 3s to 27; Finger to nose; Diastolic below 100 mmHg; Pulse less than 100; Resp. less than 24; Oximetry greater than 94%; EKG = NSR; Temp between 97 and 99.5°F

Exclusion Factors must be done at five minutes exit!

The above are norms. Assessments that do not meet this criteria must be evaluated.

Figure 12-10 *Confined space medical status sheet.*

The prevention and isolation of personnel as it pertains to safety is the main objective of the entry officer. This starts with the knowledge of the incident and the ramifications of the rescue. The fact that the responsibility of safe operations ultimately lies with this individual boils down to the training of the rescue team. The central point of the operation hinges on the ES and his or her ability to effectively command his or her segment of the rescue.

Tactical Officer

It is the entry officer's job to deal with *when* entry is made. The tactical officer's (TO) duty is to consider how entry will be made. The tactical officer looks at the tasks and assigns the personnel who best suit the job. For example, if the confined space responded to has high ambient temperatures and if the TO knows that two crew members are more acclimated to a hot environment than the rest of the team, the officer will be more inclined to send those individuals in before someone who is less acclimated to a hot environment. Heat acclimatization, task skill level, medical knowledge, education on the space, and technical knowledge are just a few of the considerations the tactical officer must make on who will make entry into the particular space.

After deciding who will enter, the TO must next decide how. Will rope operations have a hand in the performance of the rescue? Will there be a need for bilingual communication? How about the size of the rescuer? It is the TO's responsibility to decide on the best short-range plan for the rescue operation. This whole process is continuous. It starts with the initial scene size-up and continues toward a maintained evaluation. Response criteria and support should be a part of this decision-making process.

The TO will brief the rescuers on the preplan of the site if known. If it is a non-preplanned space, the TO will reference the space or have it referenced to the best of his or her ability. Prior knowledge of the space is a necessity before entry is made. This includes all hazards (chemical, structural, mechanical, electrical, etc.). By knowing the space prior to entry, the plan can be discussed with the entry team and questions about the operation can be asked. This in itself can lower the anxiety felt by many rescue workers.

Once the initial scene size-up and assignment has been accomplished it is time to plan for future considerations. This is normally regarded as the strategy, with the tasks of the operation, the tactics. In the confined space operation, the initial medical strategic goals are:

Controlling the adverse factors, which in turn stabilizes the incident.

Accessing the victim for removal.

Preventing further injury to the victim and rescue personnel.

The initial tactical goals would be:

Referencing the container for a hostile environment.

Making entry utilizing appropriate PPE and air monitoring.

Stabilizing the victim for final removal.

The tactical and entry officers must work closely together to ensure that a safe yet efficient operation progresses to a positive outcome (Fig. 12-11).

There are many variables to consider at the scene of a confined space operation. Above all, safety will be ensured through constant communication. All facts must be discussed and agreed upon before the rescue attempt is made. Pre-planning is obviously a time saver when it comes to this type of operation.

Preplanning in the form of standard operating guidelines can for the most part decrease the reference time required. By having SOG and training on the SOG, the team is prepared for the tactical objectives beforehand.

Medical personnel normally work in an aggressive mode. Even the most experienced rescue worker will find it difficult, if not impossible, to move slower during the initial stages of the operation. By giving the medical personnel the task of cursory medical you have occupied their time in an useful manner, giving the air monitoring and reference team time to organize the appropriate level of information.

Patient Stabilization

Patient care is what the rescue attempt is all about. In the initial phase of the operation chaos will increase until the incident command structure is in place. As a priority the hazards are being identified and system organization constructed.

Access to the patient is the most critical. However, several obstacles must be overcome. First the air monitoring of the environment. Second medical surveillance and suitup of the rescue team. Third is the briefing and entry itself.

Once access and arrival at the patient has occurred, then initial patient assessment can occur. At this point the emergency care priorities take place—triage if more than one patient, ACBC of each victim, and movement towards packaging.

During this initial phase, if within a hazardous or potentially hazardous environment, the patients must receive a personal respirator. In atmospheres that involve a flammable component, pure oxygen is not the gas of choice. Although it would benefit the patient, the degree of hazard may prevent oxygen delivery. In atmospheres that are not explosive, consider oxygen therapy, however by giving the patient O_2 you may create an oxygen-enriched atmosphere.

Initial assessment should involve patient assessment, both primary and secondary, correcting and stabilizing any life-threatening conditions. Stripping the patient before stabilization may be a consideration to eliminate the possibility of the saturated clothes being placed outside of the hot zone.

There are several considerations when discussing the medical approach to confined space. When making entry into the space take only the bare necessities. Although needed by your patient, oxygen may create a hazard within the space. Electrocardiograms, pulse oximetry, suction devices, and anything that utilizes a

! SAFETY
● Although needed by a patient, oxygen may create a hazard in a confined space because electrocardiograms, pulse oximetry, suction devices, and anything that utilizes a battery can be an ignition source.

CONFINED SPACE ENTRY WORKSHEET

Incident Location: _____ Incident # _____
Arrival Time:_____ Entry Time:_____ Exit Time:_____ Total Team Time:_____
A Team:_____ ; _____ Safety Officer:_____
B Team:_____ ; _____ Tactical Officer:_____
Decontamination Team: _____ : _____ : _____ : _____

	ENTRY TEAM		DECON TEAM
	A	B	
All personnel briefed on procedures?	____	____	____
Medical surveillance completed?	____	____	
Entry team equipment secured and checked?	____	____	NA
Decontamination setup?	NA	NA	____
Air monitoring completed?	YES		
Communication net in place and tested?	____		
Entry area secured?	____		
Isolation to space identified?	____		
Lockout/Tag-Out in place?	____		
Ventilation provided?	____		
PPE/SCBA > 300 ft	____		
Hazards checked (elect. mech. engulf)	____		

	Time	Level	Time	Level	Time	Level	Time	Level
OXYGEN LEVELS: 19.5%–22.5%	____	____	____	____	____	____	____	____
Lower explosive limits less than 10%	____	____	____	____	____	____	____	____
Toxicity greater than PEL	____	____	____	____	____	____	____	____
Carbon monoxide < 35 ppm	____	____	____	____	____	____	____	____
Hydrogen sulfide < 10 ppm	____	____	____	____	____	____	____	____
Sulfur dioxide < 2 ppm	____	____	____	____	____	____	____	____
Ammonia	____	____	____	____	____	____	____	____
Chlorine	____	____	____	____	____	____	____	____
Cyanide	____	____	____	____	____	____	____	____
Gasoline	____	____	____	____	____	____	____	____
Arsine	____	____	____	____	____	____	____	____
Toluene	____	____	____	____	____	____	____	____
Dusts—vision less than 5 ft	____	____	____	____	____	____	____	____
Other	____	____	____	____	____	____	____	____

Considerations
Responsible party _____
Space permit/MSDS_____
Number of victims_____
Scene schematic:

Type of Space
Tank ____
Pipe ____
Other ____

Support
Equipment needed_____
Personnel needed_____
Air Monitoring Officer_____
Hoisting Officer _____
Ventilation Officer _____
Air Supply Officer_____
Medical Officer _____
PIO_____
Staging Officer _____

Figure 12-11
*Confined space
entry worksheet.*

battery can be an ignition source within the confined space. If these pieces of equipment are absolutely necessary you must have the space properly ventilated and monitored to confirm the ventilation process. Never assume a space is safe; confirm with ventilation and air monitoring.

Decontamination

Decontamination is the process by which physical or chemical substances are removed from exposed persons and equipment. By removing clothing in most cases between 60% and 80% percent of the contaminates can be removed. In the confined space operation by removing the clothing (gross decontamination) at the time of patient packaging you are assured of at least a 60% decon even before leaving the space. Coupled with the decontamination process a complete decontamination can occur fairly rapidly (depending on the chemical involved).

■ NOTE

Same procedure utilized within a hazmat incident: Decon of a chemically contaminated patient.

The process is simple. If on a backboard, place your patient into the first stage of decontamination. Scrub the patient gently to avoid any mechanical irritation. Because of the purity, a nonscented soap is suggested so that the chance of a secondary reaction is minimal. Take time to wash the areas behind the ears, about the nose and mouth, arm pits, groin (wash the male and female genitalia well), and nail beds. Always wash from head to toe. Avoid using any pressure when washing the patient. Water pressure can irritate the skin and force chemicals into the pores. By performing decon utilizing this technique you are ensuring removal of the chemical away from the airway. Limit the exposure of the chemical by avoiding contact with the runoff. Decontamination personnel also should be held to a minimum and always protected. If removal of the clothes was not done within the space, removal must occur prior to entrance into the first stage of decon.

The patient can be placed into the second stage of decon by using a scoop stretcher. Remove the initial backboard and wash the back. Place the patient on the second-stage backboard and repeat the process. Each time you will reduce the contaminate while maintaining cervical support. A three-stage decontamination corridor should be sufficient for the removal of contaminates.

Once decontamination is completed, blot the skin dry and protect the patient from the environment. Isolate the patient, limiting the number of medical personnel around the patient. Depending on the chemical, the medical personnel may be in Level B or C protection. Ideally the decontamination process has eliminated all contaminates such that the transport can occur in Level D.

During the decontamination process medical therapy can begin. However, one must realize that equipment and personnel may also have to go through decon with the patient.

Rehabilitation

The physical and mental condition of all emergency responders is a concern. While operating at any emergency scene, fatigue of personnel occurs. We all must be cognizant of the fact that deterioration of mental alertness can occur with any-one given enough time and environmental stressors. Once fatigue sets in the safety of all emergency crews are in jeopardy.

Commanders, sector officers, and supervisors must maintain an awareness of their crew's physical and mental capacities. This may sound easy but in reality it is not. All personnel should be required to undergo rehabilitation during an incident in which physical exertion has taken place. During lengthy operations, complete rotation of crews is appropriate. For example, at approximately the 4-hour mark, fatigue can be at an extreme, even without any physical activity. Although hot conditions tend to fatigue rescue workers faster than a cooler ambient temperature, any operation should have a component of rehabilitation.

The site should afford easy access to rescue crews, but should be some distance away from the "main event." The area should allow rest, so that recuperation from the physical and mental demands can occur. PPE, SCBA, and turnouts should not be allowed within this sector.

! **SAFETY**
The rehabilita-tion site should be some distance away from the "main event."

It is up to the incident commander to initiate the rehabilitation sector. For control of this sector an officer is placed in charge as a designee to the IC. From the feedback received in rehab from the medical personnel, the rehab officer can establish the readiness of crews. This allows the IC to plan for future manpower needs. If, for example, the incident is working with a small core group of personnel, once these individuals go into rehab, the need for personnel is evident. However, as the incident escalates the need for additional personnel may not be evident. By allocating personnel to evaluate rescue crews, the cycling of fresh and refreshed crews can be maintained. Through this cycling of personnel an incident can be handled without the problems of fatigue and mistakes made as a result of fatigue.

Within the rehabilitation sector such items as redehydration beverages, snacks, and a place to sit are needed to provide a relaxing atmosphere. During the rehab stay, personnel should be medically monitored by EMS crews. Vital signs and hydration are the focus. Pulse rate below 100, respirations below 18, blood pressure diastole less than 90 mm Hg, and body temperature less than 100° F should be your rehab objective medically.

All personnel who enter and exit the rehabilitation area must be recorded. Time that the crew entered, hydration fluids, vitals, and temperature must all be completed prior to exit.

Summary

The challenges of this type of rescue are many. As seen in the hazmat event, the confined space rescue breeds its own intrinsic problems. Rotation of crews, fatigue, medical surveillance and rehabilitation are some of the most challenging issues.

Over the last few years articles have been written on the "magic" time frame for rotation and rehabilitation. Some of this research has even reduced rotation and rehabilitation to formulas. On the surface these ideas look to be the most advantageous way to manage such problems, but the fact is that a human element is involved. At times individuals may be able to handle long periods of activities, while at other times, shorter durations may be more appropriate. One common idea, however, is that in order to reduce fatigue, rotation and rehabilitation should occur every 30 to 40 minutes. Rehabilitation lasting for a duration that will facilitate physiological and psychological rest will depend on the type of incident and the particular involvement of the individual.

In situations where the agency is in a standby mode, rotation from standby to active participation is anywhere between 2 and 4 hours. During this time, all standby personnel need to be briefed on the operation at set intervals. This gives two effects. First, the personnel can focus on what has been done and the direction that the incident is taking. This allows the "mental participation" into the incident. Second, the briefing gives the personnel the chance to objectively think of additional solutions away from the incident. If these individuals are then placed into the incident fresh ideas are given for scene mitigation.

This concept is not unique to the confined space rescue but can be used at any major incident. At any hazmat incidents this type of off-site systems analysis can prove beneficial to the operation.

The medical surveillance component can be handled in the same way as the hazmat event. Medical surveillance should guide the personnel availability for entry while at the same time providing a maintenance of all rescue personnel.

scenario

You and your partner are on the hazardous materials team as hazmat medical technicians. Your unit has been requested to respond to a confined space operation. You recognize the address and start to discuss the facility with your partner. She remembers that there is a 50,000 gallon container that was emptied and placed underground many years ago. Both of you try to remember the original contents of the container, but cannot.

Your prehazard plans do not describe this vessel and actually state that it has been removed per the property owner. However, the dispatcher updates the incident with a description of this vessel as the confined space. All you and your partner can remember is that it was once a container that had flammable liquids. Your discussion progresses to the weather and how hot it has been lately. The temperature is 90° F with a humidity factor of 60%.

Scenario Questions

1. Considering that this vessel may have contained a volatile chemical at one time what are your considerations?

2. How will you set up a sectorization of such an incident?

3. What are your concerns within each sector?

4. How can the incident management system assist your operation?

   • Given a small incident can you do away with the incident management system?

   • Under what conditions can you classify the confined space as a nonpermitted space?

5. List the hazard atmospheres.

6. What consideration does each hazard atmosphere present to the medical technician?

7. What are your considerations at such an incident?

appendix

A

Exposure Risk Assessment and Referencing

A hazardous materials incident can become very confusing in the initial phases of scene arrival. During the beginning stages of setting up, incident management should be the key concern to all. Injury and accidents can occur if the scene is poorly organized. This disorganization can also contribute to the magnitude of the incident in relationship to the initial amount of personnel on the scene. Until the management structure is established, stress of the incident will affect all involved.

Adding to the stress may be a lack of knowledge about the chemical(s) involved, therefore the reference sector *must* start identifying the chemical and the potential hazards as soon as possible. Delaying the information gathering can lead to an increase in secondary exposure, thus increasing patient load, needed manpower, and overall scene management. The initial amount of information that the reference sector must gather can be overwhelming. However, if the process is organized, all of the information can be gathered quickly and efficiently.

In a rescue, reference can be done quickly, gathering only the most pertinent information so that entry can be accomplished safely. In these cases there must be a risk versus benefit analysis done prior to the emergency entry. However, most scenes are conducted at a much slower pace, allowing all pertinent information to be gathered prior to entry. In either case the benefits must out weigh the risks. If the risks are larger than the benefits, one should go on to the more comprehensive research approach. Even if it means that entry is delayed, safety and positive benefit must be present.

Preplanning a hazardous materials incident is the highest level of preincident control. Through preplanning, the identification of chemical types, and the resources needed in the event of an emergency can be identified. Most U.S. communities are dynamic and ever changing. New technologies and business bring with them the increased need for the preplanning process. Communities that experience demographic changes periodically must plan for the increase or decrease in population. In view of today's changing economics and how these economics can affect the delivery of emergency service, no community can afford not to plan.

Preincident planning has become a standard for the fire service. Completing hazardous materials preplans allows a responding team to act in a more informed, efficient, and safe manner. One of the easiest ways to store and access preincident plans is to place them in a computer database. The database will assist the reference section with quick analysis of all the chemicals within the facility.

By having this information on database, accessible through chemical name, number, and location, a quick reference can be made. Unfortunately many departments that do not have the capability of building such a database must do comprehensive research.

RISK ANALYSIS

Risk analysis is an assessment of the incident and the surrounding environment. We can think of this assessment in two parts: a *monitoring assessment* and a *model* or a *graphical analysis* of the problem. With monitoring, a chemical release has occurred. Monitoring is a method by which the task operator can perform an analysis of the environment surrounding the incident and make real time decisions based on the direct air monitoring assessment. It is tangible information supported with research and known factors. See Chapter 11 for a detailed description of this technique.

Model assessment, or graphical analysis, evaluates the incident based on a group of knowns associated with a set of variable values and the assumption of the movement of these variables. Several factors are involved in this form of analysis, all of which must be evaluated with each clue and variable researched.

There are no ideal solutions to the technical problem given (that is, in the pure defined assessment of the given problem—hazardous materials release), rather the solution is a weighing of all the presented information in order to gather a viable solution. The solution should have a fundamental basis in ethics and morality. The solutions to the hazardous material incident are a quality of life decision. This decision is in essence based upon one's experience, education, and the ethics/morality one holds.

As discussed in Chapter 3, chemistry can give us clues to the chemical and physical properties and nomenclature of hazardous materials. Under the direct heading of chemical reactivity, in terms of fire potential the natural laws of chemistry and physics can give us the necessary information to make judgments based on the principles of chemistry and toxicology in order to make the best risk versus benefit decisions, however, toxicological decisions require a more comprehensive approach to analysis.

We can look at our problem and automatically make some realistic assumptions. First we know that we can have one or more chemicals released into the environment moving in one or more of three basic modes:

1. The release of a chemical from a container into the atmosphere.
2. The release of a chemical from a container into a waterway.
3. The release of a chemical from a container onto the ground.

The first release, the fastest moving condition that we can be presented with, requires an understanding of gases and their movement within the environment. Temperature, pressure, topography, and weather all have an impact on this analysis. For most chemicals it is this mode of release that we must preplan for and quickly analyze once on scene. Plume dispersion modeling can assist us in our decision making.

Although the second mode of release can happen fairly quickly, it is usually slower than a gaseous release. However, once waterways are affected, the movement of a chemical from one community to another becomes a possibility. Water sources, such as for drinking water, can become contaminated beyond control, and groundwater can be affected, thus affecting wildlife, the environment, and surrounding communities.

This leads us to the third type of release. Here the impact is long term, usually affecting more than one community and the groundwater. Runoff from any hazardous materials incident should be appropriately disposed of in order to minimize the environmental impact any of these releases may have.

Risk benefit analysis is the process by which we look at the problem, identify the known variables, and try to solve our problem while identifying the possibilities. For the most part it is easier to identify the risks and hazards than the benefits. However, it is vitally important to identify these benefits in terms of associated prioritized risks. The following detailed discussion of those variables that we can reasonably analyze is a representation of the point of in-depth research and is not necessarily a full analysis. For this reason, when one becomes involved with a hazardous materials release, one should have identified the chemical specialists within the response area so that the expertise can be tapped.

The hazard risk benefit analysis is composed of interrelated four parts that make up our initial referencing plan of action:

1. Hazard identification
2. Chemical and associated medical assessment
3. Assessment of the incident
4. Final analysis and resource contact

HAZARD IDENTIFICATION

In Chapter 1 we discussed the components of response and the theory of pre-planning. It should be apparent that the intent of the emergency response system is to identify the hazards within the community and to relate these hazards to levels of possible release. Only through a detailed analysis of the community can potential problems be identified. Community infrastructure, although not considered sensitive, can be severely crippled if affected, once the hazmat event occurs that has not been planned for.

All emergency response agencies must design procedures and practice these procedures as often as possible. Hospitals must be included in the response loop. Hospital staff are expected to help us during any large incident, yet these same individuals receive little if no training geared in system response, hazardous materials mitigation, or how to handle the chemically exposed patient.

Runoff, generation of plumes, and ground leaching, although beyond the scope of this text, must be planned for. There is a high potential for secondary exposure when the substance is thought to be harmless or is not identified. EMS, hospital, and fire departments must practice medical applications of hazardous materials. The reality is that medical personnel will be called to the incident. Once this call for help is sounded, it is assumed that the medical sectors will perform within their respective medical understanding. However, without proper training and hazardous materials insight (experience) this will not be done. Secondary exposure, although easy to manage, creates most of the low-level hazardous materials exposures.

As with any form of communications a clear and precise understanding of the chemical name is required. As discussed in Chapter 3, the difference of one letter can mean a completely different chemical (alkane vs. alkyne). Spelling the chemical using the universal phonetics is the safest way to transmit the chemical name on radio or telephone. An alkane can sound like an alkene when discussing the chemical.

Universal Phonetics

A Alpha	J Juliette	S Sierra
B Bravo	K Kilo	T Tango
C Charlie	L Lima	U Uniform
D Delta	M Mike	V Victor
E Echo	N November	W Whisky
F Foxtrot	O Oscar	X X-ray
G Golf	P Papa	Y Yankee
H Hotel	Q Quebec	Z Zulu
I Indigo	R Romeo	

CHEMICAL AND ASSOCIATED MEDICAL ASSESSMENT

Although the details of this analysis can be done by EMS, hospital, or fire department staff, all information and research must be discussed with knowledgeable individuals. It requires evaluating the chemical in terms of toxicology and chemical properties, while analyzing the scene, the situation, and the related variables.

Basic Referencing Guides

The *North American Emergency Response Guidebook*, in general, is the quickest reference book available for the emergency responder. In fact many states require every emergency response vehicle (police, fire, or EMS) to have one on board. Another useful book for emergency responders is the *NIOSH Pocket Guide to Chemical Hazards*. This book is a bit more technical however, once you are familiar with the basic organization, it can give you a wealth of information. These two books should be the minimum library carried aboard any medical unit that responds to a hazardous materials situation. Other, more advanced references and texts are listed at the end of this discussion.

Department of Transportation Emergency Response Guidebook   The *North American Emergency Response Guidebook* is the one resource that was developed (and has been maintained) for the emergency responder. Its primary goal is to identify the product and some general properties by identification numbers. From the U.N. identification number specific materials, groups of materials, and action plans can be researched. This book is excellent for the police officer, first response fire unit, and EMS unit in the field. It is even the first book used by professional hazardous materials response teams. In general it describes initial action that should to be taken in order to protect oneself, other emergency responders, and the public.

The *Emergency Response Guide* is divided into five sections, which are indicated by different colors. The white pages designate the general information and introduction to the book. This area assists the reader in how to use the book, with an explanation of its limitations. In this area the user is also introduced to the CHEMical Transportation Emergency Response Center (CHEMTREC) in Washington, D.C., and the National Response Center (NRC), both of which are manned 24 hours daily. The introduction also displays pictures of placard types, shipping paper organization, and information for Canadian shipments and shipments to the Republic of Mexico. The color representations of the placards are given so that the user can compare what she sees with what is pictured in the book.

At the back of the book are the protective action guides, which describe general actions to be taken. One point to remember is if there is no information available (placards may not be apparent due to dusk or dawn light, fire, or degradation of the placard) about the product, use guide 111.

The yellow pages are a numeric listing (U.N. number of chemical in numerical order) of specific materials or chemical families, each one followed by a guide number. The ID numbers start from 1001 to 9500 as of the last printing (1996). The guide numbers listed in this section refer to the action guides found in the orange pages. The material names identified in this section may be specific, such as chlorine with the U.N. number of 1017 or the number may identify a general commodity such as 2478 as three types of isocyanates. NOS (Not Otherwise Specified), ORM (Other Regulated Materials), PIH (Poison Inhalation Hazard), and LSA (Low Specific Activity, a radiation qualifier) are but a few abbreviations in the book. (LSA refers to packages of radioactive material and must be transported under special conditions.) When the product is highlighted, the responder should go to the green pages for the immediate action. These are initial evacuation and isolation distances and are only good for the first 30 minutes!

The blue pages list chemicals alphabetically by their name. (Chemicals have synonyms. If not found, research for the synonym must be done.) Each is followed by a guide number and the U.N. identification number. After each material name is an action guide number, and again, this refers to the action guides found in the orange section. The corresponding ID number is listed in the last column. The highlighted areas indicate additional information located in the green section for immediate evacuation and isolation.

The orange section is the action guides that are referred to by the yellow and blue sections of the *Emergency Response Guide*. Each guide number is found at the top of the page. The numbers run from 111 to 172 and they list potential hazards as well as emergency response and public safety. These are general procedures that can, and more than likely will, be expanded upon. They are a starting point for the first responder. The potential hazards, such as fire probability and explosion possibilities, as well as the health hazards are enumerated. Potential hazards identify health, fire, or explosion hazards with a public safety section identifying general safety procedures, protective clothing and evacuation. Emergency response procedures identify fire control procedures, size of the spill or leak, along with the first aid to be taken. Decontamination is not always mentioned but is always considered.

The green section gives the isolation and protective action distances including a table of initial isolation and protection distances for large and small spill. The chemicals identified in this section are those that were highlighted in the yellow and blue sections. The primary goal of this section is to identify materials that, because of their vapor production or the possibility of vapor production, can under certain conditions, produce either explosive or poisonous effects. This table is only useful within the first 30 minutes of the incident. (Remember there could have been an EMS notification delay!) The materials are listed numerically by their identification number and the name as it appears in the yellow and blue sections. Both identification numbers correspond. The very end of this section has a listing of dangerous water reactive materials with a list of toxic vapors produced. Once the vapor is identified, then the new chemical can be researched in the yellow and corresponding orange section. One primary rule about this section is if the substance is on fire or you are past the 30-minute time frame, this table should not be used. It is only good for vapors and vapor potential.

The last section which is identified by white pages has basic hazmat concepts, PPE, fire and spill control, and a helpful glossary.

The NIOSH Pocket Guide to Chemical Hazards   Produced by the National Institute for Occupational Safety and Health, this guide is more technical than the DOT guidebook. It contains more specific chemical information than does the DOT book but is quick and easy to use. The authoring services are the U.S. Department of Health and Human Services, the Public Health Service, and the Centers for Disease Control.

The introduction gives information about the service and how the data was gathered. The reader must become familiar with the variety of abbreviations used within each chemical section. Each chemical entry is very specific (unlike a few entries in NAERG). The book is limited to information on commonly found chemicals in the workplace and does not contain many of the unusual compounds you may come in contact with.

Section III discusses the use of the pocket guide with an explanation of each chemical category. Each section in the body of the book is explained in this section. The listing of chemical data begins with Acetaldehyde and ends with Zirconium compounds. Each chemical listing is read from the left-hand page through the right-hand page with a variety of information on each chemical.

A DOT number, CAS number (Chemical Abstract Service) number, chemical formula, RTECS (Registry of Toxic Effects of Chemical Substances) number, U.N. number and guide number are found in the first column.

Headings of information include Synonyms, Exposure Limits, Physical Description, Chemical and Physical Properties, Incompatibilities, Personal Protection, and Health Hazards.

Remember, in order to have true identification, the information on the chemical must come from a minimum of three sources. This applies to initial identification and referencing of the material's properties.

Advanced Referencing Techniques

This comprehensive method is composed of the following elements of research:

Chemical name	Vapor pressure or boiling point
Type of chemical	TLV–TWA–PEL–IDLH–STEL
Vapor density	Explosive limits
Suit compatibility	Solubility

With this method we look at general criteria to answer three basic questions. First, (what do we have) what are the health hazards involved with this material? Vapor pressure, boiling point, TWA, TLV, PEL, IDLH, and STEL can give this answer. Second, (what direction is it going) what explosion or fire potential do we have? Explosive limits and vapor density can provide answers to these questions.

And finally, can a offensive role be initiated or is a defensive mode more appropriate (if we did nothing at all would it change the outcome)? Through the information gathered about the chemical, suit compatibility and the characteristics of the chemical in terms of environmental impact, a tactical plan should be developed.

By looking at the more detailed referencing method it is easy to understand the reasoning toward the limited approach (presented in Chapter 1) and appreciate this more comprehensive technique. The data points that are gathered with the concentrated approach are lengthy. Preplanning in the form of identifying occupancies that present an unusually high chemical hazard, then conducting the research before an incident takes place will assist in expediting the long process when a true emergency occurs.

Chemical and Medical Properties   Within this general category, the chemical and physical properties are referenced as they relate to the incident. Medical decisions can also be made from this information.

Identify the chemical name by using the *NAERG*, shipping papers, or other intelligence. From the name use other reference material to compile a list of synonyms, chemical formula, molecular weight and typical chemical, and physical data.

For example, we have the name of a chemical, 1,2-Dichlorobenzene, but attempts to find this chemical in the *NAERG* are unsuccessful. However, from the discussion in Chapter 3 we know that 1,2-Dichlorobenzene is a synonym for o-Dichlorobenzene with the formula of $C_6H_4Cl_2$, an aromatic hydrocarbon. From this we can conclude that this aromatic may be toxic and will not dissolve in water. Once finding it in the *NAERG*, we see that it has a U.N. number of 1591. Looking up the guide number (152) shows us that inhalation of vapors can be extremely irritating and contact should be avoided. But before we make any tactical decisions, further data must be gathered to include:

■ NOTE
UN 1591—
O-Dicholorobenzene
UN 1592—
p-Dichlorobenzene

Boiling point	Vapor pressure
Flash point	Lower explosive limit
	Upper explosive limit
Exposure limits	Molecular weight
Reactions	Health hazards

By researching the chemical (o-Dichlorobenzene as an example), we can find all the synonyms and trade names. These may become important when researching the material in other books, talking to the company that processed the material, or any other research you may perform later. So for this example we find the following information:

o-Dichlorobenzene

Boiling Point: 357° F   Vapor Pressure: 1 mm

Flash Point: 151° F

Lower Explosive Limit: 2.2 %

Upper Explosive Limit: 9.2 %

Exposure Limits: TWA as 50 ppm

IDLH as 1,000 ppm

Molecular weight: 147 AMU

Reactions: Strong Oxidizer

Health Hazard: Insoluble in water, CNS depression, causing injury to liver and kidneys. If inhaled the vapor may cause respiratory depression.

Working through each of these categories we can gain insight as to the hazard we are dealing with. Boiling point, flash point, and vapor pressure all have a relationship that translates to the LEL, UEL and exposure limits.

Boiling point and flash point have a direct relationship with each other. When we see boiling point go up, flash point also increases. This has to do with size of the molecule and the polarity. In general, the larger molecules have high boiling points and thus high flash points. Conversely, the lower molecular weighted compounds have lower boiling points and thus low flash points.

Boiling point and vapor pressure have an inverse relationship. The higher the boiling point, the lower the vapor pressure and vice versa. Low vapor pressure compounds for the most part do not always pose a significant health hazard. On the other hand, most high vapor pressure (thus low boiling point) compounds must always be considered a significant health hazard, thus personnel protection should be a primary concern.

At times, air monitoring (monitoring assessment) must be performed in order to establish a perimeter. It is important to know that several contaminates in the air can alter the readings of air monitoring devices. When in doubt, work on the side of safety and dress for the worst case atmosphere.

Flammability ranges differ between compounds. General comparisons of these ranges and their relationship to flash point can be made. Compounds with low flash points develop a flammable range further away from the surface of the substance. This gives the concentrated vapors time to mix with the air. The inverse of this principle concerns chemicals with a higher flash point. Because the vapors are given off much more slowly, giving the chemical more chance to mingle with the air, the closer the flammable limits are to the surface of the material.

The size of the molecule affects the boiling point, flash point, vapor pressure and, for this discussion, the flammable limits. Remember, these are general concepts and do not hold true for all compounds. Researching the chemical is the only method that should be used for data gathering prior to entry. Following is a general list of flammable limits according to the functional grouping. The importance of this principle is the fact that toxic limits are below the flammable limits of a substance.

Functional Group	Flammable Limit Average	
	LEL %	UEL%
Esters	1	10
Alcohols	1	30
Aromatics	2	10
Ketones	2	14
Amines	2	14
Ethers	2	40
Aldehydes	3	50

Mixtures   Remember Dalton's law of partial pressure? It is the additive total of all the partial pressure that would give the resultant (total) partial pressure, when we had a mixture of gases within a container. When we have two or more gases interacting we can use the same method to figure out the possible mixture concentration for the toxic gas in question.

In Dalton's law we saw:

$$P_T = P_1 + P_2 + P_3 + ... + P_n$$

where P_T is the total pressures of all the gases, $P_1 + P_2 + P_3$ are all the gases within the container and, P_n are all the gases n = as many gases as required to measure.

OSHA and EPA have used this basic concept to calculate toxicity concentration for mixtures. Here we see the concentration divided by the exposure limit, in the same format as Dalton's law of partial pressure:

$$E_c = C_1/L_1 + C_2/L_2 + C_3/L_3 + ... + C_n/L_n$$

where E_c is the exposure concentration and C_n/L_n is the number of gases present in the form of concentrations divided by the limit value. Rewriting it we have:

$E_c = (C_n$ = on scene actual concentration)/
$\qquad\qquad\qquad (L_n$ = reported exposure limit value (TLV–TWA) + ...

When we add all the gases' concentration divided by the threshold limit value and receive a sum greater than one, the exposure limit has been met or exceeded. If the result is less than 1, then we are within the reported threshold limit values. The one problem is that the vapor pressure in relation to the ambient air temperature is not considered. For example, if our number is close to one, and all our products that we are calculating for have relatively high vapor pressures, consider that the mixture has been exceeded. If the VP is low and the temperature is high which may create vapors, again even if our number is close to one but not greater, consider it to be above one. But if our commodity has a low VP, moderate temperatures, and the number is less than one, we have approximately the reported value for each product.

Calculating Molecular Weight   The molecular weight of a compound can be derived from the chemical formula (or chemical name if you have studied chemical nomenclature). By adding the molecular weight of each element within the formula, the weight of the compound can be derived. Using the example of o-Dichlorobenzene $(C_6H_4Cl_2)$ there are 6 carbons (12), 4 Hydrogen (1) and two chlorine (35) so the molecular weight can be calculated.

$$(6 \times 12) + (4 \times 1) + (2 \times 35) = 146$$

This calculated figure is very close to the referenced figure of 147. This becomes helpful when we have to convert mg/m³ to ppm.

If the exposure limit is 300 mg/m³, how can the ppm be calculated? By using the following formula and rearranging algebraically, we can see that 50 ppm equals 300 mg/m³:

$$mg/m^3 = ppm \times molecular\ weight/24.5$$

So, for our example,

$$300\ mg/m^3 = ? \times MW\ (147)/24.5$$
$$300\ mg/m^3 \times 24.5 = ?ppm \times MW\ (147)$$
$$300\ mg/m^3 \times 24.5/147 = ?ppm$$
$$50 = ppm$$

One must be careful when referencing parts per million and parts per billion. In some literature, if the original text was from Europe, the translation may not have used the appropriate number of zeros (this is usually because someone did not take into consideration the differences between metric and English conversion).

Calculating Parts Per Million from Percent   Another means of calculating a relationship involves the LEL percent. Percent has a relationship to parts per million. Using the example of o-Dichlorobenzene, the LEL is 2%. Using a knowledge of the relationship between percent and parts per million, it can be seen that 2% equals 20,000 ppm. As you can see in this example, the toxic levels for o-Dichlorobenzene are well below the LEL. In this case (as with most) we must transverse through the toxic range well before the LEL is reached (or even 10% of the LEL which would be .2 % or 2000 ppm)!

Defining Toxic Levels   Defining toxic levels is extremely difficult. Many factors are involved with the toxic value process. So how can the information be utilized to aid in the decision-making process?

First the toxic levels must be identified through reference. The parameters identified in the reference material are items such as TLV-TWA, STEL, TCL, TDL, IDLH, and TLV-c. The testing methods used to determine these values are taken from a study group in which animal weight, diet, and breeding were controlled. Sensitivity and adaptive reactions are not well documented in most of the literature. However, when researching the chemical, look first for any human exposure. Some reference books will list the human event with the level of toxicity found. If human exposure is not documented, determine what type of animal was tested. Broad spectrum testing is the best if the human experience is not available. At times the test animal is noted during the discussion of the chemical or in parenthesis under the toxic value. Apes, monkeys, and large mammals are the next choice. From there general mammals, large dogs, cats, and rabbits will give fairly accurate information.

When comparing the information after it has been gathered (utilizing three *different resources*) look for the most toxic level and base your action plans on these numbers. Many textbooks will use the same documents to establish their levels. Make sure the information is not a repeat of what has been already found. For example you may look on the MSDS and note a TWA of 5 ppm with OSHA as the reference source. As you continue your research you find in another book value of 4 ppm with OSHA as the reference source. Take the lower value as your toxic level. Most all chemicals have a 10% safety margin and for most this will suffice. However, if the chemical presents unmanageable concerns, 50% of the lowest reported value would give the rescuer a higher than reasonable safety margin. Always err on the side of safety.

ASSESSMENT OF THE INCIDENT

Calculating Vapor Density

Vapor density can be calculated by utilizing the following formula when the compound is small:

$$MW/29 = Vapor\ Density$$

	MW	Calculated VD	Actual VD
Methane	16	0.5517	0.554
Ammonia	17	0.5862	0.597
Hydrogen	2	0.0689	0.069
Butane	58	2.0000	2.046
Sulfur Dioxide	64	2.2068	1.450
Hydrogen Sulfide	34	1.1724	1.190

As the examples show, the heavier compounds do not indicate accurate calculations.

As previously calculated, o-Dichlorobenzene = 146. It has an estimated VD of 146/29 = 5.03 (calculated). The actual VD is 1.30. However, light gases when in a high humidity environment can react similarly to gases that are heavier than air. This calculation process would be very beneficial in establishing those densities during the research phase. In some textbooks, the value is enumerated as specific gravity for the gas. For our purposes, specific gravity$_{(gas)}$ and vapor density can be considered the same.

Another application is the light gas in the presence of high humidity or water vapor. Here the vapor density can be calculated with water (18 being the molecular weight of water) as a component of the molecule:

$$MW + (18)/29 = \text{Vapor Density}_{(\text{within a high water vapor environment})}$$

Particle Size

The size of the chemical (particle size) has an impact on the hazard component of the incident. If we look at this part of the problem, we can see that weather conditions (ambient air temperature, humidity, temperature dew point), chemical and physical properties (vapor pressure, boiling point, element/compound size) can influence our scene dramatically. Containment processes and/or confinement are an integral part of the dispersion of a chemical into the environment.

■ NOTE
Plume dispersion modeling utilizes mathematical equations to estimate patterns.

Plume Dispersion Modeling   The phenomena of plume dispersion is a relatively new science in the hazardous materials field. Most of the data pre-1987 were generated from the military and some major chemical refining accidents. These data points were not done with an experimental model in mind, rather they were a compiled list of observations. This discussion is an introduction to the plume dispersion theory, presented to review the important elements as they pertain to medical engineering control systems.

Most of the present experimental models interrelate several factors to generate a computer-assisted, graphic representation of the mathematical equations. With a generation of a plume, some assumptions are made when calculating the dispersion. Most models are based on the assumptions that:

1. The plumes will release their total quantity into the surrounding atmosphere.

2. Wind speed is 3.4 mph, and the sky is clear.

3. The level of concern is calculated to be 1/10 of the published toxicity value.

4. The terrain is flat and level, without obstructions in plume dispersion models.

The reasoning of these assumptions is quite clear. As a vapor moves across the ground, wind speed tends to increase. Because of natural obstructions (caused by topography), the turbulence is not homogeneous. A wave pattern, similar to the waves' turbulence seen in the ocean as they break along the beach, is observed. This is partly due to the vertical height in which the bottom of the cloud has interference with the ground surface (friction). The center line that is observed with plumes that are emitted from a smokestack, as an example, is relatively straight; with ground dispersion there are no true center lines. This tends to make the mathematics complicated. Temperature inversions, surface friction, and obstructions all tend to create eddies in the moving vapor cloud. With this in mind, the assumptions were developed in order to give an exaggerated estimation of the plume and not a true representation.

The question that we must answer and discuss is the release management problem in general. What are the chemical and physical properties that the material possesses? How does this information effect/affect our model and management goals?

We can answer these questions with our knowledge of basic chemistry. Particle size can be discussed on separate but related levels. The first is the molecular level, and the associated chemical and physical properties; such things as vapor pressure, orientation in space, and the polarity of the molecule. The second is the information that will impact the decision-making process—the size of the "vapor" and the "type of the vapor."

- *Dust.* Solid particles that do not diffuse but tend to settle. The mouth and upper pharynx are the usual places they are found after exposure.
- *Fumes.* Solid particles that have been generated by condensation. They tend to go deeper into the respiratory system but only to the bronchi.
- *Mist.* Suspended liquid particles that can make their way into the lungs, past the bronchi and into the bronchioles, and sometimes into the alveolar space.
- *Vapors.* A gas generated by a solid or a liquid. Here vapor pressure is important to know. Anything above 760 mm Hg is a respiratory hazard. Can get into the alveolar space and the bloodstream.
- *Gases.* A formless body of material that occupies the space within a container. For our purposes the difference between vapor and gas is negligible.

In discussing any of the above "vapors," the question of how a health hazard is determined is often asked. The form that the material is in will dictate the mitigation methodology.

Some of the reference books represent the level of concern as milligram per meter cubed (mg/m³). Most of us have a difficult time understanding what this means. However, by placing our mg/m³ in the formula and algebraically rearranging the formula we can gain the parts per million, placing the number in a reference frame that we can understand. If parts per million are hard to visualize, then convert again to percent.

$$(\text{mg/m}^3) = \frac{\text{PPM} \times \text{MW}}{24.5} \quad \Rightarrow \quad \text{PPM} = \frac{(\text{mg/m}^3)(24.5)}{\text{MW}}$$

Airborne Hazard Ratio

We can look at vapor pressure differently—as the degree of airborne contaminates. As we have already discussed, the vapor pressure is the force exerted by the chemical(s) once airborne. We can apply the gas laws and evaluate chemicals based on the degree a chemical goes toward the airborne state.

1. $P = nRt/V$
2. $V = (P) \times (1 \times 10^6)/ 760$ mm Hg
3. Airborne Hazard = Vapor Pressure $_{in\ ppm}$/ Threshold Limit Value $_{in\ ppm}$

By mathematically manipulating the formula we can have a ratio formula that can be used to calculate the degree of hazard if presented with two or more chemicals that are releasing a toxic inhalation hazard and have similar physical characteristics. Basically the greater the number, the higher the hazard. For example:

Toluene VP 20 mm Hg V = $20 \times (1 \times 10^6)/760$ mm Hg = 26316

Xylene VP 7 mm Hg V = $7 \times (1 \times 10^6)/760$ mm Hg = 9210

Toluene 26316 ppm/100 ppm = 263

Xylene 9210 ppm/100 ppm = 92

In this example the airborne hazard is greater with the toluene. However, the toxicity of the chemical may have an impact. This is fairly obvious by viewing the vapor pressures and comparing. However, this procedure is distinctly beneficial when looking at chemicals with extremely small vapor pressures (.01–.1 ranges) and the relatively same TLVs. This whole procedure is very useful with aromatics and their derivatives (volatile chemical compounds). Remember to take into consideration the toxic effects that different functional groups may have on the aromatic hydrocarbon.

DECONTAMINATION

When donning protective equipment there will be times that duct tape may be necessary. The downside of this operation is that the garment will have some of the adhesive (especially on hot days) left on the suit when the tape is removed. The area that has adhesive can become a medium for the attachment of chemicals. If the taping procedure is employed, after decontamination every attempt should be made to remove the remaining adhesive.

Solubility plays an important role in the decontamination process. Decontamination of patients and entry team personnel should be done with the dilution method or utilizing a light detergent solution. Solubility in water has a relationship with polarity of the material. If, for example, the contaminate is not polar, then the chemical in question will not be miscible in water, necessitating the use of a detergent. On the other hand, if the chemical is polar, then it will be miscible in water. This general rule of thumb applies to chemicals with lower molecular weight. Always research the solubility of the chemical and its miscible potential in water prior to decontamination. The following is a listing of the polar functional groups:

Alcohols

Aldehydes

Esters

Ketones

Organic Acids

During decontamination of equipment and personnel several facts must be kept in the minds of the decontamination crew. By utilizing water in the decontamination process, the chemical characteristics will not change. The concentration of the material will decrease, however, the strength of the material will remain the same. Therefore, all runoff must be confined and tested prior to disposal. The chemical found within the decontamination corridor is a lower concentration of the chemicals found in the hot zone. The warm zone must be treated with the same respect as the hot zone.

The runoff fluids must be tested and compared to the EPA's recommendations of toxicity. If the runoff is below this standard, then the solutions may be disposed of in a "normal" manner. Contact local EPA authorities before disposal and handling of any chemical. If the levels are above the recognized level or it is felt that it may endanger the public or environment, the material must be disposed of as a hazardous commodity.

Antidotal treatment may be a consideration at the scene of a chemically exposed patient. Drugs as well as treatment modalities must be reviewed periodically. By using the reference cards pictured in Fig. 6-9 the treatment of specific hazmats can be managed easily.

Rehabilitation must accompany all hazardous materials operations. Within this area, medical surveillance and the surrounding issue must be incorporated with SOG, hydration being one of the biggest issues. Weighing before and after and again periodically within rehab will ensure that personnel have regained the lost fluids.

FINAL ANALYSIS AND RESOURCE CONTACT

Once all of the research has taken place and the appropriate sectors have received the information, the next step is to notify or contact any additional resources needed to mitigate the incident. These contacts may include the chemical manufacturer or transport company and, in some cases, a cleanup contractor. Other contacts that may be made are to agencies such as the Department of Environmental Protection, Department of Health, Department of Transportation, or to local chemical engineers. Contacting your department's public information officer or utilizing a police PIO will help keep the news media informed about the incident. Providing concrete information to the community will lessen fears generated from a major incident.

REFERENCE TEXTS

The following is a list of reference texts currently available, each of which can provide valuable information. However, like any practiced skill, familiarity with the text is suggested prior to needing the information on scene. Some of the texts require an in-depth knowledge of chemistry, whereas others are "friendlier." The reference technician must be familiar with current abbreviations and nomenclature.

Rapid Emergency Reference Guides

North American Emergency Response Guidebook, (DOT, 1993)

NIOSH Pocket Guide to Chemical Hazards (NIOSH, 1990)

Emergency Care for Hazardous Materials Exposure (Bronstein and Currance, 1988)

Handbook of Poisoning (Dreisbach and Robertson, 1987)

Poisoning and Overdose (Olsen, 1988)

Handbook of Medical Toxicology (Viccello, 1993)

Medical Texts or References

Hazardous Materials Toxicology—Clinical Principles and Environmental Health (Sullivan and Krieger, 1992)

Clinical Toxicology of Commercial Products (Gosselin)

Pesticide Fact Handbook (EPA)

Handbook of Toxic and Hazardous Chemicals and Carcinogens (Sittig)

Medical Toxicology, Diagnosis and Treatment of Human Poisoning, (Ellenhorn)

Poisoning, 5th Ed. (Arena et al.)

Clinical Management of Poisoning and Drug Overdose (Haddad et al., 1990)

Patty's Industrial Hygiene and Toxicology, Vol. II, A, B, and C (Cralley and Cralley, 1985)

Industrial Toxicology (Williams et al. Van Nostrand-Reinhold Publishers)

Toxic and Hazardous Industrial Chemical Safety Manual (International Technical Information Institute)

Goldfrank's Toxicological Emergencies (Goldfrank et al., 1986)

Industrial Medicine Desk Reference (Tver et al.)

Effects of Exposure to Toxic Gases, First Aid and Medical Treatment (Stopford and Bunn)

Manual of Toxicologic Emergencies (Noji and Kelen)

Medical Management of Radiation Accidents (Mettler et al.)

Toxicological Emergencies (Myer and Rumach)

Comprehensive Review in Toxicology (Bryson)

Handbook of Pesticide Toxicology (Hays and Laws)

Documentation of the Threshold Limit Values (ACGIH)

Threshold Limit Values for Chemicals and Physical Agents and Biological Exposure Indices (ACGIH)

Nonmedical Care Reference Texts

The Common Sense Approach to Hazardous Materials (Fire, 1986)

Chemical Data Notebook (Fire, 1994)

Emergency Handling of Hazardous Materials in Surface Transportation (AAR)

Farm Chemical Book

Rapid Guide to Hazardous Chemicals in the Workplace (Sax)

CHRIS Hazardous Chemical Data U.S. DOT/U.S. Coast Guard

Dangerous Properties of Industrial Materials, 8th ed. (Sax and Lewis, 1992)

Merck Index *Encyclopedia of Chemicals, Drugs, and Biologicals* (S. Budavari, ed.)

Chemistry of Hazardous Materials 2nd ed. (Meyer, 1989)

Hazardous Materials: Managing the Incident (Noll et al., 1988)

The *Condensed Chemical Dictionary* (Hawley, 1993)

Computer Databases

Tomes Plus, Micromedex, Denver, CO

Poisondex, Micromedex, Denver, CO

RTECS from NIOSH (Registry of Toxic Effects of Chemical Substances)

HSDB (Hazardous Substances Database) National Library of Medicine

OHM/TADS from EPA

TRI (Toxic Release Index) EPA

CD-ROM and disc

Hawleys Condensed Chemical Dictionary, Van Nostrand Reinhold Publishers

Dangerous Properties of Industrial Materials, Van Nostrand Reinhold Publishers

ATSDR's Toxicological Profiles, CRC Press

NIOSH Pocket Guide to Chemical Hazards, NIOSH

Appendix B

Looking toward the Future

The use of electronic retrieval systems is becoming almost a necessity in emergency management. Storing and retrieving information concerning facilities, chemical information, regulations, and so forth requires some type of electronic database. The only limitations involve the cost, the skill needed to operate such a system, and the imagination to use it. The functions of hazardous materials response teams (HMRT) and their support staff (this can include police, ambulance, poison control, hospitals, fire, and emergency management) include a myriad of service activities. These activities increase in range as time and technology continue. In order to manage such on-scene operations, information must be easily retrieved, which requires a system that not only stores this vast base of information, but also produces legible and intelligent data.

Over the past decade, the computerized informational systems have become a piece of equipment that all emergency services providers cannot afford to do without. Emergency responders have been slow to accept the computer although momentum is gaining daily.

When PCs entered the market, we saw programs that emergency response groups could use with direct application (Computer-Aided Management of Emergency Operations; CAMEO). More recently we have seen the availability of data networks, CD ROM, and the storing of information (prehazard planning) on optical storage disks. Emergency response personnel have come to rely on this informational system and the advantages of rapid information assessment by the use of cellular phones, radios, and apparatus-based computer systems and the Internet.

An informational network, such as the one maintained by the International Association of Fire Chiefs (ICHIEFS), is an electronic bulletin board providing the fire service manager with an extended informational resource list. Utilizing the superhighway, virtually everyone can interact. The applications appear endless. The motivation, the ingenuity, and, of course, money are the restrictive factors. The following is a brief list of computer applications currently being used:

Laptop Computers

Fire investigation

Prefire/prehazard planning

EMS response reports

Hazardous materials database link to mainframes

Hazardous materials reference databases

Databases

Human resource management

Relational databases for health maintenance

Informational links to bulletin boards and/or mainframe databases

Service statistics

CAD programs

Drawing packages for prefire/prehazard planning

Hazardous materials locations and quantities (Tier II report planning)

Water mains, gas lines, electrical stations and grids, water supply, and storm drains

Street locations

Population densities and incident locations

Tactical operations in the field and classroom

If a list were to be compiled that included all of the information needed at a hazardous materials incident, the document would be inches thick. However, the goal is not how much information can be gathered, but rather, how much good meaningful information can be used. Information for prehazard plans or the incident itself is dependent on the user's understanding of the resource. Whether that resource is a book or a computer program, it still requires education and familiarity.

PREHAZARD ANALYSIS

Prehazard/critique analysis starts with the following four basic categories:

1. Exposure hazard analysis
2. Hazardous materials Tier II reports

3. Quality improvement

4. Relational database

Exposure Hazard Analysis

Emergency responder exposures are on the increase. These exposures are the result of chemical and biological agents found in the environment frequently visited by emergency workers. With the number of communicable diseases within the homeless population, the number of nonordinary combustibles such as plastics, common household chemicals, and many other exotic compounds, the potential for exposure is high. With this variety of exposure and the potential vulnerability, the need for documentation and the maintenance of such documentation is ever pressing.

Some states have adopted cancer-presumption legislation, which make it imperative for agencies and departments to track and record exposure to diseases and hazardous materials (through the use of a rational database). Under this type of legislation (and some areas of litigation) worker's compensation will be paid only to those individuals who can demonstrate a relevant cause and effect of exposure.

With these topics in mind, it behooves all emergency responders to preplan the hazardous materials event. Dispatch centers, for example, can place the known occupancies of hazardous materials on the dispatch screen or on dispatch cards. Calls for multiple patients at one occupancy should send up red flags to the dispatch operator of a potential hazardous materials event.

Hazardous Materials Tier II Reports

Just as exposure documentation is new to emergency services (in some areas of the country, biologicals have been reported and documented since the 1980s, however chemical exposure does not have the same level of documentation) so is the federal mandated legislation of location and quantity of hazardous materials (Tier II Reports). SARA Title III has provided the fire service (and to some extent has imposed on current available resources) with forced hazardous materials preplanning. This has placed pressure on the local emergency response groups to preplanning hazardous materials potential events. Ideally, this method of preplanning would identify all hazards within a community. However, the manhours needed to immediately accomplish this goal is beyond the means of many services. A few areas around the country are gaining resources from their communities by forming an alliance between the business/industrial sector and the emergency services.

The LEPC has the ability to respond to the needs of the emergency services, given the financial resources which few have. However, the community service businesses such as EMS, fire, and hospitals can work together in order to provide training and record keeping in order to help meet the goals of medical surveillance, preplanning, and the requirements of SARA Title III.

Relational Database

The diversity of information that the present and future manager must analyze ranges into volumes. Information management is a big factor in the future of the HMRT and the services that it provides. A database that is able to correlate information into workable and definable data is a foreseeable trend within hazardous materials.

The following is a list of standards and regulations that should be reviewed:

NFPA Standards

471	Responding to Hazardous Materials
472	Standard for Professional Competence of Responders to Hazardous Materials Incidents
473	Competencies for EMS Personnel Responding to Hazardous Materials Incidents
1500	Fire Department Occupational Safety and Health Program
1521	Fire Department Safety Officer
1561	Fire Department Incident Management System
1581	Fire Department Infection Control Program
1981	Self-Contained Breathing Apparatus
1993	Standard on Support Function Protection Garments for Hazardous Chemical Operations
1999	Standard on Protective Clothing for Emergency Medical Operations

Code of Federal Regulations

29 indicates OSHA (Occupational Safety and Health Administration)

40 indicates EPA (Environmental Protection Agency)

49 indicates DOT (Department of Transportation Regulations)

29 CFR 1910.120	Defines Hazardous Materials
29 CFR 1910.20	Maintenance of Medical Records and Testing
29 CFR 1910.132	General Standards for Personal Protective Equipment
29 CFR 1910.134	PPE for Respiratory Protection
29 CFR 1910.146	Confined Space Standard
40 CFR 311	Defines Hazardous Materials

Relational databases correlate different pieces of information and produce useable information, provided that the information gathering was done correctly and the information input was sound. The problems are that one must know the questions prior to system design. Relational databases have been used by data analyses groups within the science disciplines, business, and industry for many years. Unfortunately, not much attention has been given to the hazardous materials field and its health concerns.

Quality Improvement

Quality improvement (QI) has become an important aspect of any successful business. In the broad sense QI is a process that, when established properly, is a directed action into producing quality products. In the emergency services the product under improvement is the delivery of service. This service improvement is appraised by the citizens. Furthermore, it is our responsibility to educate the public on the needs and the requirements of safer operations. It is also our responsibility to monitor for the unnecessary exposure level of our workforce. This whole concept impacts liability costs in terms of litigation reduction.

Citizens, local government, and regulating agencies look at the emergency services from a consumer aspect. How will the system work when it is needed? How qualified are the responding individuals? Will they act appropriately? More importantly, have they been trained to the level required to handle such an incident? These and countless other questions are but a few viewpoints used by the consumers, that is, the citizens.

Emergency services managers have a somewhat different viewpoint. They must ensure that the design of the system not only answers these questions, but also that the end product (system delivery) is the same on every shift, at every station, and on every scene. This is generally accomplished through strong educational goals and standard operating guidelines and procedures. The education of the emergency services must mirror the currently used protocol that has been developed.

Several ideas and designs exist in the private sector as well as in the public sector. Different philosophies exist across the country. However, if a complete understanding approach is used, evaluating operations from an educational standpoint rather than a punitive one, quality improvement can increase our service potential in emergencies.

THE NEED FOR A QUALITY IMPROVEMENT PROGRAM

Quality can be defined as a degree of excellence. When applied to community evaluation, human resource management, and cost containment, this unit of measurement is sometimes not realistic. Measurement in the area of service delivery is difficult, if not impossible, to evaluate. However, this measurement can be thought of as the comparison of training standards, field evaluations, and systems protocol tracking.

The variables in emergency management sometimes turn into lists of nonobtainable objectives. One must realize that the objectives must also meet a level of acceptable standards. One of the goals must include an increase of the performance level in total. Nonetheless, the quest for excellence and the avenues to achieve it are a result of pride and willingness to approach and initiate a most difficult program. Quality improvement's main goal is to identify weak areas while capitalizing on the strong areas of system delivery.

The demand for a quality improvement program begins with the administration's observance of the reliability, accuracy, and timeliness of the provided services. The same basic principles that apply to industry (once manipulated to meet the emergency service's needs) can be used in the emergency services.

As with most new programs, the first question is, what are the costs of the program? When reviewing costs versus the benefits that can be obtained, the actual dollar amount spent is nominal. Input must be structured in order to have efficient and effective meaningful output. From the gathering of information, to performance, to quality of personnel, an overall systems response analysis can be evaluated and improved.

Every industry, including the emergency services, has objectives that must be accomplished over a certain period of time. These objectives are assigned criteria by which managers are able to measure the progress taking place and compare the desired goals with the actual achievements. Because emergency services consist of providing service in disastrous circumstances, it may be difficult to measure the service provided. However the quality of service must be built into the everyday operations of service delivery and consistently reevaluated.

The purposes of a quality improvement program are to: (but not limited to)

- Identify the strengths and weaknesses in current service delivery
- Identify potential weaknesses with viable solutions
- Develop solutions directed toward long-term affect
- Review the operational procedures of the system
- Evaluate the need for new SOG
- Inspect equipment for serviceability
- Inspect vehicles for inventory compliance
- Identify development programs for job enhancement and enrichment
- Review incident reports for trends
- Serve as a resource panel for other agencies
- Evaluate new equipment and technology.

The scope of evaluation involves the total emergency response system. The emergency system should be reviewed as an entity that can handle all emergencies within the grasp of available resources. Furthermore, the system should include fire, police, and hospital services, along with support agencies such as poison control centers, college chemists, toxicologists, and chemical manufacturers.

Education

A quality improvement program can help to establish a model for continuing education, while providing an atmosphere that will have a positive focused effect on:

- Individualized instruction
- Learning for competency
- Performance-based instruction

Evaluation against current standards such as the NFPA 472, 473, DOT EMT, and paramedic curricula, which would include, but not be limited to:

- A specific program of systems' strengths and weaknesses
- Methods to ensure problem resolution
- Clearly defined tasks to be accomplished
- Evaluation of individuals within the system through competency-based programs

Statistical Evaluation

The purpose of statistical evaluation is to justify and understand the available resources. The knowledge base required for the successful management of a hazardous materials response is difficult to obtain if it is not planned for. Therefore, a well-established statistical gathering system, once started and maintained, can give in-depth understanding of problem areas. Utilizing this type of system can give rise to new ideas and concepts, in both EMS and hazardous materials.

Report Evaluations

The purpose of report evaluations is to establish a format that will identify problems within the reporting system. Its primary goal is to identify problems in order to circumvent possible future litigation. Postincident critiques with lessons learned is the easiest way to manage such a task.

Quality Critique Circles

All emergencies are different, but most can be placed into general categories. We can all learn from the experiences of others while establishing and redefining policy, procedures, and guidelines acceptable to all personnel. Asking the players to identify problem areas and come up with realistic solutions enables emergency workers to buy in to the program at large.

Acronyms

Much of the jargon in the Emergency Medical Services and the Hazardous Materials field are words which are formed by the initial letters of each word in the phrase or clinical terminology. The following is an alphabetical listing of such acronyms and jargon used in emergency response.

ABC	Airway, Breathing, Circulation
A/C	Air-conditioning
ACGIH	American Conference of Governmental Industrial Hygienists
ADA	Americans with Disabilities Act
ALD	Average Lethal Dose
ALS	Advanced Life Support
APIE	Analyze, Plan, Implement, Evaluate
APR	Air Purifying Respirator
ATP	Adenosine triphosphate
AV	Arterioventricular
BEK	Butyl ethyl ketone
BLS	Basic life support
BMR	Basal metabolic rate
BOCA	Building Officials and Code Administrators
BVM	Bag Valve Mask
C	Centigrade or Celsius
CAMEO	Computer-Aided Management of Emergency Operations
CAS	Chemical Abstract Service
CBC	Complete blood count
CDC	Centers for Disease Control
CFR	Code of Federal Regulations
CGI	Combustible Gas Indicator
CHEMTREC	Chemical Transportation Emergency Center
CISD	Critical Incident Stress Debriefing
CL	Ceiling level
CN	Chloracetephenone
CNS	Central nervous system
CO	Carbon monoxide
COHg	Carboxyhemoglobin
COPD	Chronic obstructive pulmonary disease
CPAP	Continuous positive airway pressure
CS	Chlorobenzalmalonitrile
DDT	Dichlorodiphenyltrichloroethane
DECON	Decontamination
DER	Department of Environmental Resources
DMA	Dimethylamine
DMSO	Dimethyl sulfoxide
DO	Designated officer
DOE	Department of Energy
DOT	Department of Transportation
DRI	Direct reading instruments
DUMBELS	Diarrhea, Urination, Miosis, Bronchospasm, Emesis, Lacrimation, Salivation
EC	Effective concentration

ED	Emergency department
EDC	Emergency dispatch center
ECG	Electrocardiogram
EEL	Emergency exposure limit
EL	Excursion limits
EKG	Electrocardiogram
ERG	Emergency Response Guide
EMS	Emergency medical services
EMT	Emergency Medical Technician
EMT-P	Emergency Medical Technician–Paramedic
EPA	Environmental Protection Agency
ES	Entry supervisor
ET	Effective temperature
F	Fahrenheit
FD	Fire department
FDA	Food and Drug Administration
FEV	Forced expiratory volume
FID	Flame ionization detector
FP	Fire point
FVC	Forced vital capacity
GRS	German Research Society
H	Humature
HBO	Hyperbaric oxygen
HBV	Hepatitis B virus
HEPA	High-efficiency particulate air
HF	Hydrofluoric
HIV	Human Immunodeficiency Virus
HM	Hazardous materials
HMIS	Hazardous Materials Information System
HVAC	Heating, ventilation, and air-conditioning systems
ICS	Incident command system
IC	Incident commander
IV	Intravenous
IVP	Intravenous push
IDLH	Immediately dangerous to life and health

IMS	Incident management system
IP	Ionization potential
IT	Ignition temperature
IUPAC	International Union of Pure and Applied Chemistry
LC	Lethal concentration
LC-low	Lethal concentration–low
LCD	Liquid crystal display
LD	Lethal dose
LEL	Lower explosive limit
LEPC	Local Emergency Planning Committee
LIPE	Life safety, incident stabilization, property conservation, environmental protection
LPG	Liquid petroleum gas
LR	Lactated Ringers
MAC	Maximum allowable concentration
MAK	Maximum allowable concentration
MEK	Methyl ethyl ketone
MAST	Medical Antishock trouser
MetHg	Methemoglobin
mm Hg	Millimeters of mercury
MSDS	Material safety data sheets
MW	Molecular weight
NAERG	North American Emergency Response Guidebook
NA	North America
NFPA	National Fire Protection Association
NRC	National Response Center
NIOSH	National Institute for Occupational Safety and Health Administration
OC	Oleoresin capsicum
O_2Hg	Oxyhemoglobin
ORM	Other regulated material
OSHA	Occupational Safety and Health Administration
PaO_2	Partial Pressure of Oxygen
PCB	Polychlorinated biphenyl
PCC	Poison control centers

PCP	Phencyclidine	SLUDGE	Salivation, Lacrimation, Urination, Defecation, Gastrointestinal, Emesis
PEL	Permissible exposure limit		
PEEP	Positive end expiratory pressure	SMAC	Sequential multiple analysis chemistry
PFT	Pulmonary function test	SOC	Standard of care
PID	Photoionization detector	SOG	Standard operation guidelines
PIO	Public information officer	SOP	Standard operation procedures
PK	Peak value	STCC	Standard transportation commodity code
PPD	Purified protein derivative		
PPE	Personal protective equipment	STEL	Short-term exposure limit
PPM	Parts per million	STP	Standard temperature pressure
PRN	*Pro re nata*, Latin for "as needed"	TEPP	Tetraethylpyrophosphate
PT	Prothrombin time	THI	Temperature humidity index
RAD	Radiation absorbed dose	TCL	Toxic concentration, low
REM	Reontgen equivalent man	TDL	Toxic dose—low
RD	Respiratory depression	TLV	Threshold limit value
REL	Recommended exposure limit	TLV-c	Threshold limit value—ceiling
SaO_2	Saturation of oxygen	TLV-s	Threshold limit value—skin
SAR	Supplied air respirators	TNT	Trinitrotoluene
SARA	Superfund Amendments and Reauthorization Act (1986)	TWA	Time weighted average
		UN	United Nations
SERC	State Emergency Response Commission	UEL	Upper explosive limit
SCBA	Self-contained breathing apparatus	USCG	United States Coast Guard
SG	Specified gravity	VD	Vapor density
SLUD	Salivation, Lacrimation, Urination, Defecation	VP	Vapor pressure
		WBGT	Wet bulb globe temperature

Glossary

Absorption   The incorporation of a material into another, or the passage of a material into and through the tissues. The ability of a material to draw within it a substance that becomes a part of the original material.

Acclimation   The ability of the body to adapt to an environment.

Acetylcholinesterase   An enzyme that hydrolysizes (washes) acetylcholine

Acid   Chemicals with a hydrogen ion concentration and a pH of less than 7.

Activation energy   The amount of energy that is required for the reactants to produce products. It is the "hill" greater or smaller that a molecule must cross in order for a reaction to take place.

Acute   Refers to short term. Can be used to explain either a short-term exposure or the rapid onset of symptoms.

Additive   The effect that is noted when two or more toxins are combined.

Adipose tissue   Fat cells.

Administrative controls   The controls, namely policies and procedures, that are employed to minimize injury.

Adsorption   A process that will take up and hold a gas, liquid, or dissolved substance on the surface of another substance.

Aerobic metabolism   Cellular metabolism utilizing oxygen.

Aerosol   Suspended nebulized particles within a gas.

AIDS   Acquired immune deficiency syndrome.

Air purifying respirator (APR)   Filtration mask that filters outside air.

Alkali   A chemical with a concentration of hydronium ions and a pH of greater than 7.

Alkanes   Saturated hydrocarbons with single bonds.

Alkenes   Unsaturated hydrocarbons with double bonds.

Alkynes   Unsaturated hydrocarbons with triple bonds.

Alveoli   Microscopic air sacs in the lungs.

Alveolar wall   The semipermeable membrane that makes up the walls of the alveoli.

Anion   A negatively charged ion.

Antagonism   A combined effect that cancels the effect of one of the toxins.

Anticoagulant   A chemical product that prevents the clotting of blood.

Appearance   The physical state of matter, or the physical form of a chemical.

Asphyxiant   A chemical that displaces oxygen in the air or interferes with the use of oxygen in the body. An asphyxiant may not have any toxic effect in or of itself.

Atom   The smallest particle of an element. It consists of nucleus and electrons that revolve around the nucleus. In the nucleus are the proton and neutron.

Autoignition temperature   The minimum temperature required for a material to spontaneously ignite and maintain combustion.

Basal metabolic rate (BMR)   The heat production of an individual through normal metabolism while at rest but not asleep.

Biochemistry   The chemistry of biological processes.

Biohazards   Communicable diseases that are a threat to the life and health of emergency response personnel unless precautions and protective measures are exercised.

Biotransformation   Chemical reactions that occur during metabolism.

Bleb   A weak area of the lung that may burst causing a pneumothorax or subcutaneous emphysema.

Blepharospasm   The involuntary closing of the eye that occurs when the eye is irritated and/or injured.

Boiling point   The temperature at which a liquid will move into the gaseous phase.

Bronchospasm   A spasmodic contraction of the bronchioles due to the inhalation of an irritant.

Calibration   A controlled exercise in which the full range of an instrument is analyzed.

Calorie   The amount of heat required to raise the temperature of one gram of water one degree Centigrade.

Canthus   The angle at the medial and the lateral junction of the eyelid.

Capsicum   A lacrimating agent made from hot peppers. "Pepper gas."

Carbamate   A synthetic organic insecticide.

Carbohydrates   A major class of foods derived from plants.

Carboxyhemoglobin (COHg)   Hemoglobin bound with carbon monoxide.

Catalyst   A chemical that lowers the activation energy.

Cations   Positively charged ions

Cell lysis   The destruction of cells.

Cellular respiration   The use of food and oxygen to form energy (glucose + oxygen = energy [ATP] + heat + carbon dioxide + water).

Cellular hypoxia   A lack of oxygen within the cell.

Chelating agents   Chemicals that are used to combine a toxic complex for biochemical elimination.

Chemical asphyxiant   A chemical that acts in the body to interfere with the transportation of oxygen, or hamper its use on the cellular level.

Chemical bond   The attachment between elements to form molecules.

Chemically induced pulmonary edema   Noncardiogenic pulmonary edema. Pulmonary edema stimulated because of injured lung tissue and not increased pulmonary blood pressures.

Chemical properties   A description of matter and how it reacts with other substances. Used in hazardous materials to describe the hazard risk potential.

Chilblain   A mild form of cold injury characterized by dry, red, rough skin that sloughs off during the healing process.

Chronic   Refers to a long time or of long duration. Can be used to explain either a long-term exposure or a long or late onset of symptoms.

Cilia   Small hairlike structures found in the external respiratory system that sweep contaminants to the upper part of the respiratory system so they can be swallowed or expectorated.

Clandestine drug laboratory   A laboratory where illegal drugs are manufactured.

Coagulation necrosis   The formation of a tough layer of tissue as a result of acid exposure. The process of transforming a liquid based product into a solid or semisolid mass.

Cold zone   Also called the safe zone or support zone; the area where equipment and personnel directly involved in an incident are located.

Compound   A substance composed of two or more elements, which once combined into its proportions cannot be separated by physical means.

Concentration   The relative percent of the monitored gas in the atmosphere.

Conduction   Heat transferred to a substance or object in contact with the body.

Conjunctiva   A thin layer of epithelial tissue that is found over the globe of the eye that extends to the undersides of the upper and lower lids.

Contamination   The process whereby a person or piece of equipment has contact with a toxin.

Convection   The circulation of colder air around the body.

Cornea   The clear portion of the anterior eye where light enters.

Corrosive   Either an acid or an alkali capable of dissolving or wearing away metal or other substances.

Covalent bond   Electrons shared between atoms to make a molecule. The force that holds the atoms together.

CPAP   Continuous positive airway pressure. A continuous positive pressure within the airways of the lungs throughout the respiratory cycle provided by an outside source.

Critical pressure   The pressure required to liquefy a gas if its critical temperature is attained.

Critical temperature   The temperature above which a gas will not liquefy with an increase in pressure.

Cryogenics   Gases that are cooled to very low temperature rather than by using pressurization to make the gas into a liquid.

Cyanogenic glycoside (Amygdalin)   The product found in many plant seeds that can form cyanide under the proper conditions.

Cytochrome oxidase   An enzyme responsible for the movement of oxygen within the cell during cellular metabolism.

Decontamination   The removal of a chemical or biological agent from a person or equipment.

Density   The mass of a substance per a given volume.

Dermis   The second layer of skin that contains blood vessels, nerve endings, and hair follicles.

Digital opening of the eye   Forced opening of the eyes with the fingers.

Dilution   Reduction of the concentration of a chemical by adding another. Water is widely used as the agent for dilution with hazardous materials.

Dose response   The range of effects observed within a population.

Dosimetry   The measurement process of radiation.

DUMBELS   An acronym used to describe the symptoms found during an organophosphate poisoning. Diarrhea, Urination, Miosis, Bronchospasm, Emesis, Lacrimation, and Salivation.

Ebola virus   A devastating virus thought to have started in Africa that causes bleeding through all orifices of the body.

Electronegativity   An atom's desire to gain electrons.

Element   A substance that cannot be further broken down into smaller substances by chemical means.

Emergency management   Describes a function responsible for planning, coordinating the mitigation efforts, and recovery from disasters. An all-encompassing group of public safety, local, state, and federal committees that all participate in preparing for and controlling the situation in the event of a large emergency.

Endothelium   The basement membrane of the cornea.

Endothermic   Characterized by the absorption of heat during a chemical reaction.

Engineering controls   The physical controls employed to reduce potential injury.

Epidermis   The top layer of skin made up of four or five layers: Stratum corneum, stratum lucidum (in high friction areas), stratum granulosum, stratum spinosum, and the stratum basale.

Epithelial tissue   The surface covering of the cornea that becomes part of the conjunctiva.

Evaporation   Loss of heat due to the vaporization of water from the body surface.

Exclusion factors   Predetermined factors that enable the medical sector to release from rehab those individuals who do not meet the criteria for working within the hot or warm zone. A criteria used during the cursory medical exam to evaluate medical status of personnel.

Exothermic   Characterized by the evolution of heat during a chemical reaction.

Exposures   Persons who are in the area of a chemical release. These persons may or may not be contaminated.

External respiratory system   That part of the respiratory system that is open to the outside environment, above the dividing line of the alveolar membrane.

Fasciculations   Uncontrolled muscle tremors.

Fats   Stored energy-rich food that may be assimilated in the future by the animal.

Flammable limits (Explosive flammable range)   The range of percentage in which fuel vapors exist in air. The lower and upper ranges of a vapor in an air mixture. When the vapor and air mixture is between the lower explosive limit (LEL) and the upper explosive limit (UEL), an explosion and/or fire can occur.

Flash point   The minimum temperature at which a liquid will give off vapors to form an ignitable mixture in air, but not enough vapors to sustain combustion.

Fire point   The lowest temperature at which a liquid will produce enough vapor to flash and continue to burn with an outside source of ignition.

Frost point   The temperature at which atmospheric water freezes.

Glycogen   The material stored by the body that can be broken down into glucose when needed.

Graded response   The gradual increase of bound receptor sites.

Groups   The vertical columns of elements within the periodic chart. Each column presents the elements with similar properties. The naming of each group is dependent on the author and organization.

   Group IA—Alkali metals or Group 1

   Group IB—Copper family or Group 11

   Group IIA—Alkaline earth metals or Group 2

   Group IIB—Zinc family or Group 12

   Group IIIA—Aluminum family or IIA or IIIB or Group 13

   Group IIIB—Scandium family or IIIA or Group 3

   Group IVA—Germanium family or IVB or Group 14

   Group IVB—Titanium family or IVA or Group 4

   Group VA—Nitrogen family or VB or Group 15

   Group VB—Vanadium family or VA or Group 5

   Group VIA—Sulfur family or VIB or Group 16

   Group VIB—Chromium family or VIA or Group 6

   Group VIIA—Halogen family or VIIB or Group 17

   Group VIIB—Manganese family or VIIA or Group 7

   Group VIIIA—Noble gases or VIII or Group 18

   Group VIIIB—Iron family or platinum family or VIIIA or Group 8, Group 9, Group 10 respectively.

Hazardous materials   According to the EPA, a material that may be potentially harmful to the public's health or welfare if it is discharged into the environment, and according to the DOT, any substance or material in any form or quantity that poses an unreasonable risk to safety, health, and property when transported in commerce.

Heat of combustion   Heat evolved when a definite quantity of a product is completely oxidized or complete combustion takes place.

Heat of reaction   The measurement of enthalpy or the chemical reaction.

Heat stroke   A condition that leads to the complete failure of the thermoregulatory mechanism.

Hemoglobin   A molecule found in the blood that is responsible for transporting oxygen from the lungs to the cells.

HEPA   High efficiency particulate air. Usually refers to an extremely efficient filter mask.

HIV   Human immunodeficiency virus. The virus that causes AIDS.

Hot zone   The zone immediately surrounding the chemical release. This zone extends far enough to prevent adverse effects to personnel.

Hydrophilic   Chemicals that attract water molecules.

Hydroxide   The combination of a metal and a hydroxide radical.

Hyperbaric oxygen therapy (HBO)   Placing a patient in an oxygen environment that is higher than atmospheric pressure for medicinal treatment.

Ignition temperature   The minimum temperature at which a material will ignite and sustain combustion without an outside source of ignition.

Incident command system (ICS)   A system used by the emergency services to coordinate a large number of responders performing a wide variety of tasks. Reduces the span of control into no more than seven divisions, groups, or persons. Commonly called the incident management system.

Ingestion   The incorporation of a material into the gastrointestinal tract.

Inhalation   A means of gaining access into the body via the respiratory system.

Inhalation exposure   The taking in of a chemical through the respiratory system that causes harm to the body.

Inhibitors   Chemicals that control a reaction or delay the reaction.

Injection   The forced introduction of a substance into underlying body tissue (the major components are not carbon based)

Inorganic chemistry The study of substances that do not contain carbon.

Inorganic peroxide The combination of a metal and a peroxide radical.

Interference gases Gases similar to those being tested that the instrument may misinterpret.

Internal respiratory system That part of the respiratory system below the alveolar membrane. Made up of the oxygen and carbon dioxide transport system and cellular respiration.

Interval history A periodic questionnaire that gives the medical surveillance program detailed information about lifestyle changes in the emergency worker.

Intramuscular injection A means of injecting a substance into the muscle. Usually used as a way of giving medicines, but can occur with a deep puncture wound from a sharp contaminated object.

Intraperitoneal Gaining access through the peritoneal cavity, normally a means of treatment for kidney failure, i.e., peritoneal dialysis.

Intravenous A means of gaining access by a chemical into the venous system, usually via a needle or catheter.

Ion Atoms that are positively or negatively charged by having lost or gained electrons.

Ionic bonding Electrons that are transferred between atoms. A negative and a positive end is the result. This electrostatic force holds the atoms together.

Ionization The process of an element becoming an ion.

Ionization potential (IP) The minimum energy required to release an electron or a photon from a molecule.

Iris detail The fine lines of the iris that are noticed during close examination.

Irritant A chemical that causes inflammation to tissues.

Isotopes One or more forms of an element that have the same number of protons but differ by the number of neutrons.

Kinetic molecule theory The hypothesis that states that elements are in constant motion. This motion has a direct relationship with the state of matter and the temperature of the matter.

Lacrimatory agents Chemicals that stimulate increased tear production.

Laryngeal edema Swelling of the larynx/vocal cords.

Laryngeal spasm A spasm of the vocal cords.

Law of conservation of mass The law that states that matter cannot be created or destroyed but rather changed.

Legionella An airborne bacteria that caused lung disease that killed several American Legion members.

Level I incident Potentially hazardous situation involving a small population segment.

Level II incident The start of a potentially hazardous situation involving a large population segment.

Level III incident Has high probability of becoming a serious health hazard to human life and/or will affect the environment.

Lewis structures A graphical representation of the outermost electron shell of an element.

LIPE An acronym for Life safety, Incident stabilization, Property conservation, and protection of the Environment.

Lipophilic Attraction or absorption of substances within fat or oils.

Liquefaction necrosis The liquefying of tissue through the action of an alkali, causing death of the tissue.

Locker Room A trade name used for butyl nitrite, misused to elicit a high.

Lockout/Tag-Out Method used to eliminate electrical and mechanical hazards in confined areas.

Margin of safety An arbitrarily assigned separation of a toxic quality and the harmless quantity.

Medical surveillance The process by which the health of an emergency worker is observed, maximizing the long-term health benefits while minimizing the risks.

Melting point The temperature at which a solid becomes a liquid.

Mercaptan A chemical group of organosulfur compounds sometimes referred to as thiols, found in crude petroleum and sulfur compounds. Sometimes used as an odorant in odorless gases.

Metal oxide The combination of a metal element and oxygen.

Metal salt   The combination of a metal element and a nonmetal element.

Methemoglobin (MetHg)   Hemoglobin that has had the iron atom converted from ferrous iron to ferric iron and is unable to carry oxygen.

Molecule   A stable configuration of elements by which a structural unit is made that has its own characteristic chemical and physical properties.

Morgan therapeutic lens   A contact type lens that provides irrigation to the globe of the eye.

Necrosis   Death of tissue.

Neurotoxin   A toxin that affects the nervous system as the target organ.

Neutralization   The process of bringing an acid or alkali back to a pH of 7.

Neutron   One of the particles that makes up the center of an atom without electrical charge.

Nonpolar bonds   Nondiscernible attraction between atoms.

Nucleus   The center area of the atom that contains the neutrons and protons. The only exception is hydrogen, which has no neutrons.

Octet   The rule for having eight electrons in the outermost shell of an atom.

Olfactory fatigue   Paralysis of the olfactory sensors causing a loss of smell.

Opacification   A discoloration of the cornea due to a chemical exposure. To become or make opaque.

Oral   Ingestion. The taking in by mouth.

Organic chemistry   The study of carbon compounds.

Organophosphates   Phosphorus-containing pesticides that inhibit synaptic response.

Outage   The percentage of the container that is left short of being full. The area left for expansion.

Oxygenated salt   The combination of a metal and an oxygenated radical.

Oxyhemoglobin (HgO$_2$)   Hemoglobin bound with oxygen.

Parasympathetic nervous system   The side of the autonomic nervous system that is mediated by acetylcholine release.

PEEP (Positive end expiratory pressure)   A positive pressure at the end of the expiratory cycle when the interthoracic pressure is normally equal with the ambient pressure.

Perilimbal circulation   The circulatory system that provides a blood supply to the eye globe.

Periodic table   The arrangement of the elements in a tabular form that denotes chemical and physical characteristic of the elements.

Personal protective equipment (PPE)   The equipment used to protect a worker from the effects of a hazardous chemical. Usually classified as Level A, B, C, or D.

Phase I   A metabolic process by which the body tries to convert a polar chemical site into a lipophilic compound.

Phase II   The second part of the metabolic reaction, which converts the substance into a water-soluble product, enabling the body to eliminate it.

Photophobia   Intolerance or sensitivity to light.

Physical properties   A description of how matter reacts, the condition in which it is found, and the qualities it possess.

Physics half-life   The amount of time it takes a radioactive isotope to lose one-half of its radioactive intensity.

Physiologic saline   0.9% Sodium chloride, normal saline.

Plume dispersion   A mathematical representation of how a vapor cloud may move given a certain set of conditions.

Pneumoconiosis   A disease of the lungs that develops from the chronic inhalation of dust particles.

Polar   Pertaining to molecules that carry a negative and positive charge, much like a magnet.

Polarity   The possession of two opposing tendencies. The existence of a negative and a positive end on a molecule.

Polymerization   A chemical reaction in which two or more smaller chained molecules combine to produce a larger molecule. A runaway reaction can occur on its own without the addition of inhibitors. It is a chain reaction in which long molecules are formed.

Positive hydronium ions   The pH scale based on the relationship of hydronium ions to hydroxide ions. Acids are scaled from 0–6.9 and 7.1–14 indicates bases. Seven is considered neutral.

Potentiation   The increased effect of a chemical.

Precursors   Initially purchased chemicals used to make illegal drugs.

Preplanning   The act of identifying target hazards within a response district, forecasting possible emergency situations, and projecting the needs of these situations.

Primary wave   The pressure wave as it moves through the air.

Products   The end result after chemical combination.

Proteins   Naturally occurring complex molecules that are assimilated by animals.

Proton   One of the particles that makes up the center of the atom and carries a positive charge.

Pulmonary fibrosis   Unnatural growth of fibrous tissue in the lungs.

Pyloric valve   A valve found in the base of the stomach that regulates the quantity of food or fluid that reaches the bowel.

Quantal Response   An observed all-or-none toxic event.

Radiation   The loss of heat in the form of infrared rays.

Reactants   Two or more substances that combine during a chemical process.

Reagents   Chemicals that are used to change precursors into illegal drugs.

Regulated waste   The materials that are potentially infectious, such as liquid or semiliquid blood, body fluids, and/or tissue.

Rehabilitation   Eating, fluid intake, and resting.

Rush   Isobutyl nitrite. Used to elicit a high when sniffed.

Sclera   A tough white fibrous membrane that makes up the "white" of the eye.

Secondary contamination   Contamination from a previously contaminated person or object that occurs away from the initial scene.

Secondary wave   A wave created when the surrounding air around the blast rushes in to fill the vacuum created during the original explosion.

Self-contained breathing apparatus (SCBA)   A closed breathing system that contains a limited air supply used in hazardous atmospheres.

Significant exposure   Exposure by a contaminated needle stick, sharp, or other instruments in which blood-to-blood contamination takes place.

Simple asphyxiants   Chemical gases that displace oxygen from the air.

Size-up   Gathering and weighing all of the known facts about an incident so competent decisions can be made. Size-up continues throughout the incident for future decisions.

SLUD   Acronym used to describe the symptoms of organophosphate poisoning: Salivation, Lacrimation, Urination, and Defecation.

SLUDGE   Acronym used to describe the symptoms of organophosphate poisoning: Salivation, Lacrimation, Urination, Defecation, Gastrointestinal, and Emesis.

Slurry   A pourable mixture of a solid (semisolid) and liquid.

Specific gravity   The weight of a solid or liquid as compared to an equal volume of water.

Specific heat   Heat movement measured as a calorie per gram. The ratio of the heat capacity of a substance to the heat capacity of water.

Solubility   The ability of a material to blend uniformly with another material.

Solute   The substance being placed in the solvent.

Solution   A mixture of substances. The even mixture of molecules of two or more substances.

Solvent   The liquid in which a substance is dissolved; chemicals used to dissolve precursors and reagents.

Standard atmospheric pressure   14.7 pounds per square inch. Equals 760 mm Hg and one atmosphere.

Stereochemistry   The spatial arrangements of molecules that affect their chemical and physical properties.

Stimulating emesis   The act of stimulating vomiting by giving a cathartic.

Stroma   The middle layer of the cornea that makes up 90% of the thickness.

Subcutaneous injection   A means of gaining access of a substance into the subcutaneous tissue. This is a normal way to give medicines with a needle but can also occur as a result of a sharp contaminated object puncturing the skin.

Subcutaneous tissue   The tissue located under the dermal layer.

Sulfhemoglobin   The chemical combination of sulfur with hemoglobin that is unable to carry oxygen.

Supplied air respirator (SAR)   A breathing apparatus that is plumbed into a large reservoir of air, either bottled air or a breathing air compressor.

Surfactant   A substance that lowers the surface tension of a liquid.

Synergistic effect   The combined effects of chemicals.

Table I materials   Materials determined to be the most hazardous according to the Code of Federal Regulations.

Target organ   The organ system effected by a chemical regardless of route of entry.

Tear film   A film of fluid that covers the anterior portion of the eye. It is made of a saline solution and covered with a lipid layer to provide lubrication and moisture to the eye.

Testing   Verification of the basic operation of an instrument.

Tier II reports   Forms that describe certain chemical quantities and locations within an occupancy.

Thermodynamics   The interconversion of heat flow.

Threshold Limit Value (TLV)   That level of exposure that starts to produce an effect.

Toxic   The condition of a substance's harmful ability.

Toxic exposure   The concentration of a dose that causes a response.

Toxicity   The degree of a substance's ability to cause injury or harm.

Toxin   A chemical that will do harm if sufficient quantity is present and exposure takes place.

Trenchfoot (Immersion foot)   Common wartime injury where the feet stay wet and cold over extended periods. The wound is characterized by a lack of circulation to the tissues.

Turbinates   Bony protrusions in the nasal cavity that serve to swirl air so better filtration can occur.

Universal precautions   Measures taken when approaching the treatment of all patients as if they are infected with a contagious disease.

Valence number   The whole number given to an atom that describes the ability to gain or lose electrons.

Vapor density   The weight of a vapor or gas as compared to an equal volume of air.

Vapor pressure (VP)   The pressure a material exerts against the sides of an enclosed container as it tries to evaporate or boil.

Vasodilatation   The dilating of the blood vessels.

Virulence   The amount of a microorganism that can cause an effect in a host organism.

Viscosity   A measure of flow.

Visual acuity   The ability to see clearly.

Warm zone   A buffer area surrounding the hot zone where decontamination of entry personnel, victims, and equipment takes place at a hazardous materials incident.

Bibliography

The authors have made lengthy attempts to include all resources for the production of this book. Any reference not included in the list below was as an oversight and not intentional.

ACGIH. 1992–1993. *Threshold Limit Values for Chemicals and Physical Agents and Biological Exposure Indices.* American Conference of Governmental Industrial Hygienists. Cincinnati, Ohio.

ACGIH. 1993. *Documentation of the Threshold Limit Values.* American Conference of Governmental Industrial Hygienists. Cincinnati, Ohio.

Adams, Donald R. 1993. Seeing is believing. *Rescue-EMS Magazine,* July–August, pp. 38–40.

Albert, A. 1970. *Selective Toxicity.* John Wiley and Sons, New York.

Allied Signal. 1990. *Hydrofluoric Acid Exposure Recommended Medical Treatment,* Allied Chemical Company. Morristown, NJ.

American Academy of Clinical Toxicology. 1989. *Veterinary and Human Toxicology.* Publication Office, Comparative Toxicology Laboratories, Kansas State University, Manhattan, Kansas, pp. 243–246.

Ashford, Nicholas, A., and Claudia S. Miller. 1991. *Chemical Exposures, Low Level and High Stakes.* Van Nostrand Reinhold, New York.

Bevelacqua, Armando. 1992. *Prehospital Documentation, A Systematic Approach.* Brady Books (Prentice Hall), Englewood Cliffs, NJ.

Borak, Jonathan, Michael Callan, and William Abbott. 1991. *Hazardous Materials Exposure.* Brady Books (Prentice Hall), Englewood Cliffs, NJ.

Bronstein, A. C., and P. L. Currance. 1988. *Emergency Care for Hazardous Materials Exposure.* C.V. Mosby Co.

Brown, Ralph L. 1989. Dehydration in emergency operations. *Response Magazine,* Summer, pp. 20–22.

Cashman, John. 1988. *Hazardous Materials Emergencies Response and Control.* Technomic Publishing.

Carder, Thomas A. 1993. *Handling of Radiation Accident Patients by Paramedical and Hospital Personnel.* CRC Press, Boca Raton, FL.

Carr, David K., and Ian D. Littman. 1993. *Excellence in Government, Total Quality Management in the 1990s.* Coopers & Lybrand.

Carson, Rachel. *Silent Spring.* (1962)

Center for Labor Education and Research. 1992. *Emergency Responder Training Manual for the Hazardous Materials Technician.* Van Nostrand Reinhold, New York.

Code of Federal Regulation, Title 29, 1993. 1910.120.

Code of Federal Regulation, Title 29, 1993. 1910.146.

Code of Federal Regulation, Title 40, 1993. Part 311.

Coleman, Ronny J., and John A. Granito, 1988. *Managing Fire Services.* International City Management Association.

Confined Spaces. 1989. Virginia Regulation.

Connellan, Thomas. 1975. *Management by Objectives in Local Government: A System of Organizational Leadership.* International City Management Association.

Cooke, Robert. 1996. A Plague on All Our Houses. *Popular Science* (January 1996) pp. 50–56.

Cralley, Lewis J., and Lester V. Cralley. 1985. *Patty's Industrial Hygiene and Toxicology.* Vol. III, parts A & B. John Wiley and Sons, New York.

CRC. 1993. *Handbook of Chemistry and Physics.* Lewis Publishers, CRC Press, Boca Raton, FL.

Emergency Resource Inc. 1992. Surviving the Hazardous Materials Incident. *OnGuard,* Fort Collins, Colorado.

EPA. 1985. *Chemical Emergency Preparedness Program.* United States Environmental Protection Agency.

EPA. 1990. *Occupational Medical Monitoring Program Guidelines for SARA Hazardous Waste Field Activity Personnel.* United States Environmental Protection Agency.

EPA. 1992. *Standard Operating Safety Guides.* United States Environmental Protection Agency.

Evan, James R., and William M. Lindsay. 1989. *The Management and Control of Quality.* West Publishing Company.

FEMA. 1990. *Fundamentals Course for Radiological Monitors.*

FEMA. 1992. *Emergency Incident Rehabilitation.*

FEMA. 1993. *Hazardous Materials Operating Site Practices.*

DiPalma, Joseph, R. 1971. *Drill's Pharmacology in Medicine.* McGraw-Hill Book Company, A Blakiston Publication.

Dreisbach, Robert, H., and William O. Robertson. 1987. *Handbook of Poisoning.* Prentice Hall, Englewood Cliffs, NJ.

Fairbridge, Rhodes W. 1987. *The Encyclopedia of Climatology.* Van Nostrand Reinhold, New York.

Fire, Frank. 1994. *Chemical Data Notebook, A User's Manual.* Fire Engineering Books and Videos, New York, NY.

Fire, Frank, Nancy Grant, and David Hoover. 1990. *SARA Title III, Intent and Implementation of Hazardous Materials Regulations.* Fire Engineering Books and Videos, New York, NY.

Fire, Frank. 1986. *The Common Sense Approach to Hazardous Materials.* Fire Engineering Books and Videos, New York, NY.

Fernandez, Jack. 1982. Organic Chemistry: An Introduction. Prentice Hall, Englewood Cliffs, NJ.

Gad, Shayne C., and Rosalind C. Anderson, 1990. *Combustion Toxicology.* CRC Press, Boca Raton, FL.

Garrett, Alfred, W. Lippincott, and Frank Verhoek. 1972. *Chemistry, A Study of Matter.* Xerox Corporation.

Gleason, Marion N., Robert E. Gosselin, Harold C. Hodge, and Roger P. Smith. 1993. *Clinical Toxicology of Commercial Products.* Williams and Wilkins, Baltimore, MD.

Goldfrank, L. R. 1986. *Goldfrank's Toxicologic Emergencies.* Appleton-Century-Crofts.

Guyton, Arthur C. 1976. *Textbook of Medical Physiology.* W. B. Saunders Company, Philadelphia, PA.

Haddad, Lester M., and James F. Winchester. 1990. *Clinical Management of Poisoning and Drug Overdose.* W. B. Saunders Company, Philadelphia, PA.

Hawley, Gessner G. 1993. *The Condensed Chemical Dictionary,* Van Nostrand Reinhold, New York.

Houghton, David D. 1985. *Handbook of Applied Meteorology.* John Wiley and Sons, New York.

International Association of Fire Fighters. 1991. *Training for Hazardous Materials Team Members.* IAFF, Washington, DC.

International Society of Fire Service Instructors. 1993. *Hazardous Materials Technician Program.* ISFSI, Ashland, MD.

International Society of Fire Service Instructors. 1988. Safety perspectives, time out for rehab. *Rekindle.* ISFSI, July, pp. 15, 16.

Jay, Gregory D. 1991. Pulse oximetry. *Emergency Medical Services* 20(5, May):40–42, 75.

Johnson, Kevin W. 1990. EMS in the hot zone. *Firehouse Magazine,* July, pp. 46, 47, and 119.

Kamrin, Michael A. 1988. *Toxicology.* Lewis Publishers.

Keenan, Charles, Jesse Wood, and Donald Kleinfelter. 1976. *General College Chemistry.* Harper & Row.

Kurhl, Alexander. 1994. *Prehospital Systems and Medical Oversight.* National Association of EMS Physicians, Mosby Lifeline. Mosby Yearbook, St. Louis, MO.

Levesque, William R. 1991. Blasting vindicator. *Lakeland Ledger* (November) Lakeland, FL.

Lewis, Robert J. 1991. *Hazardous Chemicals Desk Reference.* 2nd ed. Van Nostrand Reinhold, New York.

Lauwerys, Robert R., and Perrine Hoet. 1993. *Industrial*

Chemical Exposure, Guidelines for Biological Monitoring. Lewis Publishers.

Leikin, J. B., D. Daufman, and J. W. Lipscomb. 1990. Methylene chloride: Report of five exposures and two deaths. *American Journal of Emergency Medicine* 8(6, Nov.):534–537.

Luciano, Dorothy, Arthur Vander, and James Sherman. 1983. *Human Anatomy and Physiology, Structure and Function.* McGraw-Hill.

Manahan, Stanley E. 1992. *Toxicological Chemistry.* Lewis Publishers.

Maslansky, Carol J., and Steven P. Maslansky. 1993. *Air Monitoring Instrumentation.* Van Nostrand Reinhold, New York.

McKay, Charles A. Jr. 1990. Prehospital chemical and radiation exposure. *Emergency Care Quarterly* 5 (Feb., 4): 17–28; Aspen Publishers, Inc.

McMullen, M. 1994. Medical Monitoring of Hazardous Materials Responders. A seminar delivered at Hazardous Materials Response Teams Conference, Fairfax, Virginia.

Memmler, Ruth L., and Dena Lin Wood. 1977. *The Human Body in Health and Disease*, 4th ed. J. B. Lippincott, Philadelphia, PA.

Merck Index, 1989. *An Encyclopedia of Chemicals, Drugs, and Biologicals*, 11th edition. Merck and Co. Inc.

Meyer, Eugene. 1989. *Chemistry of Hazardous Materials.* Prentice Hall, Englewood Cliffs, NJ.

Ness, Shirley A. 1991. *Air Monitoring for Toxic Exposures.* Van Nostrand Reinhold, New York.

NFPA 471. 1992. *Recommended Practice for Responding to Hazardous Materials Incidents.* NFPA, Quincy, MA.

NFPA 472. 1992. *Professional Competence of Responders to Hazardous Materials Incidents.* NFPA, Quincy, MA.

NFPA 473. 1992. *Competencies for EMS Personnel Responding to Hazardous Materials Incidents.* NFPA, Quincy, MA.

NFPA 1500. 1992. *Fire Department Occupational Safety and Health Program.* NFPA, Quincy, MA.

NFPA 1581. 1991. *Fire Department Infection Control Program.* NFPA, Quincy, MA.

National Institute for Occupational Safety and Health (NIOSH), Occupational Safety and Health Administration (OSHA), United States Coast Guard (USCG), United States Environmental Protection Agency (EPA). 1985. *Occupational Safety and Health Guidance Manual for Hazardous Waste Site Activities.* United States Department of Health and Human Services.

National Institute for Occupational Safety and Health (NIOSH). 1990. *NIOSH Pocket Guide to Chemical Hazards.* DHHS/CDC.

National Response Team. 1987. Hazardous Materials Emergency Planning Guide, NRT of the National Oil and Hazardous Substances Contingency Plan, Washington, DC.

Newman, Benjamin G. 1982. The hazards of taking the heat. *Fire Service Today*, August.

Nimble, Jack B. 1986. *The Construction and Operation of Clandestine Drug Laboratories.* Loompanics Unlimited, Port Townsend, WA.

Noll, Gregory G., Michael S. Hildebrand, and James G. Yvorra. 1988. *Hazardous Materials, Managing the Incident.* Fire Protection Publications, Oklahoma State University.

Orlando Fire Department Hazardous Materials SOPs, Orlando, FL, 1993.

Ottoboni, Alice M. 1984. *The Dose Makes the Poison.* Vincente Books.

Paget, G. E. 1962. Toxicity tests: A guide for clinicians. *Journal of New Drugs* 2:78–83.

Pepi, John W. 1987. The summer simmer index. *Weatherwise Magazine*, June, pp. 143–145.

Pettit, T., and H. Linn. 1987. *A Guide to Safety in Confined Spaces.* NIOSH, DHHS Publication 87-113.

Piantadosi, C. A. 1987. Carbon monoxide, oxygen transport, and oxygen metabolism. *Journal of Hyperbaric Medicine* 2(1):27–41.

Porter, R. S., M. A. Merlin, and M. B. Heller. 1990. The fifth vital sign. *Emergency Magazine*, March 1990, pp. 37–41.

Powell, William. 1971. *The Anarchist Cookbook.* Barricade Books, Secaucus, NJ.

Reimann, H. O. 1963. Biorhythms and disease. *Journal of the American Medical Association* 183:879.

Rekus, John F. 1994. *Complete Confined Space Handbook*. Lewis Publishers.

Sax, Irving. 1992. *Sax's Dangerous Properties of Industrial Materials*, 8th ed. Van Nostrand Reinhold, New York.

Spenctor, W. S. 1956. *Handbook of Toxicology*. Vol. 1, Acute Toxicities. Saunders Publishing.

Stager, Curt. 1987. Killer lake, silent death from Cameroons. *National Geographic*, September 1987.

Staten, Clark. 1992. Could this happen to you? *Emergency Medical Services* 21(5):28–35.

Sullivan, J. B., and G. R. Krieger. 1992. *Hazardous Materials Toxicology—Clinical Principles of Environmental Health*. Williams and Wilkins, Baltimore, MD.

Teele, Bruce W. 1993. *NFPA 1500 Handbook*. NFPA, Quincy, MA.

Tintinalli, Judith E., Robert J. Rothstein, and Ronald L. Krome. 1985. *Emergency Medicine, A Comprehensive Study Guide*. American College of Emergency Physicians, McGraw-Hill, New York.

Tokle, Gary. 1993. *Hazardous Materials Response Handbook*. NFPA, Quincy, MA.

Tortora, Gerard J., and Nicholas P. Anagnostakos. 1987. *Principles of Anatomy and Physiology*. Harper and Row, New York.

Trendelenburg, U. 1963. Supersensitivity and subsensitivity to sympathomimetic amines. *Pharmacology Review* 15:225–276.

Tuve, Richard. 1976. *Principles of Fire Protection Chemistry*. NFPA, Quincy, MA.

United States Department of Health and Human Services. 1992. *Managing Hazardous Materials Incidents, Vol. I, Emergency Medical Services, A Planning Guide for the Management of Contaminated Patients*. Agency for Toxic Substances and Disease Registry.

United States Department of Health and Human Services. 1992. *Managing Hazardous Materials Incidents, Vol. II, Hospital Emergency Departments, A Planning Guide for the Management of Contaminated Patients*. Agency for Toxic Substances and Disease Registry.

United States Department of Health and Human Services. 1993. *Managing Hazardous Materials Incidents, Vol. III, Medical Management Guidelines for Acute Chemical Exposure*. Agency for Toxic Substances and Disease Registry.

Upfal, M., and C. Doyle. 1990. Medical management of hydrofluoric acid exposure. *Journal of Occupational Medicine* 32(8, Aug.):726–731.

Verdile, Vincent P., and Robert A. Full. 1990. EMS in the haz mat response. *Emergency Magazine*, September 1990. Warren, Faidley and Weatherstock.

Viccello, Peter. 1993. *Handbook of Medical Toxicology*. Little Brown and Company.

Virginia Beach Fire Department. Special Operations Hazardous Materials Team SOP. Virginia Beach, Virginia.

Wallance, Deborah. 1990. In the Month of the Dragon. Toxic Fires in the Age of Plastics. Avery Publishing Group.

Watrous, R., and B. Olson, 1959. Diethylsilbestrol absorption in industry, a test for the early detection as an aid in prevention. *Journal of American Industrial Hygienist Association*.

Weinstein, Mark I. 1986. *Introduction to Civil Litigation*. West Publishing Company.

Windisch, F. C. 1990. Hydration prevents heat exhaustion, *Fire Engineering Magazine*, February 1990, pp. 75–76.

Wingard, Lemuel B., Theodore M. Brody, Joseph Larner, and Arnold Schwartz. 1991. *Human Pharmacology*. Mosby Year Book.

Index

DATE DUE
